Springer Biographies

The books published in the Springer Biographies tell of the life and work of scholars, innovators, and pioneers in all fields of learning and throughout the ages. Prominent scientists and philosophers will feature, but so too will lesser known personalities whose significant contributions deserve greater recognition and whose remarkable life stories will stir and motivate readers. Authored by historians and other academic writers, the volumes describe and analyse the main achievements of their subjects in manner accessible to nonspecialists, interweaving these with salient aspects of the protagonists' personal lives. Autobiographies and memoirs also fall into the scope of the series.

More information about this series at http://www.springer.com/series/13617

Michael Wiescher

Arthur E. Haas—The Hidden Pioneer of Quantum Mechanics

A Biography

Springer

Michael Wiescher
Department of Physics
University of Notre Dame
Notre Dame, IN, USA

ISSN 2365-0613 ISSN 2365-0621 (electronic)
Springer Biographies
ISBN 978-3-030-80605-7 ISBN 978-3-030-80606-4 (eBook)
https://doi.org/10.1007/978-3-030-80606-4

Translation from the German language edition: *Arthur E. Haas—Ein verborgener Pionier der modernen Physik, Eine Biographie* by Michael Wiescher, © LIT Verlag, Vienna 2021. Published by LIT Verlag, Vienna. All Rights Reserved.

This Springer imprint is published by the registered company Springer Nature Switzerland AG
The registered company address is: Gewerbestrasse 11, 6330 Cham, Switzerland

To our sorrow much has vanished that was well
And the most beautiful shows the shortest spell.

Heimito von Doderer (1896–1966)

Foreword

With his biography of the Austro-American physicist Arthur Erich Haas (1884–1941), Michael Wiescher presents an undiscovered piece of scientific history that begins in the period around 1900, Fin de Siècle Vienna, and continues to the 1930s and 1940s when Haas was appointed professor of physics at the Catholic University of Notre Dame in the USA, a position he held until his early death.

The German-American physicist Wiescher works and teaches at Notre Dame as Freimann Professor of Physics in the field of experimental astrophysics and is one of the outstanding experts in this branch of research. With his numerous works on the history of the natural sciences, Wiescher's professional interests go beyond his narrow field of expertise and his publications on the history of physics predestine him to be Arthur Haas' biographer. A special constellation of disciplines is conducive to this task: Wiescher, an experimental astrophysicist, works in the position of Haas' professional successor at the University of Notre Dame. Further, Haas organized the first conference dedicated to physical cosmology at Notre Dame in 1938, which can be considered the birth of this then still more-than-exotic discipline.

But it is not only Wiescher's professional closeness to Haas that predestines the author for this extensive biography. Wiescher is in the fortunate—indeed, for any historian of science, enviable—position of being able to draw on extensive, hitherto inaccessible archival material, which Haas' descendants have generously granted him access to, including photographic material that lends the book a special liveliness of presentation.

These personal documents, especially Haas' extensive correspondence with his family, enable Wiescher to give a dense account of Haas' life, which was marked by turbulence. In addition, Wiescher draws a panorama of Austrian cultural and its social and political life from the disintegration of the Danube Monarchy to the 1930s, the period of Haas' professional, scientific and public activity as an academic teacher, author and lecturer, first in Austria and Germany and then in the USA. Born into an assimilated Jewish family of industrialists, academics and lawyers, Haas belonged to the bourgeoisie of the last years of the Danube Monarchy. Privileged by wealth, this younger generation broke away from the primarily economic activities of their fathers' generation to turn to scientific interests.

The arc of this biography ranges from Haas' childhood in Brno (now Brno, Czech Republic), to his years of teaching and travelling at German and Austrian universities, to his—it must be said—timely move to the United States in 1935. The year 1935 was already marked by the flight of many academics from Germany, from 1933 onward after Hitler's seizure of power, and this emigration soon also extended to Austria in the aftermath of the country's annexation by the German Reich in 1938. Wiescher's work is thus also an eminent contribution to exile and emigration research in the history of science.

To this day, physicists and historians of science associate the name Haas with the search for a physically based explanation of the quantization of the energy of thermal radiation postulated by Max Planck in 1900. In publications of 1910—three years before Niels Bohr's work on the later so-called Bohr atomic model—Haas succeeded in combining the radiation theory formulated by Planck with the model of atomic structure that was common at the time. Haas' outstanding physical achievement and historical merit consists in the conceptual connection of "real" atoms (the hydrogen atom) with Planck's abstract quanta, which leads to the first "quantized" understanding of atoms and thus to the estimation of atomic quantities.

Wiescher's work now provides an overall view of Haas' life and work that goes far beyond previous knowledge of Haas as a pioneer of quantum physics and reveals previously unknown sides of his work. Particularly revealing is Wiescher's detailed account of Haas's professional and private cross-connections to many protagonists of scientific and cultural life in the 1920s and 1930s. Here, Wiescher provides the reader with extensive biographical notes on the people in Haas' life, opening up many sources that are often difficult to access. Historians of science will appreciate this with gratitude.

It can probably be described as a trick of history that many natural scientists, who had to flee Germany and Austria before the persecution measures of the National Socialists, were able to achieve outstanding scientific achievements in the positive acculturation conditions of Anglo-Saxon countries, in England and the USA. These include—as one of many—the Vienna-born physicist and 1998 Nobel Prize winner for Chemistry, Walter Kohn, who was able to flee Vienna on a Kindertransport, as were the children of Arthur Haas' brother, Otto Haas. Like Walter Kohn, Arthur Haas can rightly be named as one of those for whom the expression "Hitler's gift" applies.[1]

With his monographic work on the life and work of Arthur Erich Haas, Wiescher has closed a hitherto hidden gap in the history of physics in the twentieth century. Only with the presentation of this comprehensive volume has it become possible to enter the included information so accurately into the consciousness of all those who have learned to understand the history of the natural sciences within its complex

---

[1] Jean Medawar & David Pyke, *Hitler's Gift : The True Story of the Scientists Expelled by the Nazi Regime* (New York: Arcade Publishing, 2001).

historical contexts. Our view of the history of scientific life on both sides of the Atlantic has been enriched through many facets with the presentation of this book.

April 2021

Wolfgang L. Reiter
University of Vienna
Vienna, Austria

Preface

When I came to the University of Notre Dame more than thirty years ago, as a young assistant professor to develop my research programme in nuclear astrophysics, I was told that fifty years earlier an Arthur Haas from Vienna had come to the university with similar intentions of establishing a new research programme in the field of cosmology. Despite the enormous differences in time and motivation, I regarded ours as the comparable fates of two emigrants seeking to establish their research and roots in a foreign country. Years later, partly due to my interest in the history of science, I followed up on this idea and began to explore what brought Arthur Haas to the USA and how his life developed here—interestingly enough, on the same street where I had my house.

What fascinated me about Haas, besides these quasi-personal associations, was his view regarding the close connection between the microcosm and the macrocosm—between the world of atoms and the world of stars and galaxies, respectively—which he believed followed corresponding principles. My own research focuses on the origin of the elements, in the interior of stars and the chemical evolution of the universe, reflecting a similar perspective in science.

Typically, biographies of this nature concentrate on scientific successes and on the successful and, therefore, focus on widely known names and personalities in science. However, given the large number of well-known researchers, there are an even larger number of scientists who have not or who have only partially succeeded in making a breakthrough. These scientists have done important work as contributors and idea bearers in the network of the scientific community. Arthur Haas is among these for whom social and political circumstances prevented the longed-for achievement of scientific success and recognition. And yet, he has nevertheless contributed to the development of new ideas from atomic physics to cosmology and, more importantly, has made an essential contribution to the popularization and dissemination of these ideas. Therefore, it seems important to revisit the career and destiny of Arthur E. Haas within the network of the scientific community and in light of the political developments of the early twentieth century.

My early research on Haas first led to my publication in the journal *Physics in Perspective*, in which I essentially follow the development of Haas' scientific thinking. The support of the family has motivated me to continue in my efforts to create a more complete biography of the life of Arthur Haas, including a background of the political and social developments in Austria during his time. Therefore, the first section of the book covers the period up to the end of the Habsburg Dual Monarchy. For this section, I follow Haas' unpublished autobiographical notes written in 1937. I supplement these and explain his situation, providing background information on the historical context as well as on the physics question that he discusses in these notes. In the second section, I follow Haas' life in post-World War I Austria. This includes the first five years of his career as he navigates through the times of inflation, his wedding in September 1924, and his acceptance of first employment at the Vienna Academy of Science.

The third section concentrates on Arthur Haas' years with the academy and his desperate attempts to find an academic position in Vienna or abroad. The section describes the resistance that he encountered, from right-wing conservative forces at the university, to his employment and his attempts to pursue both a science career and a career as a science communicator. These events are portrayed within the context of his personal life in Vienna and the support that he received from his young German wife, Emma Beatrice née Huber. The section also follows the political developments in Austria and how these finally forced Haas to leave Austria and immigrate to the United States. The fourth section follows the development and successes of his renewed scientific career in the USA over the short period between his arrival and the time of his death, including his enormous efforts to help Jewish and other persecuted colleagues in Austria to obtain the necessary papers for emigration.

Much of the background information, I have added as footnotes. These details about colleagues and acquaintances whom Haas mentions in his autobiographical notes, in his letters and in other written papers do not contribute significantly to the continuation of the story but do serve to illuminate Haas' world. Much of this information has been taken from Wikipedia or other lexical documents and does not contain any evaluation or interpretation on my part. As far as possible, I have complemented this with additional new information and results from my own research on this topic. Translation of original German texts and letters into English was primarily done by the author, unless otherwise referenced.

Notre Dame, USA Michael Wiescher

Acknowledgements

Numerous people have contributed to this work. I have received enormous support from the descendants of Arthur Haas and especially from his granddaughter Patricia Haas-Cleveland and her sisters Theresa and Susan, who gave me free access to the family archives. The Haas family has provided me with personal correspondence generated over forty years of Arthur Haas' life, a collection that fills two filing cabinets and gives fascinating insight into the changes in Austrian life that developed over the first decades of the twentieth century.

These letters also give an overview of both the possibilities and the opposition that Jewish scientists experienced during that period in Austria's history; Haas was not the only scientist to navigate the numerous issues confronting a Jew while, at the same time, remaining dedicated to a scientific life. Klaus Taschwer from Vienna helped me to understand the coordinated resistance against talented young Jewish colleagues by the mostly conservative world of professors. I received much help from Stefan Sienell, archivist of the Austrian Academy of Sciences in Vienna, who sent me the Academy's numerous documents on Haas from 1917 to 1937 and was a great resource in understanding the role of Arthur Haas in the Vienna Academy. In addition, I received support and motivation from other Austrian sources, in particular from Prof. Dr. Walter Kutschera of the University of Vienna, Dr. Peter Schuster, the unfortunately, recently deceased director of the Austrian Museum for History of Physics in Pöllau, and Undersecretary and Honorary Professor Dr. Wolfgang Reiter, a leading specialist in questions of Austrian science before and during the Third Reich. Pavel Šišma, Professor of Mathematics at Masaryk University in Brno helped me with information on the history of the German Technical College in Brno, and Joanna Behrman of Johns Hopkins University, now the American Physical Society helped me with background information on Wilson College and other American women's institutions in the 1920 and 1930s. I also received advice and support for historical-philosophical considerations from Professors Friedrich Stadler of the University of Vienna and Leopold Stubenberg formerly of the University of Notre Dame. I would also like to thank Professor Ani Aprahamian for helpful discussions and insight on the internal configuration of heavy atomic nuclei, a topic Arthur Haas was working on towards the end of his life. Joe Smith from the Notre Dame Archives was extremely helpful in providing me material on Arthur Haas and his time at the university. Particularly helpful in that aspect were also Fr. Thomas Blantz, C.S.C. for his 2020 published *The University of Notre Dame—A History* and his interest and comments. I also want to thank Dr. Christopher S. Temple, who provided me with a copy on his yet unpublished Ph.D. thesis on *Fostering Elite Science at an American Catholic University*, an in depth analysis of the period when Arthur Haas joined the faculty. I express my gratitude to the Coomes family for providing me unpublished and personal material on Prof. Ed Coomes at Notre Dame. Finally I would also like to thank Brigitte Roos for her help in transcribing many of the letters for Chap. 2. Special thanks, however, to Elizabeth Jo who painstakingly struggled through the manuscript to improve my English, untangle long sentences and correct the comma position. Last but not least, I want to thank Bob Bernhard as the director of the Office of Research at the University of Notre Dame for his support for this work.

Contents

About the Author

Michael Wiescher is Freimann Professor and Director of the Institute for Structure and Nuclear Astrophysics at the University of Notre Dame, and he is currently also Adjunct Professor at Michigan State University, Visiting Professor at the University of Surrey in the UK and at the Goethe University Frankfurt in Germany.

Wiescher studied history and physics in Münster and received his doctorate in 1980 on a topic in nuclear astrophysics. In the following years until 1986, he was a postdoctoral fellow at Ohio State University, at the Institute for Nuclear Chemistry in Mainz, and at the Kellogg Radiation Laboratory at CalTech. In 1986, he joined the faculty of the physics department at the University of Notre Dame. As the director of the Nuclear Science Laboratory and the Joint Institute for Nuclear Astrophysics, Wiescher developed a worldwide research programme in nuclear astrophysics with research projects in the USA, Brazil, Europe, South Africa and Japan.

Wiescher has also developed a programme in archaeometry, in the analysis of art historical and archaeological materials using atomic and nuclear physics methodology, and is involved in various topics in the history of science.

Michael Wiescher is Fellow of the American Physical Society and the American Association for the Advancement of Science and Elected Scientific Member of the Academiae Europaeae. In 2003, he was awarded the Hans Bethe Prize of the DNP & DAP of the American Physical Society and 2018 Laboratory Astrophysics Prize of the American Astronomical Society for his research. He received the Alexander von Humboldt Foundation Research Award in 2007 and the Heraeus Award at the University of Frankfurt in 2017.

Part I
Youth and Career in the Habsburg Empire 1884–1919

Chapter 1
Introduction

Arthur Erich Haas (1884–1941) provided new insights and developed new approaches during the initial period of quantum theory. In the early years of his physics career, he added significantly to the discussions about deeper meaning within the context of atomic physics. Haas devoted many years to the question of cosmology, both from a philosophical as well as a physical perspective. He rejected the idea of an infinite universe but, like many of his peers, he was attracted by the similarities between the atom and the planetary systems, looking for a deeper meaning and correlation in the fundamental constants of nature. Toward the end of his life, Haas developed a new methodology for explaining the internal structure of the atomic nucleus, an approach which today is reflected in the cluster model and is widely used to describe many features in nuclear structure, nuclear decay, and nuclear reactions. His contributions are largely forgotten today, only mentioned in works on the history of science and then only marginally as an example of an early scientific approach to quantum mechanics[i] or cosmology.[ii] Haas' approaches were recognized at the time but often not fully appreciated, thoughts and ideas having been, perhaps, too early for their reception.

Haas came from the bourgeoisie of the Fin de Siècle of the Austrian-Hungarian Monarchy. He always considered himself an integral part of this society and it is, therefore, important to understand its historical, political and social background and developments. Of equal importance is seeing how young Haas developed within the framework of this, his social environment which drastically changed over his lifetime. As a member of the Strakosch sugar dynasty,[iii] he initially lived a bohemian life in which his academic studies were merely an expression of his intellectual interests. He chose to study physics and was the last student of Ludwig Boltzmann (1844–1906) who had shaped the development of physics for half a century. Haas was witness to and effected by his time, by the Great War and the dissolution of the multi-national Habsburg Monarchy together with the collapse of bourgeois society. The resulting political, economic, and social developments shaped his life, moving him from his status as a coveted member of society to a Jewish outsider. As a Jewish representative

M. Wiescher, *Arthur E. Haas—The Hidden Pioneer of Quantum Mechanics*, Springer Biographies, https://doi.org/10.1007/978-3-030-80606-4_1

of a vanishing liberal world, he fell victim to the blame and hatred generated by an increasingly radicalized petty-bourgeois society.

Arthur Haas' further fate fell victim to Austria's development, from its defeat in 1918 through the establishment and failure of parliamentary democracy and the emergence of authoritarian conservativism, often described as clerical fascism. His fate and his work also reflect the dynamic development of modern science, which was emerging from Austria and revolutionizing the world. Franz-Serafin Exner (1849–1926) and Stefan Meyer (1872–1949), the founders of the Radium Institute in Vienna, were Haas' teachers. Among his fellow students were Erwin Schrödinger (1887–1961), a leading representative of quantum physics and founder of wave mechanics, and Lise Meitner (1878–1968), a representative of the new science of atomic nuclei that provided the revolutionary explanation for nuclear fission.

Studies and thinking were shaped by Haas' observations and impressions; he observed his contemporaries and colleagues and he described them not without arrogance. This arrogance, coupled with great scientific intellect and financial independence, shaped his early life before the war and caused his vacillation between bohemian and scientific desires. Then although his first scientific thoughts and insights on cosmology and the new science of atomic physics were fully formulated, the work mostly remained unrecognized. He was, for example, the first to apply the new concept of quantum theory to explain the spectral observation of light, a work that was largely mocked but then taken up and continued by others.

When war and subsequent inflation deprived Haas of his fortune, he sought, perhaps too late, a serious scientific career in Vienna. But it was by then a changed, dark Vienna where the bourgeoisie existed only in fragments and anti-Jewish resentment in the Catholic state prevented his scientific career. Economic hardship became a new factor. He had thoughts of emigration but a position in the scientific administration of the Vienna Academy of Sciences and his wedding with the young charismatic Emma Beatrice Huber prevented him from taking this step. From the scientist, he developed into the voice of science. He made the latest findings, which his colleagues were in the process of uncovering and developing, accessible to the public. In this way, Haas became the "Carl Sagan" of his time, known as a science communicator and popularizer of leading scientific developments. As such he became internationally known through lectures and lecture tours as well as through his books; the latter being translated into many languages.

The increasing difficulties in the last years before the Anschluss,[1] with unrest and growing fear in Austria, drove him to America. Haas emigrated with the help of Albert Einstein and subsequent events in Austria validated the necessity of this step. With immigration to America, he was transformed from victim to helper, working tirelessly to help and support members of his family as well as other Jewish scientists to escape. He organized visas, housing, and jobs as the necessary first steps for those

[1] Anschluss refers to the annexation of Austria into Nazi Germany on March 12, 1938. The idea of an Anschluss began after the unification of Germany excluded Austria and German-Austrians from the Prussian-dominated German Empire in 1871. After the end of the Habsburg Empire, reunification with Germany became an important goal for parts of the Austrian population but was opposed by the conservative Christian-oriented part of society.

beginning new lives in the United States. His fate reflects the fate of many, not only Jewish scientists, who had to leave Europe to build a new life and a new career in a new unfamiliar environment.

Arthur Erich Haas was a witness and a victim of his society and his time; the beginning of the twentieth century was a time characterized by scientific and social revolution and upheaval. He studied with the leading minds of physics and witnessed the great moments of physics—the development of the theory of relativity, the birth of atomic physics, and the formulation of quantum mechanics—during his studies in Göttingen and Vienna. In this, he played a small but authoritative role that is reflected in his books and letters. But, he was also witness to political and scientific decline in German-speaking countries. This is reflected in his observations and comments about the nationalist-conservative movement and its representatives, from the time of his student days in Göttingen to his eventual emigration. As an Austrian Jew who was a representative of the Vienna Academy of Sciences, Haas was responsible, even after 1933, for direct contact between German-speaking scientific associations and the academies. As such, he observed the exodus of science from Germany. The consequences of this drain, which began with the dismissal of all Jewish scientists from German universities, are reflected in this small anecdote: Bernhard Rust, the new Minister of Science, asked the famous mathematician David Hilbert, sitting next to him at an official luncheon, how the institute for mathematics was doing after the "removal of Jews and Jew friends," and Hilbert replied: "The institute?—that does not exist anymore!".

Haas left fragments of an autobiography that he had started writing during his early days in America. However, while occupied with building a new scientific career, helping his fellow victims to escape from Nazi terror, and with his early death, he was prevented from completing this work. Although there are isolated reports on his scientific accomplishments[iv] and on his role in scientific development, there is no complete account of his complex life in complex times. It will now be attempted here to create a picture of his life and time on the basis of his autobiography, his scientific publications, his books, and his countless letters and written documents, which he brought from Europe to America and which slumbered for decades in family archives. Biographical footnotes refer to various persons—friends, acquaintances, teachers, classmates, and colleagues—who reflect the close social and scientific network in which Haas lived, shaping his early life and studies as well as the connections that he tried to maintain in his later life.

His records describe the environment of a wealthy upper-middle-class family in the last decades of Austria-Hungary. They offer insights into domestic luxury with several servants, who are often mentioned only casually, the daily celebrations and amusements offered by the capital of Vienna, and the extensive summer resorts in the surrounding countryside or on distant coasts. However, the text also provides insight into the dark side, into and inexplicable resistance that this rich and well-connected son faced in his quest for academic success and fame. This resistance can probably, in retrospect, be traced back to his Jewish origins, perhaps to his wealth, and possibly also to the aforementioned social and intellectual arrogance with which he confronted his teachers. However, the latter may have only been reflected in his personal notes while being less obvious in the undoubtedly polished manners of young Haas.

The rejection and malicious criticism that he received for his groundbreaking work may have been due not only to the short-sightedness of his contemporary physicists but also to the socio-emotional, probably even anti-Jewish attitude with which these fellow scholars confronted Haas. This may be speculation but further developments at the universities—during the years of the young Republic of Austria, reinforced in the clerical state of the Federal Chancellors Engelbert Dollfuß (1892–1934) and Kurt Schuschnigg (1897–1977)—demonstrate a growing hostility toward young Jewish scientists. Arthur Haas was perhaps an early target of these emerging developments.

Before we come to the tale of the young Haas let me provide a brief account of his origins and family, which are closely linked to the Strakosch sugar dynasty in Austria. His autobiographical account shows close ties to, and mutual support within, the Haas and Strakosch families. This expressed itself in frequent mutual visits and close business networking. Networking is also reflected in the history of and connections between the Haas family and Moravian Jewry, ties that can be traced back to important rabbinical families and Jewish thinkers of the Middle Ages. Although he himself did not identify with active Judaism in the Fin de Siècle of Vienna, Haas was proud of these origins, of the rabbinical and intellectual background of his family that he refers to in the introduction to his autobiographical notes. These are, however, only hints, specific remarks in which the personal pride in his origins is reflected; these origins need to be considered in more detail in order to understand the world of thought from which Haas emerged as the man who is described here.

Therefore in the first part of this biography, I will annotate the autobiographical and very subjective notes of the young Arthur Haas with more objective historical information about his family, his environment and his time. I do this in order to better understand and explain the development of and influences on his world of thought. His origins were the basis for his self-confidence and the arrogance with which Haas viewed and confronted his environment. The Fin de Siècle of the Habsburg Dual Monarchy, though doomed to extinction, formed the intellectual basis for the world of thought of the twentieth century. This also formed Haas' world of thought and he wanted to be part of it, to contribute actively to it. While he was not afraid to represent this claim openly and sometimes aggressively, his desires led to negative reactions especially when he questioned authority and was openly or even covertly critical.

Notes

i. Armin Hermann, *The Genesis of Quantum Theory (1899–1913),* (Cambridge: MIT Press, 1971), 87–10.
ii. Helge Kragh, *Matter and Spirit in the Universe: Scientific and Religious Preludes to Modern Cosmology* (London: Imperial College Press, 2004).
iii. Werner Kohl and Susanna Steiger-Moser, eds., *Die österreichische Zuckerindustrie und ihre Geschichte(n) 1750–2013* (Wien: Böhlau Verlag, 2014).
iv. Armin Hermann, *The Genesis of Quantum Theory (1899–1913)*, 87–10.

Chapter 2
Origins and Childhood

In his autobiographical remarks, Arthur Haas mentions with just a few words his origins and family but they had shaped him much more extensively and more deeply than he acknowledges. While his comments avoid deep connections to the Jewish spiritual and cultural world, he does mention the intellectual leaders in his ancestorial line and does not fail to point out, with hidden pride, the historical significance of his birthplace.

> On the Great Square now Freedom Square of the Moravian capital of Brünn, rises a five-story Renaissance building crowned by a Mother of God statue, which in many art histories is described as an architectural masterpiece. The city of Brünn had given it to a citizen Raduit de Souches, originally from France, to thank him for his help in successfully defending Brünn against the besieging Swedes in the Thirty Years' War. On April 30, 1884, I was born as the son of the "Moravian-Silesian provincial lawyer" Dr. jur. Gustav Haas in an apartment of this building, just before the last fifteen years of the 19th century began.

The fact that the city of Brünn resisted the Swedes was, in itself, a world sensation at the time. Olmütz (today Olomouc) capitulated to the Swedes after four days, Iglau (today Jihlava) within a day, and Znaim (today Znojmo) surrendered without any defense of its own. News of Brünn's stance spread to Italy, Germany and France. The defenders of the city—everyone, including citizens, masters, journeymen and even students—were richly rewarded and promoted. And so also was Raduit de Souches who, nevertheless, still had to pay 60,000 guilders for his house in the Great Square shown in Fig. 2.1, a house that included a garden and the adjacent city gate. The house remained in the possession of the de Souches family for almost ninety years. Around 1884 the house belonged to the colonial-goods dealer Franz Komarek and the Haas family was renting a floor.

> The story of my family, which in former times was called Auerbach, dates back to the Middle Ages. In a document from the year 1497, an early, demonstrable ancestor is identified who had close connections with the court of the then ruling bishop of Regensburg.[1] The family later moved to Vienna. One of my ancestors [the father-in-law of an ancestor in the male

[1] Bishop from 1492–1507, Rupert II was son of Count Palatine Friedrich von Sponheim. The bishop expelled the Jews from Regensburg but made an exception for the rich court Jew, Moshe

M. Wiescher, *Arthur E. Haas—The Hidden Pioneer of Quantum Mechanics*, Springer Biographies, https://doi.org/10.1007/978-3-030-80606-4_2

Fig. 2.1 The Great Square in Brünn. The house where Haas was born is the fourth towering Renaissance building on the right side of the street. Today the gable of the house has disappeared. A history of the building can be found on the website of the descendants of the former German-speaking inhabitants of the city of Brünn (today Brno), accessed January 31, 2021, http://www.bruenn.org/www/bruna/de/haus-des-vergessenen-fremden.php

> line] was Dr. Leo Lucerna, who died on April 26, 1635 and whose grave is still preserved in Vienna; he was, as his epitaph says, doctor of three faculties. My family left Vienna in the second half of the 17th century and from then on, my ancestors lived in different Moravian cities. My very great-great-grandfather [2] enjoyed the special friendship of the then Moravian governor, a Count Zierotin.[3]

Arthur Haas was proud of his origins and of his contributions to physics, and he did not shy away from underlining these assertions with references to the scholastic achievements of his ancestors. In his writing, young Haas traced his genealogy far into the Middle Ages (Fig. 2.2), to the Jewish rabbinical dynasties of the Auerbachs, Lurias, and Lucernas who were each famed for their important Talmudic and Kabbalist thinkers. Moses Auerbach (~1462–1540) was in Regensburg and is considered one of the progenitors of the family. He moved with his wife Mushkat, from the

von Auerbach: Auerbach was allowed to stay in the city. He refused, however, and followed his brothers in faith.

[2] This is probably Emanuel Haas (1791–1850) who will be discussed later in more detail.

[3] Franz Josef von Zierotin (1772–1845), Councilor of the Moravian Gubernium. From 1817, he was administrator of the family property in Blauda, Chromče, Meseritsch, Prussia and Brno; he was also a scholar and acquired a large library. From 1828, Zierotin was a member of the board of the Economic Association of Moravia and Silesia, and later director of the Moravian Insurance Company. In 1835 he became a privy councilor and during the coronation celebrations for Bohemian King Ferdinand I in 1836, Zierotin was knighted by St. Wenceslas and awarded the Leopold Order.

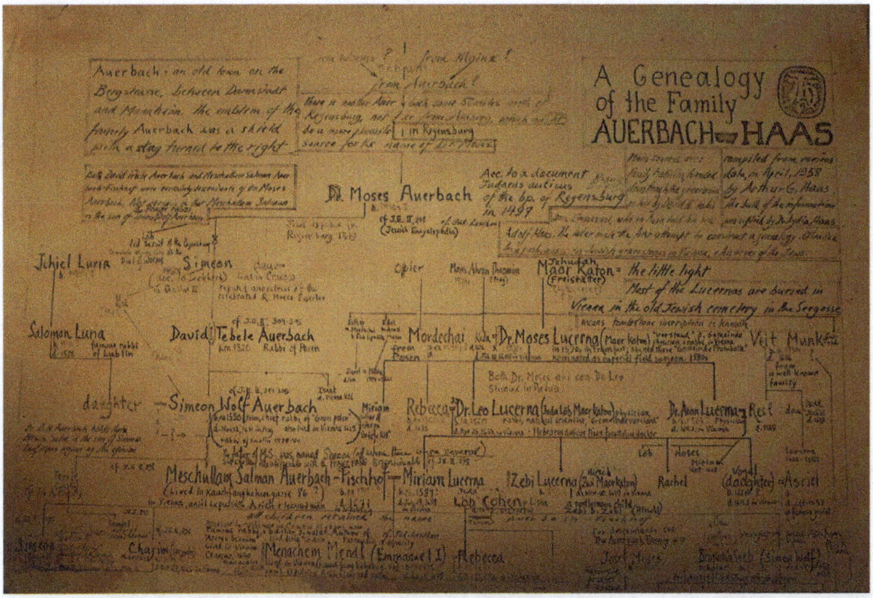

Fig. 2.2 The family tree of the Auerbach family as developed by Arthur Haas and drawn by his oldest son, the historian Arthur Gustav Haas in 1958. This shows the first five generations after Moses Auerbach from Regensburg. The origin of the family is traced back to the small town of Auerbach on the right bank of the Rhine south of Darmstadt

Lazara-Isserlee family, to Cracow after the expulsion of the Jews from Regensburg.[i] The eldest son Israel was appointed to important rabbinical posts in Posen (today Poznan) and Cracow, while the second born, Simon Wolf Auerbach founded a branch of the family in Frankfurt. The youngest son David Tebele Auerbach (1512–1575) became rabbi in Vienna and is considered the progenitor of the Viennese branch of the Auerbach family.[ii] He was married to Krondel Luria, the daughter of the Rabbi of Lublin, Shlomo Luria (~1510–1573). Shlomo Luria traced his origins through the Spira and Treves families[iii] to the well-known Rabbi Rashi-RAbbi SHlomo Itzhak (1040–1105), who in turn claims his origins from the family of David, which refers to Noah as the original progenitor.[iv]

The couple had several sons. The eldest, Simeon Wolf Auerbach served as rabbi in Polish and Bohemian cities, lastly in Prague in 1631. David Tebele Auerbach's second son Meshulam Salman Fischoff Auerbach (~1565–1677) married Miriam Lucerna (1584–1654), the daughter of Dr. Leo Lucerna (1560–1635) who had studied medicine in Padua, as Jews were not allowed to attend German-speaking universities at the time. Despite the economic difficulties during this period of the Thirty Years' War, it was a golden age for the rapidly growing Jewish community in Vienna (Fig. 2.3). The community was under the special protection of Emperor Ferdinand II who assigned them their own district, today's Leopoldstadt, as a residential area providing protection against the less benevolent citizenry.[v] Meshulam

Fig. 2.3 Assembly of traditional Jews in a painting of the late nineteenth century in Vienna

Salman Fischoff Auerbach and Miriam Lucerna had four sons, who in turn held important municipal functions as well as rabbinical functions in Moravia, such as in Eibenschütz and Nikolsburg. Yet even in relocation, the family stayed connected and maintained its relationship with fellow believers in Vienna.

Emperor Ferdinand II's successor, Emperor Leopold I, ordered the expulsion of the Jews from Vienna in 1670 at the instigation of his Spanish wife who blamed the Jewish population for her miscarriage. Leopold took this action despite the high tax revenues that Vienna and the imperial court had received from the Jewish community. As an official pretext, the Emperor and his councils accused the Jews of having collaborated with the Protestants, the Hungarian insurgents, and the Turks during the Thirty Years' War.

When the Jews were expelled from the city, some moved westward but most settled in areas near the border in Moravia. This was only possible, however, in places where the local nobility regarded the Jews and their trade as a welcome source of tax revenue, while many other Moravian towns did their best to slow down the influx of displaced Jews. Therefore, some large groups moved further to Poland, Galicia and Russia where influential Jewish communities had been forming for years.[vi]

The expulsion affected Meshulam Salman Fischoff Auerbach and his family, the remaining sons and grandsons. He moved to Nikolsburg in South Moravia where the then largest Jewish community existed. Nikolsburg belonged to the dominion of the Princes of Dietrichstein, who welcomed the Jews in exchange for high taxes and other duties. Prince Walter Franz Xaver von Dietrichstein (1664–1738) is said to have taken in eighty families in Nikolsburg alone, which developed rapidly thereafter. Of course, the Jews had to make extremely high payments for this reception. Meshulam Salman Fischoff Auerbach died there in 1677 but his son Benjamin Wolf Auerbach (1625–1705) remained and became the elder of the Jewish community of Nikolsburg. His descendants appear in later records under the name Fischhof but the connection of the Auerbach family with the Haas family remains obscure.[4]

Arthur Haas ignores these details and only noted that the family had lived in several Moravian towns. In fact, there were countless smaller Jewish communities in Moravia, of which Nikolsburg was one of the largest. However, little is found in the records[vii]; Arthur Haas himself was not quite sure about the exact generational sequence because, as late as 1937, he was still trying to create one, as he wrote to his brother Otto:

> However, the two or three intermediate generations [which lead from the Auerbach family to Haas] are missing, but maybe you can find them.[5]

This never happened; only a few months later, Germany invaded Austria and the search for ancestral connections was no longer relevant. Only when Arthur's eldest son Arthur Gustav took up this endeavor was a genealogy finally compiled of the Auerbach-Haas family, seen in Fig. 2.4.[6]

Menachem Mendel Fischhof Auerbach, the brother of Benjamin Wolf Auerbach, is mentioned in 1660 as a rabbi of the Jewish community in Raussnitz,[viii] however, he did not stay there but accepted a rabbinical position in Krotoschin in Poznan, Poland, where he died in 1689. In addition, only one member of the Haas family is mentioned, Jakob Haas as owner of a Jewish house in Kanitz.[ix] However, the mention of Count Zierotin in Arthur Haas' records points to the town of Meseritsch (today Velké Meziříčí) in the Wallachia region of eastern Moravia (Fig. 2.5). In the nearby market town of Krasno, which at that time was still an independent village, the Jew Marek Haas had acquired, in 1745 the rights to a distillery from the Zierotin dominion, for 1100 florins a year. This distillery remained in family ownership for generations.[x]

The name of the Jewish Haas family goes back to the cloth merchant Salomon ben Meier, who is mentioned in 1530 as the owner of the "Haus zum Hasen" in Frankfurt's Judengasse (Jewish Ghetto).[xi] His descendants lived in the Judengasse for centuries, often adopting the names of their respective houses according to given custom.[7] The Haas family had businesses in Amsterdam and Mainz, resulting in the

[4] In 1782, 1783 the Jews in Austria were forced to leave—often because of their resistance to adopting fixed family names—and this makes it difficult to track families in the Moravian Diaspora.

[5] AEH to Otto Haas (South Bend, March 24, 1937).

[6] Unfortunately, these family charts do not include proper documentation and bibliographical information. Arthur G. Haas, "Haas Family Tree," (unpublished manuscript, 1958).

[7] Salomon ben Meier Haas is therefore also considered the progenitor of the Grotwohl, Schuch, Kann, Stern, and Beer families who were named after their respective houses in Frankfurt's Judengasse. A detailed list of the Frankfurt families in the Judengasse can be found in the list of graves in

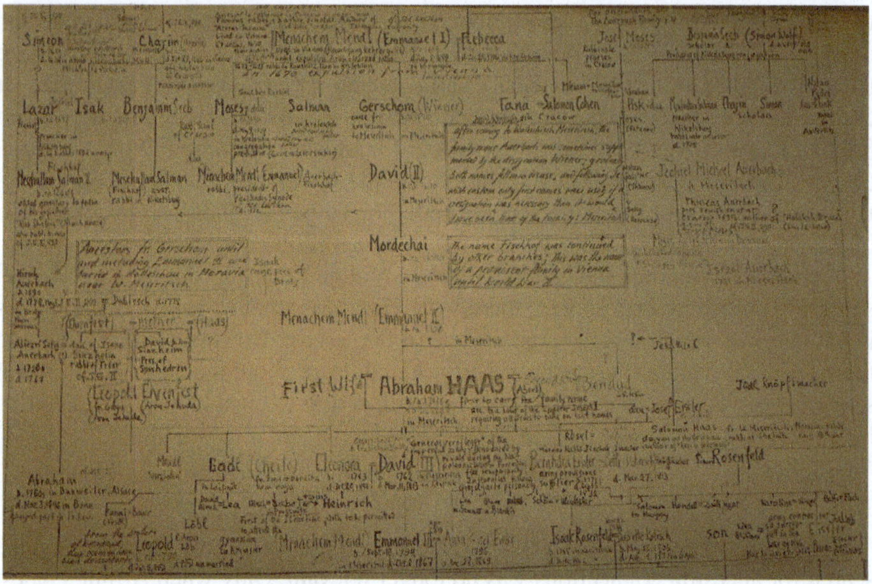

Fig. 2.4 The different branches of the Auerbach-Fischhof genealogy according to the notes of Arthur Gustav Haas created in 1958, including the first appearance of the name Haas in the family succession as descendants of Menachem Mendel Fischhof Auerbach

Fig. 2.5 Typical Jewish quarter in a South Moravian town in the nineteenth century

development of a Dutch branch of the Haas family as well as a southwest German branch.

How and when Marek or Markus Haas came to be in Moravia is not known.[8] Imperial and local decrees restricted the influx of non-native Jews to Moravia in the eighteenth century. It was probably the lucrative business of making potash, an important local product for filtering the popular local Slivovitz schnapps, that attracted Marek Haas. Producing potash was a flourishing business in Moravia and, together with the distilling of schnapps, mainly run by Jews. They were, however, required to pay large sums of money to the lessees, such as the Counts of Zierotin. However, since Jewish tenants ran the businesses of the local Counts of Zierotin, the latter helped in circumventing local restrictions. Presumably Marek Haas became a part of this enterprise, for which he must have had to invest a considerable initial capital.

The Edict of Tolerance issued by Emperor Joseph II in 1780 improved the situation of Jews (and Protestants) in the Habsburg territories, which had been expanded just ten years earlier, by the third Polish partition, to include Galicia with its Jewish population. The subsequent lifting of travel and trade restrictions led to a huge upsurge in economic activity. The Edict of Tolerance thus laid an essential foundation for the modernization and industrialization of the Austrian-Bohemian-Hungarian lands at the turn of the nineteenth century.

While it took some time in coming, the Haas family was well prepared for this change. Their local distilling business provided resources for immediate investment in these new business opportunities.

The son of Marek Haas, Mendele Abraham Haas (1746–1810), paid 2200 guilders annually to Count Zierotin in 1781 and for the lease of a trading room, he paid another 250 guilders. These prices indicate how lucrative the distillery was and demonstrate how lease of the distillery was a very profitable undertaking for the owners of the manor, relative to other sources of income. In comparison, the lease of a paper mill brought in only 250 guilders annually and the lease of a glass factory brought in 220 guilders annually.

But the Jewish tenants also seem to have earned well. Abraham's son, David Haas (1770–1830) was a rich man; despite the enormous rent and tax payments, he could afford to buy a distillery in Meseritsch, in addition to the distillery in Krasno. The seller was Bernard Eissler, who is mentioned as the owner of a distillery in Meseritsch at the beginning of the nineteenth century. The reason for the sale was probably a

the Jewish cemetery. See Shmul Ettlinger, *Ele Toldot 1241–1824*, which can be viewed in Frankfurt at the Institute for City History.

[8] In the eighteenth century, the name Haas first appeared in Bohemia and Moravia and, in the nineteenth century, spread to neighboring Galicia. In neighboring Holleschau (today Holešov) several Jewish births are registered under the name Haas, the oldest of which is Selig Haas (1740–1809) who may have been a son of Marek, born in Holleschau and who formed a separate branch of the family there.

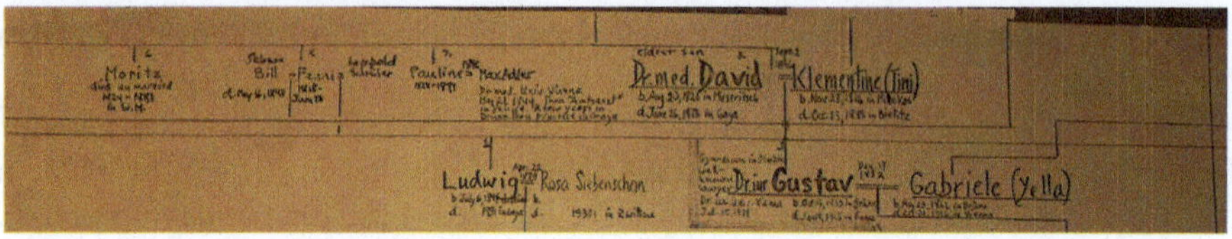

Fig. 2.6 The descendants of Emanuel Haas in Meseritsch according to the notes of Arthur Gustav Haas in 1958

marriage contract, since Eissler's daughter Anna married David Haas' son Emanuel Haas (1791–1850)[9] in 1821.

Emanuel Haas was, as his great-grandson had accurately observed, a rich man. He owned not only two distilleries but also leased a tavern in the town hall and ran a stall at the cattle market. From Emanuel's marriage with Anna, the sons Moritz (*1825), Adolf (1823–1906) and David (1822–1875) were born in Meseritsch (Fig. 2.6). Moritz Haas continued his father's business in Meseritsch. Adolf Haas, however, first moved to Olmütz with his young wife Josephine Wagner, from where he moved with his family to Vienna.

The third brother, the eldest son David Haas studied medicine in Vienna and finished his studies in 1844. The following year, in 1845, he married the young Clementine Rosenfeld (*1824) from the small Moravian town of Koritschan (Fig. 2.7). She was the daughter of Isak Rosenfeld (*1795) and granddaughter of Simon Rosenfeld (~1755–1831), a family that had lived in the town since the early seventeenth century.

> My father's father, Dr. David Haas had obtained doctorates in medicine and surgery at the University of Vienna and he worked as a doctor, first in Venice, then in Brünn and finally in the Moravian town of Gaya, where he died in 1875. As the third of his eight children, my father was born on October 15, 1850 in Brünn. My father had four brothers and three sisters. An older brother Ludwig, to his father's grief, became a merchant since his eyes did not allow him to study. Of the three younger brothers, Arnold was a pharmacist, and two others, Max and Fritz, became lawyers like my father. My father graduated from the First German Gymnasium in Brünn and then obtained a doctorate in law from the University of Vienna. Since my grandfather did not have a generous income because of his large number of children, my father had to rely on himself during his university studies. As a doctor, he entered a very prestigious law firm in Brünn, as an apprentice lawyer; it was the office of a Doctor Heschel, who had taken it over from the Baron von Giskra who instead became a leading liberal statesman in Austria in the mid-nineteenth century. After several years of work as a clerk, my father then took over the law office of Dr. Heschel, shortly before his marriage (on December 17, 1882), and was very successful from the very beginning.

David Haas had taken advantage of the relaxed admission regulations for Jews that occurred in 1780 at Austrian universities.[10] He studied medicine in Vienna and

---

[9] The early Haas family is well established on the genealogical websites, which can be easily followed into all family branches, e.g., https://www.geni.com/people/Emanuel-Haas/6000000023910361466?through=6000000038116898940.

[10] It was not until the revolutionary year of 1848, with the abolition of the ghettos, that the complete civic equality of Jews was legally established.

Fig. 2.7 Dr. David Haas as a well-established doctor in Gaya around 1870 and his wife Clementine, usually called Tini, who was born Rosenfeld. The traditional representation of the family in these oil paintings underlines their successful rise to the Austrian bourgeoisie

then, after completing his studies, he worked from 1845 as a doctor in Venice for two short years. He then returned to Brünn where his first son Ludwig Haas (1847–1891) was born in 1847. David Haas remained there as a doctor for the next ten years and three more children were born: Adolfine Haas (1849–1901)[11] was followed three years later by Gustav Haas and Arnold Haas (1852–1915).

Around 1855, David Haas moved with his family from Brünn to the small Moravian town of Gaya (now Kyjov) in South Moravia. The reasons for the move are not known but he established a new medical practice there. He had considered immigration to America at the time, as his notes for the year 1857 show. These notes detail a carefully worked schedule of steamship routes from Bremen to New York, including a detailed cost analysis. He may not have made the trip because of the associated expense and because besides another son, Max Haas (1855–1895), three additional children were born in Gaya, Selma Haas (*1858),[12] Hedwig Haas (1860–1906),[13] and the youngest son Friedrich Haas (1862–1937).

The social rise of grandfather and father was a typical phenomenon in the Austro-Hungarian-Bohemian multi-ethnic state after the lifting of all social restrictions and the full legal equality that Jews shared with other citizens following the revolution of 1848. Prince Metternich and the restorative policy of the previous thirty-five years had been swept away; a new era seemed to be dawning. In the Haas family, sufficient

[11] Adolfine Haas later married Josef Rosenfeld (1836–1920) from Koritschan in Moravia.

[12] Selma Haas later became the wife of Dr. Bernhard Fried in Gaya.

[13] Hedwig Haas married the headmaster Leopold Blüh (~1850–1899) who came from Bielitz in the neighboring Austrian part of Silesia (today Bielsko-Biala in Poland). The marriage remained childless.

Fig. 2.8 Dr. Gustav Haas in a painting from the turn of the century

capital had accumulated and although David Haas was not in good financial shape, his academic title gave him a considerable social advantage in Austria.

He was probably aware of this as it enabled his sons, Gustav (Fig. 2.8), Max and Friedrich to study law and for Arnold, a career in pharmacy. David's daughters Adolfine and Selma also found their role among the upper-middle-class with their marriages to Jewish academics. Thus within one generation, the Jewish merchant family had become a family of academics who sought their place in the bourgeoisie of the Austro-Hungarian Empire. This was accompanied by the disintegration of the traditional Jewish communities in Gaya and other South Moravian towns, a development that more traditionally minded rabbis desperately tried to stop.

Yet, traditionally minded Jews did remain with the community. One of these was Gustav's older brother Ludwig, the only one who, according to Gustav, had not studied because of an eye disease. As a merchant and trader, Ludwig remained faithful to the Jewish community even after his father's death. He married Rosalie Siebenschön from neighboring Straßnitz (today Strážnice) and was later listed as a factory owner and member of the Jewish community council of Gaya.

Ludwig Haas is never mentioned again in Haas' remarks; one can imagine that the differences in commitment and perspective led to an intellectual debate between the two brothers or even caused them to drift apart. His siblings moved to Brünn and Vienna, into the burgeoning intellectual and business worlds of the new age. This was especially true of Gustav Haas who married into one of the richest families in Austria, a connection that Arthur Haas particularly emphasized.

My mother, Gabrielle Strakosch, called Jella, came from a Moravian industrialist family, of whom, later, some were ennobled with the predicate "von Feldringen". My maternal grandfather, Bernhard Strakosch,[14] who died at the late age of eighty-eight in 1908, was one of twelve brothers, six having come from each of my great-grandfather's two marriages. With the exception of one, all the brothers became successful and very wealthy industrialists. The son of one of them, Eduard Strakosch, emigrated as a youth to South Africa, became very wealthy in the gold industry and then moved to London, where he was raised to peerage as Sir Henry Strakosch.[15] He repeatedly represented the British Empire as a financial expert in the League of Nations.

My mother was the eldest daughter of my grandfather Bernhard Strakosch. She was born on May 20, 1862 in Brünn; she had a younger sister Laura and two brothers, of whom the older Felix became one of the most important sugar manufacturers in Austria (+1931 in Vienna), while the younger one, Oskar, who earned a law degree and was a trainee with my father, died at the age of twenty-nine. Moreover, my mother had a younger stepsister Ida.

My grandfather was the only one of my four grandparents whom I knew; I was usually his guest on Sunday, and he often led me through his textile and cloth factory. In the summer we spent a lot of time in his villa in the Schreibwald [a large forest] near Brünn. His mother-in-law, my great-grandmother,[16] whom I only lost when I was twenty-one years old, also held me with heartfelt love. Being herself from a wealthy industrial family, named Flesch, she had married a leather manufacturer. Later she had the satisfaction of seeing all her brothers ennobled. Her two sons were also very important industrialists and both were consuls, one to the Netherlands,[17] the other to Turkey.[18]

Arthur Haas was rightly and very obviously proud of the fact that he came from a successful family dynasty. The Strakoschs, like the Haas family, had developed from limited Jewish beginnings in South Moravia into one of the most esteemed and prosperous families of the Austro-Hungarian Empire. [xii]

The Strakosch family came from the small South Moravian town of Butschowitz (today Bučovice and referred to in Jewish literature as Budespitz), east of Brünn. This was one of the oldest Jewish communities in the country, its population dating back to when Jews were expelled from Brünn in 1454.[xiii] The population list of 1673 already mentions a Lazar Straka as a Jewish merchant. The Jews were mainly employed as merchants of rags and clothes but they also had among them many tailors who had a special agreement with the local Christian tailoring guild.

The oldest officially registered members of the Strakosch family were Nathan Strakosch (1749–1820) and his father Dominik Strakosch (1718–1778).[xiv] Nathan Strakosch and his wife Judith had two sons: Salomon Strakosch (the elder) (1764–1843) who married Caroline Popper, the daughter of a local clothier in Butschowitz,

---

14 Bernhard Strakosch (1828–1908).

15 Henry Strakosch (1871–1943) was indeed an influential personality in British economy and politics. As a young immigrant to South Africa, he gained considerable wealth followed by considerable influence. In 1891 he entered banking in the city of London, where he played an important role in structuring colonial economies and banking. He published several books on economic issues and served as an important advisor for Winston Churchill, also providing fiscal support for the statesman who was always in need of it. https://en.wikipedia.org/wiki/Henry_Strakosch.

16 Babette Grünfeld née Flesch (1823–1906).

17 Wilhelm Grünfeld (1847–1921).

18 Arnold Grünfeld (1848–1919).

Fig. 2.9 A miniature portrait of Solomon Strakosch, the successful cloth merchant from Butschowitz, who moved his company to Brünn after 1848 and continued to expand there

and Löeble or Leopold Strakosch (1770–1841) who married Paulina Boehm in 1794. Leopold's son, Solomon Strakosch (1795–1867) married his direct cousin Anna Johanna Strakosch, the daughter of Salomon the Elder, in 1818. From this marriage Solomon had six sons, Jonas, Moritz, Isidor, Siegmund, Bernhard, and Eduard, and from his second marriage to Rosalie Wolf he had another seven sons, Simon, Dominik, Nathan, Matthias, Jakob, Julius, and Leopold.

Solomon Strakosch developed into an extraordinarily successful cloth merchant and entrepreneur whose portrait appears in Fig. 2.9. With the liberalization and dissolution of the ghettos after the revolution of 1848, he moved with his family to nearby Brünn and opened a cloth weaving and spinning mill. The close social network of Jewish families made it easy to provide necessary capital and to invest in the purchase of the latest spindles and looms. Families also invested in the construction of the railways that were rapidly developing at the time and established production facilities in favorable locations relative to these rail lines, ensuring rapid transport and trade. The six sons from Solomon's first marriage signed, in 1855, a partnership agreement to found their own factory which started operations in 1856 (Fig. 2.10).

The first report on the company's success declared,

> the factory enterprise of the Strakosch brothers is one of the better ones in the Crown land of Moravia and even in all of Austria and is worthy of the protection of the highest state administration in terms of its size, solidity and the excellence of its products, which have been well received and sold also on foreign markets.[XV]

Fig. 2.10 The six brothers Strakosch: Jonas, Moritz, Isidor, Siegmund, Bernhard, and Eduard. Individual brothers are not identified as investors except Jonas, who as founder of the company is in the center of the picture

With their success, the brothers already had plans to expand their business portfolio. With the founding of the Hohenau sugar works in Lower Austria, they entered a completely different line of business.

The brothers knew about the importance of sugar in Austria and also the resulting necessity to produce their own. As importing cane sugar from overseas became expensive and increasingly difficult, they invested in the cultivation and processing of beet sugar. As in the case of wool weaving, the Strakosch brothers drew up a partnership agreement to finance this new branch with each of the brothers investing 20,000 guilders. The factory was again conveniently located on the railway line to ensure fast transport.

Construction of the factory buildings was finished in 1868; the workers and skilled laborers came from domestic Moravia and also Slovakia, but the local population and agriculture benefited from the factory as well.[xvi] The sugar factory was an extraordinary business success and the brothers and their sons quickly rose to economic leadership in the Austro-Hungarian Empire. They had become business representatives of cloth weaving and the sugar industry, agricultural experts, and politicians, and they would soon represent Austrian culture as opera singers and impresarios. And so, over the following generations, the family was closely intertwined with Austrian society.

The next generation of brothers had already converted and were fully assimilated into Austrian social life. They no longer felt connected with the religion of their Jewish ancestors and they all considered themselves to be representatives of and advocates for German cultural life. The history of this closely intertwined family is described in detail elsewhere.[xvii] In the following, we will concentrate only on the family of Bernhard Strakosch (1828–1908) who married Regina Grünfeld (1843–1871), the daughter of Maximilian Grünfeld (1810–1881) and Babette Flesch (1823–1906) from Neu Raußnitz (today Rousínov), in 1861. The latter was the beloved Great-Grandmother whom Arthur Haas especially mentions in his memoirs.

We know nothing about the origin and origins of Maximilian Grünfeld but the Flesch family can be traced back to Rabbi Gershom Yehuda Flesch (*~1540) and thus to origins in the early sixteenth century in the Judengasse of Frankfurt.[19] His son Abraham Flesch (~1580–1640) moved to Vienna after marrying the daughter of the Prague rabbi Samuel Phoebus Teomin. After the expulsion from Vienna in 1670, the family settled in Neu Raußnitz under the name Rabbi Menachem Mendel Fischhof Auerbach. So it is quite possible that Arthur Haas had the stories of his Urgroßmutter (great-grandmother) in mind when he made this connection, referring to his origins in the Auerbach family.

Bernhard Strakosch and his wife had four children: Felix Strakosch (1865–1931) who became a partner in the Hohenauer Zuckerwerke and developed it into one of the leading companies in the sugar industry; Oskar Strakosch (1869–1898) who was applying to be a lawyer when he died early, at the age of twenty-nine; Laura Strakosch (1867–1919) who married Gustav Brum (1854–1907) and finally, as the oldest sister, Gabrielle Strakosch (1862–1916) who was married in December 1882 to the lawyer Dr. Gustav Haas. From Bernhard Strakosch's second marriage to Rosalie Rosenberg, which took place in 1874 in the Vienna City Temple, came another daughter, Ida Strakosch (1876–1943).[20] Ida Strakosch married the physician Dr. Louis Gomperz (1863–1922) in Brünn, who, as we will see later, maintained close contact with Arthur Haas and thus played an important role in his career.

Arthur Haas was born on April 30, 1884 as the eldest son of Gustav Haas and Gabrielle Strakosch, a fact he mentions only indirectly in his autobiography. The birth was recorded in the birth registers of the Israelite Cultus community in Brünn (Fig. 2.11). Despite his official membership in the community, there is no evidence of this in Haas' personal records. Accordingly, it is clear that he did not see himself as a member of the Jewish community but, rather, as part of the German population of Brünn. This corresponded with the perspective of the assimilated Jewish bourgeoisie who at that time increasingly distanced themselves from their traditional origins.[xviii]

---

[19] The family is named after the House of the Flasche (Bottle) in the Jewish ghetto. The house was built in 1530, for rabbi and schoolmaster Jakob of Prague, next to the House of the Spiegel (Mirror), cf. Alexander Dietz, *Stammbuch der Frankfurter Juden*, 80–82.

[20] Ida Strakosch married a doctor, Dr. Louis Gomperz and after his early death in 1922, she remained in Brünn. When the National Socialists took over in 1941, she was sent to the Theresienstadt ghetto as a sixty-five-year-old woman. Two years later in June 1943, she was transferred to the Auschwitz extermination camp where she was murdered in August 1943.

Geburts-Zeugnis.

Cultus-Gemeinde BRÜNN

Geburts-Zeugnis.

Fig. 2.11 Copies of Arthur Haas' birth certificate, provided in 1890 and 1904 by the Israelite community in Brünn

Arthur Haas grew up in the close family circle of the Strakosch family, as Gabrielle had a particularly close relationship with her father. His upbringing was in traditional middle-class ways. The religious influence of the local Israelite community can be considered negligible and, again, is nowhere in his later writings. These were carefree childhood days as depicted in Fig. 2.12.

Brünn was a pleasant, small town at the time and the Strakoschs were part of its upper society, through business interests in the wool-weaving mills and the sugar factory. Their position provided and required numerous social connections and obligations. As a child, Arthur was at leisure to spend many summer days at his grandfather's the villa in Schreibwald near Brünn (Fig. 2.13). This was a popular excursion destination for Brünn citizens and, "in childhood memories, [it] seemed to be the epitome of natural splendor, joy of celebration, forest darkness, mountain climbing, coffee enjoyment."[xix] His was a close circle of relatives and the families shaped the young Arthur Haas for the entirety of his life, in his liberal thinking, his bourgeois conservative views, and his cultural interests.

> During my childhood, I was raised in a completely carefree environment. My father was content with having a steady increase of income throughout his life and his income was always well above the cost of his increasingly comfortable life. My hometown of Brünn

Fig. 2.12 Yella Haas, née Strakosch with her first son Arthur, as a one or two-year-old in around 1885, and as a three or four-year-old in around 1887

Fig. 2.13 The summer resort Schreibwald near Brünn, on an old postcard. This is where Bernhard Strakosch had his retreat during the summer months, often with young Arthur

Fig. 2.14 Postcard view over the city of Brünn to the fortress Spielberg

belonged to Austria until 1918 and was at that time still a predominantly German city; only the lower classes of the population spoke Czech at that time. Almost all my relatives, in one form or another, were actively involved in maintaining the German character of their life and town.

Brünn was often called the "Austrian Manchester" because of its very important industry; but art and literature also flourished in Brünn, which at the time had about 80,000 inhabitants. The splendid city theater saw, for example, the rise of three artists who later became world famous: the later Viennese castle actor Josef Kainz,[21] the opera singer Maria Jeritza,[22] and the opera singer Leo Slezak.[23] Brünn was a beautiful city: in the center, there rose the historically famous Spielberg, surrounded by splendid parks (Glacis and Augarten) in the town itself (Fig. 2.14). To the north of the city was the "Moravian Switzerland" with magnificent mountains and wide forests.

When I was a little over two years old, my father moved from Grosse Platz 22 to a house a little further away from the city center, Alleegasse (now Legionärsstrasse) 19. There we lived on the second floor in a six-room apartment and had access to a very large garden. In this garden as well as in the neighboring public park, the Augarten, a large part of my sunny childhood took place.

---

[21] Josef Gottfried Ignaz Kainz (1858–1910) was an Austrian actor. He was regarded as one of the greatest German-speaking theater actors. In his honor, Gottfried was awarded the Kainz Medal in Austria each year from 1958 to 1999; this honor is bestowed for the significant achievements of actors and directors.

[22] Maria Jeritza (1887–1982) was a Moravian soprano singer; she became famous and sang for many years at the Vienna State Opera (1912–1935) and the Metropolitan Opera in New York (1921–1932 and 1951).

[23] The Moravian-born singer and actor Leo Slezak (1873–1946) was one of the most popular Austrian tenors in the first half of the twentieth century.

Fig. 2.15 The residence of the Dr. Gustav Haas family at Alleegasse 19, (today Kapitána Jaroše 19) in Brünn. The family lived on the second floor of the building

Thus, the family lived in a three-story manorial building at Alleegasse 19 (today Kapitána Jaroše 19); built in the style of the late nineteenth century, during the so-called founding period, it was situated on a tree-lined avenue and was bordered on the right and left by newly built tenement houses. The building is still preserved today, as seen in Fig. 2.15. Opposite and only separated by the two rows of trees on the avenue is the first Czech Grammar School of Brünn, opened in 1867 at Alleegasse 14 (now Kapitána Jaroše 14). In this school, through contact with Czech-speaking pupils, Haas became familiar with the bilingualism of his home town.

The Augarten (today Lužánky Park) is a twenty-two-hectare großer public park in Brünn that was opened and made accessible by Emperor Joseph II. With its meadows, groups of trees and streams, it forms an ideal playground. Nannies and educators took little Arthur and later probably also his brother Otto for frequent walks on the park paths, winding around sculptures, water fountains and streams.

When I was three years old, my younger brother Otto was born as the second child of my parents. I was closely attached to him throughout my childhood, despite frequent quarrels.

Fig. 2.16 A postcard depicting the Augarten in Brünn at the end of the nineteenth century

> Shortly before the birth of my brother, a Miss Klara Peche from Prussian Silesia joined the household as teacher. She stayed with us for over ten years and still maintains regular correspondence with us today (1937). We always spent the summers in the country; as a child of several months, I was in Karthaus near Brünn; as a one-year-old, in the Schreibwald near Brünn; as a two-year-old in Steindorf on Lake Ossiach in Kärnten [Carinthia]; as a three-year-old in Mondsee; later in Baden, Marienbad, Mondsee, Heringsdorf and finally for many summers in the Dutch vacation resort Scheveningen at the North Sea.

In fact, Arthur Haas and his brother Otto remained closely connected throughout their lives. They had similar interests and probably spent their childhood together in the Augarten shown in a contemporaty postcard in Fig. 2.16, sharing experiences that were only intensified by long summer hiking tours. The much younger sister Margaretha was only seen by her brother as a latecomer and hardly mentioned in the autobiography presented here.

The summer retreat was a tradition in the Haas family, inherited from Bernhard Strakosch who took yearly summer retreats with his family even if it was only to his villa in the Schreibwald. Later, holidays to the Austrian lakes in Carinthia and the province of Salzburg were added as well as to Baden near Vienna, where Bernhard died in 1908.

> When I was six years old, I entered elementary school; it was nearby, just a few houses away from our apartment. It was a school where the young educators from teachers' school were being trained. I still think of all my teachers there with great appreciation. My handwriting was so beautiful, at that time when my notes were sent to Vienna for an exhibition. When I was ten years old, I passed the entrance exam to Gymnasium.

Fig. 2.17 Augartenstraße with the teachers' training college on a contemporary postcard

Although Haas doesn't talk much about his experiences at elementary school, it is interesting to know that, by chance or luck, he had come to a special school that served as a training school for the local teachers' training college shown in Fig. 2.17. The school was located on Augartenstraße, close to the apartment and the playground, and could easily be reached by walking alone or with the nanny.

Teacher-training colleges were a new concept and, following the thought and concept of Humboldt's educational reform in Prussia, were incorporated into the Imperial and Royal State Schools in 1869 with the enactment of the new elementary-school law. With this law the elementary school, or primary school, was assigned a special significance as the foundation of the entire educational system. The concept places a special expectation on the teacher whose aim is to educate students to think independently and critically. The teacher's special training is necessary for his or her profession in two ways,

> firstly, in the material to be taught and in the inclusion of the variables that he or she is found with, through the undeniable demands of the times, including the realities (geography, history, nature, natural history, form), especially with regard to the new law on elementary schools. No less important, however, is the training of the teacher according to the form of teaching, that is, through acquaintance with the significant advances in methodology and didactics, as well as the more humane spirit in the discipline that, with the removal of the drastic means and the unjustified and unreasonable compulsion of earlier times, strives to develop the natural faculties of the child in a benevolent and reasonable direction, all-round and independently.[xx]

This represented a fundamental change from the education that, until then, had been primarily based on use of the switch and had sought to impart education through memorization and drill.

In this new endeavor, future teachers were not only given special training but also pupils to teach, specifically the pupils at the training school from whom teachers gained experience in a supervised environment. In addition to the classical subjects of grammar and arithmetic, importance was attached to training in literature and natural history. Therefore, the teacher was required to not only provide basic but also formal education and its reproduction, up to the fundamental of mathematics.

> The knowledge of elementary algebra should provide the student-teacher with a secure basis for understanding arithmetic in special numbers, whereby the latter must then be methodically practiced and brought to mechanical skill. The teaching of geometry must not be "lecturing" but teaching, i.e., heuristically and genetically developing, in order to advance its influence as practical logic. Continuous participation of the students in both arithmetical and geometrical teaching, a well-constructed set of questions that step by step instructs and illuminates the whole mental process leading to the development of truth, is as interesting for teachers and students as it is indispensable, since with this process the whole value and success of mathematical teaching stands and falls.[xxi]

These were the instructions provided to young teachers on how to arrange the lessons in elementary school. These ideas were revolutionary at the time and aimed to cultivate in both teachers and students an open and critical thinking, a thinking that would prepare the next generation for life in a modern and increasingly industrialized world.

This was far from the rule for the whole of the k.u.k.[24] educational system, but it was a stroke of luck for the young Arthur Haas who was able to profit directly and who eagerly absorbed the knowledge imparted to him, learning to look at his environment with frank criticism.

Notes

i. Rainer Krämer, *1000 Jahre Judentum in Regensburg* (Berlin: EPubli Verlag, 2016), 279–295.
ii. Siegfried M. Auerbach, *The Auerbach Family* (London: Perry Press Productions, 1957).
iii. Esriel Hildesheimer, "The Treves Families," *Avotaynu* 5 (1989): 17–22 and "Missing Treves Genealogy," *Avotaynu* 5 (1989): 39.
iv. Dan Rottenberg, *Finding Our Families* (New York: Random House, 1977), 10–14.
v. Ludwig Bato, *Die Juden im alten Wien* (Wien: Phaidon-Verlag, 1928), 11–13.
vi. Hugo Gold, *Die Juden und Judengemeinden Mährens in Vergangenheit und Gegenwart* (Brünn: Jüdischer Buch- und Kunst Verlag, 1929), 8–10.
vii. Ibid.

[24] A traditional short form for "Kaiserlich und Königlich" (imperial and royal), referring to the fact that the Habsburg empire was considered a dual monarchy ruled by the Emperor of Austria and the King of Hungary.

viii. Ibid., 407.
ix. Ibid., 275.
x. Ladislav Baletka, "Juden in der Geschichte der Stadt Valašské Meziříčí," Pressglas-Korrespondenz 2009-4-04 (6.12.2009), 1–46.
xi. Alexander Dietz, *Stammbuch der Frankfurter Juden von 1349—1849* (Frankfurt am Main: J. St. Goar, 1907), 135–138.
xii. See Maria-Theres Arnborn, *Friedmann, Gutmann, Lieben, Mandl, Strakosch—Fünf Familienproträts aus Wien vor 1938* (Wien: Böhlau Verlag, 2002), 105–135.
xiii. Hugo Gold, *Die Juden und Judengemeinden Mährens*, 173–176.
xiv. Jakob Baxa, *1867–1967: Hundert Jahre Hohenauer Zuckerfabrik der Brüder Strakosch* (Wien: R. Spies, 1967).
xv. Maria-Theres Arnborn, *Friedmann, Gutmann, Lieben, Mandl, Strakosch*, 114.
xvi. See Brigitte Semanek, Herbery Liska, Karl Schubtschik, Ernst Springer, Rudolf Springer, "Die Hohenauer Zuckerfabrik der Gebrüder Strakosch 1867 bis 2005," in Werner Kohl and Susanna Steiger-Moser, eds., *Die österreichische Zuckerindustrie*, 223–254.
xvii. Maria-Theres Arnborn, *Friedmann, Gutmann, Lieben, Mandl, Strakosch*, 105–135.
xviii. See William O. McCagg, *A History of the Habsburg Jews 1670—1918 (The Modern Jewish Experience)*, (Bloomington: Indiana University Press, 1989).
xix. Bertha von Suttner, *Memoiren* (Stuttgart: Hofenberg, 1909), 17. (Original text: "Kindheitserinnerungen als der Inbegriff von Naturpracht, Festesfreude, Waldesdunkel, Gebirgsbesteigung, Kaffeegenüssen.")
xx. Annual Report of the k.k. Lehrerbildungs-Anstalt und Übungsschule zu Brünn (Brünn 1871), 1–2.
xxi. Ibid., 6.

Chapter 3
High School Years

It is clear from his testimony that young Haas' experiences and opportunities in elementary school set the stage for his later Gymnasium education at the German Gymnasium in Brünn shown in Fig. 3.1:

> The First German Gymnasium in Brünn, which I attended from 1894 to 1902, is one of the oldest high schools in all of former Austria; it was originally founded in 1578 as a religious school but had already been turned into a state school at the time that my father studied there. I think of my old high school with the highest esteem, especially with regard to three very excellent teachers who had a great influence on my spiritual development: my mathematics and physics professor Josef Travniczek, my history professor Julius Miklau who unfortunately I only had until the sixth grade, and my Greek professor in the upper classes, Josef Wagner. It is of course inevitable that among the teachers there were also numerous who were very mediocre.
>
> The focus of the curriculum was the study of classical languages. From the first to the last grade, we had Latin for six to eight hours every week and from the third grade to the very end, we had Greek for about five or six hours a week. German language and literature also played an important role, but modern foreign languages were completely absent from the obligatory syllabus. The lessons in mathematics and physics were very good, but astronomy and chemistry were almost completely absent. In history, the emphasis was on the ancient and the history of the nineteenth century was neglected because of the political danger that such discussions posed; the same was the case for the history of recent literature. Geography was almost exclusively taught at the lower level. Hygiene, art history, economics, and political science were missing completely, while we were fed to the point of being bored on lectures about the Fatherland. A remnant of ancient tradition, offered in the last two years, was philosophical introduction, which was confined to logic and psychology, without any explanation of the great philosophical systems. Religious instruction was obligatory. I have always felt that the absence of any teaching in geometry and geometric drawing was a major shortcoming of my education. In general, we had five or six lessons a day, mostly in the morning and afternoon.

The subject matter described here is the typical material offered to students enrolled in classical grammar school. Learning the classics and the classical languages, such as Latin and Greek and in some cases Hebrew, was considered the basis of all education. Mathematics and German were also taught, and physics was considered a subject of importance. However, the latter more because of its classical

M. Wiescher, *Arthur E. Haas—The Hidden Pioneer of Quantum Mechanics*, Springer Biographies, https://doi.org/10.1007/978-3-030-80606-4_3

Fig. 3.1 The former German Grammar School in Brünn, today the Janáček Academy of Music and Performing Arts, where Arthur Haas spent his high school years

origin and less because of its influence and importance in the modern mechanized world, as with the neglect of chemistry. Modern languages had also lost their importance in school teaching. This is indicative of the increasing importance placed on German culture; in the nineteenth century, German art, German thinking, and German literature were increasingly considered superior as compared to the "French decadence" and "Anglo-Saxon merchant spirit." This was also reflected in the importance of "patriotism" in the Imperial and Royal Republic of Austria where the German language was considered the most important language.

It should also be borne in mind that in this imperial and royal multi-ethnic state, the German population had primacy. This led to perpetual tensions in the structure of the Habsburg Empire. Austria with its Bohemian, German, and some Italian possessions was regarded in the first half of the nineteenth century as a permanent part of German imperial unity, a unity that had to be reestablished after the collapse, in 1807, of the thousand-year-old Holy Roman Empire of the German Nation. The victory of the German states under Prussian leadership and the exclusion of Austria over France in 1871—with the subsequent founding of the Small German Empire under Prussian leadership—had only slightly diminished the feeling of being German. This was not only true for the German population in the multi-ethnic state but was especially true for the Jewish population who saw in the Empire their protection and the certainty of their equal rights. The Jewish people were regarded as one among the many peoples although the eventual collapse of the Habsburg Empire transposed them back into a dreaded minority status.[1]

> I was an unusually good high school student, always a privileged student and, with the exception of the first semester of the first year in which I was the second best, always Primus. I was the best in the class, usually with a very long lead over the second best.

[1] This was a fear and also a realized observation, especially reflected in the works of the two Austrian-Jewish writers Joseph Roth and Elias Canetti.

> Almost without any exceptions, I had credentials that were excellent in all subjects (apart from gymnastics and drawing in the lower classes). In the lower classes, my strength was especially in the study of Latin and Greek.
>
> After I had been at the high school for a few months, the provincial school inspector visited class. Repeatedly I had to read Latin texts aloud to him because he claimed that he had never heard such a beautiful pronunciation of Latin from a student. Already in the lower classes I demonstrated great interest in the topic of history. The lower classes at the Gymnasium [Middle School] ranged from the ages of ten to fourteen years.

These lines, like so many others in these records, express Haas' great desire to be the best and he did not shy away from showing his ambition and pointing to his successes. He wanted to impress his father as well as his wider environment and, as the anecdote about the state-school inspector shows, he was successful. He shone in all disciplines, which brought him success at school but later, during his time at university, he experienced difficulties and problems with fellow students and professors.

> Family life continued to be serene and carefree. When I was about thirteen years old, Miss Peche left us and we were first given a French governess for a few months. We then had, for a few years, a French tutor from Lausanne, Monsieur Lecultre, who later became a language teacher in Brünn. With his instruction, I soon mastered French as well as German. At home, I also took regular violin lessons. In sports I did skating and swimming, although the latter I learned late, as a thirteen-year-old at the summer resort in Millstadt [Kärnten - Carinthia].
>
> In November of 1897, our family celebrated the great event of the birth of a younger sister, Margarethe [Grete], to whom the two older brothers were always very tender. There was no lack of pleasures, of all kinds; in the Augarten I played with friends and contemporaries but always under the supervision of the "Miss" or some other warden.
>
> Numerous were the trips to the surroundings of Brünn. The summers we mostly spent in Scheveningen. We were also once at Millstätter See and once at Wörther Lake (Pörtschach), both of which are in Kärnthen. I am especially remembering fondly a fortnightly carriage ride through Tyrol, which was linked to the summer resort in Millstadt (1897).
>
> My father always gave the impression of a very satisfied man; he took great interest in our high school studies, and I had the opportunity to admire the depth of his general education. My mother also made a very happy impression, especially in Scheveningen which she loved above everything else.

Here, Haas describes his youth as the son of a bourgeois family, protected by a governess and court master. With special language instruction supplementing what was missed in school, he received the final polish of offspring who were being prepared for the future life of their class.

The summer retreat was an equally important component of upper-middle-class life. No one of "any good standing" wanted to stay in the hot city, whether Brünn or Vienna. In the summer these citizens fled to the countryside, either to the mountains or to the cooler climate of the Austrian lakes, as did the Haas family who traveled to the Millstätter See or the Wörther See. The newly built seaside resorts on the Baltic and North Sea coasts had also become fashionable. Scheveningen, situated on the Dutch coast, not only developed into a popular seaside resort for the Haas family but was generally regarded as the preferred destination, offering beach and bathing opportunities for Habsburg citizens from Bohemia and Moravia as shown in Fig. 3.2.

Fig. 3.2 Scheveningen's beach and spa, in the late nineteenth century

For those who appreciated a slightly less German-national bathing atmosphere such as the Pomeranian Baltic seaside resorts of Heringsdorf, Scheveningen was considered more neutral and international. Since 1884 there has been a large hotel and spa in the Renaissance style found along its beach, offering enough space and pleasure for a steadily growing number of guests. In the following decades, numerous new hotels were built along the coast for further accommodation.

As one can see in the two contemporary pictures, even at that time there were no deserted, unspoiled beaches during the Scheveningen holiday season; old and young people strolled along the beach, only to dive into the cooling waves from time to time in their high-necked swimsuits.

In this time from 1870 to the 1930s, it was the fashion of wealthy middle-class youth to wear a sailor suit, a custom that even extended to young Arthur Haas pictured in Fig. 3.3.

In the years of upper high school, from the age of fourteen to eighteen years, the change from boy to young man took place. Despite all this, my degree of independence was still very low and much, much less than that of my peers. In the fifth and sixth grades I was most interested in mathematics and history, which my excellent teacher Miklau captivatingly taught me, albeit with a strong German-nationalistic coloring. In the seventh and eighth grades, my interest in world history took a back seat when Professor Miklau left Brünn and in his place a rather mediocre teacher took over the history lessons. With dedication, I then devoted myself to the highest classes, in addition to the mathematics of physics.

Most of all, however, I studied Latin and Greek literature. Actually, I read all of the classical works of ancient literature as so-called "private reading," about which one could take examinations in school. I especially loved Homer, Plato, Sophocles, Aeschylus, Euripides, and the Greek poets, as well as Ovid and Catullus. I loved the Roman classics less than the Greek, but I took quite a particular dislike to Vergil, Cicero and Livy.

German literary history as well as the purely descriptive sciences, I considered rather boring and uninteresting topics in high school. From the electives, I only studied stenography and won, as a sixteen-year-old, a prize in a large public competition. We continued with the French language in school and when I was about seventeen years old, I also began to study English at home, with an old English lady who made teaching as boring as possible. Later, I always deeply regretted that in my education English played a much less important role than French.

Fig. 3.3 August Ernst Haas in a sailor suit, from around 1894

Arthur Haas valued his teachers and valued the subject matter, at least the subjects that were taught to him in a captivating way. He was not necessarily a critical and challenging student, although the first tendencies toward this are evident in his report.

He appreciated the mathematics and physics teacher Josef Travniczek, who probably looked after Haas throughout his secondary school years and kept his interest in mathematical sciences alive. Travniczek thus became the determining influence for young Haas' later career. Unfortunately, not much could be learned about the origin and fate of such a talented teacher except that he later switched to the k.k. Gymnasium in Vienna's third district. According to his name, he came from Bohemia or Moravia. His main interest was probably in geometry since he published textbooks for secondary school in 1910 and 1911, on the geometry of plane and space. In an earlier publication of 1889, Travniczek also discussed, in an historical sketch, the problem of circle measurement before Archimedes—an approach that looked at modern problems from the perspective of ancient thinkers. This approach was also extensively used by his student Arthur Haas.

Things were different for Julius Miklau, the history teacher, who privately published historical articles and writings dealing especially with forgotten figures of Austrian literary history from the age of Empress Maria Theresa. Miklau left the Brünn Gymnasium in June 1900 and transferred to the State Gymnasium in Marburg, Austria, and from there, later in 1905, to the II State Gymnasium in Graz, Styria. He was held in high esteem and the Brünn Gymnasium announced his transfer:

> Professor Julius Miklau had belonged to the local Gymnasium since 1888, i.e., for twelve years, and he was held in high esteem because of his open, honest character. As a teacher,

> he was able to attract the attention of the students and keep their interest alive through his fascinating lectures.[i]

He was replaced by Max Hansmann from Bregenz. This teacher was obviously less appreciated by Haas who did not even consider Hansmann worthy of being mentioned by name.

Arthur Haas remained devoted to the classics; he especially appreciated classical Greek literature but less the literature of the Romans. This may have been due to the literary prejudices that were widespread in his humanistic Gymnasium or due to his natural tendency toward philosophical classical literature, which he expressed throughout his life.

The criticism of German literature classes, which he describes as simply boring, is surprising. His perspective may have been due to the special teacher who was not further named; it also may have been a retrospective impression since these notes were written after his emigration from Austria which had in the meantime been incorporated by Germany. We don't know the detailed curriculum and lesson plans regarding German literature but throughout his academic career, Haas loved to refer to Goethe in particular and to quote Goethe whom he held in high esteem. It can therefore be assumed that the German lessons were not entirely in vain.

> Of course, I had a lot of friends among my high school colleagues. Among them were, above all: Paul Blum, later a secondary school professor in Brünn; Leo Branczik, later a poet and civil servant; Josef Hroch, later a lawyer and counselor in Vienna; Ottokar Stoklaska, later a lawyer; and Leo Lederer, later a lawyer and journalist. To my circle of friends and colleagues, I now also added acquaintances with the young girls from Brünn. From the age of fifteen, I regularly attended dance lessons with my brother and was also invited to small youth gatherings.
>
> Even during the summer, we did a lot of socializing. Scheveningen remained the preferred summer resort. Besides this, we once spent part of the summer in Blankenberge, a Belgian seaside resort, and once a summer on the Semmering. We also visited many foreign cities in the summer, including the Hague, Amsterdam, Brussels, Antwerp, Munich, Nuremberg, Prague, Berlin and Hamburg. We frequently visited Vienna, where we always stayed with our Aunt Laura,[2] the sister of our mother. More and more often we went to the theater, where my parents subscribed to a box.

Haas lists here a series of liberal and mostly Jewish friends who later, as officials and writers, carried a certain importance in the city of Brünn during the Fin de Siècle. Still others remained in Vienna after the war and made careers as officials and journalists; they are therefore mentioned with respect. The tragedy of the fall of the Habsburg Empire for his fellow Jewish citizens is reflected in their stories.

The Expressionist writings of Leo Branczik had already been published at the turn of the century, along with his comments on Kierkegaard and Wedekind, but these have been largely forgotten. Branczik disappeared in the turmoil of the First War.

The later lawyer and teacher Ottokar Stoklaska (*1883) was the son of the Moravian poet and Brünn German teacher of the same name, who was also the director of the Brünn Girls' Lyceum. The father determined literary and cultural life in Brünn

---

[2] Laura Brum née Strakosch (1867–1919).

at the turn of the century. Like his older brother, the son wanted to become a lawyer; he served as an imperial and royal lieutenant during the war but after the end of the war, he remained in Brünn, succeeding his father as the director of the Lyceum. With the collapse of the Habsburg Empire, his star also sank into oblivion.

Dr. Josef Hroch became a ministerial official in the Austrian Ministry of Economics and in the course of his later career, he was mainly concerned with questions of the railway and the growing motor vehicle traffic.

During the war, Dr. Leo Lederer (1883–1946) was the Bucharest correspondent for the Berlin daily newspaper and sent Haas numerous postcards from all parts of the world. After the end of the war, in the 1920s, he became the Viennese correspondent for the same newspaper, before emigrating to Paris. He was arrested there by the Gestapo in 1940 and remained in custody until 1944 when units of the American Army liberated him. After the war, he became the New Austria correspondent in Paris.[ii]

Haas continued to correspond with all of these friends even after his time in Brünn and thus remained informed of their progress as well as about the social life in Brünn.

Besides socializing with these friends, young Haas was also part of the social life of the bourgeois society of Brünn. In particular, he went for the dance lessons that were a social obligation organized by the Gymnasium. This basic requirement was meant to strengthen the bonds within the society and also to motivate acquaintances among the young people, which under careful observation and guidance should lead to marital unions. The small receptions and invitations, as well as the concerts and theater visits, were intended to introduce young people to society. These activities usually took place among the upper classes of the Gymnasium and only students of Prima and Secondary classes could take part.

> In the best tradition, I ended my high school career with a final exam, the so-called Matura. For the final grades in all subjects I was given without exception the best mark, namely "Excellent," and in Greek I was given the addition, "under emphasis of the particularly successful initiatives in private reading." In Latin, it was very petty-minded to omit the same addition simply because, absent-mindedly, I let the diction of a fourth-declination word end in "u" instead of "ui" in my written exam.
>
> The Matura, which was generally regarded as a very severe test at the time, was divided into a written and an oral exam. The written exam took five half-days and included a mathematical section with four problems: a German essay, a translation from German into Latin, one from Latin into German, and one from Greek into German. We also had to interpret a quote from Goethe as a subject topic in German, as well as a section from Suetonius and one from Isocrates in the translations.
>
> In the mathematical work, for which one had five hours, the professor was most surprised when I returned the work after half an hour; he did not want to believe that I had solved all of the problems correctly, since solving a Diophantine equation[3] alone requires at least one hour. But I had solved this with just a stroke of the pen—against all expectation of the professor. The task was: Which numbers divided by 3, 7, 11, 13 give the remainders 1, 6, 10, and 12? The solution according to the method taught to us would indeed have been very

[3] A Diophantine equation (named after the Greek mathematician Diophantos of Alexandria, around 250) has the form $f(x_1, x_2, x_3, \ldots x_n) = 0$, where f is a given polynomial function with integer coefficients that must be determined when solving the equation.

> complicated. In this special case, however, I could simply write down the solution in the following form: "Let's call the number **Z**, then **Z** + 1 must be a multiple of 7, 11, and 13, which is a multiple of 7x11x13 or 1001. Since 1,000 divided by 3 gives the remainder 1, 1000 + 3003 x **n** is the general solution."
>
> The oral exam usually included Latin, Greek, mathematics, German literary history, history and physics. But the best students were exempt from taking the last three subjects, and so I was liberated. I had to translate something in Latin from Tacitus, and I impressed the superintendent of the school board when I was able to answer his question of whether French literature (which we did not study in high school) shows counterparts to the "Ars Poetica" of Horace. In Greek, I got to translate something from Plato, and in mathematics I had a relatively easy problem to solve.
>
> I took the written Matura in early June 1902 and the oral I passed on July 2, 1902. Following the Matura was the usual farewell party. During the party my teacher, Prof. Travniczek, told me, in good humor, that what impressed him the most about me was that I received excellent grades in all subjects during the whole eight years of the Gymnasium but never received a really good grade in "behavior." Ten years after graduation, in 1912, we returned to Brünn for a reunion. When the 25th anniversary was celebrated, I was in New York but the following year we gathered again when the 350th anniversary of the old venerable Gymnasium was celebrated in Brünn.

Josef Travniczek and Julius Miklau were the two teachers who most supported and influenced young Haas. Especially Travniczek succeeded in fascinating him so much with mathematics and physics that Haas chose these for his course of study at a time when they were still unpromising endeavors. These teachers did not correspond at all with the picture that Arthur Haas' contemporary and distant cousin Stefan Zweig (1881–1942) drew in his memories, *Die Welt von Gestern* (The World of Yesterday). In this work, Zweig describes a much different relationship between teacher and pupil: "They were sitting on the top of the lectern and we below, they asked, and we had to answer, otherwise there was no connection between us."[iii]

Notes

i. Benno Imendörffer, Karl August Schwertassek, *Jahres-Bericht des ersten deutschen k.k. Gymnasiums in Brünn: für das Schuljahr 1900–1901* (Brünn: Rudolf M. Rohrer, 1901), 28, 59.

ii. Werner Röder, Sybille Claus, Daniel S. Niederland, and Beatrix Schmidt, eds., *Biographisches Handbuch der deutschsprachigen Emigration nach 1933–1945* (Munich: De Gruyter Saur, 1983), 423.

iii. Stefan Zweig, *Die Welt von Gestern* (Stockholm: Bermann-Fischer, 1944).

Chapter 4
Years at the University

At the university, young Haas was confronted for the first time by life outside of the bourgeois atmosphere in which he had been protected by parents and prosperity. His hopes of being able to continue his academic studies quickly and purposefully, admired and encouraged by his professors, failed. Unlike in high school, the structures of the university and university life offered little direct guidance and supervision. The aim of university education was to promote independence, independent thinking, and goal setting. This seems to have caused Haas difficulties. His student endeavors showed little coordination or direction and he seems to have been distracted and unsteady—a young man full of self-confidence and the will to show and shine, but without the basic awareness of which discipline he wished to pursue or in what direction he saw his future. He did not want to follow his father's plan and take over the legal practice; he did have in mind the fascinating prospects of the natural sciences:

> As harmonious as my Gymnasium times and studies were, my studies at the university were unstable, disorganized and aimless. Above all, I wanted to study physics, or at least an exact science. My dad had a tacit desire for me to become a lawyer and, as such, his successor, but since he was not only a lawyer, but also involved in the cement and sugar industries, he agreed that I should become a chemist. So I decided to go for that discipline and it was only natural that I move to the university where my grandfather and father had earned their doctorates. In October 1902, I began my studies at the University of Vienna. I lived in lodgings provided by an old lady who also included meals, in Wickenburggasse 16.

Fatherly support was necessary in choosing a course of study and it is understandable that his father would have preferred that his son study law. A bourgeoisie future would be assured in the practical application of studies such as medicine and law, while the natural sciences offered few practical applications at that time. University positions were rare and offered as little prospect for increasing the family wealth as teaching positions did in the schools. Chemistry at least offered hope for a career in the sugar industry that was very important to the family, both near and far. Gustav Haas was an influential partner in the Strakosch factories and their growth as well as the growth of his law practice were very dear to his heart, as the following will show.

Gustav supported his son, although apparently somewhat reluctantly, and helped him find his accommodation near the university, in a quiet area between the

M. Wiescher, *Arthur E. Haas—The Hidden Pioneer of Quantum Mechanics*, Springer Biographies, https://doi.org/10.1007/978-3-030-80606-4_4

Votivkirche and the town hall. This four-story apartment building, from the Wilhelminian era, was where Haas would live during his first year at the university. From there he could easily reach the Institute of Physics at Türkenstraße 3, an inadequate and outdated building. Before 1904, professors Josef Loschmidt, Josef Stefan, Ludwig Boltzmann, and Franz Serafin Exner taught there until their call for a new building was answered. The young student Lise Meitner, who had enrolled to study physics a year earlier, described the building in the following way:

> At that time the Institute of Theoretical Physics was located in a very primitive apartment building, converted for this purpose, at Türkenstraße. The entrance looked like the entrance to a chicken house, so I often thought, "If a fire breaks out here, very few of us will come out alive."[i]

Arthur Haas seems to have been less worried, for he does not mention the structural shortcomings of the institute and only the shortcomings of the scientific teaching he received:

> In the beginning, I attended the usual beginner's lectures: Experimental Chemistry (with Lieben[1]), Experimental Physics, Differential and Integral Calculus (with Escherich[2]), History of Philosophy (with Jodl[3]), and Practical Philosophy (with Laurenz Müllner[4]). I also went on a daily basis, for many hours, to the chemistry laboratory for first-year students. Lieben's lecture was quite good and careful, but completely dispassionate; Jodl's lecture was interesting while Müllner's was captivating; Escherich's lecture was certainly excellent, but much, much too difficult for a beginner.

---

[1] Adolf von Lieben (1836–1914) was an Austrian chemist who worked in Vienna, Heidelberg and Paris. He received his doctorate under Robert Wilhelm Bunsen. After his habilitation in Vienna, he taught at the universities of Palermo, Turin and Prague before he became a full university professor and chairman of the II Chemical University Laboratory in Vienna from 1875 to 1906. He developed new organic-synthetic methods using, among other things, the theory of atomic chaining in carbonyl compounds and hydrocarbons. Von Lieben introduced a more comprehensive systematic approach to research by being the first Viennese professor to assign related problems as dissertation topics to his students, leading to more effective research. Cf. Mathias Luger, "The development of the chemical institutes of the University of Vienna in the twentieth century," (diploma thesis, University of Vienna, 2011).

[2] Gustav von Escherich (1849–1935) was an Austrian mathematician. After studying at the University of Graz, he taught at the universities of Graz and Czernowitz before being called to Vienna in 1884. He was Dean of the Faculty of Philosophy from 1892 to 1893 and Rector of the University of Vienna from 1903.

[3] Friedrich Jodl (1849–1914) was a German philosopher and psychologist. Jodl studied history and philosophy in Munich and subsequently taught at the Bavarian War Academy in Munich and the German University in Prague before being appointed professor of philosophy in Vienna in 1896. Among his students were the writer Stefan Zweig and the cultural philosopher Egon Friedell, both fellow students of Arthur Haas.

[4] Laurenz Müllner (1848–1911) was an Austrian philosopher and theologian. In 1876 he received his doctorate in philosophy at the University of Innsbruck. From 1880, Müllner lectured on philosophical and theological propaedeutics for the Faculty of Theology in Vienna, where he became associate professor of Christian philosophy in 1883. He was one of the moral and intellectual supports for young Haas during the time of his academic difficulties and doubts.

> The beginner lectures in physics were a scandal. At that time there were two professors for experimental physics. One of these was Franz Exner,[5] who did not lecture during almost my entire time as a student, for some or the other reason. The other professor, a zero, inflated to bizarre fictitious size, was Viktor, Baron von Lang[6]; either at that time or later, he was a member of the upper house in parliament and president of the Academy of Sciences. (When much later a memorial was unveiled at the University, the keynote speaker was unable to mention a single scientific achievement by Lang in the list of countless honors.) I had enrolled in Lang's lecture on experimental physics; about 400 listeners gathered for the first lecture, while the hall only held half. Lang appeared and, while he turned his back on the audience the entire time and looked at the blackboard, he grunted all sorts of things into his beard, which nobody understood. For the second lecture only four students showed up, including me, and afterwards I gave it up. Since there were no other possibilities offered, I attended a rather crazy but at times interesting lecture on experimental physics, held by a nerd, the private lecturer James Moser.[7]

All in all, a devastating assessment of the physics professors by a student who had hoped for personal guidance and instruction. Instead, Haas had to share lecture halls with hundreds of other young students and in return, he was only able to view the almost unattainable academic leaders and professors from a distance.

---

[5] Franz-Serafin Exner (1849–1926) came from one of the most important academic families of the Habsburg Monarchy. He studied in Vienna and Zurich and was awarded his doctorate under August Kundt. An associate of Wilhelm Conrad Röntgen, Exner habilitated at the University of Strasbourg and, at the instigation of his Viennese mentor Viktor von Lang, was called to Vienna in 1879. There, Exner succeeded Johann Josef Loschmidt as head of the Second Physical Institute in 1902. In 1908, Exner was appointed rector of the University of Vienna. In contrast to the negative evaluation by Haas, Franz Exner was described by his students as a versatile, highly educated and cultivated physicist with strong visions. On the other hand, his student Lise Meitner was not very enthusiastic about his lectures on experimental physics, which lulled her into a nap. However, in research Exner was a pioneer in many areas of modern physics. Early on he dealt with the radically new topics of radioactivity, spectroscopy, electrochemistry and electricity in the atmosphere. The physics of the early twentieth century in Austria was determined by Exner's students. Among the most famous are: Stefan Meyer as head of the Radium Institute; Victor-Franz Hess, the discoverer of cosmic radiation; the quantum physicist Erwin Schrödinger, as well as Lise Meitner, the co-discoverer of nuclear fission. Exner probably never found the right connection to Arthur Haas, stating: "A self-confident young man, with less talent for experimentation than for philosophizing about the intellectual origins of physics." This was probably not quite where Exner's interests were focused.

[6] Viktor von Lang (1838–1921) was co-founder and pioneer of the study of crystallography. He studied in Vienna and received his doctorate in 1859 on "Physical Conditions of Crystallized Bodies" at the University of Giessen. After two years as an assistant to Gustav Robert Kirchhoff and Robert Wilhelm Bunsen in Heidelberg and Henri Victor Regnault in Paris, he returned to Vienna in 1861 and habilitated on "The Physics of Crystals." After teaching at the Kensington Museum in London and at the University of Graz, he was appointed to the chair of physics in Vienna in 1866. His scientific field was crystal physics as well as optical and electrical aspects of crystal bodies. In contrast to the assessment made by the young student Haas, the *Austrian Biographical Lexicon 1815–1950* emphasizes Lang's great didactic abilities and explains, "The extraordinary spatial limitation of the physical institution and the meager equipment at his disposal allowed only a modest addition of experiments to his lectures, but this did not in the least detract from the clarity and intelligibility of his lecture." *Austrian Biographical Lexicon 1815–1950* vol. 4 (Wien: Verlag der Österreichischen Akademie der Wissenschaften, 1969), 444, 445. This is probably a case of historical whitewashing because colleagues, including the private lecturer James Moser, confirmed young Haas' perspective.

[7] James Moser (1852–1908) studied physics in Berlin where he received his doctorate in 1876. In 1884 he habilitated in physics, including physical chemistry, at the University of Vienna, served

Fig. 4.1 Left: Viktor, Baron von Lang (1838–1921) . Right: Franz-Serafin Exner (1849–1926). These luminaries of modern physics developed the disciplines of crystal and radiation physics, precursors of today's solid state and nuclear physics

At the time of Ernst Mach (1838–1916) and Ludwig Boltzmann (1844–1906), Vienna was considered a Mecca of modern physics and, in the German-speaking world, only Berlin and Göttingen could approach its rigor. These were the theoretical physicists and philosophers of science whose famous disputes over the existence of the atom would determine science. In addition to Mach and Boltzmann, there were also renowned experimental physicists who worked on the forefront of modern physics. Victor von Lang and Franz Serafin Exner, shown in Fig. 4.1 belonged to this group of leaders, playing a critical role in the pursuit of disciplines such as X-ray and radiation physics in Vienna.[ii]

Therefore, the verdict of young Haas seems harsh and probably had, at least in part, less to do with a well-founded assessment and more to do with his personal disappointment in the lack of accessibility, a degree of accessibility to which he had grown accustomed during his days with the gymnasium professor Josef Travniczek. Young Haas still had to learn to study from a place of personal responsibility and not in order to please his teachers.

---

as assistant to Johann Josef Loschmidt until 1891 and continued to represent him as a private lecturer in some lectures. In Vienna, Moser repeated Heinrich Hertz's fundamental experiments on electromagnetic waves. As a lecturer, Moser was popular in the university world—for example, in physics lectures for doctors and philosophers—as well as in popular-scientific circles. Moser confirmed Arthur Haas' description of lecture conditions with Viktor von Lang in a letter to Heinrich Hertz. In the letter, Moser describes how, although Lang had an enviably large number of students enrolled for his lecture, up to 500, the number of listeners rapidly dropped below ten. Christa Jungnickel, Russell McCormmach, *The Now Mighty Theoretical Physics, 1870 to 1925*, vol. 2 of *Intellectual Mastery of Nature. Theoretical Physics from Ohm to Einstein* (Chicago: University of Chicago Press, 1986).

At first, I devoted myself only to studying, from morning to late evening, but I was by no means satisfied. There was no one from whom I could learn Physics. In the chemical laboratory, I was by no means among the best because I lacked the manual skills for analysis, so that even female students got better results than me, which hurt me very much. In addition to the laboratory, there were no classes in which I could have distinguished myself as I once had in high school. This depressed me a great deal and certainly did not lessen my complete loneliness, which was only occasionally interrupted by visits to the theater with my father, who often stayed in Vienna. Only occasionally did I meet with high school colleagues who also studied in Vienna; I also spent many hours with the medical student Ernst Weiß[8] who also came from Brünn. Weiß later became a doctor and a well-known poet, and he often used my family as a model for his novel, *Die Galley*.[9]

This is an amusing passage that reflects the attitude of young Haas who could not accept that anyone was superior to him, especially not the young female students. He blames his difficulties on a lack of manual skills, probably alluding to the "natural" cooking skills that presumably gave the women an advantage in chemical analysis.

But this cannot be considered a real excuse, particularly when one considers the female students who were taking the practical courses in chemistry with him. Women were not admitted to study in Vienna until 1897 and then mainly so that they could serve as teachers in the public schools.[10] Practical courses in chemistry were compulsory for all students of the natural sciences. Two-thirds of the female students came from the Jewish bourgeoisie and played a leading role in the women's movement.[iii]

It is likely that one of the female students who denied young Arthur Haas the leading role in chemical internships was Lise Meitner, later co-discoverer of nuclear fission, who began her studies in 1901 and received her doctorate from Exner in 1906; another was Selma Freud (1877–1962) who also received her doctorate from Exner in 1906 with a thesis on the photoelectric effect; there was also Emerike Volkmann (*1878) who worked on the separation of thorium isotopes and received her doctorate on this topic in 1908 after only four years, as well as the somewhat older Hermine Hug von Hugenstein (1871–1924) who received her doctorate in 1909 on radioactive precipitation in the electrolysis of radioactive liquids and who later made a name for herself as a psychologist. In addition, Olga Steindler (1879–1933) was the first woman to receive a doctorate in Vienna, in 1903 with the thesis, "On the Temperature Coefficients of Some Iodine Elements" under Franz Exner and Ludwig Boltzmann. She married her fellow student Felix Ehrenhaft in 1908 and continued to

[8] Ernst Weiß (1882–1940) studied medicine and was a doctor in Bern, Berlin and Vienna. After a meeting with Franz Kafka, Weiß decided to become a writer. He committed suicide in his hotel room after the Wehrmacht invaded Paris in 1940.

[9] A somewhat strange comment since the family of the novel's young physicist and radiologist, Erik Gyldental had a disturbed relationship with both his father and his mother. The story does not offer a harmonious example of bourgeois family life, with Gyldental raising his hand against his mother; certainly a more expressionistic than realistic family drama. See Ernst Weiß, *The Galley* (Berlin: S. Fischer Verlag, 1913).

[10] Between 1903 and 1909, seven female students graduated in chemistry and six in physics from the University of Vienna, http://scienceblog.at./frauen-den-naturwissenschaften-die-ersten-absolventinnen-der-universit%C3%A4t-wien-1900-1919#.WtQg9rgTWJU.

campaign for women's access to higher education in Austria. In addition, there were Marie Buchmayer (*1874) from Prague and Rosa Uzel (1878–1967), who received their doctorates in 1908 on an unknown subject. Apart from Lise Meitner, none of these talented students embarked on a career in physics; they became teachers and usually pursued other disciplines.

Arthur Haas was, however, according to his own statements very interested in young women. This interest was limited to his social class and did not extend to fellow female students, whom he apparently regarded as "older blue-stockings" who due to their better chemical-cooking skills stirred his competitive spirit. And so apart from old school friends such as the medical student Ernst Weiß, Haas had little opportunity to expand his social connections. His reference to his friend's first novel, *The Galley* and its connection to the Haas family is therefore also significant.

The novel describes both the research and the love challenges of a young medical physicist who successfully experiments with strong X-rays. But he is less successful in his relationships with young women, which is psychologically explained by strong parental ties. His relationships often manifest themselves in emotional outbursts and violence. The young researcher suffers from these circumstances and from his inability to communicate with women, but he eventually succumbs to radiation death.[11] To what extent the Haas family was reflected in the novel published in 1913 remains unclear but Haas may have given his friend some physical background on X-rays and their effects on the human body.

Nevertheless, Haas was clearly lonely in his early years of study:

> The loneliness caused me, after a few months, to take a very ill-considered step, to which my father reluctantly gave his consent; I became a fraternity member. Not that I think badly of my former friends, but the loss of time that this fraternity affiliation caused was considerable. My involvement kept me far too often from Viennese cultural and social life and, also, the original claims and ideals of the fraternity were too far from my own expectations and ideals.
>
> The fraternity, Fidelitas, was founded in 1876 as part of the German Academic Association (later fraternity). The members wore dark green caps and green-white-green ribbons. The group was well organized and required saber fencing events, which were called Bestimmungsmensuren [Introduction Rites]. Fidelitas identified itself as national liberal, as the term was called. (Already a few months earlier I had become a member of the German Academic Reading Association in Brünn, which had black-green-gold bands.) In contrast to Germany, drinking and eating habits in the Fidelitas were very moderate.
>
> Of my former fraternity brothers, I especially like to remember: Ludwig Ferneböck,[12] later a Hofrat [German honorary title which might be translated as court councilor] and a lawyer; Willi Hellmann,[13] who was at the time a factory director and my original Leibbursch [mentor], and who later, like his father, became an important Bohemian textile-industrialist;

---

[11] The latter is an interesting ending that clearly underlines the fact that the physics and medical communities were already aware of the dangers of X-rays and other radioactive radiation in 1910.

[12] Ludwig Ferneböck (1882–1942) came from a wealthy Jewish family of factory owners. After attending a traditional grammar school, he studied law. He then became a civil servant and ended his career as a ministerial councilor before moving to the supervisory board of the Länderbank. He was a member of the Social Democratic Labor Party and was denounced and arrested by the Gestapo in 1942. He was murdered after three months in the Mauthausen concentration camp.

[13] Willi Hellmann (1883–1964) was a Bohemian silk manufacturer who, after the end of the war, moved back to Vienna where he lived with his wife Daisy née Steiner. After the Anschluss of

> the second person of greatest assistance to me was Robert Fischer, who later became a PhD and chemical manufacturer in Berlin; and furthermore, the two future physicians Oswald Hesky[14] and Walter Sterz were helpful to me. Of older Fidelitas members, I was especially friends with Oskar Scheuer,[15] later a physician and author of several books and publications on the history of student life, and Eduard Beyer who later became a general physician in the Austrian Army.

Instead of describing his personal impressions as a fraternity member, Haas again takes the opportunity to provide names and his connection to fellow members who later became well known and regarded. These young men were mostly from Jewish families—not all of them could be identified—which shows that Fidelitas was a more liberal fraternity as compared with more German nationalist fraternities that would not allow Jews as equal members. Nevertheless, the classical traditions of fraternities, such as drinking and sabre fighting, were preserved. Haas found these aspects quite unpleasant but was attracted to friendship and recognition and so he accepted these traditions, trying to take a leading role in them.

> That first Christmas, I spent the holidays in the Brünn family home, carrying my green cap everywhere with childish pride. And so, the first year of study was largely filled with fraternity life but I still studied more than the average student. My parents often visited me in Vienna, especially my father for whom business affairs kept him in the capital; he always took me to the theater or a concert. At the end of the first year, I took, without being urged by my parents, an exam in organic chemistry with Hofrat Lieben. I spent part of the summer of 1903 in my parents' Brünn home and part of my time in Alt-Aussee, in Salzkammergut, where we saw Emperor Franz Josef with the Romanian king,[16] from near-by.

Despite the difficult political conditions in the Habsburg Empire, Vienna at the turn of the century was a Mecca for music and the arts. A wide variety of concerts and theater performances—in the most diverse directions of art, music and theater taste—were offered daily. There is of course a question regarding where the liberal-conservative father led young Haas. The cultural life of the city was largely influenced by so-called Viennese Modernism[iv] which, in these final years of the monarchy,

---

Austria he fled, leaving his art collection behind. He escaped via Czechoslovakia to Paris and from there, in 1941, by way of Spain and Portugal to Brazil. In 1950 the couple returned to Vienna. Cf. Gertraud Diendorfer and Herbert Pichler, *Kunstraub—Raubkunst, Informationen zur Politischen Bildung*, vol. 20 (Bozen: Studien Verlag, 2004), 144–145.

[14] Oswald Hesky (1882–1942) studied medicine and pharmacy. The date of his death is not known but he was probably killed as his brother, Dr. Egon Hesky was in 1942, murdered in the Polish transit-camp Izbica or in one of the subsequent extermination camps.

[15] Oskar Franz Scheuer (1876–1941) was an Austrian dermatologist and student historian. He held the view that Heinrich Heine was thrown out of the Göttingen fraternity not because of his Jewish origin but because he had violated the prevailing chastity law. It can be noted that after the annexation of Austria in 1938, Scheuer lost his work permit as a doctor, being a Jew. In addition, he had to sell his student-history collection; this collection formed the basis for the Institute for Student History in Vienna, which was founded in 1939. Scheuer was arrested with his wife in Vienna, in October 1941, and taken to the Litzmannstadt concentration camp (Łódź) where he was probably murdered. Friedhelm Golücke, *Verfasserlexikon zur Studenten—und Hochschulgeschichte* (Köln: SH-Verlang, 2004), 288.

[16] Karl Eitel Friedrich Zephyrinus Ludwig von Hohenzollern-Sigmaringen (1839–1914) was from 1866, Carol I. Prince and after 1881, King of Romania.

once again produced intellectual excellence in the city's creative culture. New art and stylistic associations, such as the Vienna Secession and the Wiener Werkstätten (workshops), shaped craft and artistic creation.

The historical architecture of the founding period, that had dominated until then, came to a rapid end. Painters such as Gustav Klimt, Oskar Kokoschka, and Egon Schiele set new goals and directions, sharply distinguishing themselves from the naturalism that had dominated art. But most notably, the music and the kinds of concerts offered to father and son Haas changed rapidly. These new developments thrived under the influence of the late-romantic tonal symphonies of Anton Bruckner and Gustav Mahler, as well as the tone pieces and radically new twelve-tone music from Arnold Schönberg to Anton Webern. The theater was no different; the stages offered plays from the conservative Hugo Von Hoffmansthal of wealthy Jewish-Moravian heritage to Arthur Schnitzler[17] who, under the influence of Siegmund Freud, brought psychologically motivated plays to the Viennese stages. It was a cultural revolution and young Haas, exposed in concerts and at the theater, was certainly influenced in his thinking, and he was probably compelled to attend lectures in literature and design as well.[v]

In addition to this interesting distraction from his declared goal of chemistry, as vowed to his father, Haas fell increasingly under the influence of the physicist Ludwig Boltzmann, the leading figure in classical physics. From his work in thermodynamics and statistics, Boltzmann had derived, as a natural philosopher, the real and objective existence of atoms. Boltzmann's work stood against the views of his very skeptical colleague Ernst Mach who fundamentally doubted the existence of atoms and declared energy to be foundation of matter. While Haas dreamed of becoming a natural philosopher like Mach, he felt attracted by the personality of Boltzmann and notes:

> I spent the first semester of my second year in Vienna. During this semester, I was completely overwhelmed by the genius who by-far outshone everyone else and highlighted their insignificance in Physics: Ludwig Boltzmann.[vi] I heard two lectures from him; the first, attended by many hundreds of students, was about natural philosophy and the second was on a topic of pure physics, Maxwell's theory of electricity. The first lecture was easily understandable but I had tremendous difficulty understanding the second. And no wonder, one had to be at least a PhD to participate in and understand these lectures,[18] which later appeared as a book.[19]

---

[17] Arthur Schnitzler bought his villa at Sternwartestraße 76 in 1910 and thus became a neighbor of the now noble family of Siegfried Strakosch von Feldringen, one of the directors of the Hohenau sugar factory. A close relationship developed between the two families. The eldest son, Heinrich Schnitzler, married later after Schnitzler's death to Lilly Strakosch, the daughter of Siegfried Strakosch. See Georg Markus, *Unter uns gesagt: Begegnungen mit Zeitzeugen* (Vienna: Amalthea Verlag, 2008).

[18] Lise Meitner had also enrolled as a student in 1902 and attended the same lecture cycle. She writes: "In my memory, his lectures are the most beautiful and stimulating I have ever heard. [...] He himself was so enthusiastic about everything he taught us that one went away from every lecture with the feeling that a whole new and wonderful world was opening up." David Rennert and Tanja Traxler, *Lise Meitner—Pioniering des Atomzeitalters* (Salzburg-Wien: Residenz Verlag, 2018), 30.

[19] Here young Haas must have missed something because Ludwig Boltzmann published his two-volume textbook on Maxwell's theory of electricity and light as early as 1891, while in Leipzig.

> At home I diligently studied Lommel's textbook on experimental physics and Christiansen's introduction to theoretical physics. There were no lectures at the University of Vienna that would have introduced me to theoretical physics, despite a plethora of faculty. In addition to Boltzmann's lectures, I took several other classes in higher mathematics (theory of functions), physical chemistry, philosophy, and, needlessly, two on the literature of romanticism and aesthetics.

Arthur Haas had probably just finished his chemistry studies, after passing his examinations with Professor Lieben, and he turned completely to physics in this time of scientific revolution.

Even before 1905 and the *Annus Mirabilis* (the year of miracles in physics),[20] new editions of textbooks were being published every one to two years in order to record and pass on newly acquired knowledge. Eugen von Lommel's *Lehrbuch zur Experimentalphysik* was published in its first edition in 1893 by Leipzig's Johann Ambrosius Barth Verlag, whose publications reflected the world and the groundbreaking discoveries of the nineteenth century from mechanics to electricity, magnetism, optics, and thermodynamics. Lommel's book encompassed the world of industrialization and the engineering sciences; it was thought that everything was understood and little could be added in this new emerging century.

Then, in 1895, came the second edition and just one year later, in 1896 the third and fourth were published, followed in 1897 by the fifth edition. With such rapid change, it can be assumed that Arthur Haas used for his studies the sixth edition of 1900 or the seventh edition of 1902, which differed fundamentally from the first edition in that it contained all of the new discoveries in the field of spectroscopy and X-ray physics. These discoveries provided the observational basis for the emerging science of the twentieth century.

It is not clear whether the young student realized this rapid development; the textbooks on theory, with the detailed, mathematically complex formulas that he preferred, followed the history of discovery more slowly, since the new phenomena could not yet be explained theoretically nor substantiated mathematically. The change in structure, presentation and content presented in the textbooks on theoretical physics came, therefore, with a delay. This may well have influenced the education and thinking of Haas, a young student who was oriented toward theory.

The books by Boltzmann that are mentioned by Haas probably refer to the complete works, edited by the student Hasenöhrl and published in 1909 after Boltzmann's death. Therefore, in 1904, Haas probably used Christian Christensen's *Elemente der theoretischen Physik*. The first German edition, published in 1894 by Johann Ambrosius Barth Verlag in Leipzig, was still fundamentally influenced by the formalism of classical physics, with only the chapter on electromagnetic light theory[21] pointing beyond it. The second edition was published in 1903 and this is the

---

Boltzmann's publisher was the then leading scientific publisher in the German-speaking world, Johann Ambrosius Barth.

[20] So named because in 1905, Albert Einstein published his three revolutionary writings on Brownian molecular motion, the photoelectric effect, and special relativity and he formulated the famous equation for the equivalence of mass and energy $\mathbf{E = mc^2}$.

[21] A topic on which Haas later intended to write his first doctoral thesis.

one that Haas is likely to have used. Compared to the first edition, however, it does not show any significant innovations as the authors themselves admit in the second preface:

> When editing the second edition of *Elements of Theoretical Physics,* no significant changes have been made in the arrangement and presentation. A section on diffusion and osmosis has been added, as well as a section on electrolysis. In addition, in the first section, some information about the calculation of moments of inertia is included, and the principle of virtual velocities is added at the end of the same section. In thermodynamics, the considerations of circular processes and entropy are extended; the concept of free energy and thermodynamic potential are introduced. Also included was a somewhat expanded chapter on the mathematical treatment of light theory.

Only the third edition of 1910 shows fundamental changes with the introduction of vector theory—a later favorite of Arthur Haas—and a fundamental revision of electricity and magnetism, up to the introduction of electron theory and the disappearance of light theory. However, this edition was not accessible to young Haas at the time of his studies and, despite his growing interest, he found himself distracted by and wrestling with more mundane concerns:

> The fraternity proved more and more to be an annoying obstacle to study and cultural education. Before Christmas 1903, I had a fencing event in which I remained unharmed but provided my opponent with a huge "pull-through" on the cheek. In January 1904, I had a duel the origins of which had nothing to do with the fraternity but, rather, originated at a Brünn dance affair. My opponent was a young man from Brünn, of shady reputation, who later ended up in prison. We inflicted each other with many saber wounds and as a consequence I now have to wear, for the rest of my life, numerous scars as a sign of Central European idiocy.
>
> Eventually I became so tired of the so-called student life that I decided to change from the University of Vienna to another place. Unfortunately, I only considered universities within the German-language. (How blessed would it have been for me if I had gone to Cambridge in England!) Since in the fields of physics and physical chemistry, as well as mathematics, Göttingen was at the top, I decided, in agreement with my father who very much approved the election, on the University of Göttingen.

Physics had won. His social experiment in the fraternity was a failure that left him with an everlasting mark, much as the scars on his cheek and forehead. He had probably been seeking female acquaintance in the Viennese dance establishments, indicating that his personal loneliness still existed and weighed on him despite his declaration to a life of physics à la "l'art pour l'art."

He had chosen the University of Göttingen to continue his studies; but while Göttingen had emerged as a center for theoretical physics, as an academic institution it had already been called into question by Heinrich Heine in his travel description of 1826, *The Harz Journey*:

> In general, the inhabitants of Göttingen can be sorted into students, professors, philistines, and cattle; these four groups are nothing less than strictly separated. The cattle faction is the most important one. To count the names of all the students and all the professors, both ordinary and disorderly, here would be too broad; nor do I remember all the names of the students at this moment, and among the professors there are some who have not yet been named.[vii]

Nevertheless, Göttingen had become a Mecca for theoretical physics, in the eighty years following this characterization, and it was to remain so for another thirty years. Haas had not been unwise in his choice of academic training but he was not prepared for the student life prevailing in Wilhelmine Germany at the time.

> In the middle of April 1904, I arrived in Göttingen via Leipzig where I stayed overnight. I was much too early, two weeks before the semester began. According to custom at the time, I moved into a two-room apartment, directly next to the University, Weenderstrasse 82. Göttingen gave the impression of being a beautiful ancient city, with nice villa districts in a gracefully wooded, hilly landscape that was unabated.
>
> But then came the horror; the most disgusting manure one can imagine spilled out over the city, a flood of color carrying student bodies, the various corps and fraternities. From morning to night and night to morning, heavily drunk and further publicly quaffing, nosily brawling and quarreling with indescribable brutality, bloated beer faces filling all lecture halls and streets and houses and theaters with an unbearable stench of guzzled and again vomited beer, of unwashed bodies and Prussian homosexuality. I could barely breathe with disgust, but I decided to endure the semester.
>
> On the other hand, the level of the training in natural sciences at the University of Göttingen towered far above that of Vienna. I immediately studied, with great zeal, with Nernst[22] in electrochemistry, with Riecke[23] in experimental physics, with Voigt[24] in theoretical physics,

[22] Walther Hermann Nernst (1864–1941) was awarded the Nobel Prize in Chemistry in 1920 for his work in the field of thermochemistry. Haas attended Nernst's lecture on physical chemistry; this was probably Haas' last lecture in chemistry as he increasingly leaned toward physics. Nernst habilitated at the University of Leipzig in 1889 with a treatise on the electromotive effectiveness of ions. One year later he went to the University of Göttingen. Here Nernst initially worked as a private lecturer and associate professor, then from 1894 as a full professor of physical chemistry. In 1905 he moved to the University of Berlin. In 1906, his research in Göttingen led to his important discovery: the Nernst heat theorem, better known as the third law of thermodynamics, which he applied to the calculation of chemical equilibria.

[23] Eduard Riecke (1845–1915) was a German experimental physicist. He studied physics at the Polytechnic in Stuttgart, at the University of Tübingen, and at the University of Göttingen where he received his doctorate in 1871 and was appointed professor shortly afterwards. In 1873 he became associate professor and in 1881 full professor. Together with his colleague Voigt, he was responsible for the building of the new Physics Institute in Göttingen, which was completed in 1905 when Haas was studying there. Riecke carried out experiments on electrical conduction in metals, for which he tested a model of electron conduction developed by Paul Drude, for its consistency. Among other things, the model was able to explain the decrease in conductivity due to temperature increase. Later he worked with, among other things, the conduction of electricity in gases.

[24] Woldemar Voigt (1850–1919) was a theoretical physicist and director of the Department of Mathematical Physics at Göttingen. His successor in 1914 was Peter Debye, who was succeeded by Max Born in 1921. Voigt worked on the physics of crystals, thermodynamics and electro-optics. His main work was *Lehrbuch der Kristallphysik* (The Textbook of Crystal Physics) published in 1910. He also made music and published writings on musicology, becoming known as a Bach expert.

in applied mechanics with Prandtl,[25] and I also eagerly visited physics laboratories, which had not existed in Vienna.

So, the first semester was quite successful and, under the influence of some new acquaintances, I decided to add another term, dismissing my general disgust. At that time, I was mainly with a Viennese, Karl Jellinek,[26] who was Nernst's private assistant and who later became professor of physical chemistry at the Gdansk technical college, and with his closest friend Paul Schüler,[27] and also a medical student Walter Kersten. Together with Karl Jellinek, I also frequently visited a sympathetic bachelor, the professor of physical chemistry Kötz.[28]

Haas quickly befriended Jellinek; it was not only their Viennese origins that fostered this friendship but also their common interest in the great philosophical and psychological questions of the time, the debate between religion and science. These were pressing questions particularly for members of the young Jewish generation who were asking the question of meaning, having lost the religion of their forefathers and seeking to support the rationality of the natural sciences. Haas attended special lectures with Jellinek, such as that of Max Verworn[29] on "body and soul." This led both students to the physiological origins of thought as a way of overcoming the dualism between body and soul.

Another event that was to have a profound influence on Haas was a lecture by Julius Baumann,[30] professor of philosophy at Göttingen, on the subject of the poetic

[25] Ludwig Prandtl (1875–1953) was an engineer and pioneer in the development of mathematical analyses for the study of flow and aerodynamics. He began his studies at the Technical University of Munich in 1894 and completed with a doctorate after six years. After a short stay in industry, he was appointed Professor of Fluid Mechanics at the Technische Hochschule in Hannover in 1901. Thanks to his scientific achievements, Prandtl became head of the Institute for Technical Physics at the University of Göttingen in 1904. In 1925 he became director of the Kaiser Wilhelm Institute for Fluid Dynamics in Göttingen.

[26] Karl Jellinek (1882–1971) was a physical chemist. He came from a Jewish family in Vienna (baptized Protestant), studied in Vienna and Göttingen and received his doctorate in 1905 with a thesis on reaction kinetics. From 1908, Jellinek worked at the TH Danzig. In 1922 he became professor there and in 1933 director of the institute. In 1937 he was prematurely retired, emigrated to England in 1939 and lived in Switzerland after 1957.

[27] Paul Schüler (1881–) studied chemistry in Halle from 1901 to 1903 and then moved to Göttingen where he received his doctorate in 1906 with a thesis on organic chemistry under the supervision of Arthur Kötz.

[28] Friedrich Arthur Kötz (1871–1944) studied chemistry at the University of Leipzig and received his doctorate in 1894. In 1898 he habilitated at the General Institute of Chemistry of the University of Göttingen and became a private lecturer for the Faculty of Philosophy. In 1903 he was awarded the title "Professor," and in 1909 he was appointed Associate Professor of Chemistry and Chemical Technology. In 1921 he became head of the Chemical-Technological Collection (also known as the Chemical-Technological Institute) of the University of Göttingen and was given a teaching assignment for Technical Chemistry. From 1922 to 1935 he was full professor at the Faculty of Mathematics and Natural Sciences.

[29] Max Verworn (1863–1921) was a developmental psychologist who had studied medicine and natural sciences under Ernst Häckel in Berlin. From 1901 to 1910, he was professor in Göttingen. See Max Verworn, "Die Mechanik des Geisteslebens" in *Jahrbuch des Freien Deutschen Hochstifts* (1907): 132–153 and *Die Mechanik des Geisteslebens. Aus Natur und Geisteswelt, 200th volume* (Leipzig: Teubner, 1907).

[30] Julius Baumann (1837–1916) was a secondary school teacher in Berlin and Frankfurt am Main in 1863 after studying and obtaining his doctorate in Berlin. In 1869 he was appointed professor

and scientific view of the world.[viii] In this, Baumann examined whether poetry and poetry as truth should be placed next to or above scientific truth. The topic was examined using the work of many writers, from Shakespeare to d'Annunzio, but Baumann was particularly concerned with Goethe's view that experience is the only true science and the reflection of this in Goethe's *Faust*. This is a topic that appears again and again in Haas' later writings, whereby he repeatedly compares his scientific statements with and against *Faust*.

Through these lectures and their common interests and discussions, Jellinek and Haas developed within a short time a close friendship that lasted for many years. In their discussions, they sought to understand the origin and unification of all things, especially in questions of science and the humanities, philosophy and religion. The physiology of Verworn was a trigger and they continued to follow the natural philosophical writings of Hans Driesch (1867–1941) regarding the soul as an elementary natural factor. These themes on the search for unification and salvation were offered in 1920, while still under the shock of the war, as a lecture by Karl Jellinek in Danzig (today Gdansk). The lecture, and the book resulting from it, formulated the claim that:

> we must come back to harmonious education, to the tremendous synthesis between all natural sciences and all humanities and, further, the synthesis between science, philosophy and religion must be dared.[ix]

These thoughts,

> which in detail summarize and bring together the realm of unorganized matter, the realm of the living body, the realm of the soul, and the realm of the spirit…

from astronomy to physiology, from psychology to epistemology, from history to culture, are the result of discussions between these two friends.

Haas' stay in Göttingen was also made easier by family visits, excursions and trips during the summer holidays, which were easily arranged by way of Göttingen.

> In July 1904, my brother visited Göttingen for a few weeks. At that time, he was about to finish his last high school year and he buried himself in Greek and Latin classical studies. On the way to Scheveningen in the middle of July, my parents and my sister and her tutor came through Göttingen. My father, who from childhood hated the Prussians with a passion, was speechless with horror when he saw Göttingen's student life. He nearly chocked with disgust, barely took a bite, and left Göttingen as quickly as possible.
>
> My brother and I later followed to Scheveningen. Not for my parents but for me, this was the last time that I stayed in this wonderful summer resort. On the way back from Scheveningen to Brünn, I spent a few days in Göttingen and then visited, with my brother, Wartburg (which we had already seen with our parents fourteen years earlier) and Weimar.

Whether or not Arthur Haas used the summer with his parents and brother to talk with them about his decision to give up the Jewish faith, leave the religious

---

of philosophy at Göttingen University. He is regarded as a representative of Realismus which was influenced by the natural sciences and he was very concerned with questions of the philosophical interpretation of scientific concepts such as time and space as well as approaches such as mathematics in the natural sciences.

community and join the Lutheran confession, is unknown. The decision to change his religion was a consequence of the relationship and discussions with Karl Jellinek but Arthur's visit to the Wartburg, as the historical monument for the Lutheran, faith may have fortified his decision. After returning from Brünn for the autumn semester, Arthur Haas was baptized on November 20, 1904 in Geismar, a small parish south of Göttingen. The baptism, performed by Pastor Konrad Hinrichs, took place in the small St. Martin's Chapel of the Protestant congregation shown in Fig. 4.2.

In addition to his friend Jellinek, the high school teacher Erich Kreisle from Vienna served as baptismal witness but appears nowhere else in the records of Arthur Haas (Fig. 4.3).

Despite being very unhappy with the social environment, Haas found enjoyment in the study of physics and his friendship with Jellinek, and these were the decisive elements for continuing to stay in Göttingen.

After his initial decision to study in Göttingen, the range of lectures presented was oriented toward the study of modern physical concepts and had a motivating effect. Especially the experimental practical courses in physics offered by Professor Riecke were highly appreciated by Arthur Haas who saw the need to compensate for his fascination with theory and philosophy. He therefore complemented his preference for mathematical physics theory with practical application. He did this extensively as his textbook entries demonstrate in Figs. 4.4, 4.5, 4.6, 4.7 and 4.8.

During this first semester Haas had turned completely away from traditional chemistry. He concentrated on physical chemistry or electrochemistry in Nernst's lecture, but his main focus was on experimental physics. This included several exercises and practical courses as well as a lecture on applied physics, specifically electrical engineering by Hermann Theodor Simon.[31] This was a purely technical lecture on electrical circuits in electrical engineering, radiation, and illumination phenomena and may have given Haas insights for his later work on light theory. The choice of this lecture might have been based on interactions with Simon since, given Haas' description, he seemed to like Simon personally. Despite his preference for historical-philosophical and theoretical approaches to physics, Haas made at least one attempt to obtain a doctorate with Simon on a topic in applied physics.

> In October 1904, I returned to Göttingen to continue my studies. I now lived at Bühlstrasse 19 with the wife of a grammar school professor who himself resided in a mental health asylum, but who had three sons and two daughters. I eagerly continued my studies, gradually turning from chemistry to physics. With Riecke, I took additional experimental physics classes and

[31] Hermann Theodor Simon (1870–1918) studied in Heidelberg and Berlin, received his doctorate in 1894 with a thesis on UV radiation and habilitated in Erlangen in 1896. In 1898, Eduard Riecke called Simon to Göttingen as his assistant. Easter of 1900, Simon went to Frankfurt am Main as a lecturer in physics at the Physical Society and director of the Physical Institute. In autumn 1901 he returned to Göttingen and became an extraordinary professor and director of the department of applied electricity. In December 1905, the new physics building was completed and the department was transformed into a separate institute. In 1907, Simon became full professor. Around 1909, he suggested the founding of the Radioelectrical Experimental Station for Navy and Army in Göttingen, which was then established by Max Reich after Simon's death.

Fig. 4.2 The Protestant St. Martin's Church in Geismar, built in 1743 in the baroque style; the baptismal room is now equipped with modern furnishings

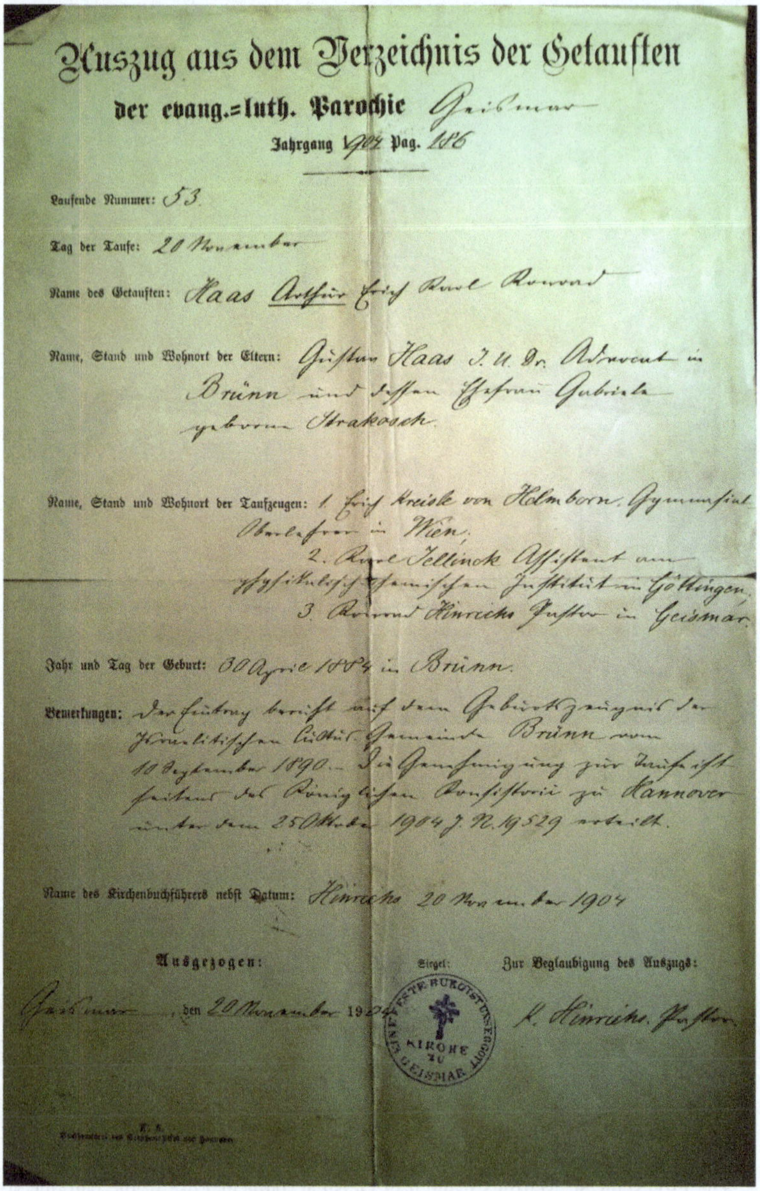

Auszug aus dem Verzeichnis der Getauften
der evang.-luth. Parochie Geismar
Jahrgang 1904 Pag. 186

Laufende Nummer: 53

Tag der Taufe: 20 November

Name des Getauften: Haas Arthur Erich Karl Konrad

Name, Stand und Wohnort der Eltern: Gustav Haas J. U. Dr. Advocat in Brünn und dessen Ehefrau Gabriele geborene Strakosch.

Name, Stand und Wohnort der Taufzeugen: 1. Erich Kreidl von Holmborn, Gymnasial Oberlehrer in Wien,
2. Karl Jellinek Assistent am physikalisch chemischen Institut in Göttingen.
3. Konrad Hinrichs Pastor in Geismar.

Jahr und Tag der Geburt: 30 April 1884 in Brünn.

Bemerkungen: der Geburtstag beruht auf dem Geburtszeugnis der Israelitischen Cultus Gemeinde Brünn vom 10 September 1890. Die Genehmigung zur Taufe ist seitens des Königlichen Konsistorii zu Hannover unter dem 25 Oktober 1904 J. N. 19529 erteilt.

Name des Kirchenbuchführers nebst Datum: Hinrichs 20 November 1904

Ausgezogen: Geismar, den 20 November 1904

Siegel: FESTE BURG IST UNSER GOTT KIRCHE ZU GEISMAR

Zur Beglaubigung des Auszugs: K. Hinrichs, Pastor

Fig. 4.3 Baptism record for Arthur Erich Haas in the church book of St. Martin's parish in Geismar near Göttingen

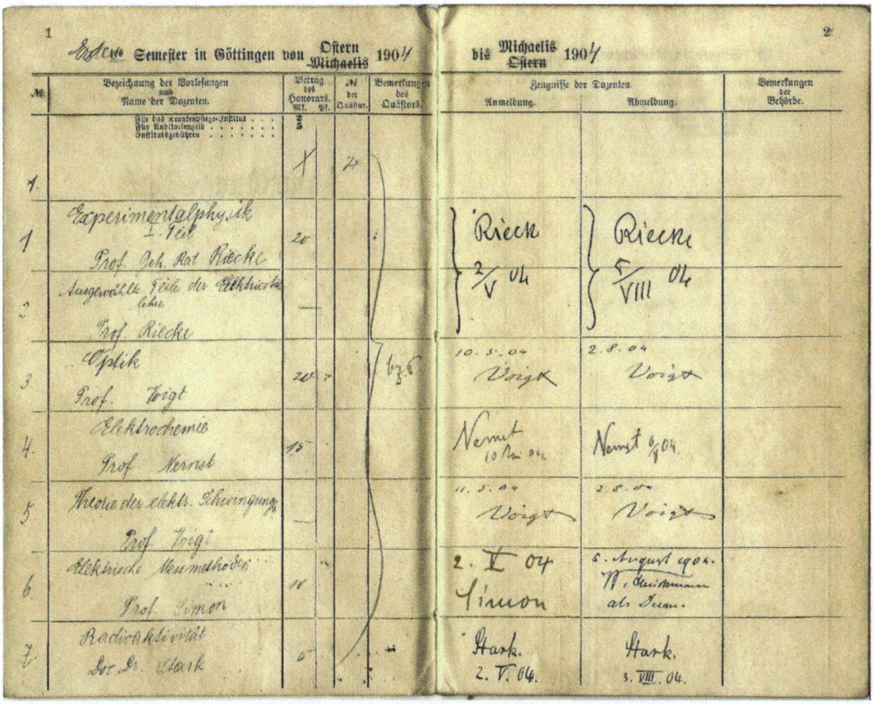

Fig. 4.4 Entries in Arthur Haas' study book for the 1904 summer semester at the University of Göttingen. Besides classical experimental physics, which included an internship with Eduard Riecke, Haas took part in theory lectures by Nernst and Voigt. For a short time, Haas was also interested in questions regarding measuring techniques and methods, as read by Simon. Johannes Stark's lecture on radioactivity provided a long-term, formative element for Haas

a seminar on electron theory. Voigt's lectures covered almost the whole field of theoretical physics. I also attended physics lectures by Prandtl, Simon, Bose,[32] and also Stark[33] (the later

---

[32] Emil Bose (1874–1911) studied and habilitated under Nernst in Göttingen. In 1905 he was a private lecturer for physics and physical chemistry there but was subsequently appointed to a chair in Danzig. In 1909, he was appointed to the newly founded University of La Plata in Argentina. He died there in 1911 after a short stay, from typhoid fever.

[33] Johannes Stark (1874–1957) studied physics in Munich from 1892 to 1897 and received his doctorate in 1897 with a thesis on soot. After two years as an assistant, he moved to the University of Göttingen where he habilitated in 1900. He remained with Eduard Riecke as a private lecturer and assistant at Göttingen University until 1906, then became an associate professor of physics at the Technische Hochschule Hannover. In 1909 he moved to Aachen and in 1917 to the University of Greifswald. In 1919, Johannes Stark received the Nobel Prize in Physics for "the discovery of the Doppler effect in canal rays." In 1920, Stark was appointed to Würzburg as successor to Wilhelm Wien but left the university after only two years due to differences with the faculty. After Hitler came to power, Stark became president of the Physikalisch-Technische Reichsanstalt Berlin from 1933 to 1939. Because of his anti-Jewish publications and his membership in the NSDAP, Stark was put on trial after the war. Von Laue, Heisenberg and Sommerfeld testified against him. On July

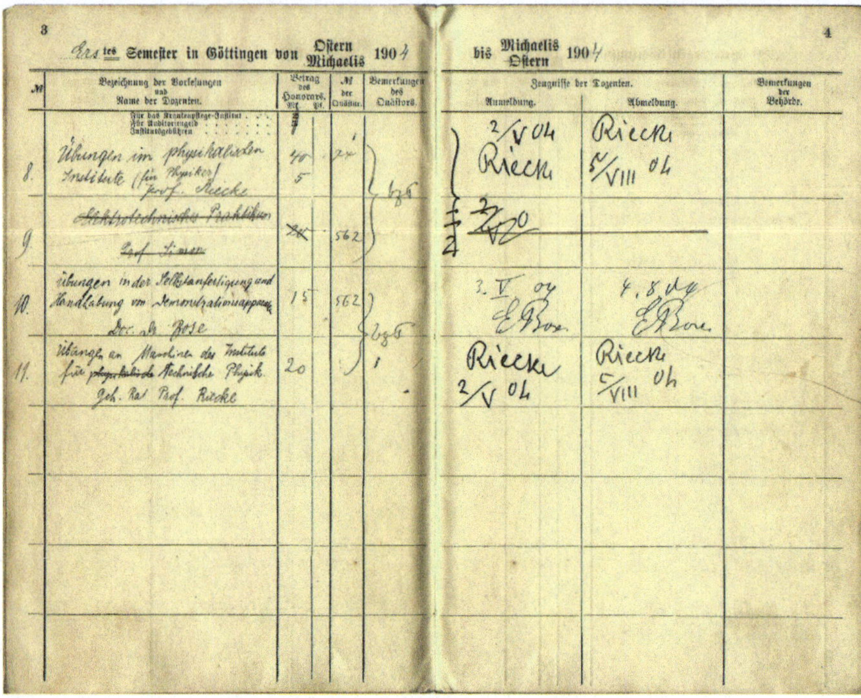

Fig. 4.5 Entries in Arthur Haas' textbook for the 1904 summer semester at Göttingen University. Further events were practical laboratories and exercises by Riecke and Bose

joker and Nazi president of the physical-technical Reichsanstalt). In addition, I continued my participation in physics and specialty laboratory classes.

Haas' winter semester courses are noted in Figs. 4.6 and 4.7. Arthur Haas took a course with Johannes Stark on the phenomenon of radioactivity which had only been discovered a decade earlier. Haas was extraordinarily impressed by the topic and, despite his later criticism, also by the lecturer. At the time, Stark had founded a new journal, *Jahrbuch der Radioaktivität und Elektronik* (Yearbook of Radioactivity and Electronics), a publication that was able to quickly spread the latest information about the new phenomenon. In the first issue of 1905, Stark presented both the experimentally observed phenomenon of radioactivity and the mathematical formalism of the laws of decay, as well as providing an independent article on a number of possible explanations based on his lecture on the given topic.[x] It is notable that

20, 1947, the court found him guilty of the main crime and sentenced him to four years in a labor camp. The sentence was later reversed.

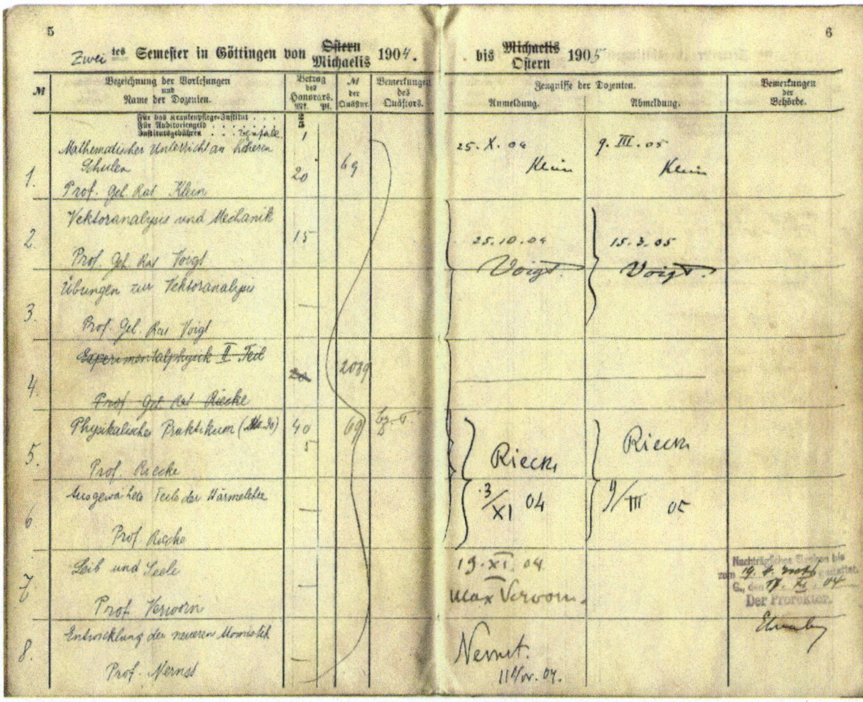

Fig. 4.6 Entries in Arthur Haas' course book for the winter semester 1904/05 at the University of Göttingen. In this semester, Haas concentrated more on mathematical-theoretical and philosophical lectures, without neglecting his practical courses in physics. The lecture on vector analysis formed the basis for his own work in the field

in these papers, he frequently quoted Albert Einstein and did not show any of the anti-Semitic sentiments that characterized his later career.

It is interesting that classmate and later Nobel Prize winner Max Born (1882–1970) expressed a completely different opinion about Stark, perhaps understandably. As with Haas, Born was surely influenced by Johannes Stark's later personal developments but during his time as a student, his impressions were clear:

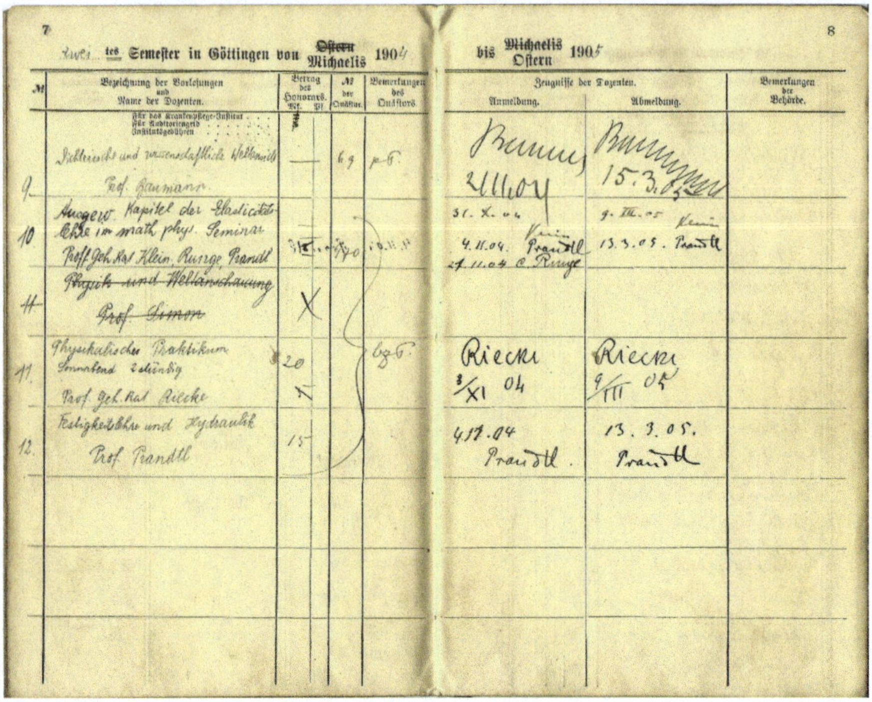

Fig. 4.7 Entries in Arthur Haas' course book for the winter semester 1904/05 at the University of Göttingen. For unknown reasons Haas cancelled the lecture by his mentor Simon, "Physik und Weltanschauung."

> The private lecturer Johannes Stark had announced a lecture on radioactivity [...] but what Stark presented seemed to me to be pure dogma. He did not show experiments but described them in rather unclear terms, and the results were formulated like articles of faith without any attempt at explanation. After a few lectures I came to the conclusion: If this is modern physics, then I don't want anything to do with it. Stark [...] was a genius in the operation of devices and in related gadgets.[34]

[34] Born's mathematical-theoretical education, during which he had admittedly minimized the experimental component of physics as much as possible, was entirely based on the mathematically deductive method of his teachers. With this, he showed little understanding for the approach of experimental physicists. Especially in the case of new phenomena such as in this case radioactivity, physicists first describe the physical behavior phenomenologically in mathematical terms in order to extrapolate predictions, which in turn provide the basis for understanding the phenomena. That is not a deductive but a descriptive mode for a new phenomenon. Stark provided experimental results at a time when radioactive decay, although observed, was still far from physical interpretation and mathematical description. Such a phenomenological approach must necessarily have been boring for Max Born. Cf. Max Born, *Meine Leben: Die Erinnerungen des Nobelpreisträgers* (München: Nymphenburger Verlagshandlung, 1975), 133.

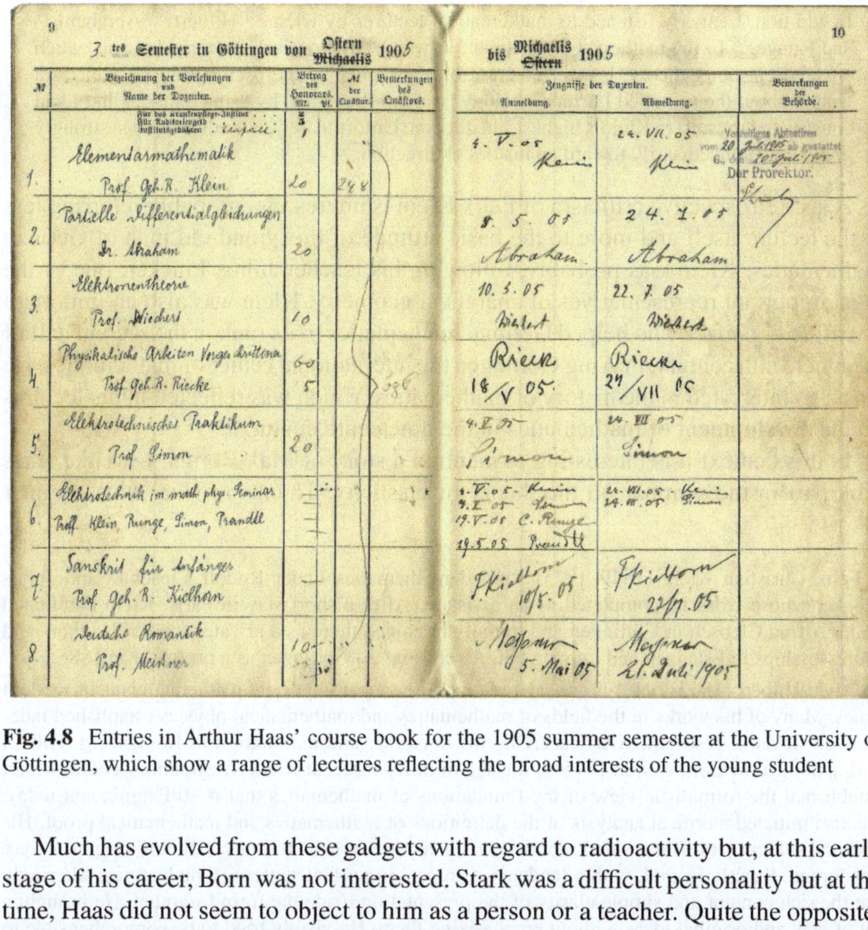

Fig. 4.8 Entries in Arthur Haas' course book for the 1905 summer semester at the University of Göttingen, which show a range of lectures reflecting the broad interests of the young student

Much has evolved from these gadgets with regard to radioactivity but, at this early stage of his career, Born was not interested. Stark was a difficult personality but at the time, Haas did not seem to object to him as a person or a teacher. Quite the opposite, the phenomenon of radioactivity occupied Haas throughout his life and had a lasting influence on his scientific and religious thinking. Many of Stark's original thoughts and ideas flowed into these considerations as will be discussed later.

In addition, I eagerly listened to mathematical lectures by Klein,[35] Hilbert,[36] Abraham,[37] and Runge.[38] In the mathematical seminar, in which Minkowski[39] participated and which covered elasticity theory, I gave a lecture on the theory of the oscillograph as well as a lecture about the so-called Larmor Analogy (which I illustrated by using a model that I had constructed myself) to Klein's highest satisfaction. Unfortunately, Klein's influence strongly guided me unnecessarily toward an historical direction.

Haas' retrospective criticism of Felix Klein is interesting, probably referring less to the lecture itself and more to the basic attitude of this grand old man of German mathematics. Klein was probably, following his teacher Julius Plücker, one of the most important representatives of analytical geometry. Klein was also an important scientific organizer who helped German mathematics to its rank in the second half of the nineteenth century, making Göttingen the international center of mathematics. He was also interested in the history of mathematics, which was reflected in his lectures on the development of mathematics in the nineteenth century.[xi]

In this context it is interesting to mention a story of Max Born's, who like Haas took part in the seminar on the theory of elasticity. Thanks to Haas, this seminar

---

[35] Felix Christian Klein (1849–1925) studied mathematics under Rudolf Lipschitz and Julius Plücker whose work he completed as an assistant. After a short stay in Paris, Klein habilitated under Alfred Clebsch in Göttingen and initially remained there as a private lecturer. He then held professorships in Erlangen and Leipzig before returning to Göttingen as a professor in 1886.

[36] David Hilbert (1862–1943) is considered one of the most important mathematicians of modern times. Many of his works in the fields of mathematics and mathematical physics established independent fields of research. After receiving his doctorate and habilitation in Königsberg, Hilbert was appointed to Göttingen in 1895, at the age of thirty-three, at Klein's instigation. There Hilbert established the formalistic view of the foundations of mathematics that is still significant today. He also initiated a critical analysis of the definitions of mathematics and mathematical proof. His doctoral student Otto Blumenthal reported on his lecture style as follows: "Hilbert's lectures were unadorned. Strictly factual, with a tendency to repeat important sentences, he lectured falteringly but the rich content and simple clarity of the presentation made the form forgotten. He brought a lot of new and original ideas without emphasizing them. He visibly tried to be comprehensible to all; he read for the students, not for himself." Arthur Haas will have profited greatly from this.

[37] Max Abraham (1875–1922) was a student in Berlin and after his doctorate in 1897, he was an assistant to Max Planck. After habilitation, Abraham taught as a private lecturer in Göttingen from 1900 to 1909. He received a call to the University of Illinois but returned to Europe after only a few months. He received a professorship in Milan but had to give it up because of the First World War. He remained as professor of physics in Stuttgart until 1921. One year later he died, only forty-six years old, of a brain tumor.

[38] Carl David Tolmé Runge (1856–1927) grew up in Havana where his father was Danish consul. He studied in Berlin under Karl Weierstrass and was awarded the doctorate there in 1880; six years later Runge was called to Hanover. His work concentrated mainly on the applications of mathematical methods in physics and astronomy. In 1904 he was called to Göttingen at the instigation of Felix Klein, where Runge remained until his retirement in 1925.

[39] Hermann Minkowski (1864–1909) was born in what is now Kaunas (Lithuania). After studying in Berlin and Königsberg, he received his doctorate in Königsberg in 1885, where he was a close friend of David Hilbert. After teaching in Bonn, Königsberg, and at the Polytechnic in Zurich where Albert Einstein attended his lectures, Minkowski moved to Göttingen in 1902 at the instigation of his friend David Hilbert. Minkowski researched the structure of quadratic equations with n variables as well as the geometric properties of n-dimensional spaces. Around 1907, Minkowski developed the

had a considerable influence on Born's further development, as he describes in his memoirs:

> I was not very interested in elastic problems and decided to listen to Klein and Runge's lectures and discussions at the seminar without playing an active role myself. But I couldn't avoid participating completely. For each problem a speaker and a co-lecturer were appointed, the latter mainly in case the speaker himself was prevented by illness from giving his lecture. Now I became a co-lecturer for a student named Arthur Erich Haas and our problem was the elastic line or *clastika*, and its stability. I read about the topic in the usual textbooks, but left the study of the special literature to Haas.
>
> About a week before the lecture was due, Haas came to my room and explained that he could not give his seminar lecture because he was overworked and ill. In fact, he did not look well; but I had my doubts about his illness and thought that it might have been some kind of stage fright—a very common occurrence at performances under the direction of the terrifying Felix Klein. I tried to get some information about literature out of him but found his knowledge as poor as mine.
>
> Then a hectic week followed. I presented my findings to the seminar, where they were received with some interest; but Runge regretted that, as a result of Haas' collapse, the seminar audience was deprived of hearing about the many interesting cases found in the literature.[xii]

This is an interesting story that Armin Herman uses as an example of when Haas was given an opportunity to prove himself but shunned the energetic step to success.[xiii]

Participation in all of these mathematical lectures was driven by Haas' desire for practical application. The lecture on vector analysis and classical mechanics[xiv] by Voigt particularly influenced Haas but later, in his own books on this topic, he followed a different formalism than had been presented. Voigt introduced mechanical laws in his presentation after first introducing the basic units—using points, lines, the volume elements of solids and liquid bodies—while Haas had a more mathematical approach, using scalar, vector and tensor formalisms for the description of mechanical properties.

In addition to these lectures, Haas also heard a lecture on electron theory from Professor Emil Wiechert (1861–1928) who was the first to develop an electron theory of X-rays[xv] when he was a young assistant in Königsberg, before moving to Göttingen in 1897. There he became director of the newly founded Institute of Geophysics and developed into one of the leading seismologists of his time. However, Wiechert is largely forgotten today despite his numerous scientific achievements in physics.[xvi] In this development, but also in his broad scientific interests, he is comparable to Arthur Haas. Wiechert's considerations were an integral part of his lecture:

> In view of today's natural science, we will probably have to give up the idea of going into detail and discovering the last foundations of the world at some point, and I believe we can do so with a light heart. The world is "infinite" in all directions, not only upwards, into the great, but also downwards, into the small. If we follow the content of the world further and further from our human point of view, then in both directions we finally reach misty distances in which first the senses and then also the thoughts let us down.

---

idea of the four-dimensional space-time continuum—a thought that directly influenced Einstein's general theory of relativity. Minkowski died at the age of forty-four of a ruptured appendix.

Arthur Haas later opposed these considerations with the demand of finiteness in his conception of cosmology. However, Wiechert's basic idea of free ether as a carrier of electromagnetic waves had to be abandoned only ten years later, in the face of modern developments in the theory of relativity. Yet, despite this research, Wiechert did not give up his ideas about ether and therefore is regarded as a model for the fictitious physicist Victor Jakob, in Russell McCormmach's tale *Night Thoughts of a Classical Physicist*.[xvii]

But let's return to the memories of Arthur Haas, in which he also describes his participation in lectures on the humanities and literature as shown in the course book entries displayed in the Figs. 4.8 and 4.11.

> For no apparent reason, I also studied Sanskrit with Kielhorn,[40] German literary history with Meißner,[41] aesthetics with Vischer,[42] and philosophy with Baumann. In the sixth semester, I began to work independently as a scientist with Riecke, but without success; Riecke assigned me to study magnetic substances, with which he had worked decades earlier and in which he had still kept an interest. I was supposed to determine if the magnetic constants of the substances had changed over time. I soon found that these changes had to be smaller than the limits of the observation errors, so the investigation went nowhere.

"For no apparent reason" probably referred to Haas' aspired goal of becoming an applied physicist. For the physicist, as for the engineer, the more philosophically based scientific questions were considered secondary at that time. This line of thinking undoubtedly had serious negative consequences for physics in Germany when, just one year later, Albert Einstein formulated the special theory of relativity with mathematical arguments followed only a few years later by the general theory of relativity. The mathematical construct itself was instinctively rejected by many experimental physicists and was further labeled as Jewish physics.

Applied physicists and engineers often turned to ideas that corresponded more with their intuition, ideas based more on "classical physics" than on direct observation. These alternative approaches even included obtuse theories such as the "world

---

40 Franz Kielhorn (1840–1908) was a German Indologist who did research in particular in the fields of Sanskrit grammar and Indian epigraphy. After studying in Göttingen, Breslau, and Berlin, Kielhorn stayed for a year in Oxford in 1866. From there he moved to India where he taught as professor of oriental languages at Deccan College in Poona. In 1881 he took over the chair of Indology at Göttingen University where he remained until his death.

41 Georg Paul Rudolf Meißner (1862–1948) was professor for old Germanic languages after having studied at Göttingen and Berlin. After obtaining his doctorate and postdoctoral qualification in Göttingen, he remained there from 1896 to 1906 as a private lecturer in German language and literature. He married the daughter of Robert Vischer and followed a call to Königsberg in 1906. After 1913, he was a professor of Germanic philology and later rector of the University of Bonn.

42 Robert Vischer (1847–1933) studied art history and philosophy at the universities of Zurich, Heidelberg, Bonn, Munich and Tübingen. In 1872 he received his doctorate at Tübingen. He moved two years later to the Vienna Academy of Art and in 1878 to the University of Munich, where he habilitated in 1879 and taught as a private lecturer. In 1882, he was called to Breslau where he was initially offered an extraordinary professorship for new art history. This was followed in 1885 by a call to Aachen, where he was full professor of general art history and aesthetics until 1893. Finally, in 1893, Vischer moved to the University of Göttingen where he taught in the same capacity until his retirement in 1911.

ice" and the "hollow world" theory, which were presented by technically pre-trained engineers and were propagated and supported particularly in the German-speaking world during the time of the Third Reich. Especially for the older physicist, classical physics was considered German physics. This view necessarily turned against the new theories of the twentieth century, such as the theories of relativity and quantum physics. These were considered abstruse since they brought only uncertainty and doubt into the solid predictable functioning of the world. A reflective representation of this perspective of the older physicist can be found in the already mentioned, *Night Thoughts of a Classical Physicist*. While the theories of relativity and quantum physics are experimentally confirmed today and belong to the standard framework of our physical knowledge, 100 years ago many physicists were still in doubt.

Haas himself had a broad interest and he considered natural philosophical questions to be an important aspect of his field. His experimental interest had been aroused by exactly these kinds of questions—such as the thesis postulating that things are only "reality" if they can be measured. This is what drove him into the experimental field where he wanted to find the reality and truth that he did not see in abstract theory. However, experiments have as little or as much to do with reality as theoretical models; experiments are designed to verify or falsify the theoretical predictions of models for natural phenomena. Experiments do not describe reality but only identify a carefully selected effect through a sophisticated, well designed experimental mechanism formulated to verify a specific model prediction.

Gottfried Keller had already described this endeavor rather poetically, and also accurately, in the first chapter of the novel *Das Sinngedicht* (The Poem of Sense), written in 1855:

> When the sun rose a few spans, he again closed the windows to the beautiful world, with everything that lived and weaved in there, and let just a single ray of light into the darkened room, through a small hole he had drilled in the shutter. When this ray was carefully stretched to order, Reingart began his day's work without further hesitation, picked up paper and pencil and looked inside to continue where he had left off the day before.[xviii]

The development of the experiment as described in this way—by the reduction of disruptive effects and uncertainties—requires a certain talent and it is this talent that makes a successful experimental physicist. With Haas, this ability was less apparent despite all of his efforts.

Nonetheless, in this second year Haas established social connections that made his life in Göttingen easier.

> At that time, I also came into social contact with several families of faculty in Göttingen, such as Professor Klein, Professor Tschackert,[43] and university librarian Lutz[44]; these were

43 Paul Tschackert (1848–1911) was a Protestant theologian and church historian born in Freystadt, Silesia. Tschackert studied history, theology and philosophy at the University of Halle and continued his education at the University of Göttingen in 1873. In 1875 he moved to the University of Breslau. In 1877 he became a private lecturer at the University of Halle and from 1884, professor at the University of Königsberg. In 1890 he moved to Göttingen, where he remained until his death.

44 Emil Lutz (1849–) studied law. After receiving his doctorate in 1877 he became a volunteer in Königsberg, but in 1879 he moved to the University of Göttingen where he became curator of the legal department in 1883 and chief librarian from 1897.

all families with young daughters. I was also often invited to the home of the Vischer family. He was a son of Friedrich Theodor Vischer, the well-known writer and professor of art history and aesthetics, and married to an unusually sympathetic Viennese lady. They had a very beautiful, already married daughter[45] whom I later often heard in Vienna as a concert singer. For a long time, I was also close friends with a Viennese engineer, Eduard, Knight of Grimburg[46] who wanted to educate himself in Göttingen; his father,[47] President of the State Railway Company in Vienna, was a friend of my father. My acquaintances also included Max Born,[48] later a professor at the universities of Breslau [today Wroclaw] and Edinburgh,

[45] Eleonore Meißner née Vischer (1880–1948) was married to the much older Rudolf Meißner. She was a concert singer and sculptor in Vienna.

[46] Eduard Grimus von Grimburg (–1918).

[47] Rudolf Grimus von Grimburg (1839–1917). After studying and working abroad, he was professor of mechanical engineering at the Technical University in Vienna from 1868. In 1890 he became director of the Austrian network of the State Railway Company, whose performance and profitability he increased with great success until nationalization in 1908.

[48] As a theoretical physicist and mathematician, Max Born (1882–1970) played a major role in the development of quantum mechanics. After his studies in Breslau, Born arrived at the University of Göttingen in 1904, where he studied with the famous mathematicians Felix Klein, David Hilbert and Hermann Minkowski. Born wrote his doctoral thesis on "Stability of the Elastic in Surface and Space" and received the prize of the Faculty of Philosophy of the University for this work. Arthur Haas probably met Born in the seminars associated with this topic. In 1905, Born began research into special relativity with Minkowski and subsequently wrote his habilitation thesis on the Thomson model of the atom. Born remained in Göttingen as a private lecturer but with the beginning of the war, he was called to Berlin to serve on the Artillery Examination Commission for research work important to the war. In 1921 he returned to Göttingen. Under Born, Göttingen became one of the world's leading centers for physics. In 1925, Born and Werner Heisenberg formulated the matrix-mechanical representation of quantum mechanics. In the following year, Born formulated the interpretation of the probability density function for $\psi * \psi$ in the Schrödinger equation, for which he received the Nobel Prize in 1954. In January 1933, the NSDAP came to power in Germany. Born, who was Jewish, was suspended. He emigrated to Great Britain where he was appointed Tait Professor of Natural Philosophy at the University of Edinburgh in October 1936. He became a British citizen on August 31, 1939, one day before the outbreak of the Second World War. He remained in Edinburgh until 1952. He retired to Bad Pyrmont in West Germany and died in a hospital in Göttingen on January 5, 1970.

Györö Zemplén,[49] later a professor at the University of Budapest, the young Hungarian mathematician König,[50] and a young Indian, Zia Ahmad.[51]

Based on these societal connections Haas built his network, remaining in correspondence with many of these young scientists. By the time Haas wrote his biography, his former acquaintances were all well established and he probably considered them worth mentioning.

Haas also appreciated his social contact with the daughters of the various professors; he was looking for a firm bond that could support him in his development and his world of thought. This was well known to his friends, as Leo Lederer expressed with sarcastic sneers in a letter to Haas:

> What do you intend to do now besides slipping around the parquet floor of various Salons? Apparently, you are still the old one with the eternal longing for happiness with women, Aryan connections and salon successes.

Haas obviously had no long-term success with the Göttingen professors' daughters. This may have been due to the art of Prussian competition which, despite his long stay, still managed to get his blood boiling.

> With my increasing number of social contacts, it was inevitable that I also met young law clerks and young representatives of the Prussian Herrenkaste [derisive expression for young military noblemen], terrible all of them (though I had not yet been in contact with the worst of them). They had a combination of those characteristic ill manners that are, in their entirety, marked by the word Prussianisms: ever-rampant, big mouthed, bragging and boasting that arises from their inner cowardice; dog-like, slave-like with a complete lack of dignity and with servility toward the presumed superior, coupled with brutal vulgarity against the supposedly lower ones; a cynically scornful contempt for all that is sacred to others, mingled with insincere hypocrisy; parvenus pride concealing a lack of any true culture, with a constant mendacity and readiness to find a higher "moral" ground for every broken word and dishonor—in short, all those features that found their highest representative in the then ruler of Germany.[52] Indescribable was the deification of militarism. Mental giants

[49] Győző Zemplén (1879–1916) was a Hungarian physicist working in the field of hydrodynamics and gas kinetics. After a research stay in Göttingen from 1904 to 1905, he returned to Budapest to take up a professorship there. At the beginning of the World War, he reported to the front and died in 1916 during battle against Italy.

[50] Dénes Kőnig (1884–1944) was born in Budapest as son of the mathematician Gyula Kőnig. He spent his first four semesters at the University of Budapest and his last five in Göttingen, where he studied with the famous mathematicians József Kürschák and Hermann Minkowski. In 1907 he was awarded a doctorate in geometry from the University of Budapest, with a dissertation on geometry. He remained there until 1944 as a member of the faculty, first as a private lecturer and then as a professor. In October 1944, the National Socialist Arrow Cross Party took over the country. Nine days later he committed suicide to escape persecution by the Nazis as a Hungarian Jew.

[51] Sir Ziauddin Ahmad (1878–1947), born as Ziauddin Ahmed Zuberi. He received his degree in mathematics from Cambridge University. There he was the first Indian to be awarded the Isaac Newton Scholarship in 1904. Afterwards he moved to Göttingen to earn a doctorate in mathematics. After his return to India, he became professor of mathematics at the Muhammadan Anglo-Oriental College which was later transformed into the Aligarh Muslim University under his direction. He served the university for three terms as vice chancellor and was also rector. During his career in India, he was a mathematician, logician, natural philosopher, politician, parliamentarian, political theorist, educator and scholar. For his merits he was ennobled by the British King.

[52] The German emperor, Kaiser Wilhelm II. von Hohenzollern.

like Klein died in awe of a young lout in lieutenant's uniform and they proudly showed, when visited, new cannon models that they had been given.

It was the time of German militarism and general worship of the Prussian and now the German Army which had defeated France in the war of 1879/71. Victory had given the Germans self-confidence but also military arrogance and a feeling of superiority. This explains the dominant role of the military in German society at the turn of the century. Eventually this commanding military presence led to the catastrophe of the First World War, which would fundamentally change Europe as well as the European bourgeoisie, intrinsically altering the life of Arthur Haas. The resulting loss of Arthur's homeland—which translated into personal retrospective identification with life in the Habsburg Empire, at the time considered the "home of the Jewish bourgeois"[xix]—together with the loss of "the world of yesterday,"[xx] explains the bitterness and contempt that he still expressed thirty-five years later in his autobiographical notes.

Often I went to concerts in Göttingen, many were very good. I also went to the theater, although I restricted myself to serious plays because the others were frequently disturbed by the rabble-romping of the corps students who, if it pleased them, would vomit from the balcony in the auditorium without anyone daring to intervene. I played the violin with zeal; I also did a lot of sports, especially swimming and ice skating. I also continued with what I had started in the upper school classes, tennis and cycling. Strangely enough, though I had too much pocket money, I never thought of going on trips to other cities. Only once did I travel to Bremen, when I had to report to the Austro-Hungarian consul for military service.

In July 1905, I returned to my parents, having previously given my father great pleasure by sending him a reprint of my first publication; it was a short treatise on the theory of physical mass systems published in the *Physikalische Zeitschrift*.

This essay about the measurability of reality was his first, albeit very short, scientific publication.[53] The title reflects his already compelling interest as motivated by Voigt and Simon, which would only be reconciled with the introduction of reliable unit systems and measuring scales.

With this work, Haas proposes to introduce velocity v as a further basic unit, in addition to the unit of length l. He suggests that the various powers of velocity itself give the dimensions of many quantities, namely the most important ones of electricity, so that a system of measurement with only one basic unit would suffice for all of them. As a unit for velocity, he suggests the speed of light c. With this he integrates the electromagnetic quantities into a mechanical unit system. Today this is called the natural or Planckian system,[xxi] primarily used in relativistic physics where the speed of light is set as dimensionless quantity $c = 1$.

However, after this initial scientific success, a number of changes took place for the family. His father decided—as legal representative of the Strakosch business and its rapidly growing sugar empire, as well as a member of the board of directors of various other industrial companies, such as the Portland Cement Factory Company in Galicia—to move from familiar Brünn to the capital city of Vienna.

[53] A. E. Haas, "Über ein Masssystem, das die Längeneinheit und die Lichtgeschwindigkeit als Grundeinheiten enthält," *Physikalische Zeitschrift* 6, no. 7 (1905): 204–205.

At that time a tremendous event took place, the move of my family from Brünn to Vienna (moreover, in July 1905, my brother passed his Matura exam at the First German Gymnasium in Brünn with a certificate that was in no way inferior to my own). For many years, my father had had the desire to move his law firm from Brünn to Vienna; this desire was quite understandable, as my father's sphere of activity grew constantly. In addition, he had begun to play an important role in various industries, since he was constantly dealing with the highest legal authorities in Austria, who of course had their seat in Vienna.[54]

The fulfillment of my father's wish had failed time and again due to the rigid will of my [maternal] grandfather. My mother did not want to part with her father and therefore only wanted to move with him to Vienna; but her father did not want to hear about moving, despite the fact that he had another married daughter[55] in Vienna and the promise that after a cure in Carlsbad and Gastein, he would find his accommodations from Brünn to be unchanged in Vienna. After years of back and forth, my mother finally decided to move to Vienna without her father but, until his death, she spent a day every week in Brünn with him.

My father rented a very spacious apartment that summer, on the mezzanine of a modern house on the Ringstraße—one of the most distinguished streets in Vienna—it was Stubenring 14. His office was connected with the apartment. Of the eight large street-rooms, one served my sister and her tutor, the next was my parents' bedroom, followed by a beautifully paneled dining room, an oval drawing room, the gentleman's room, my brother's room (where I sometimes lived), my father's study, and my uncle Fritz's study, with whom my father had a joint law office. Of the other smaller rooms, one served as a hall, one as a buffet room, one as a cloak room, and some as chancery rooms. The rooms were most tastefully decorated with new furniture and in September 1905 the family could move in.

The Stubenring was undoubtedly one of the most desirable addresses for upper-class families in the city (Fig. 4.9). The Ringstraße, around the old center of Vienna, had been developed in the second half of the nineteenth century as the signature and center of urban renewal, reforming the wide strip of the former and now demolished fortifications of the city. Ringstraße was a development with public buildings and large multi-story rented palaces in the style of the Gründerzeit (so called, "founding period" architecture) and Art Nouveau. The buildings reflected new ideas about urban life and the expectations of the bourgeois class.[xxii]

The corner house, Stubenring 14 with its tower-like corner rounding was built in 1904/05 by Jakob Gartner. Gartner was a well-known Jewish master builder in Vienna who built various synagogues, called temples in Vienna, in addition to residential and commercial buildings. Gustav Haas had probably obtained this modern apartment for a rather modest rent through his connections to the Viennese Jewish community.

The interior decoration of the apartment and the law office was in the fashionable dark style, with heavy furniture to demonstrate conservative dignity and prosperity as shown in Fig. 4.10. Heavy old German furniture in the dining room and study was alternated with more comfortable furniture in the salon or living room, from the Biedermeier period and probably owned by David Haas. Arthur Haas loved these pieces of furniture and did his best to keep them in his possession; some of these pieces are still owned by the family today. So, the apartment reflected the conservative taste

[54] Gustav Haas was an important industrial lawyer. He not only played an important role as legal representative of the Strakosch Group, in its various legal matters, but also extended his legal consulting activities to other industries, especially the cement industry.

[55] Laura Brum née Strakosch (1867–1919).

Fig. 4.9 The apartment building at Stubenring 14 as it is today. The apartment and office of Dr. Gustav Haas were located in the mezzanine, i.e., on the somewhat lower first floor of the building

of the Viennese bourgeoisie. Uncle Fritz Haas, on the other hand, had a completely different style called Viennese modernism.

The move to Vienna and the associated work did not, however, prevent the family from travelling to escape summer-hot Vienna, so that they could relax on the annual summer retreat.

> We spent the summer of 1905 in Baden and Alt-Aussee, and in October 1905, I returned to Göttingen.

Baden near Vienna and Altaussee in Styria were typical holiday resorts for Viennese citizens. Being imperial summer residences, Baden and Altaussee attracted Habsburg citizens and nobility. However, the stay in Baden was probably more dedicated to visiting grandfather Bernhard Strakosch, who spent his summers there; he died there only three years later in August 1908. Altaussee was also a meeting place for writers and other artists, from Arthur Schnitzler to Gustav Mahler, who also left the hot city to spend the summer in alpine silence. The silence did not last long as this constant immigration led to the construction of new hotels and mansions. So,

Fig. 4.10 Clockwise, the dining room and study of Gustav Haas and two views of the living room in the large apartment that stretched around the corner of the building

the Haas family followed the general trend, although Scheveningen remained the preferred holiday resort.

But after this recovery in the Habsburg homeland, Arthur Haas went back to cold unloved Prussia where he settled back in Göttingen to resume his studies:

> The following seventh[56] semester of my studies was completely miserable. I did not go to many lectures because I was mainly concerned with the completion of my doctoral thesis. But with that I had bad luck, though not through any fault of my own. I had decided to do my dissertation with Professor Simon, who taught Applied Electricity, because he had a favorable opinion of me and distinguished me with his special kindness. Unfortunately,

[56] This was the winter semester that began in September 1905 and officially ended Easter 1906. It was the seventh semester since Haas had started his university life in Vienna and the fourth semester spent in Göttingen.

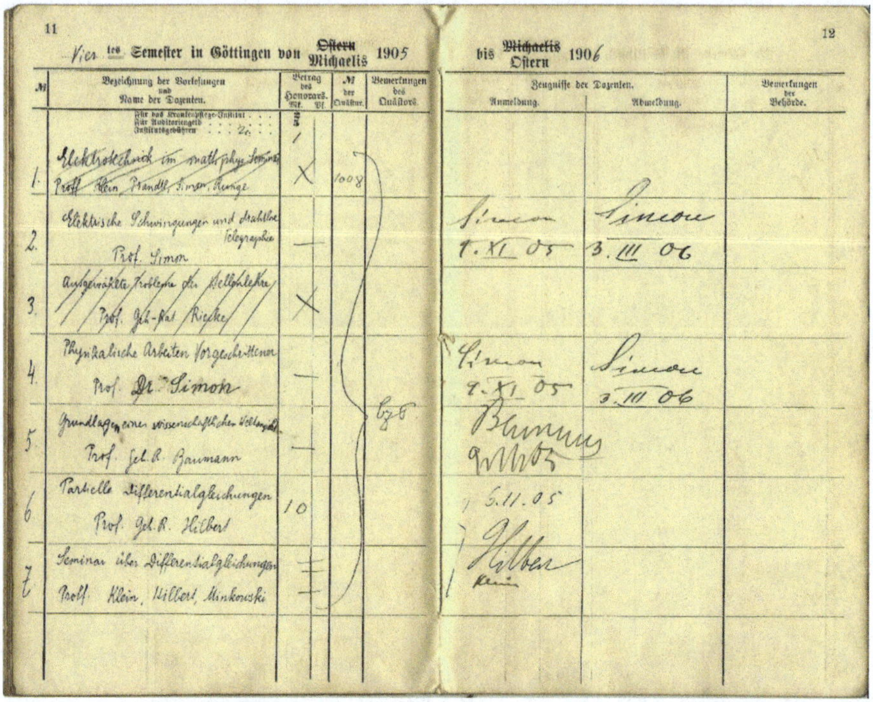

Fig. 4.11 Entries in the course book of Arthur Haas for the winter semester 1905/06 at the University of Göttingen. During this fourth semester, Haas devoted himself to the study of partial differential equations in addition to his studies of electricity. He also attended another lecture with Baumann on the foundations of a scientific view of the world

> as a dissertation topic he gave me acoustic interference experiments with singing electric-arc lamps.[xxiii] Acoustic intensity measurements at that time were a complicated matter and, in any case, required continuous constancy of pitch; this did not, however, work in the singing arc because the urban alternating current that I had to use had considerable frequency fluctuation. Therefore, I could not do my work.

In fact, the course book records only practical research studies under the guidance of Hermann Theodor Simon, which were probably intended to provide the experimental data for Haas' doctorate (Fig. 4.11). However, he also attended David Hilbert's lecture on partial differential equations which also play a crucial role in experimental physics. Haas had remained interested in questions of the philosophy of science and so, he also attended another lecture on this topic by Julius Baumann.

Haas seems to have lost his motivation due to the experimental difficulties with his doctoral topic. It was the early days of electrification—today's standard frequency of 50 Hz (Hz) for alternating currents (AC) had just become the norm in Germany—and the stability of the AC frequency depended directly on the power consumption on the grid, the so-called grid load. With increasing grid load, caused by increased power consumption in the evening, the speed of the power generators slows down, followed by a reduction in the grid frequency. This can also occur if the current feed-in decreases unexpectedly due to the failure of one of the previously unstable generators. Conversely, the main frequency increases if the current feed is higher than expected or if the load is lower. Typical in the early days of electrification, these instabilities led to the fluctuations[57] that caused Haas difficulties in his precision measurements, spoiling his appetite for experimentation and measurement.

> This discouraging scientific mishap was combined with an unbearable, growing mental and physical disgust with Göttingen University life. So I decided to simply cut the Gordian knot. I pretended that there were urgent family matters that were forcing me to return to my parents' home immediately. I said goodbye to Simon, abandoned the dissertation, and received the full bill for the semester fees despite early termination at the beginning of February 1906. (When I met Simon in Cologne two and a half years later, he told me that it was a pity that I had not completed the interesting dissertation; another candidate had successfully continued with and completed the work immediately afterwards. I asked where I could read the printed dissertation; Simon replied that it was not yet done—I never saw the work.)[58]
>
> When I crossed the border again on the trip home, I felt as if a tremendous nightmare had been taken from me. Getting to know Prussia, in Göttingen, turned me from an unsuspecting adolescent into a man who despised mankind—this persisted through three decades, until America gave me back my faith in compassion and in humanity.

Apart from the growing annoyance with Prussia, the decisive factor was that he was alone again in unloved Göttingen. His friend Karl Jellinek had received his doctorate in the summer of 1905 and had followed his doctoral supervisor Walter Nernst to Berlin as his assistant.[59] Thus, Haas was alone when he returned to Göttingen in September 1906. Personal discouragement about his lack of success in his scientific endeavors as well as his growing loneliness abroad were probably the real motivations that drove Haas back to familiar Austria.

> Understandably, my father was not particularly pleased when I arrived in Vienna in the middle of the semester, but he did approve of my decision when he learned of the circumstances. My only goal at that time was to complete a doctoral thesis as soon as possible in order to complete my university studies in good time, after eight semesters (although, in Germany at that time, ten semesters was the norm).
>
> Therefore, I immediately went to Boltzmann. I was shocked when I did not find the Boltzmann of earlier days but, rather, an old man with heavily damaged health. He recommended that I pursue an historical-critical analysis of the second law of thermodynamics and said

57 Today this problem is regulated by frequency stabilization which compensates for fluctuations in performance, supplying or removing so-called balancing energy.

58 Perhaps Haas is not quite fair in his judgement; his teacher Simon published in 1906 and 1907 some theoretical works on the effects of light arcs. Th. Simon, "Zur Theorie des selbsttönenden Lichtbogens," *Physikalische Zeitschrift* 13 (1906): 433 and Hermann Th. Simon, "Theorie des singenden Lichtbogens," *Electrical Engineering* 28 (1907): 804.

59 Jellinek stayed in Berlin for a year and then moved into industry. Following a brief period in industry at Bayer Farbenwerke in Leverkusen, in 1908 he returned to an academic career at the Technical University of Danzig, thanks to the mediation of Carl Duisberg (1861–1935). Until the end of his life, Jellinek remained interested in the ideological issues that he had discussed so intensively with Haas. Rudolf Steiner was one of his admirers and Jellinek followed Steiner to Dornach in Switzerland, where Jellinek wrote a five-volume work on the mystery of man, in which he sought to create a synthesis of natural science and anthroposophy.

that he would discuss the details with me in a few days. This conversation did not happen, because of his illness, and I had to find my own way.

Ludwig Boltzmann had a brilliant scientific career and received enormous scientific recognition during his lifetime, but throughout his life he suffered from severe depression and anxiety.[xxiv] He vacillated between high spirits, as he was in 1902 when Haas first met him, and depression and despair, as when Haas met him again in 1906. The last creative years of Boltzmann's life, after he returned to the University of Vienna in 1902, were mainly filled with the controversy over the question of atomism, which caused him considerable professional and psychological strain.

In the summer of 1905, Boltzmann travelled to California to lecture at Berkeley and he wrote a very amusing report on the journey of a German professor to Eldorado, ending with the words, "But by far the most beautiful thing about the whole journey, that is the moment when you are back home again."[xxv] But he was also weakened by the effort of travel. In the following months, Boltzmann's eyesight began to get weaker and weaker; he often suffered from severe headaches, had to give up lecturing, and could only dictate his work.

In May 1906, the dean of the university reported to the ministry that Prof. Dr. Ludwig Boltzmann was suffering from severe neurasthenia and had to abstain from any scientific activity, according to medical orders. It was during this time that Haas had wanted to discuss with Boltzmann the doctoral thesis topic, the critical analysis of the second law of thermodynamics.

The second law of thermodynamics presents the *direction of processes*, postulating *an ever-increasing entropy of the (closed) system* in which the process takes place. Boltzmann expressed the concept of entropy as: the scale of the order or disorder of a closed system, expressed by the (now famous) formula $\boldsymbol{S = k \cdot ln\ W}$, where $\boldsymbol{k}$ is a universal constant (named after Boltzmann) multiplied with the logarithm of $\boldsymbol{W}$, with $\boldsymbol{W}$ corresponding to the number of possible modes of excitation (energy states) that atoms and molecules within the system can assume.[60] If there is great disorder in a system, the numerical value of the states and thus the entropy is great. On the other hand, the more ordered a system is, the smaller is the numerical value of the states and the entropy. Boltzmann himself described it in the following way:

> The kinetic energy distributed to the individual molecules of a body always passes from a less probable distribution state to a more probable one, but not vice versa. If, for example, all air molecules are initially in one corner of a room, they are distributed evenly throughout the room: entropy increases. It is practically impossible, however, that conversely the evenly distributed molecules all accumulate in one corner of the room.[xxvi]

Haas had been planning to do his doctorate on a theoretical extension and substantiation of these statements but Boltzmann's illness prevented the meeting. Thus, Haas was left with the topic of heat theory, a topic that he would later write about in his scientific and popular scientific publications.

---

[60] American physicist Eric Johnson has contrasted this system of ordering systems, as developed by Boltzmann, with the personal disorder in Boltzmann's psyche in order to interpret Boltzmann's suicide. Eric Johnson, *Anxiety and the Equation—understanding Boltzmann's Entropy* (Cambridge: The MIT Press, 2018).

In 1906, however, Haas had no patience for finding a new adviser with whom he could start this new project; he wanted a fast conclusion to his studies and he looked for alternative topics:

> I realized that there was no longer sufficient time for an experimental project and a theoretical thesis without Boltzmann's guidance would not suffice. But, in Göttingen I had developed a strong interest in the history of science, so I decided, especially as my Greek and Latin were still exquisite, on a very peculiar theme: Ancient Theories of Light.

This was indeed an unusual topic from the history of science, but it combined the historical interests of the young doctoral student—including his preference for and appreciation of classical antiquity, whose languages he still mastered and cultivated—with his preferences for electromagnetism and the resulting theories of light, which he had studied so extensively with Professors Voigt and Simon in Göttingen. The topic was obviously chosen with a view to optimizing the project without the need to enter new areas of learning in theory or experiment.

It was not a work of physics but, rather, a more natural-philosophical presentation of the various ancient theories of light. He focused on the question of seeing or, as Haas put it, on how the effect of the visible object on the eye can be explained. Even today the work is a pleasure to read and his praise of the philosophers, whom Haas clearly admires, is only too justified.

Haas first presents the various classical theories regarding this question. Then, thanks to his enormous knowledge of the subject matter, he presents a detailed analysis of the respective arguments both for and against the various theories, including the resulting development in classical thought. One of the most detailed theories that Haas discusses is that of the rays of sight, a concept that is actually being applied today in radar technology and astronomy. According to this theory, the eye contains an inner source of energy; from the inner fire of this energy, the rays of sight emerge in a cone shape and are reflected back to the eye by surrounding objects. This idea was particularly popular with the Pythagoreans since the straightness of the visual rays results in mathematical optical laws. An example is the law of reflection, in which the angle of incidence must be equal to the angle of reflection.

However, there were objections; for example, Aristotle asks: Why, with constant inner eye fire, is one only able to see during the day and not at night? Others ask: Why are the fine rays of vision not deflected by wind and water? Or, how much inner-fire must one have in the eyes in order to be able to see as far as the stars? These criticisms and doubts gave rise to alternative theories. For example, the Epicurean theory of images is based on the opposite viewpoint: the eye is only a perceptive object; all bodies constantly send out signals that penetrate the pores of our sense organs and trigger sensations of all kinds. Regarding the nature of these signals as images of the perceived body, various interpretations developed, these leading to the formation of various schools. Democritus' theory of "air imprints" suggests that the images do not go directly into the eye but create an imprint in the air, which is then reflected in the watery eye. The Platonic theory of Synaugia[61] combines both models, whereby

[61] Plato formulated the theory of Synaugia in his dialogue *Timaios*, a script that consists mainly of a monologue by the title character Timaeus of Locri about nature and man.

the process of seeing is the result of the mixing of the rays of vision from the inner eye with the imprints of the objects.

Aristotle postulated a new idea, which most closely corresponds to the understanding of the early twentieth century: The process of seeing consists in the action of the object on the eye through the medium in between. Aristotle refers to this medium as the transparent, which in subsequent interpretations emerged as the fifth element or ether. This interpretation shaped the idea of ether as the transport medium of light rays or, more generally, of electromagnetic waves, a theory that was maintained until the early twentieth century.

The dissertation presented by Haas was thus a descriptive and interpretive work based on exceedingly careful study and interpretation of the ancient literature, but it was not based on an experimental or theoretical analysis of a physical—in this case optical—system, as Haas himself admits:

> I spent months in zealous study of the ancient classics. And, although I was otherwise withdrawn from everything, I did once in a while visit my old Fidelitas connection of which, in due turn, I became a distinguished member (without having gone through the regular fraternity student career).
>
> In around May, after only a quarter of a year's work, my dissertation was finished. It seems to have been very good in its way. The famous Viennese connoisseur of ancient philosophy Theodor Gomperz[62] praises the thesis in his book *Greek Thinkers*, saying that the work is "excellent." I also heard much additional praise for the work—but it was not a physics thesis. It was published, with a length of about forty pages, in the *Archive for the History of Philosophy*, this being already my fourth published work.[63] The already mentioned first work was followed, also in the *Physical Journal*, by an article on the relationship between Newton's Law and Coulomb's Law,[64] and the *Bibliotheca Mathematica* had printed an historical essay on the medieval physicist Johannes Philoponus, shortly thereafter.[65]

These two subsequent publications are thematically very different, the first being a somewhat critical commentary on a proposal for a new fundamental constant, which Haas exposes as a misinterpretation of physical units and scales—a topic of interest to him throughout his life.

The second work, on the other hand, reflects his historical interests but was, in actuality, merely a response to a lecture by Emil Wohlwill (1835–1912), who spoke on the importance of Philoponus as Galileo's predecessor in the sixth century. The lecture was given by Wohlwill at the meeting of the Society of German Natural

[62] Theodor Gomperz (1832–1912) came from Brünn, the home town of Arthur Haas. He studied ancient humanities first in Brünn, later in Vienna. After obtaining his doctorate in 1867 at the University of Vienna, he became a private lecturer and in 1873 was appointed professor of classical philology. Gomperz later resigned from his professorship in Vienna to devote all of his energy to his opus magnum, the three-volume work *Greek Thinkers*, the first edition of which appeared in 1893. The commentary on Arthur Haas' theories of light must have appeared in a later edition.

[63] Arthur Erich Haas, "Antike Lichttheorien," *Archiv für Geschichte der Philosophie* 20 (1907): 345–386.

[64] Arthur E. Haas, "Die Beziehungen zwischen dem Newtonschen und dem Coulombschen Gesetze," *Physikalische Zeitschrift* 7 (1906): 658–660.

[65] Arthur E. Haas, "Über die Originalität der physikalischen Lehren des Johannes Philoponus," *Bibliotheca Mathematica* III (1906): 337–342.

Scientists and Physicians in Merano in 1905. In his essay, Haas denies Philoponus this significance as an intellectual link between antiquity and the Renaissance—again with a somewhat youthfully arrogant polemic toward Wohlwill, who was almost fifty years older—and assigns the importance to Hipparchus on the basis of his interpretation of free fall. It was a classic scholarly dispute about the primacy of an idea, about which neither the old electrochemist Wohlwill nor the young doctoral student Haas could claim to be a true expert, with regard to the ancient intellectual world.[xxvii] The discussion did, however, provide a motivating element for Haas, giving him a direction and an opportunity to deal with the ancient world of ideas that he found so compelling.

> The dissertation was indeed fully recognized and even awarded the grade "excellent." Before the summer holidays, I took, with professors Jodl and Müllner, the minor exam for philosophy, which was obligatory for all doctoral students of the philosophical college. I chose ancient philosophy for the subject of the examination (it was not necessary to master the whole field of philosophy), and made a brilliant impression on the examiners, who of course gave me "honors." Then I moved alone, without my parents, into a quiet summer resort—first Schloss Aue and then Schottwien, both in the Semmering area—and studied for the main exam.
>
> I took the exam in mid-October with physics as a major and mathematics as a minor. My examiners were the physicists Lang and Exner and the mathematician Escherich. I also passed this major challenge with distinction but not without the ever grumpy and mischievous old man, Lang, trying to torment me in the meanest way; he grunted and, to the most difficult questions that he could have chosen, he responded the same to each of my answers: "that was correct for a change," although I never missed an answer. Boltzmann was no longer among the examiners.

The final result of the *viva voce* (rigorosum) is displayed in Fig. 4.12. The exam included the final defense of the theses that has been made in the doctoral dissertation as well as examination of the candidate on his expertise. The first part was, especially with the philosophers, a self-evident success. But the physicists, who thought little of ancient speculations on nature and who had little knowledge or interest in the subject, concentrated on physics questions concerning the modern interpretation of the theory of light, i.e., electromagnetism and electromagnetic waves. Professor Lang was an expert in crystallography and light refraction and Exner was a protagonist of the modern physics of radioactivity, of which Haas only knew through the lecture by Johannes Stark. These examiners knew the candidate well and used the opportunity to not only test Haas on his knowledge but also, annoyed by his youthful self-confidence and arrogance, remind him of his status in the lowest level of the academic rankings. No wonder he missed Boltzmann, who was famous and popular for his understanding and support of young students. But Boltzmann was gone, leaving a huge gap for Haas personally:

> At the beginning of September, Austria was shaken by the news of the terrible tragedy that had ended his life; his family had found him hanged on a window cross when they came home unsuspectingly. With him the heyday of the Vienna School of Physics was buried, which had found in him, Boltzmann as well as in Doppler, Stefan and Loschmidt its immortal representatives. What remained, after Boltzmann, did not exceed the level of modest mediocrity. How it hurt me later that I was only allowed to be a pupil of Boltzmann's

PHILOSOPHISCHE FAKULTÄT DER K. K. UNIVERSITÄT WIEN.

D.-Z. 2127 Wien, am 20. April 1910

1906

Rigorosumzeugnis

Herr Arthur Haas, geboren zu Brünn in Mähren, hat am 22. Oktober 1906 die zweistündige strenge Prüfung aus Physik in Verbindung mit Mathematik mit dem Teilbeil, "per vota majora mit Auszeichnung approbiert" abgelegt.

Der Dekan der philosophischen Fakultät:

K. k. Universität Wien
Dekanat der philosophischen Fakultät

Fig. 4.12 Confirmation of the rigorosum, passed with distinction by Arthur Haas

to such a modest extent. He was a genius, and only a few like him are possible within any given century.

Boltzmann's suicide was indeed a great loss for Vienna and for the entire physics community. With his death, the classical physics of the nineteenth century came to an end. Although he was one of the leading proponents of atomism, his arguments were in the field of natural philosophy and he neither mathematically nor physically dealt with a possible explanation of the theory. Lise Meitner, who admired him unreservedly, noted that Boltzmann did not discuss the theory of thermal radiation, which was fundamental to quantum physics, in the manner of a Max Planck. Nor had Boltzmann discussed Albert Einstein's three groundbreaking works of 1905, even though Einstein's formalism on the Brownian motion of molecules and atoms provided the theoretical underpinning for Boltzmann's thermodynamics and atomism. Thus, leaving Meitner to comment, "I have often wondered why Boltzmann never said a word to us about this."[xxviii]

Boltzmann left a gap that could only be filled in Berlin and Göttingen by the representatives of twentieth century physics, modern relativity and quantum physics. Arthur Haas' Göttingen teacher, the theorist Woldemar Voigt, opened the obituary in the *Physikalische Zeitschrift* with this statement:

> Boltzmann did not leave us in the midst of the richest creative output; for years there had been signs of a decline in production, which at times mounted to total silence. But the interspersed receptiveness and creative joy that showed under favorable conditions in between, left room for hope that this slackening would only be temporary. It was not meant to be. With Boltzmann, one of the most distinctive leaders among the physical researchers of all nations passed away. Just as he had many a strange habit in his daily life, he often followed lonely paths in research. For a long time, he was admired without actually being fully appreciated.

Voigt further honored Boltzmann's achievements in the fields of thermodynamics, electrodynamics and atomistics, with which he had opened a new era.[xxix]

In the next month, young Haas passed his exams as was probably expected by both sides, examiner and examinee. Nothing more stood in the way of his doctorate and the whole family probably participated at the graduation event:

> On October 31, 1907, I received my doctorate, at noon in the banquet hall of the University of Vienna, in the usual solemn manner to become a Doctor of Philosophy. It happened coincidentally that my cousin Maximilian Grünfeld[66] (who had studied chemistry and later died in the World War) was my immediate neighbor in the midst of the alphabetically ordered row of candidates. Much later, I had my philosophical doctoral degree framed with the law degree of my father and the medical and surgical degrees of my grandfather, so that all four serve as a wall decoration of my study.

Arthur Haas had succeded in receiving his doctoral degree or PhD as documented in Fig. 4.13. He had joined the academic tradition of his family. It was a success that he had long sought, to prove himself to his father and dispel his doubts about the practicality of an academic degree as opposed to a career-oriented direction such as law or medicine. Yet to complete this last step, Arthur Haas needed a permanent position or better still a professorship, which in the German-speaking world, despite the relatively low income associated with it, was considered the pinnacle of the bourgeois social hierarchy.[67]

---

[66] Maximilian Grünfeld (1883–1916) was the son of Wilhelm Grünfeld who was the brother of Regina Grünfeld, the grandmother of Arthur Haas. Maximilian fell on June 4, 1916, the first day of the Brusilov Offensive on the Galician front, when the Austrian Army was overrun by Russian forces.

[67] With the exception of social groups, such as nobility, church, and the military.

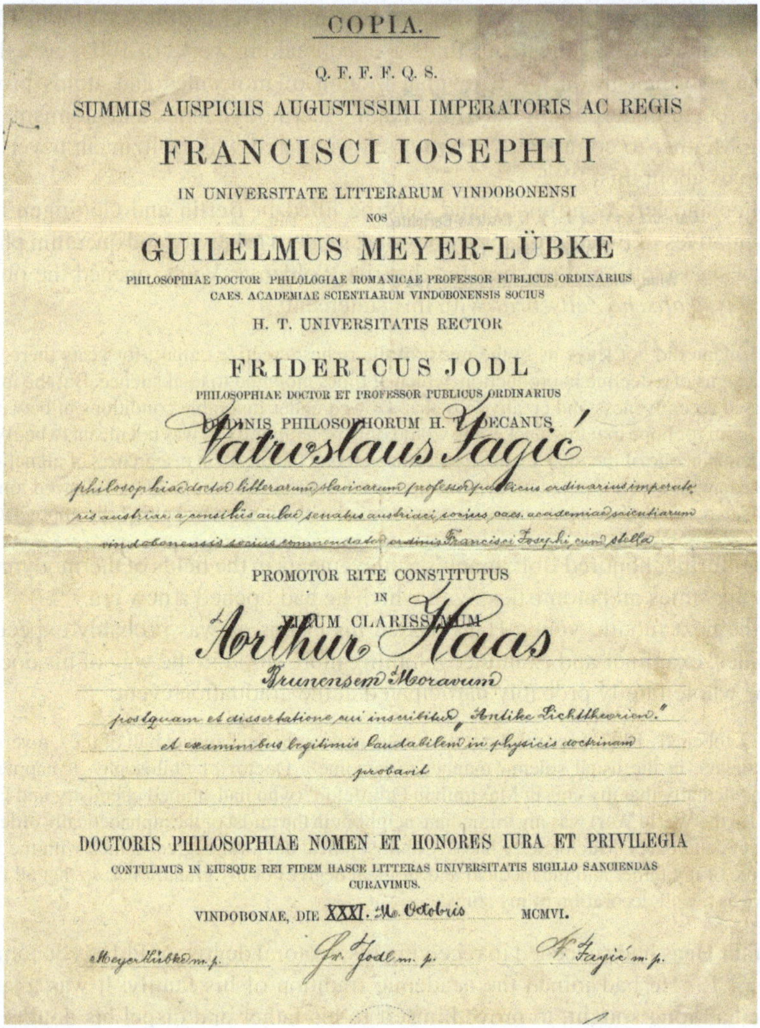

COPIA.

Q. F. F. F. Q. S.

SUMMIS AUSPICIIS AUGUSTISSIMI IMPERATORIS AC REGIS

FRANCISCI IOSEPHI I

IN UNIVERSITATE LITTERARUM VINDOBONENSI

NOS

GUILELMUS MEYER-LÜBKE

PHILOSOPHIAE DOCTOR PHILOLOGIAE ROMANICAE PROFESSOR PUBLICUS ORDINARIUS
CAES. ACADEMIAE SCIENTIARUM VINDOBONENSIS SOCIUS

H. T. UNIVERSITATIS RECTOR

FRIDERICUS JODL

PHILOSOPHIAE DOCTOR ET PROFESSOR PUBLICUS ORDINARIUS

ORDINIS PHILOSOPHORUM H. T. DECANUS

Vatroslaus Jagić

philosophiae doctor litterarum slavicarum professor publicus ordinarius imperatoris austriae a consiliis aulae senatus austriaci socius caes. academiae scientiarum vindobonensis socius commendator ordinis Francisci Josephi cum stella

PROMOTOR RITE CONSTITUTUS

IN

VIRUM CLARISSIMUM

Arthur Haas

Brunensem Moravum

postquam et dissertatione cui inscribitur „Antike Lichttheorien."

et examinibus legitimis laudabilem in physicis doctrinam

probavit

DOCTORIS PHILOSOPHIAE NOMEN ET HONORES IURA ET PRIVILEGIA

CONTULIMUS IN EIUSQUE REI FIDEM HASCE LITTERAS UNIVERSITATIS SIGILLO SANCIENDAS CURAVIMUS.

VINDOBONAE, DIE XXXI. M. Octobris MCMVI.

Meyer-Lübke m. p. Fr. Jodl m. p. V. Jagić m. p.

Fig. 4.13 The doctoral certificate of the now Dr. Arthur Haas, from October 1906, written in Latin

Notes

i. Lise Meitner and Dietrich Hahn, *Erinnerungen an Otto Hahn* (Stuttgart: Hirzel S. Verlag, 2005), 83.

ii. Wolfgang L. Reiter, *Aufbruch und Zerstörung* (Wien: LIT Verlag, 2017), 421–423.

iii. Michaela Raggam-Blesch, *Zwischen Ost und West. Identitätskonstruktionen jüdischer Frauen in Wien* (Wien: Studien Verlag, 2008).

iv. Carl E. Schorske, *Fin de Siècle Vienna* (New York: Alfred A. Knopf, 1979).

v. Stefan Zweig, *Die Welt von Gestern* gives a literary account of how Viennese life at the turn of the century presented itself to young people.
vi. See Carlo Cercignani and Roger Penrose, *Ludwig Boltzmann: The Man Who Trusted Atoms* (Oxford: Oxford University Press, 1998).
vii. Heinrich Heine, "Die Harzreise," in *Reisebilder. Erster Teil* (Hamburg: Hoffmann und Campe, 1826).
viii. Julius Baumann, *Dichterische und wissenschaftliche Weltansicht—mit besonderer Beziehung auf "Don Juan," "Faust" und die "Moderne"* (Gotha: F.A. Perthes, 1904).
ix. Karl Jellinek, *Das Weltgeheimnis—Vorlesungen zur harmonischen Vereinigung von Natur—und Geisteswissenschaft. Philosophie, Kunst und Religion* (Stuttgart, Enke, 1922), Preface.
x. Johannes Stark, ed., *Jahrbuch der Radioaktivität und Elektronik* 1 (Leipzig: S. Hirzel, 1905): 1–12 and 70–82.
xi. Felix Klein, *Vorlesungen über die Entwicklung der Mathematik im 19. Jahrhundert* (Berlin: Springer, 1926), 1979.
xii. Max Born, *Mein Leben*, 150.
xiii. Armin Hermann, "Arthur Erich Haas—Eine Biographie," unpublished first version later published in abbreviated form as *Arthur Erich Haas—Der erste Quantenansatz für das Atom* (Stuttgart: Ernst Battenberg Verlag, 1965).
xiv. Woldemar Voigt, *Elementare Mechanik—als Einleitung in das Studium der theoretischen Physik* (Leipzig: Veit & Co., 1901).
xv. Emil Wiechert, "Die Theorie der Elektrodynamik und die Röntgen'sche Entdeckung," *Schriften der Physikalisch-Ökonomischen Gesellschaft zu Königsberg* 37 (1896): 1–48.
xvi. Joseph F. Mulligan, "Emil Wiechert (1861–1928): Esteemed seismologist, forgotten physicist," *American Journal of Physics* 69 (2001): 277.
xvii. Russell McCormmach, *Night Thoughts of a Classical Physicist* (Cambridge: Harvard University Press, 1982).
xviii. Gottfried Keller, *Das Sinngedicht* (Stuttgart—Berlin: Cotta'sche Buchhandlung, Nachfolger, 1919).
xix. Marie-Theres Arnborn, *Damals war Heimat—Die Welt des Wiener jüdischen Großbürgertums* (Wien: Amalthea Verlag, 2014).
xx. Stefan Zweig, *Die Welt von Gestern.*
xxi. See Siegfried Valentiner, *Die Grundlagen der Quantentheorie in elementarer Darstellung* (Braunschweig: F. Vieweg & Sohn, 1914), 51–52.
xxii. Carl E. Schorske, *Fin de Siècle Vienna.*
xxiii. A. Blondel, "Über den singenden Lichtbogen," *Éclairage électrique* 12 (1906): 4, 1–58, and *Journal der Physik* 5 (1906): 77–97.
xxiv. See Dieter Flamm, ed., *Ludwig Boltzmann—Henriette von Aigentler* (Wien: Böhlau, 1995), 15–78.
xxv. Ludwig Boltzmann, "Die Reise eines deutschen Professors ins Eldorado," in *Populäre Schriften* (Leipzig: J.A. Barth, 1905).

xxvi. Ludwig Boltzmann, *Vorlesungen über die Gastheorie* (Leipzig: J.A. Barth, 1895).

xxvii. See Michael Wolff, *Fallgesetz und Massebegriff—Zwei wissenschaftschaftshistorische Untersuchungen zur Kosmologie by Johannes Philoponus* (Berlin: De Gruyter, 1971).

xxviii. David Rennert and Tanja Traxler, *Lise Meitner—Pioniering des Atomzeitalters*, 31.

xxix. Woldemar Voigt, "Ludwig Boltzmann †," *Physikalische Zeitschrift* 7, 649–650.

Chapter 5
After the Doctorate

Students usually see the doctorate as their ultimate goal and the gateway for entering academic paradise. Their older predecessors on this career path know, however, that it may be the first step to academic hell, a hell that one will have to survive for years without any or with very little income, while always hoping for the permanent position that often never comes. This is as true today as it was a hundred years ago, and the options available to young Haas were in principle the same as those available to any physics student after successfully completing his PhD thesis. It is true that Haas had the personal advantage of being from a wealthy home, but this kept him financially dependent and each decision required his father's input and support.

> I was supposed to now, as a young man of twenty-two years, enter the professional life. I had two options: either join one of the industrial companies in which my father was involved or pursue an academic career, with the first objective being to get a position as a private lecturer [Privatdozent] at the University of Vienna. I asked my father's approval to try the latter and he gave it. But the path I took to reach that goal was full of missteps. There were a huge number of obstacles that stood in my way, or that I unnecessarily and unwisely created for myself.
>
> At the University of Vienna, a clique economy of the worst kind prevailed. A group of closely knit families actually decided on the fulfillment of various faculty positions, paying particular attention to the interests of their own sons, sons-in-law, nephews and other relatives. Professorial positions and chairs were trade objects. In addition, there were the uncontrollable influences from higher forces outside of the university. Far more important than scientific achievement was the constant kowtowing to important professors, so that one was invited to their homes to entertain them as a musician or similar things. I was quite reluctant to work using such means. Of course, there was the option of choosing another university city, instead of Vienna, but a deep love for Viennese life awoke in me at that time and the unpleasant memories of Göttingen kept me from taking such a step.

This analysis is somewhat harsh but certainly not quite incorrect. Professors saw it as one of their tasks to provide successful students with career chances that led to a professorship. Having often worked closely for years with these students, professors used personal connections to achieve successes for their would-be colleagues. This not only helped their specific students but also helped in the further dissemination of their own ideas and scientific conceptions. The German term *Doktorvater* reflects this

M. Wiescher, *Arthur E. Haas—The Hidden Pioneer of Quantum Mechanics*, Springer Biographies, https://doi.org/10.1007/978-3-030-80606-4_5

view of professors as academic fathers who are expected to provide for their academic and intellectual offspring. Loners stiving within this academic-family community, such as Arthur Haas who was floating between specific research disciplines without a mentor, had and have little chance of long-term academic success.

This mentor-based approach exists to support one of the most important requirements for academic success—the creation of a social network in one's academic field, through which one can be recognized and known. This was the case in Vienna but is also true in every other academic community. Therefore, the loner Haas found positioning himself within the academy to be quite difficult, and for a while he was critical of the structure itself, aimlessly skulking about. But eventually he worked to achieve his goal by building up relationships with the Viennese philosophers who were more positively disposed toward him. In this time of emotional and professional uncertainty he was certainly helped by the financial support of his father, support that many of his professional colleagues did not have.

> So I spent the first few years after graduation in scientific isolation, especially since there was nothing left in Viennese physics after Boltzmann's death that could somehow have attracted me. My financial independence and self-confidence were actually a hindrance to my professional development. I lacked a strong desire to work, which is usually driven by fiscal need.
>
> It did not help that at the time I was far too interested in Viennese social life, which led to a tremendous waste of my time. But most of all, it hurt me that I had been such a good student in high school, in all subjects. With my still excellent knowledge of Greek, I continued my interest in Greek physics but, due to my strongly developed historical sense, I was far too interested in the history of physics.
>
> The high esteem that I enjoyed with the philosophy professor Laurenz Müllner always led me to the most challenging philosophical problems. I took an active part in the sessions of the Philosophical Society at the University of Vienna[1] and also gave some lectures in the field. The goal that I had in mind at the time was a professorship as Ernst Mach once had, namely natural philosophy and the history of the inductive sciences. Thus, I diligently immersed myself in a wealth of books that dealt with the history of the sciences and in numerous works that were considered to be on natural philosophy.

This passage certainly reflects a degree of self-criticism as Haas considers his long-gone student days from the perspective of middle age, with a better understanding of where he had gone wrong.

In order to be able to obtain a professorship, a habilitation is expected in the German-speaking system.[2] The habilitation requires further scientific work and a public lecture, which must be evaluated and approved by the entire faculty. It is, therefore, another hurdle; the aspiring academic must persuade all professors to give their agreement to one's thesis in order to obtain the *Venia Legendi*, the right to give lectures. This obviously requires remaining on generally good terms with the professors of one's field. This path of least resistance was not necessarily a strength

[1] Illustrious representatives of the natural sciences were Philipp Frank (1884–1966), Ernst Mach (1838–1916) and Stefan Meyer (1872–1949). Cf. R. Reininger, ed., *Geschichte der Philosophischen Gesellschaft an der Universität Wien 1888–1938* (self published, Vienna, 1938).

[2] Still today, habilitation is indispensable in the successor states of the Habsburg Dual Monarchy.

of the young Haas. Being very sure about his superior knowledge and abilities, young Haas inevitably faced the problems that have been described in this chapter.

> During the first period after attaining my doctorate, I published treatises on "Aesthetic and teleological aspects in ancient physics" [in *Archiv für die Geschichte der Philosophie*], on the problem of matter ["Annalen der Naturphilosophie"], and a paper regarding the philosophical conclusions emerging from the second law of thermodynamics [in *Archiv für Systematische Philosophie*]. In 1908, I devoted myself entirely to the completion of my habilitation thesis, which appeared in December 1908 as a book under the title *Die Entwicklungsgeschichte des Satzes von der Erhaltung der Kraft* [The History of the Development of the Law of Energy Conservation]. The book was produced at my father's expense by a Viennese publisher and was then sold by the Viennese bookseller Alfred Hölder.
>
> In September 1908, I took part in the annual gathering of German physicians and naturalists, which then met in Cologne. I nearly exclusively participated in the session on the history of medicine and the natural sciences, and I delivered a lecture on the "Historical Analysis of the Energy Principle." With that I won over the heart of the (then fifty-five-year-old) grand old master of the discipline of medical history, the Privy Councilor [Geheimrat] Karl Sudhoff, who served as Professor of the History of Medicine at the University of Leipzig; he was from Frankfurt and overflowing with temperament. In 1908/09 I also completed the very significant work on the history of ancient dynamics, which was published in the then newly founded *Archiv für die Geschichte der Naturwissenschaften und Technik*.

These years between doctorate and habilitation were a learning process for young Haas. He was not lazy and unmotivated, although he had claimed these as his problems. Rather, during these years he managed to step out of his purely descriptive approach to the history of science, which he had used in his doctoral thesis, and he began to recognize and analyze new connections between the old and the modern. He developed a systematic approach to the history of ideas and in doing so considered the theory of heat, which had been seen as pure sensation in antiquity. The theory of heat had, however, been formalized in nineteenth century classical physics by no one other than his early mentor Ludwig Boltzmann.

Haas' work on the problem of matter[3] also demonstrates that his interests had developed. Descriptions of ancient physics and its history had progressed to discussions of the emerging and developing fundamental modern physical principles—such as the theories of energy and entropy—as derived from ancient ideas. This work was, however, still motivated by Haas' original conversation with Boltzmann about the dissertation project and reflects the lasting influence of the latter. On the other hand, this new work was the first in which Haas formulated thoughts about cosmology and used the argument of the heat death of the universe as a necessary consequence of the second law of thermodynamics.

On this subject, he followed with another paper outlining in more detail the philosophical and cosmological consequences of the second law of thermodynamics.[4] This is the development of a fundamental idea that Haas was to follow throughout his life. It is the first work in which he reflects on the effects of the second law as it

---

[3] Arthur Erich Haas, "Die allgemeinsten Gesetze des physikalischen Geschehens und ihr Verhältnis zum zweiten Hauptsatze der Wärmelehre," *Annalen der Naturphilosophie* 6 (1906): 20–30.

[4] Arthur Erich Haas, "Die Physik und das kosmologische Problem," *Archiv für Systematische Philosophie* 13 (1907): 511–525.

relates to the formation and development of the universe. He pleads for universally valid and unchangeable laws of physics and its fundamental constants. From this he argues for the necessity of a finite universe:

> We thus arrive at the conclusion that, according to the laws of physics, the world must only have existed over a finite time and this period, between its beginning and the present, would be a shorter time than **Z**.[5]

A key experience for Haas, in the development of both personal and academic work, was his participation in the conference of doctors and natural scientists in Cologne. His lecture and the resulting essay on the "Historical Analysis of the Energy Principle"[6] were part of his preparation for habilitation. The important thought in this paper was the idea of permanence. The ancient idea of eternity or the conservation of matter—coupled with the ideas coming from antiquity about the eternity of the world as reflected in its astronomical movements, and the atom as the smallest particle—arose from the idea of the conservation of energy, which finally led to the idea of the equivalence of all forms of energy in "modern science."

In 1909, Haas completed his natural philosophical work in two papers exploring the influence of ancient thought models on the modern concepts and principles of science. In his publication on the topic of aesthetic and teleological aspects in ancient physics,[7] he examined the geometrical-aesthetic ideas of antiquity with respect to their impact on modern and primarily mechanical scientific concepts, from atomism to astronomy. This approach was extended in his work on the history of ancient dynamics[8] in which he considered these concepts within the framework of modern concepts such as gravity and hydrodynamics as well as the laws of motion. This was the completion of his work on ancient physics, which had begun with the theories of light and now covered most aspects of classical physics. However, this work also touched on the world of modern physics from atomism to cosmology. This evolution of thought was to be the basis for his habilitation thesis.

Regarding more personal issues, Haas had successfully developed a network for himself through the acquaintance of Karl Sudhoff, the acknowledged Nestor of German scientific history, who recognized a kindred spirit in young Haas.[9] Finding

[5] **Z** was an arbitrary time shorter than the time within which the universe would suffer the heat death.

[6] Arthur Erich Haas, "Die historische Analyse des Energieprinzips," *Annalen der Naturphilosophie* 7 (1908): 410–416.

[7] Arthur Erich Haas, "Ästhetische und teleologische Gesichtspunkte in der antiken Physik," *Archiv für Geschichte der Philosophie* 22 (1909): 80–113.

[8] Arthur Erich Haas, "Die Grundlagen der antiken Dynamik," *Archiv für die Geschichte der Naturwissenschaften und Technik* 1 (1909): 19–47.

[9] Karl Sudhoff (1853–1938) is considered one of the early founders of the history of science. He succeeded in establishing this field as a legitimate discipline for research and teaching within medical faculties. In 1905 he became the first professor of medical history at a German university. At the University of Leipzig, Sudhoff founded the Institute for the History of Medicine and dominated in the field of German medical history for decades, until his retirement in 1925. Despite his achievements, Sudhoff is still considered a controversial figure today. In 1933, to the great dismay of many of his colleagues, Sudhoff joined the NSDAP and retained his membership for the rest of his life.

Fig. 5.1 Arthur E. Haas in 1907 as a yellow Dragoon in Wiener Neustadt

such a patron was a great success for Arthur Haas. It shows that he had become well aware of the importance of scientific conferences for the presentation of his own results but also for the intellectual exchange with colleagues and potential sponsors. Through this meeting and the subsequent exchange, Sudhoff became the sponsor and mentor that Haas had originally imagined in Ludwig Boltzmann. However, this also fixed his scientific course, committing him to further work on the history of science.

The next goal was the habilitation thesis. Before this could take place, however, Haas was asked to attend the general compulsory military service of the Habsburg Empire. Figure 5.1 shows the young Haas in the uniform of the Yellow Dragoons.

> In the fall of 1907 an event took place that, while of zero impact on my life and desires, nevertheless cost me several months of wasted time. As a consequence of the general military draft that at that time existed in Austria, I joined the fifth Dragoon Regiment in Vienna on the first of October, as a so-called "one-year volunteer." The regiment was known as the Yellow Dragoons for its yellow uniform collars.

The reasons for this step are thought to be political conformity and a thirst for power. In 1933, Sudhoff again took over the Institute in Leipzig before finally resigning from its management in 1934, four years before his death. See Dominik Gross and Werner Friedrich Kümmel, "Karl Sudhoff (1853–1938) and National Socialism," Sudhoff's Archive 100/1 (2016): 2–22, and Claudia Stein, "Divining and Knowing: Karl Sudhoff's Historical Method," *Bulletin of the History of Medicine* 87/2 (2013): 198–224.

I did not fit in with the military nor did it fit me. After a few weeks spent in the training camp for one-year volunteers at Breitensee in Vienna, I became tired of the regiment and the regiment became tired of me. After an incident that resulted in a three-day barracks arrest, I reported sick. I had the life-saving fortune, considering the later war, to be declared unfit for military service and dismissed from the army, after medical examination.

What made it possible for Haas to be relieved of his one-year compulsory service as a volunteer remains uncertain. This limited one-year service was already a privilege of the bourgeois educated class. General compulsory service lasted for up to three years but if one volunteered as a citizen of higher education, the service was limited to only one year with training that led to a reserve-officer rank. However, in the following years these men were also required to regularly attend the annual reserve exercise. Many of Haas' Brünn friends from youth chose the reserve-officer career, which also brought considerable social benefit in civilian life. One of these officers was Ottokar Stoklaska who regularly served in Vienna at the War Ministry and took the opportunity to enjoy Viennese life with his friend Haas.

After his return to Vienna, Haas lived mostly in his parents' apartment at Stubenring 14, but he also had sufficient financial resources to afford his own small apartment as a private refuge.

I felt closely attached to my parents even though I had my own furnished, small rental apartment, in almost constantly changing locations, after my return from Göttingen. Yet, I went to my parents' every day for lunch, usually also for dinner. Since moving to Vienna, my father's standard of living had increased considerably. At that time, besides an English governess for my sister, we had two chambermaids (Anna and Tini, who even decades later sent cards for family birthday celebrations), a cook, a servant, and, after 1907, my father owned an automobile and employed a chauffeur. Numerous were the rather luxurious events that my parents now hosted, with some reluctance on the part of my mother who preferred a more modest life.

My father's professional career expanded constantly. He was at that time not only a lawyer but also a board member of Portland-Cement-Farbrik in Szczakowa[10] [on the Galician-Silesian border], vice-president of the sugar factory in Mödritz near Brünn,[11] a board member of the Doloplasser sugar factory in Moravia[12] and of the Upper Hungarian sugar factory.[13] His relations with the Parisian multimillionaire Baron Hirsch had taken him frequently to Paris in earlier days, and he continued to travel often to Paris, where he was once even the guest of the French Minister of Justice. Baron Hirsch had become aware of my father when the latter arranged a legal matter for him connected with a Moravian possession of the baron's. Baron Hirsch developed such a high opinion of my father that he charged him with formulating the text of his will. My father's pride was his success in negotiating the building of cartels, which not only increased the profits of the cartelized producers but also

---

[10] The Galician village of Szczakowa belongs today to the Polish town of Jaworzno. The cement plant was built in 1883 and initially developed well, surviving two wars, but fell into insignificance during the 1990s. The mighty plant complex was later demolished.

[11] The sugar factory in Mödritz on the southern outskirts of Brno was founded in 1850.

[12] The village of Doloplass (Czech: Doloplazy) is also located near Brünn. The sugar factory was built in 1851 and belonged to the Skene brothers before the company was converted into a joint-stock company in 1899.

[13] The associated sugar factories were located in Kaschau in Upper Hungary (now Kosice, Slovakia) and had been founded in the 1840s.

cheapened the product for the consumers. At that time, my father wrote a brochure[i] about the problem of an international sugar contingent, which attracted much attention.

Gustav Haas' business success is obvious; his connection to the Strakosch family had opened the door to a successful career as an industrial lawyer. His family enjoyed the luxury of their time, which was reflected in their elegant city apartment and in the choice and quality of their servants; sugar was king in Habsburg Vienna where there were several sugar refineries, some even within the city proper. More importantly, Vienna was the seat of administration for numerous other industries in the Habsburg Empire, while also holding distinction as the Imperial capital and residence city. The various sugar companies were closely connected through cartels and price fixing and it had, therefore, been important for Gustav Haas to move from Brünn to Vienna in order to handle the legal aspects of industrial communication. At the same time, new career opportunities opened for him when he became a member of the supervisory boards of several of these officially competing companies. From there, it was an easy step to move into neighboring and less-neighboring industries and companies. His was a successful legal career that culminated in Gustav's work for Baron Hirsch who was world famous at the time.

Baron Maurice de Hirsch (Moritz von Hirsch, 1831–1896) came from a distinguished family of Jewish court bankers to the King of Bavaria and continued to move in the ranks of the European aristocracy, in a manner comparable to the bankers of the Rothschild family. The baron was counted among the confidants of the Prince of Wales, later Edward VII and Austrian Archduke Rudolph. Hirsch's career began when he joined the Brussels Bank Society, Bischoffsheim & Goldschmidt in 1851. He initiated a bold program of financing for the construction of the Oriental Railway, which connected Constantinople with Europe, and secured control of the railway concession for the world-famous Orient Express. The railroad project and his often-pioneering efforts in the sugar and copper industries brought Hirsch's fortune to an estimated $100 million by 1890, earning him a reputation as an outstanding industrialist and financier.

It was during this period that Hirsch became aware of the plight of Middle Eastern Jewry and gave the Alliance of Israelites one million Swiss francs ($200,000) for the creation of schools. He eventually consolidated his donations to the Alliance into a foundation with an annual income of 400,000 francs ($80,000). He then created his own organization, the Baron de Hirsch Foundation, for educational work in Galicia. In 1891, he created the New York Baron de Hirsch Fund to support immigrants in the United States and later in Canada. In the same year he founded the Jewish Colonization Association (ICA) to facilitate the mass emigration of Jews from Russia to agricultural colonies, especially in Argentina and Brazil.

This type of colonization project prompted representatives of the Zionist movement to ask Hirsch for support, but Hirsch regarded the creation of a Jewish state as a pipe dream and rejected all support. In this way of thinking, he was completely in the tradition of the Viennese Jewish bourgeoisie which, despite its philanthropic activities, no longer wanted to identify itself as Jewish and rejected the Zionist movement. He hoped for and expected complete assimilation into the German or European bourgeoisie,[ii] with whose art and cultural ideals he identified.

The task of settling the estate of Baron Maurice de Hirsch, who was recognized as one of the five richest people but also the greatest Jewish philanthropist of his time, was without doubt a great honor. But it was also an enormous responsibility for Gustav Haas, who was required to work with various other international law firms in order to sort out the Baron's complex financial affairs. Hirsch left the ICA organization 50 million dollars, which today is worth 11.55 billion dollars. These details were not made available to the public.[iii] According to tabloid reports, the rest of the $130 million fortune went to his only son and, after his untimely death, to the Baron's illegitimate daughter.[iv]

This connection with Hirsch helped to further improve the living conditions of Gustav Haas and his family, as well as those of his brother and partner Friedrich Haas in the law firm. And, as Haas reports, his father's connections further promoted their cultural and intellectual interests:

> My father went with great passion to concerts and with my mother to the theater. My father also maintained a lively interest in the fine arts, which at that time flourished in Vienna. Of course, the automobile was used for numerous excursions in the Vienna woods and also for further, more distant travels.
>
> My brother had studied law at the University of Vienna since 1905 and he obtained the doctorate of law in 1910; four years later, he was awarded the PhD in Geology and Paleontology, as his favorite subjects.
>
> My sister attended the Vienna Girls' Gymnasium in the Rahlgasse from 1907; she spent all day with her tutor Miss Norah Barnett, who had come to us in the last year of our Brünn stay and with whom we stayed friends for decades to come.
>
> Among our Viennese relatives, my father's brother, Uncle Fritz, was the most important. Uncle Fritz had already had a magnificent apartment built on Lichtensteinstraße in Vienna and before we moved, it had been furnished by two famous artists, Urban and Lefler. Uncle's apartment was depicted, in great detail, in the English magazine *Studio*, and the artist Urban later became famous and rich as one of New York's most successful architects, building skyscrapers for the city. The apartment was the scene of many brilliant society receptions where many important artists could be met.

Like Arthur, his brother Otto Haas also had wide-ranging interests. His goal was to study geology and paleontology but his father, who had already sacrificed his first son to science, insisted that the second son should be the successor to the Haas brothers in the law practice at Stubenring 14. This forced Otto, regardless of his wishes, to study law. It was a correct decision; Uncle Fritz (Fig. 5.2), unlike his older brother, was *a bon vivant* who could not be entrusted to manage the practice without oversight, as later events have shown. However, Otto's subsequent second study in the field of genuine interest to him would prove to be life-saving later on, as we shall see.

Uncle Fritz and his wife Josephine née Singer were childless and they invested in their private life and comfort. He had given up his private law practice in the Moravian city of Proßnitz (now Prostějov) so that he could move to Vienna ahead of his brother, to cope with the increasing workload of the firm. He afforded himself a

Fig. 5.2 Fritz Haas in the office at Stubenring 14

magnificent apartment entirely in Viennese Art Nouveau style and furnished by the two important interior designers of turn of the century Vienna, Urban and Lefler.[14] The apartment was located in a four-story apartment building at Liechtensteinstraße 59 in the 9th district of Vienna, near the Strudlhofstiege and opposite the garden of the Liechtenstein Summer Palace. The details given in the magazine *Studio* demonstrate Urban's preference for exquisite material:

> Dining room in mahogany inlaid with mother-of-pearl, Library in mahogany inlaid with mother-of-pearl, old silver fittings; Boudoir with walls of purple silk and mahogany inlaid with mother-of-pearl.[v]

It was indeed a highly elegant and undoubtedly costly interior as shown in Fig. 5.3, which has largely disappeared today.

The reference to his uncle Fritz prompted Haas to also mention two of his uncles on his mother's side, the cousins of his mother, Siegfried Strakosch and Rudolf

[14] Joseph Urban (1872–1933) studied architecture at the Academy of Fine Arts in his home town of Vienna. He established himself as an architect as well as a book illustrator, exhibition designer, interior designer and stage designer, often in collaboration with his brother-in-law, the painter Heinrich Lefler (1863–1919). Urban and Lefler were co-founders of the Hagenbund, an exhibition company similar to the Secessionists. In 1912, Urban emigrated to the US at the age of forty; Lefler remained in Vienna but worked only privately. Cf. Markus Kristan, *Joseph Urban, die Wiener Jahre des Jugendstilarchitekten und Illustrators 1872–1911* (Wien: Böhlau Verlag, 2000).

Fig. 5.3 Dining room (left) and library (right) in the apartment of Friedrich Haas. Social gatherings were usually held in the dining room and the adjoining conservatory

Strakosch, who had earned academic honors in Vienna and whom Haas admired for this reason.

> Two other relatives also had large homes in Vienna: the most closely related cousin of my mother, Siegfried von Strakosch who later received an honorary doctorate from the College of Agricultural Sciences for his services to agricultural science and economics, and for authoring numerous important works on the topic; another cousin was Rudolf Strakosch who, like Siegfried, traveled a great deal, especially in the Orient, and played a major role in the Scientific Club.

Siegfried Strakosch von Feldringen (1867–1933) , son of Isidor Strakosch, was heir to the Strakosch sugar empire. He was less interested in business matters and the industrial requirements of his company, being more curious about the scientific foundations of his agricultural and ecological enterprises. He attended and gave lectures on agricultural theory and plant physiology. He took privately financed research trips to North America where he conducted light-climatic and light-physiology studies together with the Viennese botanist Julius Wiesner, investigating the influence of light on plant growth.

Siegfried continued these measurements in Sudan and Egypt in 1907 and went on to draw far-reaching conclusions about the future possibilities of agricultural development.[vi] Thus he gradually became one of the most influential advisors to the Ministry of Agriculture in Vienna and to other government agencies. Siegfried was ennobled with the title Edler von Feldringen for his services in the Austrian sugar and agricultural industries. He lived in a luxurious house at Sternwartestraße 56, seen in Fig. 5.4. His daughter Lilly later married the son of neighboring writer Arthur Schnitzler.[vii]

In Siegfried's obituary, the historian and economist Ludwig von Mises wrote:

Fig. 5.4 The villa of Siegfried Strakosch at Sternwartestraße 56 in Vienna

> Siegfried Strakosch was one of the typical representatives of the Austrian upper class, which in many respects left its mark on Viennese life in the age of Emperor Franz Joseph. He was a tireless and successful businessman. But his interests included much more. He was well acquainted with all intellectual and artistic developments. Among his friends he counted many of the most important musicians, writers and visual artists. He had the gift of understanding and appreciating creative achievements of every kind.

It is not completely clear to whom Arthur Haas is referring when he mentions Rudolf Strakosch. In the directories of the Scientific Club in Vienna there are two persons by this name, one was the owner of the wool company Rudolf Strakosch & Co. While this profession probably gave this Rudolph less opportunity for traveling afar, the second Rudolf J. Strakosch lived as a privateer in a quiet location at Brahms Platz 2 in the 4th district. This private existence may have supported the above-mentioned trips to the Orient but no further details are recorded.

In his description of social life in Vienna during the Fin-de-Siècle and the last decade before the war, Haas listed members of the Viennese society, including their titles and honors; Haas socialized with these people and was taken with their daughters.

> The number of families with whom I was befriended was rather large; it may have been at least forty. To name only those who were closer to me: Section Chief Falk with his daughters Hanna, Lisa, and Maria, of whom the eldest was Dr. phil. [Doctor of Philosophy] and unfortunately died young, while the youngest later became my sister's intimate friend; Ober-Baurat v. Emperger, a world-famous concrete engineer and architect (daughter Elga); the medical doctor von Erlach (daughters Trude and Hilda); Baron Gussich (daughters Franziska and Grete, son Paul a good friend of mine); Baron Mérode, a painter (daughter Constance);

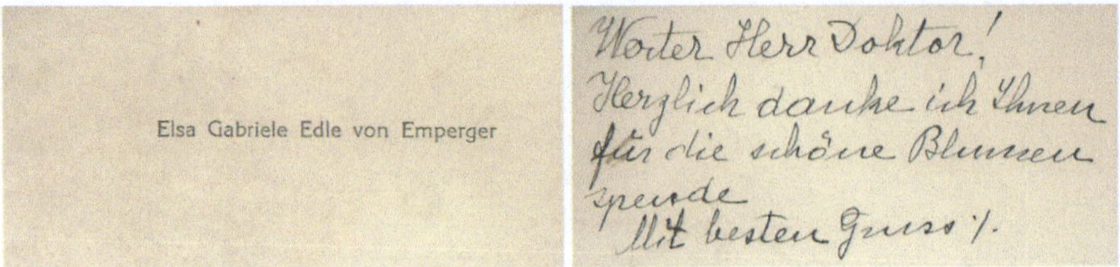

Fig. 5.5 First formal card exchange with Elsa von Emperger

> Hofrat Baron Possaner (daughters Lilly and Elly); prosecutor Baron Klingspor (daughter Dorle); pastry manufacturer Koestlin (daughter Tilly); Baroness Isbary (daughters Alice and Clarisse); Hofrat Baron Benz (daughter Helene later became Dr. phil.); Dr. Sobotka (daughter Marianne); Frank (daughter Hedwig;, her brother was my co-student and a later professor at the University of Prague, Philipp Frank); Hofrat von Beiersdorf (three daughters); Colonel Dowager of Pitter (daughter Grete); Professor von Weilen (daughters Mitzi and Helene); District head v. Fraenzel- Vesteneck (daughter Marianne); notary Wagner (daughter Melitta). I was a frequent guest of almost all and a regular for the tea hour.

The number of acquaintances was large but the extent of his personal relationship with these acquaintances remains uncertain, although some developed closer connections that resulted in longer-lasting relationships. This was especially the case with Elsa von Emperger, the daughter of the civil engineer Friedrich von Emperger (1862–1942) who lived in the same building as Uncle Fritz at Liechtensteinstraße 59.

Formal meetings, with flowers and sweets, were followed by postcard exchanges from various summer resorts (Fig. 5.5), from Karlsbad in Bohemia (today Karlovy Vary) to Westerland. The relationship intensified through 1808 when the playful exchange of poems suggests a close-dance relationship. Thus in August 1908, Elsa wrote a poem to Arthur with an image of herself mounted as a seal, shown in Fig. 5.6.

> We are well on the way
> to return from this strand
> to our homeland
> But since I won't stay
> I came to the sand
> Put on fins as my task
> Now I would like to ask
> If my dancers will mind?

However, subsequent postcards became more formal and the relationship cooler, only to end in the spring of 1909 with a "best greeting from Elsa von Emperger to the Hochwohlgeborenen [highly esteemed] Dr. Arthur Erich Haas."

Other dance relationships were even more short-lived; for instance, relations extended to only one meeting with Constance Baronne de Merode (Fig. 5.7), Marquise de Trélon, daughter of the Austrian painter Carl von Merode (1853–1909) from a Belgian-Austrian noble family.

Fig. 5.6 Poem and picture for Arthur Haas from Elsa von Emperger

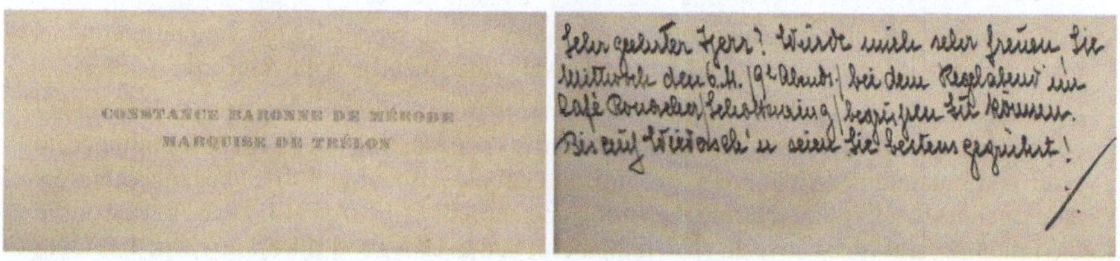

Fig. 5.7 First and only exchange with Constance de Merode

From Arthur's notes, Constance must have been the cousin of the dancer and ballerina Cléo de Merode who kept Paris in suspense with her beauty and dance during the Belle Epoque. The fate of Constance is not known.

This misfortune was followed by a lengthy letter relationship with Marianne von Fraenzel-Vesteneck, a student of geology and geography at the University of Vienna. She frequently asked for help with attending or rescheduling lectures and typically had questions about geography and the conversion of physical units. Marianne lived mainly at Schloß Altlack near Bichofslack in Upper Carniola and she asked him to send her answers there. Several letters were received but these are more concerned with her personal situation and circumstances than with young Dr. Haas. This was probably neither a friendship nor a dance relationship.

Another relationship took place with Hanna Falk but this was only expressed in short letters and thank-you notes, remaining formal. Hanna Falk studied mathematics and physics and she then turned to philosophy and history.

There is no written legacy of exchanges with other daughters from the important Viennese families that Haas mentioned in his list. It is reasonable to assume that these daughters were mere acquaintances whom Arthur Haas had met through his friends or their sisters, or perhaps in a chance meeting at one of the countless Viennese balls or other social events. While these people don't appear in his further life, the exception is Philipp Frank whom Haas already knew from his time with Boltzmann. Frank would have a considerable influence on Arthur Haas' career.

> The number of balls that I attended was extraordinarily large. In the winter months, there were always several taking place during the week and it was my habit to return home at dawn. Some of the balls I particularly liked: The Women's Home Balls (so named after the benevolent purpose they served); the Hunter Balls, with countless Comité meetings; the Jordan Evenings (hosted by Section Chief v. Jordan); the Klaudy Evenings (arranged by Baurat Claudy); the Mazur Evenings organized by the Polish aristocracy in Vienna; the Viribus-Unitis picnic, the violet picnic, etc. Fortunately, at that time the waltz was almost exclusively danced with the wonderful sounds of the great Viennese composers.
>
> I was also a very frequent guest at the various charity events. In 1908 many social events were held in connection with the 60th anniversary of the reign of Emperor Franz Joseph and a grand pageant was also organized by the Viennese society.[15] Of course, the number of friends I won at that time was numerous. With me were my two closest friends, Oskar Schmarda, who was then active in the Emperor's Supreme Court Office and later became Governor of the Prater, and Fritz Reichsritter [imperial knight] von Hohenblum, a member of the Ministry of Agriculture who unfortunately died young. As active as my social life was, at the time I was not even remotely thinking of marrying.
>
> I often went to the theater, frequently to the opera, where especially the Wagnerian operas made a strong impression on me. I went relatively infrequently to the concerts.
>
> Unfortunately, I abandoned the violin after my return from Göttingen. I also neglected, almost completely, possible excursions into the beautiful surroundings of Vienna. I continued to skate and I maintained swimming, cycling, and tennis. Also, after 1907, I began, with great passion, horseback riding. During the spring and autumn of 1907–1914, I used to ride out in the Prater several times a week. (My father, too, had been a good rider who owned a riding horse for a long time in Brünn.)
>
> As a summer resort, I first chose Baden near Vienna in 1907, where I underwent a bath cure for a rheumatic disease; afterwards I went to Schottwien. In 1908, I spent the summer at Lake Millstatt in Kärnthen, where I found very nice associates.
>
> In the middle of August, I interrupted my stay in Millstatt to go to the funeral of my grandfather in Brünn; he had died of old age in Baden. My mother had been prepared for the sad event for a long time and endured it despite her heartfelt love for her father.
>
> In mid-September I drove from Millstatt to Munich where I stayed for a few days, enthusiastically wandering through the museums, and then traveled to Cologne for the meeting already mentioned. When traveling from Mainz, I went down the splendid Rhine Valley on a steamer, but the journey was completely ruined by the overbearing presence of drunken students. In Cologne, the city authorities had organized an impressive reception for the participants of the meeting, in the wonderful town hall. Although I was still accustomed to "German nature" from my time in Göttingen, I was, nonetheless, very surprised when at this reception, where the guests were treated with Rhine wine, worthy and well-to-do full-beard fathers were rolling drunk on the floor. From Cologne I went for a short visit to Göttingen, from there to Nuremberg and then back to Vienna.

[15] The procession was designed by Joseph Urban and Heinrich Lefler.

Fig. 5.8 Arthur Haas during a garden society (left) and in his study (right), in both cases with a book in his hand

Haas describes the carefree life of a young, financially independent bon vivant; it was the first decade of the twentieth century and the vibrant life of Vienna was the pinnacle of Haas' social life and cultural experience. Even though he did not find happiness with his various female acquaintances, Haas took part in all receptions, balls and other dance events. He was in his early twenties and he wanted to enjoy life despite his unfocussed academic ambitions. While he was at odds with himself about where and how he wanted to develop, academic life still dominated and was recorded in photographs shown in Fig. 5.8. As previously noted, it was also during these years that he wrote several publications on the influence and interaction of ancient thought, noting its impact on modern ideas of science.

But at last, despite distracting pleasures, Haas completed his habilitation thesis on the history of the development of the theorem of the conservation of power. His thesis then had to be bound and printed according to the habilitation rules of the university. Arthur's father paid for these expenses since he was proud of the completed work, despite his reservations about his son's academic career. The work was published by the k.u.k. Hof-und Universitäts-Buchhändler der Kaiserlichen Akademie der Wissenschaften Alfred Hölder in 1909 and bound in green silk paper (Fig. 5.9).

The theme was the conservation law of energy as well as the idea of causality. These were both basic principles of scientific thinking in nineteenth century physics, which Haas had traced back to ancient origins. As he himself wrote:

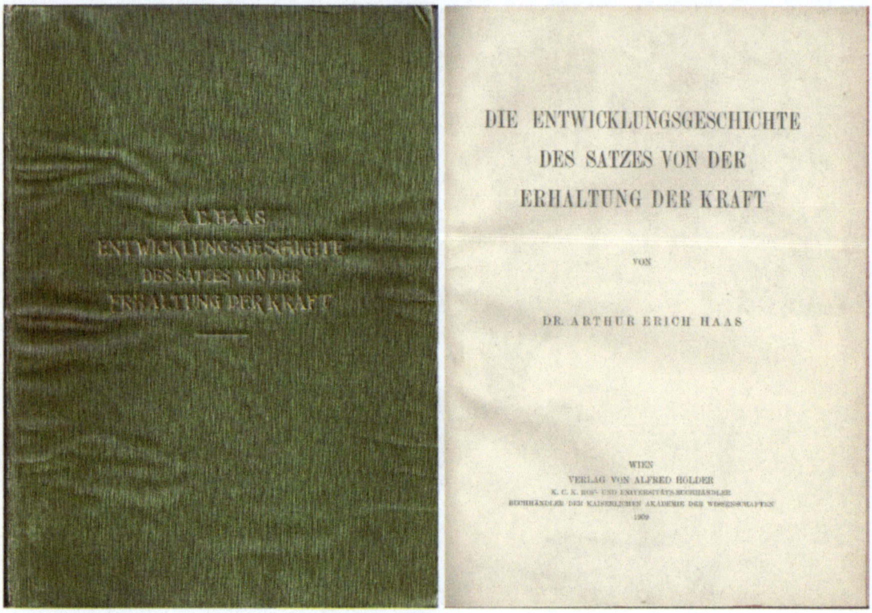

Fig. 5.9 The printed edition of Arthur Haas' first submitted habilitation thesis in Vienna

> The law of the conservation of energy ultimately reflects also the idea of causality; applied to the connection between all death and awakening in nature, causality finds its most determined and sharpest expression. The so-called discovery of the law of energy was therefore not a happy coincidence. It emerged naturally with the development of the most important physical ideas, which, through progressive convergence, were destined to finally merge into a single thought. ... The study of the historical development of these ideas thus also convinces the physicist of the truth of Goethe's words: "The history of science is science itself."[16]

It was a work of natural philosophy and history of science, a very detailed and coherent compilation of all the arguments that have been put forward over the centuries. Haas progresses from Lucretius, who links the preservation of matter with the idea of the indestructibility of motion, to Descartes who argues from the standpoint of God's perfection, which is expressed not only in the immutability of the deity itself but also in the constancy of its effects. Although Isaac Newton as well as Voltaire opposed this assumption of conservation, the principle prevailed with the introduction of potential energy by Gottfried Wilhelm Leibnitz. With potential energy, the original idea of exclusively kinetic energy was extended by a concept of bound or temporarily frozen energy, which has the potential, if released, to produce dynamic changes. This argument was followed by the development of the laws of heat and the realization that heat is a form of energy.

[16] Arthur Erich Haas, "Die Entwicklungsgeschichte des Satzes von der Erhaltung der Kraft," (habilitation thesis, Wien, 1909), 2–3.

Haas' thesis is a thorough and scholarly text that bears witness to his careful study of the historical scientific literature and ends in mathematical processing of the energetic ideas. It is not, however, a physical work in the strict sense, since no new physical findings of an experimental or even theoretical nature were developed and presented. The Goethe quotation: "The history of science is science itself," at the end of his introduction, certainly did not help him to overcome the omission.

When by Christmas 1908 my habilitation thesis was completed, I went to my mentor Prof. Müllner. He advised me to submit the habilitation application. Unfortunately, this did not have the desired success. The matter was assigned to the physicists (essentially Exner and Lang). Although they probably could not easily refuse me because my work was favorably evaluated, the peculiarity of the historical work gave them a headache. So after months, the not at all badly meant decision came: I should, since only two years had passed since my doctorate, wait a bit and, after completion of a purely physical (not historical) work, submit again in addition to the already submitted habilitation thesis. This decision annoyed me and it effected my father even more so, our immediate response was that I would turn to law school.

Arthur Haas was easy to annoy, especially when his achievements were not recognized and when resistance came from professors such as Lang and Exner, whom he did not hold in high esteem. He now regarded them as the main obstacle to his rapid career in academia. Nevertheless, the decision was correct—the subject matter had little to do with actual physics and was actually an historical analysis of a scientific thought. Such topics were dealt with by other departments of the university. His insistence on physics required that he turn to an actual physics topic, from which he had surely strayed after Boltzmann's illness. Initially Haas remained annoyed and defiant, which distracted him even further from his academic goals. This distraction was, however, well in line with his father's plans and hopes.

I had taken a precautionary measure in 1907 as a Dr. of phil. I had enrolled as an ordinary student in the law program of the University of Vienna and thereby, I had the right to attend the compulsory examination given in the middle of the law curriculum, in the fall of 1909. Due to my enrollment, no one could object that I had never actually attended any of the juridical lectures.

In the spring of 1909, I studied from books for the four subjects of the examination: Roman law, which interested me very much; ecclesiastical law, which I found interesting; Austrian imperial history, which in any case I knew extremely well; and German legal history, which I found completely boring.

Silly enough, I then spent the summer of 1909 in Vienna, going daily to the then newly opened Gänsehäufl, a beach on the Danube.

At the end of September 1909, I went back to my work, attending the meeting of doctors and naturalists that, at that time, took place in Salzburg. There I once again gave an historical lecture on ancient physics, fortifying Sudhoff's benevolence for me. I also won the affection of the leading figure in the history of the actual natural sciences, Prof. Siegmund Günther [17] of the Technical University in Munich.

[17] Adam Wilhelm Siegmund Günther (1848–1923) studied geography and mathematics at various German universities, including Erlangen, Heidelberg, Leipzig, Berlin and Göttingen. After his doctorate in 1872, he began teaching as a secondary school teacher in Weißenburg, Bavaria. He submitted a mathematical habilitation thesis in 1873 and became a private lecturer at the Munich Polytechnic. In 1876 he transferred to the University of Ansbach. From there he was appointed

In October, I passed the law exam easily. The board of examiners included the Romanist Jörs,[18] the professor of German law Schwind,[19] and the Bishop Marschall.[20] I even received an award in Roman Law due to the fact that my knowledge of Latin was still excellent. By custom, I would first translate the section of the corpus juris that I would then interpret, but the examiner stated that the outstanding quality of my translation had already clearly demonstrated that I understood everything and a further explanation was not necessary. I remained formally inscribed and I formally received the Absolutorium[21] of the Faculty of Law in 1911, although I never went back to studying law.

In the description of his law studies, Haas demonstrates a self-confidence that almost borders on arrogance, as well as making his political prejudices clearly evident. His interest in Roman legal history and his condescending devaluation of the history of the resulting German law is difficult to explain, especially when one considers that Austria was part of this empire until the dissolution of the old German Empire only one hundred years earlier. It is reasonable to assume that this devaluation of the subject of German law had less to do with his lack of interest in the subject and more to do with the adjective "German." By the time of his writing, Haas deeply hated the political rise of the right-wing extremist Nazi movement and the associated need to emigration to America.

Arthur Haas received his law degree in the autumn of 1911 as shown in Fig. 5.10, having worked hard for two years on the required lectures and seminars. However already in 1909, after spending time in the swimming pool at the beginning of his law studies, Haas had clearly turned back to his actual purpose. This began with the autumn meeting of physicians and scientists where he sought the support and advice of his mentor Karl Sudhoff from Leipzig. Haas also tried to break away from his parents, asserting his independence by getting his own apartment.

In the year 1909 another not-insignificant event took place in my life: the establishment of my own apartment. Since 1906 I had moved from one furnished flat to another, from annoyance to annoyance. I first lived in the ninth district in the Wasagasse, then in the Baumann-, Bechard-, and Rechte Bahngasse in the third district. Then in February 1909, I rented my own apartment in the third district since we had surplus furniture, partly from my grandfather and partly as a result of new Viennese acquisitions. With the consent of my father, I moved

---

to Munich where he remained until his retirement. Günther served as professor of geography and particularly distinguished himself as a geographer and mathematician, as well as an historian of mathematics and a natural scientist.

[18] Paul Jörs (1856–1925) studied history and law in Bonn. He received his doctorate in Bonn in 1882 and, after professorships in Kiel, Giessen, and Breslau, went to Vienna in 1905 where he taught Roman law.

[19] Ernst, Baron von Schwind (1865–1932) studied law and political science in Vienna. In 1888 he received his doctorate in law and habilitated in 1891 with a thesis on German law at the University of Vienna. In 1894 he was appointed associate professor of German law and Austrian imperial history in Innsbruck, and in 1897 he moved to Graz for the same subjects. From 1899 he was a full professor at the University of Vienna.

[20] Godfried Marschall (1840–1911) was vicar general and auxiliary bishop in the archdiocese of Vienna.

[21] The Absolutorium is a certificate of completion of the law program at a university; it does not correspond to a university degree. Therefore, students had to pass the actual final exams after passing the Absolutorium.

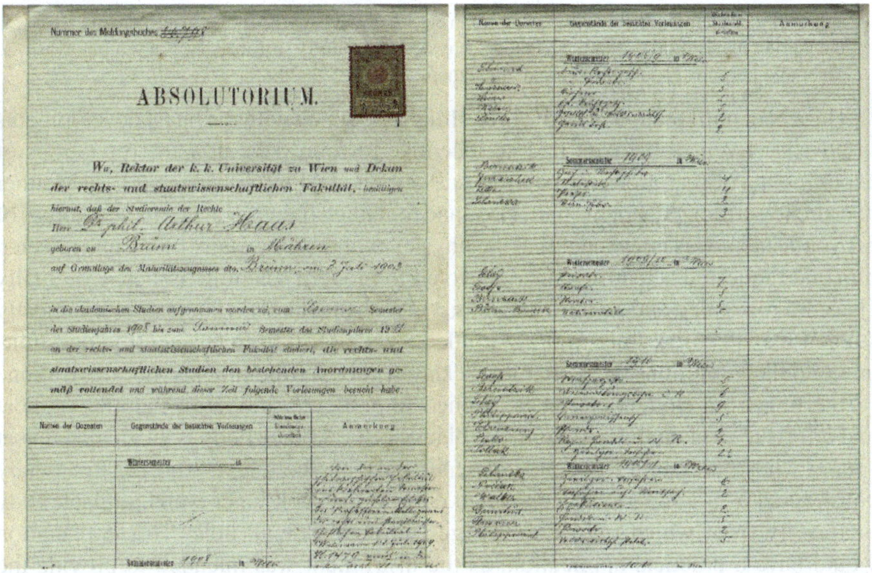
Nummer des Meldungsbuches

ABSOLUTORIUM.

Wir, Rektor der k. k. Universität zu Wien und Dekan der rechts- und staatswissenschaftlichen Fakultät, bestätigen hiermit, daß der Studierende der Rechte Herr Dr phil. Arthur Haas geboren zu Brünn in Mähren auf Grundlage des Maturitätszeugnisses dto. Brünn, am 7. Juli 1903

in die akademischen Studien aufgenommen worden sei, vom ... Semester des Studienjahres 1908 bis zum ... Semester des Studienjahres ... an der rechts- und staatswissenschaftlichen Fakultät studiert, die rechts- und staatswissenschaftlichen Studien den bestehenden Anordnungen gemäß vollendet und während dieser Zeit folgende Vorlesungen besucht habe:

Namen der Dozenten	Gegenstände der besuchten Vorlesungen		Anmerkung

Fig. 5.10 The document confirming his law degree, the Absolutorium

> to Neulinggasse 24. This remained my home until my marriage, in April 1925; I stayed there for more than sixteen years.
>
> It was a ground-floor garden apartment, consisting of two large rooms facing the garden, a walk in closet, and adjoining rooms. The garden of the house was enclosed by parks with tall trees belonging to various old palaces, so that my view in the spring and summer gave the complete illusion of living in the countryside. The apartment was only ten minutes walking distance from the Ringstrasse, and only five minutes from the city park; I was able to reach my parents' house in about ten minutes. One room was set up as a living room, with my desk and my book-boxes, as well as a big table and a divan inherited from my grandfather; the other room was furnished as a bedroom. Over the years, of course, the interior constantly improved.

With this independence, Arthur was free from paternal pressure to take over his father's law firm with his brother Otto. And so, Haas' concentration again returned to physics. He wanted to show the professors, whom he despised so much, that he was indeed capable of working out complex physical theories leading to new groundbreaking results. He regarded law only as a last resort or last alternative should he, contrary to expectations, fail to achieve a scientific breakthrough.

> By passing my state law exam, I had secured the opportunity to become a lawyer if my habilitation plans should fail. Then, in agreement with the decision of the faculty, I turned to the task of completing a pure physics thesis. I delved into study of the most recent literature and, in February, 1910, I had already succeeded in a very significant achievement, or so I believed; at any rate, it was the greatest achievement of my life.
>
> I was the first to apply the quantum hypothesis to the problems of atomic structure and I was the first to find the connection between the spectroscopic fundamental constant and the elementary quantum of action, the electron charge and the electron mass. Today, this

> achievement is universally acknowledged. Yet, the greater my life-long grief became when the intellectual limitations of the Viennese physics community suffocated this attempt toward great and perhaps groundbreaking achievement.

Haas had, until then, done little or nothing associated with the latest developments in physics. Even Boltzmann had no incentive to follow the new interpretations of experimental observations raised by Max Planck and Albert Einstein, as Lise Meitner had already criticized. For his part, Haas had remained focused on the ancient topics of his youth and it was no wonder that the physicists at the University of Vienna rejected him and requested scientific papers devoted to modern scientific issues. Eventually Haas did recognize and acknowledge this, albeit reluctantly. He then systematically studied the latest literature in order to identify the most critical scientific questions and find their solutions. He certainly lacked neither self-confidence nor dedication.

Starting from Boltzmann, Haas studied in particular the latest theories on heat and efforts to explain the radiation of a black body. A black body is an idealized heat source whose radiation frequency depends only on the temperature $\boldsymbol{T}$. Practically constructed as a temperature-controlled cube with black inner walls, a black body is a cavity from which radiation emerges through a small hole. This explanation would have a practical application, as the instrument was intended to provide the standard value for industrially manufactured light sources. It turned out, however, that although the emerging spectrum—the intensity distribution of the radiation as a function of the radiation frequency—changed significantly with temperature as predicted, the observed dependence could not be fully described by the formalism of the various theoretical models.

The Stefan-Boltzmann law, developed in Vienna, predicted—for the radiated power—a proportionality of $\boldsymbol{T^4}$. The Rayleigh-Jeans law, formulated by the two British physicists Lord Rayleigh (1842–1919) and James Jeans (1877–1946), could only describe the spectral energy distribution for large wavelengths or small frequencies, whereas the Wien Radiation Law, developed by Wilhelm Wien[22] could only describe the spectral energy distribution for small wavelengths or large frequencies, respectively. While the Wien Displacement Law did at least establish the relationship between the emission maximum of a black body and its temperature.

The first successful description was that of the Berlin physicist Max Planck (1858–1947), which he proposed to the German Physical Society in 1900.[viii] The formula developed therein, Planck's radiation law, was able to reproduce exactly the observed spectra of black bodies over the entire frequency range for all temperatures. However, Planck used a small, as he thought, mathematical trick for achieving this, in that he did not let radiation be emitted continuously, as was generally assumed, but let it be emitted in infinitely small energy packets whose energy $\boldsymbol{E}$ was directly proportional

---

[22] Wilhelm Wien (1864–1928) is primarily known for his work on electromagnetism, thermodynamics and the development of the Wien filter, in the velocity separation of moving particles, as well as for the formulation of Wien's displacement law in thermal radiation. He was awarded the Nobel Prize in 1911 and in 1919 was called from Würzburg to Munich to take over the chair in physics, as Wilhelm Röntgen's successor. Despite his very patriotic nationalistic opinions, Wien supported Jewish physicists, unlike other German-national colleagues such as Philipp Lenard and Johannes Stark. This is probably why Haas sought contact with him.

to the frequency ν of the radiation: $E = h \cdot \nu$, where the constant of proportionality h had a value of $h = 6.548\text{–}10^{-27}$ erg·sec (erg is the unit for energy, 1 erg = 10^{-7} J).

These energy packets are today called quanta or photons. Today, the proportionality constant h is simply called Planck's constant and is considered to be one of the fundamental constants of physics, which reconciles the macrocosm of our observable world with the microcosm of the quantum world of atoms and nuclei.

Haas was familiar with Stefan Boltzmann's law from his studies with Boltzmann but this development of Planck's appeared to be a new world. Haas recognized the importance of the constant of proportionality and the necessity of assigning a physical meaning to a mathematical construct. Yet, the questions remained: What is the constant? What is its physical role and meaning? He was well acquainted with the meaning of scales and constants from his studies of electromagnetism and electron theory and he was always fascinated to understand the physical basis and meaning of these constants, which seemed to appear so suddenly and mostly unmotivated in the textbooks.

Haas' goal became to investigate the actual physical role and significance of Planck's constant h. A central part of Planck's theory was the assumption of resonators in the radiating black body, which maintain the balance of absorption and emission. From the relationship between the entropy of the resonator and the total internal energy of the system, the black-body spectrum could be calculated. While entropy was also a term well known to Haas, this was another purely theoretical concept and the actual nature of these resonators had not been discussed in Planck's paper.

Ten years after the formulation of Planck's radiation law, Haas would base his own theoretical approach to this unanswered question regarding the nature of the resonators on an atomic model. He used the model developed in 1904 by the British physicist Joseph J. Thomson (1856–1940), the so-called plum pudding model. According to this model, the negative electrons were supposed to be distributed homogeneously, like raisins, within a dough comprised of an equal number of positively charged protons, thus obtaining the electrically neutral atom. However, in the same year that Haas worked on the problem, the British physicist Ernest Rutherford (1871–1937) proposed a new model based on his interpretation of alpha-scattering experiments on gold. In this model, the positive charge of the atom was located in a small core, the nucleus, while the electrons were located in a cloud surrounding the nucleus. These results were unknown to Haas at the time. However, he recognized from his reading of Thomson's publication[ix] that Thomson's atom could provide a physical meaning for the Planck resonator. Haas explained his approach in the following way:

> In the following study we consider Thomson's model of a hydrogen atom as a special case of an optical resonator. By using electron theory and Planck's radiation theory, which deal with the same problem, we should be able to find a connection between these two theories.

While the Thomson model is based on the assumption of a homogeneous distribution of electrons in positively charged pudding, Arthur Haas' calculations revealed a non-homogeneous distribution—a special case where the electrons are located in the outer sphere of a generally positively charged sphere with radius ***a***. The positive charge of the sphere was compensated by the negative charge of the electron shell, so that the total charge was zero. Comparing the potential energy of the electron in this configuration with the Planck energy quantum was a radically new approach by young Haas.

Equating centripetal force with the Coulomb force was a typical approach from classical physics, with **ω** being the rotational frequency of the body:

$$h = 2\pi \cdot e \cdot \sqrt{a \cdot m} \quad or \quad a = \frac{h^2}{4\pi^2 \cdot e^2 \cdot m}$$

However, by equating different forms of energy, as well as forces of mechanical and electrical natures, Haas found a direct correlation between the radius ***a*** of his electron sphere and Planck's constant ***h***,

$$\frac{e^2}{a} = h \cdot \upsilon = h \cdot \frac{\omega}{2\pi} \qquad m \cdot \omega^2 \cdot a = \frac{e^2}{a^2}$$

It should be noted that Haas only took the first step (left side of the system of equations) because he considered atomic dimensions to be fundamentally indivisible parameters.

Haas was too much influenced by the classical Greek concept of A-Thomos—the smallest indivisible particle—as formulated by Democritus (ca. 460 BC—ca. 370 BC). Haas had internalized this concept as undeniable fact through his work on Greek philosophies and conceptions of nature. Therefore, Haas wanted to give the Planck constant a physical meaning within the framework of what he considered to be the most fundamental constants—the dimensions of the atom as the smallest indivisible particle, measured as atomic mass and atomic radius. This was his goal and he did not see beyond it. This may have cost Haas the Nobel Prize. The second formula (on the right in the system of equations) shows the radius of the atom ***a*** as a function of the Planck constant and the atomic mass ***m***, and is directly derived by solving the left-hand equation for ***a***. This is the expression for the so-called Bohr radius, as predicted three years later by Niels Bohr (1885–1962) within the framework of the Rutherford atomic model. The step was a small conversion of an equation, but it was a giant step in the intellectual significance of the interpretation. As Bohr wrote fifty years later to Arthur Erich Haas' son, Arthur Gustav Haas:

> As you will know, he [Haas] was among the first to be interested in the interpretation of spectra on the basis of quantum theory and atomic models, however, the significance of the discovery of the atomic nucleus was not recognized, and therefore also not the radical departure from accustomed ideas ...[23]

Haas had omitted this last step because he believed in the infallibility of the Greek philosophers' smallest entity.

[23] Niels Bohr in a letter to Arthur Gustav Haas, the son of Arthur, fifty years later.

Fig. 5.11 The absorption and emission spectrum of hydrogen gases in the visible wavelength range from about 400 to 760 nm. Haas was the first to theoretically predict the position of these lines which, according to current understanding, are generated by quantum transitions in the hydrogen atom

However, the theory established by Haas not only predicted the relationship between the Planck constant and the atomic dimensions of hydrogen in the ground state but also tried to provide an explanation for other phenomenological parameters, such as the Rydberg constant in atomic spectroscopy. Haas thus offered a first physical explanation for the previously, entirely empirical Balmer formula.

The Balmer formula describes the energy distribution—or frequencies, or wavelengths—of the observed lines in the visible spectrum of hydrogen gases displayed in Fig. 5.11.[24] In the case of gas discharges, such as those generated in Plücker tubes, the Balmer lines that dominate the visible emission spectrum can be easily detected by the use of prisms. In sunlight, the Balmer lines appear as black absorption lines which are generated by filtering out the hydrogen light in the outer solar atmosphere.

Haas' theory also predicted important relationships: the mass ratio between an electron and the hydrogen atom, the two fundamental particles at that time, and the ratio of their physical sizes.

This work was quite sensational and Haas submitted it to the university in 1910 as his habilitation thesis. His tormentor Lang had retired the year before and his successor was Ernst Lecher, a former student of Lang's whom Haas did not find particularly appealing.[25]

> Lang had retired in 1909. His successor was the experimental physicist Ernst Lecher from Prague, a man who could be called with some benevolence a witty jerk (his admirers called him "boisterous"). Incidentally, he was a brother of the very influential German-progressive Member of Parliament at Brünn, Dr. Ing. Otto Lecher, who had been a frequent visitor in my parents' home in Brünn during my childhood.

[24] On the history and development of the term spectrum and its experimental acceptance and meaning in all of the natural sciences, cf. Klaus Hentschel, *Mapping the Spectrum* (Oxford: Oxford University Press, 2002).

[25] Ernst Lecher (1856–1926) went to Vienna in 1909 after holding faculty positions in Innsbruck and Prague. He succeeded Viktor von Lang as director of the First Institute of Physics. Lecher worked in the field of radio and frequency technology, so he had little knowledge of developments in modern quantum or atomic physics.

> Boltzmann's successor was a completely insignificant man of only thirty-three years, but he had the privilege of being the nephew of the immensely influential director of the Theresianum, a high school where the youth of the whole high aristocracy studied at that time. His name was Friedrich Hasenöhrl, a completely blank slate. I only remember him as a very limited man, while others had a better opinion of him. That he was a very distinguished character is, in retrospect, certain.[26]

Friedrich Hasenöhrl[27] was not only Boltzmann's student but also became Boltzmann's successor at the University of Vienna in 1907. He was the head of the Department of Theoretical Physics, where he was mainly concerned with the radiation of black bodies and the associated laws. He was therefore regarded as an expert in the scientific question considered in Haas' habilitation thesis. Hasenöhrl had a number of illustrious students in Vienna, in particular Erwin Schrödinger[28] who later received the Nobel Prize in Physics for his contributions to quantum mechanics, namely the famous Schrödinger equation. More recent research results assign Hasenöhrl a more important role than that granted to him by Haas.

In 1904, Hasenöhrl was able to predict Einstein's famous formula $\boldsymbol{E} = \boldsymbol{m} \cdot c^2$ exactly from his investigations, except for a small correction; this achievement was never fully appreciated.[x] In this regard, the experiences and disappointments of Haas and Hasenöhrl may have been similar.

> The success that I achieved in February 1910 was a very substantial and crucial part of the achievement that Bohr repeated in 1913, for which he was awarded the Nobel Prize. Bohr himself called me his predecessor, in his Nobel lecture. In addition, later, in Sommerfeld's classical work *Atomic Structure and Spectral Lines*, as well as in Reiche's *History of Quantum Theory*, my results were emphasized accordingly.
>
> However, at that time in Vienna, I found only contempt, even mockery. When I talked about my work at the Vienna Chemical-Physical Society, Lecher thought himself especially funny when he called the whole thing a carnival joke, in the public discussion. When Laurenz Müllner asked Hasenöhrl about me, the latter explained that I was not to be taken seriously because I naively confused branches of knowledge that could not have anything to do

---

[26] Here Haas refers to the attitude Hasenöhrl showed in later discussions, allowing Haas to publish his results.

[27] Friedrich Hasenöhrl (1874–1915) studied natural sciences and mathematics at the University of Vienna under Josef Stefan (1835–1893) and Ludwig Boltzmann. In 1896, Hasenöhrl received his doctorate under Franz-Serafin Exner with a thesis on the temperature coefficient of the dielectric constant in liquids and the Clausius–Mossotti formula. After a stay in Leiden, he was appointed back to Vienna, to the chair of Boltzmann. With the outbreak of war Hasenöhrl volunteered for the front and in 1915, he fell during an attack on the Italian positions at Folgaria.

[28] Erwin Schrödinger (1887–1961) studied in Vienna under Franz-Serafin Exner and Friedrich Hasenöhrl between 1906 and 1910. That was the time after Haas had returned from Leipzig but there may not have been much overlap since they were at different stages of the student career. In 1911, Schrödinger became an assistant to Exner at the time when Haas sought his habilitation. The two may have become acquainted through similar interests in philosophical questions. Schrödinger was strongly influenced by Arthur Schopenhauer and, as a result of extensive reading of Schopenhauer's works, was deeply interested in philosophy throughout his life. In his lecture "Mind and Matter," Schrödinger said that, "The world extended in space and time is but our representation." This is a repetition of the first words of Schopenhauer's main work. Haas mentioned a few occasional personal interactions with Schrödinger in his later writings.

> with each other, such as quantum theory (as something related to the theory of heat) and spectroscopy (as something related to optics).

It was precisely this step, to bridge the boundaries between the two traditionally separate areas and explain optical radiation phenomena by using the quantum mechanical approach, that put Haas ahead of his time. He was too early for recognition and it is possible that he himself did not recognize its importance. He did not want to develop a model of the atom, as he himself admitted. Haas was only looking for a definition and explanation of the physical nature of the Planck number within the framework of the classical parameters set forth in electron theory.

This lack of recognition and the resulting rejection constituted another setback and Haas began looking for alternatives. He soon found a new direction thanks to the personal connections of his father.

> Understandably, I was deeply discouraged; but there was a new hope. The excellent mathematics professor of the Brünn Technical College, Emil Waelsch[29] whom my father knew well from Brünn, advised my father that I ought to habilitate at Brünn University. The now former professor of physics at the university, Jaumann,[30] was an ingenious theorist who had anticipated some of the mechanics of waves. He was quite isolated because of the challenging nature of his articles. The chair for mechanics was held by Hamel,[31] a relatively young German who later returned to Germany.
>
> At the invitation of Jaumann [Fig. 5.12], I gave a guest lecture at the Brünn University of Applied Sciences in which I outlined my work on the quantum-theoretical significance of the spectroscopic constants. I obviously made an excellent impression on Jaumann, who invited me to submit a habilitation application. I did so in the spring of 1910 and, on the basis of the very favorable reports by Jaumann, Hamel, and the Munich professor Siegmund Günther who had been invited as external evaluator, the professors' collegium decided to allow me to apply for habilitation. With this, the matter seemed to be decided favorably; the still to be held colloquium, as well as the trial lecture, were considered by everyone to be merely unessential formalities.
>
> So I studied gladly and diligently for the colloquium and traveled on the appointed day to Brünn where, for lunch, I was my uncle's guest; Louis Gomperz[32] was married to my mother's younger stepsister. After lunch, my uncle told me that, a few hours earlier a Professor

[29] Emil Waelsch (1863–1927) worked mainly in the field of geometry. After initial lectures in Prague, he studied with Felix Klein in Göttingen and Max Noether at the University of Erlangen, where Waelsch was awarded his doctorate in 1888. After stays in Leipzig and Prague, he was appointed to the Technical University in Brno in 1885, where he remained until the end of his life.

[30] Gustav Jaumann (1863–1924) studied chemistry and physics in Prague, where he was a student and assistant to Ernst Mach. Jaumann received his doctorate in 1890 and habilitated only half a year later. In 1893, he was appointed extraordinary professor of experimental physics and physical chemistry at the k.k. German Karl-Friedrich University in Prague. In 1901, Jaumann moved to the German Technical University in Brünn as a full professor of physics.

[31] Georg Hamel (1877–1954) studied in Aachen, Berlin, and Göttingen, where he received his doctorate in 1901 under David Hilbert. Hamel was habilitated in Karlsruhe in 1903 and became a full professor in Brünn in 1905. He moved to Aachen in 1912 and in 1919 to the TH Berlin. He is best known for his work on fundamental problems as well as for his work on the axiomatic structure of classical mechanics, which represents an important contribution to technical mechanics.

[32] Louis Gomperz (1863–1922) was the husband of Ida Strakosch (1876–1942), the half-sister of Arthur Haas' mother. Gomperz worked as a lawyer in Brünn and later in Vienna. His wife Ida died in 1942 in the Brno concentration camp, a sub-camp of Auschwitz.

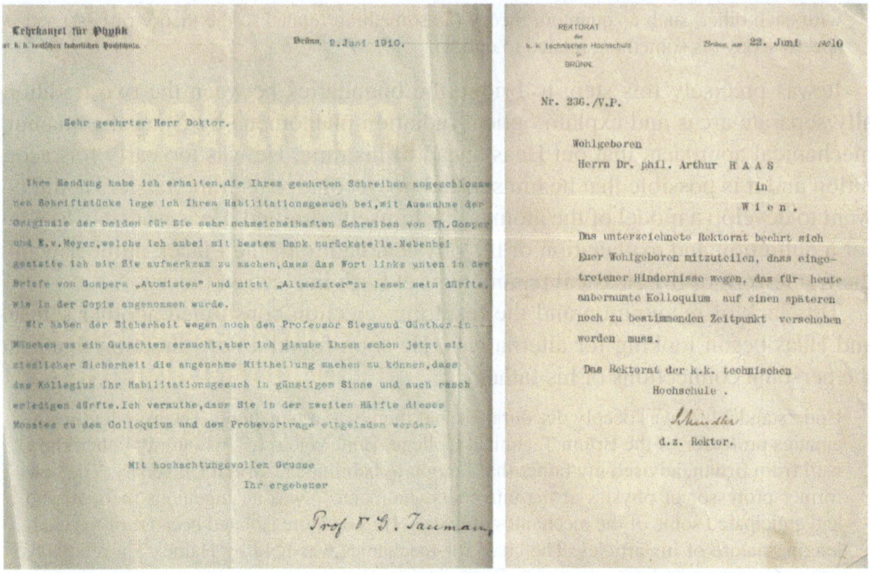

Lehrkanzel für Physik
der k. k. deutschen technischen Hochschule.

Brünn, 2. Juni 1910.

Sehr geehrter Herr Doktor.

Ihre Sendung habe ich erhalten, die Ihrem geehrten Schreiben angeschlosse-
nen Schriftstücke lege ich Ihrem Habilitationsgesuch bei, mit Ausnahme der
Originale der beiden für Sie sehr schmeichelhaften Schreiben von Th. Gomper
und E.v.Meyer, welche ich anbei mit bestem Dank zurückstelle. Nebenbei
gestatte ich mir Sie aufmerksam zu machen, dass das Wort links unten in dem
Briefe von Gompers „Atomisten" und nicht „Altmeister" zu lesen sein dürfte,
wie in der Copie angenommen wurde.

Wir haben der Sicherheit wegen noch den Professor Siegmund Günther in
München um ein Gutachten ersucht, aber ich glaube Ihnen schon jetzt mit
ziemlicher Sicherheit die angenehme Mittheilung machen zu können, dass
das Kollegium Ihr Habilitationsgesuch in günstigem Sinne und auch rasch
erledigen dürfte. Ich vermuthe, dass Sie in der zweiten Hälfte dieses
Monates zu dem Colloquium und dem Probevortrage eingeladen werden.

Mit hochachtungsvollem Grusse
Ihr ergebener

Prof G. Jaumann

REKTORAT
der
k. k. technischen Hochschule
in
BRÜNN

Brünn, am 22. Juni 1910

Nr. 236./V.P.

Wohlgeboren
Herrn Dr. phil. Arthur H A A S
in
W i e n .

Das unterzeichnete Rektorat beehrt sich
Euer Wohlgeboren mitzuteilen, dass einge-
tretener Hindernisse wegen, das für heute
anberaumte Kolloquium auf einen späteren
noch zu bestimmenden Zeitpunkt verschoben
werden muss.

Das Rektorat der k.k. technischen
Hochschule .

Schindler
d.z. Rektor.

Fig. 5.12 Letters from the Brünn mathematician Jaumann and the rector, informing Haas of the state of his colloquium lecture

> of Materials Science, Haussner had shown up and introduced himself as a spokesman and representative for a whole group of professors. He claimed that the faculty did not wish for me to become part of their group, in spite of the invitation and approval that I had received from experts in the field. Haussner maintained that my uncle should convince me to voluntarily withdraw the application for habilitation at once, otherwise they would cause me to fail at the colloquium. My uncle thought it wise to postpone the date of the colloquium. So, I drove back to Vienna without any success.
>
> I learned that the group, whose exponent was this individual named Haussner, argued against me using the vilest slander, all of which was a complete lie. This group claimed that the College of Soil Culture, where I supposedly wanted to habilitate, had received scathing and damning evaluations about me. My father immediately wanted to press criminal charges against the scoundrel Haussner, being convinced, as a lawyer, that Haussner's actions were a crime of extortion. He stopped, however, when he considered that legal action would ruin my chances in Vienna and that he would also greatly damage the position of my uncle in Brünn. So, with disdain, I withdrew my petition for a habilitation in Brünn.

This is indeed a curious story of inner-university intrigue. At that time Brünn had two universities; the German Technical University was considered the leading university in Moravia and the fifth largest polytechnic school in the Habsburg Empire. Although not Vienna, Haas would have been content to return to the city of his youth for his habilitation. The reason for resistance by some members of the faculty remains unknown. As already mentioned, all professors of the university have to agree to a habilitation but this was mostly a mere formality since professors of other disciplines did not care much about what happened in physics. The extent of the opposition, as described by Haas, was unprecedented for that time. Only after the First War did the opposition to Jewish candidates grow to significant dimensions.

The leader of the opposition Alfred Haussner (1859–1928) had studied at the Technical University in Graz. He was an assistant for mechanical technology in Prague and was habilitated in Graz in 1889, where he taught as a private lecturer for five years before being appointed professor of mechanical engineering at the Brünn University of Technology in 1894. Haussner was indeed influential and popular. He mainly dealt with technical issues of mechanical paper manufacture and published several textbooks on the subject. He also wrote a commemorative publication on the occasion of the 75th anniversary of the university in 1924.[xi] During his academic career in Brünn, he was twice elected rector of the university.[xii] From a practical point of view, one cannot imagine less of an overlap between Haussner and Haas; the reason for opposition to Haas' habilitation by the former is therefore not obvious. Anti-Jewish motives that were later so evident in the filling of university positions may, however, have already played a role at the time.[33]

> It goes without saying that the months that followed were a time of severe mental depression. Certainly I had the satisfaction that my work on the quantum mechanical interpretation of the spectroscopic constants was published in the session reports of the Vienna Academy of Sciences[34] (on the basis of the recommendation by the Prague professor of theoretical physics v. Lippich), followed by related articles in the *Physikalische Zeitschrift*[35] and in the *Jahrbuch der Radioaktivität und Elektronik.* [36]
>
> It was also gratifying that the professor of philosophy at the University of Jodl in Vienna, who according to these Brünn goons had delivered a devastating verdict on me, denounced their assertion as completely unfounded slander in a letter to Brünn Technik—but, nonetheless, my chances of a university career seemed to be completely destroyed.
>
> I decided to work during the fall as an intern in the sugar industry. I spent the summer in Schottwien and afterwards two wonderful weeks as a guest of my parents at the Lido, from where I went daily to Venice, which made a tremendous impression on me.

The rejection at Brünn did indeed seem to be the end of young Haas' physics career. That his work on the topic could be published gave him a certain satisfaction. The reason for the strong opposition is not quite obvious, since Haas followed the scientific thinking of his time in his deductions. As the science historian Armin Hermann points out:

---

33 Anti-Semitism had been successfully used as a political weapon by Austrian politician and Viennese mayor Karl Lueger (1844–1910) for years, and had thus brought about largely anti-Jewish thinking in the academic population, which can perhaps be attributed more to envy of success than to racist sentiments.

34 Arthur Erich Haas, "Über die elektrodynamische Bedeutung des Planck'schen Strahlungsgesetzes und über eine neue Bestimmung des elektrischen Elementarquantums und der Dimensionen des Wasserstoffatoms," *Sitzungsberichte der kaiserl. Akademie der Wissenschaften Wien* 119 (1910): 119–144.

35 Arthur Erich Haas, "Über eine neue theoretische Bestimmung des elektrischen Elementarquantums und des Halbmessers des Wasserstoffatoms," *Physikalische Zeitschrift* 11 (1910): 537–538.

36 Arthur Erich Haas, "Der Zusammenhang des Planckschen Wirkungsquantums mit den Grundgrößen der Elektronentheorie," *Jahrbuch der Radioaktivität und Elektronik* 7 (1910): 261–268.

> Haas' approach is therefore, and it cannot be otherwise, quite "classical," and his efforts include, among many others, deriving the law $\boldsymbol{E} = \boldsymbol{h}\cdot\boldsymbol{\nu}$ or the nature of the quantum constant from known principles, or at least classifying it through known principles. But Haas starts his considerations with the problem of the atomic constitution and therefore his "attempt" became extremely significant.[xiii]

The approach was not ridiculous, as assumed by the Viennese faculty, if one considers the broader context of the scientific work and publications of his time.

Planck's formula for black-body radiation was well known and the scientific community was equally well aware that the concept of resonators required a physical meaning behind the concept. Also, Hasenöhrl in particular was indeed very familiar with this topic. Haas used the Thompson model of the atom and Hasenöhrl was familiar with this theory. It was a theory that, despite some shortcomings, was widely accepted in the time before the publication of the results of the scattering experiments of Ernest Rutherford.

According to his notes, Haas studied the latest scientific literature, including a publication on the same subject by his teacher Stark in December 1907.[xiv] In this publication Stark developed a similar concept, identifying a negative electron in the atom as the Planck model resonator, with the electron moving in an orbit around the positive center of the atom. He writes:

> If we take the fact that in Planck's radiation law only one characteristic radiation constant of electromagnetic resonators occurs, we have to conclude that in nature there is only one kind of elementary electromagnetic resonator, the negative electron.

Stark postulates a central force to maintain the stability of the orbit against energy loss due to electromagnetic radiation. Haas most likely knew this work and combined these ideas of Stark's with those of the Thompson model, thus positioning the negative charges in the outer orbit of the positive plum pudding.

It is inertia—as a first reaction, upon the presentation of new scientific postulates—that produces rejection before one has taken a closer look at the thought. Even the new theories from Planck to Einstein were initially met with great reluctance and only became scientific common property after a delay of many years. As Johannes Stark himself writes in a subsequent publication on the motivation of his experiments:

> However, it does not seem right to me to persist in a passive resistance to it [Planck's quantum hypothesis of light] and to demand that the new hypothetically introduced characteristic be made understandable on the basis of already known properties of electromagnetic resonators; perhaps one is asking for something impossible here, insofar as it is not excluded that the new characteristic of electromagnetic resonators is a primary elementary property, such as the electric charge and mass of negative electrons.[xv]

According to this statement, Stark was very much open to the light quantum hypothesis and the possibility of a correlation to the interpretation of the resonators by electron theory.

But Arthur Haas was impatient. He saw his dreams being shattered and he decided to abandon physics for good and join the management of the Strakosch family's sugar empire.[xvi] However, such a change is difficult even with good family connections and young Haas needed to first learn the business and its operations from scratch.

Fig. 5.13 Felix Strakosch (left) and Siegfried Strakosch from Feldringen (right), the owners and managers of the Hohenau sugar factory

> In the fall, I started in the intern position at the sugar factory in Hohenau [Fig. 5.14]. The factory had been founded by my grandfather and was now owned by his son, my uncle Felix Strakosch, and his cousin Siegfried Strakosch [Fig. 5.13]. The factory was located in a small village between Vienna and Brünn. I lived as a guest with my uncle Felix and his wife Lisi, who already had three children. There, I got to know the company thoroughly and also studied several books on the sugar industry.
>
> From Hohenau, I went for a few weeks as a volunteer to the sugar factory in Mödritz near Brünn. As the son of the vice president, I was, of course, treated like a prince. During that time, I lived in a hotel in Brünn. In the evenings I spent time with my brother who was there as a one-year volunteer in a Dragoon regiment.

This seems to have fulfilled a long-cherished wish of Gustav Haas that his son should follow him into business life. With the completion of his law studies, Arthur Haas had the necessary qualifications; he also had his basic training in chemistry, which would make it easy for him to understand the production mechanisms of sugar beet processing. He was only missing the practical part, the on-site training, in order to understand both production conditions and management structures. Volunteering under the guidance of the family owners seemed the appropriate way to complete this practical part of his training.

Felix Strakosch was the younger brother of Gabrielle Haas and had taken over the shares in the sugar refinery. In 1891 he had joined the family business in the production of woolen goods and in 1895, he became a public-company partner. In 1903, he and his cousin Siegfried Strakosch applied for the deletion of this family business from the commercial register. In 1901, Felix had become a public partner in the family's sugar factory in Hohenau and was elected president in 1905. The headquarters of the Strakosch brothers' Hohenau sugar factory was moved from Brünn to Vienna in 1902, which also prompted Gustav Haas to move to Vienna.

Fig. 5.14 Entrance gate and aerial view of the Hohenau sugar factory (around 1930), which was built by the Strakosch family

Siegfried Strakosch took over business administration in Vienna while his cousin Felix Strakosch remained in Hohenau to manage the local matters of the company. Having a direct connection to production was more appealing to Felix than the life of the city, which involved many social obligations.

Arthur Haas lived in the house of his uncle Felix Strakosch and his wife Elise née Grünfeld (1876–1968) during his stay in Hohenau. He doesn't say much about his personal experiences in industry nor about the extent to which he became acquainted with the company's operations as a volunteer; he only points out that it was by studying textbooks that he learned about technical issues and processes in an approach that does not demonstrate great enthusiasm.

From Hohenau, Haas moved to the factory branch in Mödritz (today Modřice) in order to gain a better understanding of the different production conditions in a smaller refinery. In addition, he was to develop an understanding of the network that facilitated the operations of the family's rapidly growing sugar empire. Mödritz is located only seven kilometers south of the city center of Brünn. Haas preferred the freedom of hotel accommodation, rather than staying with his uncle Louis Gomperz or other more distant relatives of the Strakosch family. This allowed Arthur to continue his social life with his brother and various friends in Brünn without hindrance or obligation.

But it turned out differently than the family had imagined. Business life seemed boring to young Haas; studying production processes while analyzing and clarifying legal questions did not interest him when contrasted with the exciting world of scientific discoveries. Moreover, numerous friendly connections and incentives did not let him forget his joy in science. Besides his mentor Karl Sudhoff in Leipzig, it was also his friend and former fellow student Philipp Frank[37] who provided him with information and tried to persuade him to not give up his scientific goals. During his studies under Boltzmann and Hasenöhrl, Frank had been a member of the Viennese Circle,[xvii] in whose wider halo Arthur Haas had moved as well.[38] Both had found there, together, their interest in the philosophy of science. As Frank remembers:

> ... the domain of my most intensive interest was the philosophy of science. I used to associate with a group of students who assembled every Thursday night in one of the old Viennese coffee houses ... We returned again and again to the central problem: How can we avoid the traditional ambiguity and obscurity of philosophy? How can we bring about the closest possible rapprochement between philosophy and science?[xviii]

In Frank, Arthur Haas had found an intellectual successor and discussion partner to his Göttingen friend Karl Jellinek who, unfortunately for Haas, soon had to move on to Prague. It was through Frank that Haas found unexpected help from outside.

> During my time in industry, which I did not appreciate very much, I received a happy message. My friend and colleague Philipp Frank had gone to Göttingen to hear a series of six lectures given by the famous founder of electron theory and Nobel laureate H. A. Lorentz from Leiden University in Göttingen. The topic was "Old and New in Theoretical Physics," and Frank told me that in one of these lectures (reprinted in the *Physikalische Zeitschrift*, in excerpts), Lorentz talked about my work for a quarter of an hour.
>
> I was fortunate enough to have created something valuable, despite the failure of my habilitation attempts. Hasenöhrl was noble enough to appreciate the injustice that had been inflicted upon me and to recognize his error, especially since my work was being discussed in detail at the highly significant Solvay Congress in Brussels attended by scholars from all over the world, including Hasenöhrl. Thus Hasenöhrl asked me, in the spirit of the original decision by the faculty of philosophy, to resubmit my habilitation application for the history of physics with the same post-doctoral thesis, but he also asked me to complete yet another purely physical work.

Hendrik Antoon Lorentz (1853–1928) was one of the leaders in theoretical physics in Europe at the turn of the century. He was a Dutch mathematician and physicist who formulated the mathematical foundations on which Albert Einstein's special theory of relativity was built. Terms like *Lorentz force* and *Lorentz transformation* were named after him. He was also well acquainted with the questions of quantum physics and immediately recognized the importance of Arthur Haas' work, becoming one of his great patrons. Without Lorentz's public support in Göttingen, Haas' achievements would have remained unrecognized and unknown. He probably would have had to continue his life in the sugar factories.

---

37 Philipp Frank (1884–1966) was a physicist, mathematician and philosopher in the first half of the twentieth century. After completing his studies in Vienna in 1907 with a dissertation in theoretical physics, Albert Einstein recommended Frank as his successor to the German Karl Ferdinand University in Prague, a position that he held from 1912 to 1938. With the annexation of Czechoslovakia by Nazi Germany, Frank emigrated to the United States where he was appointed professor at Harvard University.

38 There is no evidence that Haas himself participated in the discussion rounds of the Vienna Circle, but he was befriended and acquainted with many of the members and leaders of the Circle. It is not known whether there was a specific reason for not getting more directly involved.

Fig. 5.15 Participants of the first Solvay conference in 1911 at the Hotel Metropole. Seated (L-R): W. Nernst, M. Brillouin, E. Solvay, H. Lorentz, E. Warburg, J. Perrin, W. Wien, M. Skłodowska-Curie, and H. Poincaré. Standing (L-R): R. Goldschmidt, M. Planck, H. Rubens, A. Sommerfeld, F. Lindemann, M. de Broglie, M. Knudsen, F. Hasenöhrl, G. Hostelet, E. Herzen, J.H. Jeans, E. Rutherford, H. Kamerlingh Onnes, A. Einstein and P. Langevin

As the founder of the famous Solvay Congresses, Lorentz had the opportunity to support Haas' work even in the face of his critical colleagues. The publications of Haas' theory were the subject of discussion at the first Solvay conference in Brussels in the autumn of 1911, sponsored by the Belgian industrialist Ernest Solvay. The conference was devoted to the topic "Radiation and Quanta."[xix] Many leading physicists took part, from Lorentz as chairman to Marie Curie (1867–1934), Arnold Sommerfeld (1868–1961), Ernest Rutherford, Albert Einstein, and also Friedrich Hasenöhrl from Vienna. Participants are seen here in Fig. 5.15.

Haas' model was a topic of discussion following Max Planck's lecture on the light quantum hypothesis. Lorentz noted,

> I would like to think that the model found by Haas ... deserves a slightly better evaluation than Mr. Planck gave it. ... if one stands on the ground of these ideas, one must assume the existence of effects that do not fall within the scope of mechanics ... and therefore the theorem of continuous energy distribution ceases to be valid. I will come back to this subject immediately, but first I want to point out once more that this obvious hypothesis leads us to the Haas model in a quite natural way.

Arnold Sommerfeld,[xx] one of the leading theoreticians of his time and generally not a supporter of the idea of atomic models, was not convinced. He expressed the following opinion:

Regarding Haas' hypothesis, I would like to note that the numerical relation between quantity ***h*** and the molecular dimensions can also be derived from the much more general assumptions without such specific ideas. However, the connection here is limited to a certain ultraviolet oscillation range and therefore does not appear fundamental. For my part, I prefer the general hypothesis about the quantity ***h*** over some special atom-models.

Lorentz replied in defense of Haas' approach,

Mr. Sommerfeld does not deny that there is a connection between the constant ***h*** and the atomic dimensions. This can be expressed in two ways: The constant ***h*** is determined by these dimensions or the dimensions attributed to atoms depend on the quantity ***h***. I fail to see any substantive difference in this.[xxi]

While the members of the Solvay Convergence remained ambivalent about the relevance of Haas' idea, Rutherford listened to the debate and discussed it in 1912 with his visitor, the young Niels Bohr. Only one year later, Bohr postulated his now famous Bohr's atomic model, replacing the Thomson model with the atomic concept newly suggested by Rutherford.

For Haas personally, however, an important result of this debate was that during the discussion at the Solvay Conference, Hasenöhrl recognized the possible scientific relevance of the work and withdrew his objections to the habilitation. He suggested that Haas submit a generalized paper to minimize the internal objections of the more conservative faculty members of the university.

Haas was delighted, gave up his unloved internship and turned to this new task:

After the start of the new year of 1911, I returned to theoretical physics and wrote a lengthy study of the equilibrium positions of electron groups, which was published in the session reports of the Vienna Academy.[39] Then I submitted the work again to the University of Vienna for my habilitation and admission as a lecturer. Now, without any difficulties, everything went smoothly; in the summer of 1911, the professors voted to allow me to join their ranks.

Haas' father seems to have finally given in to both losing his son to science and not finding in him a successor for the firm. Perhaps the internal assessment by the Strakosch cousins, who had noticed a lack of interest and commitment by the intern, aided Gustav Haas in his acceptance. On the other hand, he was proud of his son who had finally achieved his long-cherished goal.

Summer began and bourgeois Vienna went to the summer resort; his decision made, young Arthur Haas went with them since he no longer lacked the necessary funds.

My father was very satisfied with the favorable development and gave me a larger sum of money so that, in fulfillment of a long-cherished wish, I could spend a month in Paris. So I spent the summer of 1911, first alone in Krumpendorf on the Wörthersee. Then at the end of September, I traveled to Paris, via Munich where I spent a few days in the middle of the beginning of October Fest.

I was mesmerized by the beauty of the French capital from the first minute and the four weeks I spent in Paris are among the most beautiful of my life. I lived in a small hotel on

[39] Arthur Erich Haas, "Gleichgewichtslagen von Elektronengruppen in einer äquivalenten Kugel von homogener positiver Elektrizität," *Sitzungsberichte der kaiserl. Akademie der Wissenschaften Wien* 120 (1911): 1111–1171.

> Rue d'Antin near the Opera and with much enjoyment I took my meals in the more-simple restaurants. Every day I made a pilgrimage to the collections of the Louvre, where the Nike of Samothrace and the Venus of Milo made the strongest impressions on me. Again and again, I was captivated by the old oriental collections and the famous picture gallery. But I also studied the other museums with great zeal.
>
> I walked a lot in the Bois de Boulogne and loved to stroll along the banks of the Seine, in the Latin Quarter and in the Jardin du Luxembourg. In the evening, of course, I strolled a lot along the boulevards and enjoyed many theater visits. I remember brilliant performances of *Aida* and *Tristan* in the Grand Opera, of *Manon* in the Opéra Comique, of Molière's *Bourgeois Gentilhomme* in the Odéon, of Offenbach's *Vie Parisienne* in one Boulevard Theater and Maeterlinck's *L'oiseu* in the Théatre Rejane. Of course, I did not miss making numerous visits to the variétés and the circus.
>
> After returning from Paris, I again studied eagerly for the colloquium, which I passed with good success. My very tactful examiners were Exner and Hasenöhrl, while Lecher was completely passive without opening his mouth. By the end of the year 1911, I had practically passed the probationary lecture procedure. The Faculty of Philosophy made the decision to admit me as a lecturer but the decision still needed the "supreme" approval. Thus, the recommendation went, for months, to the Ministry of Education and Teaching and from there to the Cabinet Office of the Emperor. In August 1912, my appointment as a private lecturer was officially announced.

The prerequisites that Haas had successfully mastered were described as follows in the habilitation regulations of 1888:

> On the basis of the probationary lecture, the colloquium, the submitted scientific work and the knowledge of the morality and honorability of the applicant's character, the faculty gives its decision on the admissibility or non-admissibility of the applicant for private lecturing.

However, this decision needed to be confirmed by the Ministry of Education. The habilitation process was extraordinarily complex and habilitation regulations required not only scientific qualification of the applicants but also a personal qualification. This stipulation made it easy for the faculty and ministry to reject unappreciated applicants without further arguments regarding their scientific qualifications.[xxii] In addition, even the requirement of a printed, explicitly published habilitation thesis excluded applicants who could not afford the associated costs.

This was certainly not the case for Arthur Haas but, nevertheless, a full year passed before the official appointment as private lecturer was granted by the Ministry of Education. The reasons for this long delay are unknown but one strategy often used, in dealing with professionally qualified but unwanted habilitation candidates, was to delay the habilitation as long as possible.[xxiii] However, Haas did not get discouraged during the waiting period and used the opportunity to undertake further educational and pleasure trips:

> In the spring of 1912, I spent two weeks in Berlin for pleasure, using the remaining money that was left over from Paris. I entertained myself by watching a few plays and other performances, but it was clear that a terrible antithesis existed between the noble taste of French culture and the pretentious, perfunctory lack of culture and taste of Wilhelmine Germany! I could not believe my eyes when I saw the kitsch of the various magnificent buildings, monuments, and portals. But I did eagerly visit the museums in Berlin where, especially, the Egyptian antiquities made a great impression on me. The Natural Science Museum of Oceanography

also interested me very much; in its center was a gift from Emperor Wilhelm, a collection of models that illustrated the different types of saluting to superiors in the German Navy!

As summer resort in 1912, I again chose Krumpendorf am Wörthersee; there I received notice that I had finally been appointed as private lecturer but also, at the same time, the bitter news about the failing health of my father. Since 1906, I had always gone alone on summer vacation while my parents usually stayed with my sister on the Dutch or Belgian North Sea, in the Dolomites or in Switzerland. In the summer of 1912, my father had gone to Westende, a Belgian seaside resort. There, though not quite but certainly surprisingly, all sorts of heart complaints asserted themselves. The doctors advised him to spend the rest of the summer in the milder climate of the Adriatic.

My father therefore moved to the paradisiacal island of Brioni near Pula, where my mother and I joined him. I was much shaken when I did not find our father as strong as usual but I calmed down when his condition improved. Brioni was a pearl, both in terms of scenic charms and ancient monuments, but the grim concern for my father obscured everything. After a few weeks we departed to Trieste, in a terrible storm, and from there back home. On the advice of the doctors, my father, together with my mother, spent another six weeks in the Semmering spa, apparently ameliorating his condition. We visited him often, including on his birthday, and at the end of October my father returned to Vienna.

Notes

i. Gustav Haas, "Ein Vorschlag für die gesetzliche Kontingentierung der österreichischen Zuckerindustrie," (Brünn, 1902).

ii. William O. McCagg, *A History of Habsburg Jews,* 197–200.

iii. See Jewish Telegraphic Agency, March 11, 1926, accessed January 31, 2021, https://www.jta.org/1926/03/11/archive/ica-withholds-last-will-of-baron-de-hirsch-from-jewish-public.

iv. *The Journal*, "Little Miss Hirsch—The richest Heiress," (New York, May 31, 1896).

v. *The Studio,* "The Art Revival in Austria, Special Summer Number," (London, 1906), 56–58.

vi. Siegfried Strakosch, *Erwachende Agrarländer. Nationallandwirtschaft in Ägypten und im Sudan unter englischem Einflüsse* (Berlin: P. Parey, 1910).

vii. Marie-Theres Arnborn, *Damals war Heimat*, 123–131.

viii. Max Planck, "Zur Theorie des Gesetzes der Energieverteilung im Normalspektrum," *Verhandlungen der Deutschen Physikalischen Gesellschaft* 2 (1900): 237.

ix. Joseph J. Thomson, "On the Structure of the Atom: An Investigation of the Stability and Periods of Oscillation of a number of Corpuscles arranged at equal intervals around the Circumference of a Circle; with Application of the Results to the Theory of Atomic Structure," *Philosophical Magazine* series 6, 7:39 (1904): 237–265.

x. See Stephen Boughn, "Fritz Hasenöhrl and $E = mc^2$," *The European Physical Journal H. 38* (2013): 261–278.

xi. Alfred Haussner, "Geschichte der deutschen technischen Hochschule in Brünn 1849–1924," in *Festschrift der deutschen technischen Hochschule in*

Brünn zur Feier ihres fünfundsiebzigjährigen Bestandes im Mai 1924 (Brno, 1924).

xii. Pavel Ŝiŝma, *Zur Geschichte der Deutschen Technischen Hochschule Bünn—Professore, Dozenten und Assistenten 1849–1945* (Linz: Trauner Verlag, 2009), 74.

xiii. Armin Hermann, *Arthur Erich Haas—Der erste Quantenansatz für das Atom* (Stuttgart: Battenberg Verlag, 1965), 10.

xiv. Johannes Stark, "Elementarquantum der Energie, Modell der negativen und positiven Elektrizität," *Physikalische Zeitschrift* 8, 24 (1907): 881–884.

xv. Johannes Stark, "Neue Beobachtungen an Kanalstrahlen in Beziehung zur Lichtquantenhypothese," *Physikalische Zeitschrift* 9, 22 (1908): 767–773.

xvi. Werner Kohl and Susanna Steiger-Moser, eds., *Die österreichische Zuckerindustrie*, 224.

xvii. Friedrich Stadler, *The Vienna Circle—Studies in the Origins, Development, and Influence of Logical Empiricism Ursprung, Entwicklung and Wirkung des Logischen Empirismus im Kontext* (Wien-New York: Springer, 2001).

xviii. Philipp Frank, *Historical Background—Discussion in a Vienna Coffee House, in Between Physics and Philosophy* (Cambridge, MA, 1941) and J.T. Blackmore, R. Itagaki, S. Tanaka (eds), *Ernst Mach's Vienna 1895–1930: Or Phenomenalism as Philosophy of Science* (Dordrecht: Springer Science and Business Media, 2001), 61–84.

xix. Cf. Norbert Straumann, "On the first Solvay Congress in 1911," *European Journal of Physics* H 36 (2011): 379–399.

xx. Cf. Michael Eckert, *Arnold Sommerfeld—Science, Life and Turbulent Times 1868–1951* (New York, Heidelberg: Springer, 2013).

xxi. Translated from: Arnold Eucken, ed., *Die Theorie der Strahlung und der Quanten—Verhandlungen auf einer von E. Solvay einberufenen Zusammenkunft* (30. Oktober bis 3. November 1911), (Halle: Wilhelm Knapp, 1945), 100–105.

xxii. See Kamila Maria Staudigl-Ciechowicz, *Das Dienst-, Habilitations- und Disziplinarrecht der Universität Wien 1848–1938* (Wien-Göttingen: V&R Unipress, 2017), 249–251, 257–269.

xxiii. Ibid., 274.

Chapter 6
Life as a Private Lecturer

The *Venia Legendi*, or *Venia Docendi*—the right to give lectures free of charge and without pay or promise of a permanent position, as a private lecturer—was not only a right but also a duty. If no lectures were offered within a year, the private lecturer lost his *Venia Legendi*. The ideal of the private lecturer in the academic course of study was that,

> it should be seen as an irreplaceable nursery for professors, which always regenerates itself without the state's assistance or expenditure of money, and whose young individuals sprout precisely where they will be most needed.[i]

This statement was accompanied by a warning to the older generation,

> to maintain their activity and zeal of the professors, since they need to understand that they must be all the more careful not to be outflanked by them [the private lecturers], since the academic teachers will lose not only their audience but with them also the tuition fees [...].[ii]

From the pool of private lecturers, only the best were to be selected as professors. This was a favorable arrangement for the state and the university but a less favorable one for the candidates, especially those who had no means of their own and thus had to live only from the meager tuition funds which in turn depended on the number of students in the lecture.[iii] Arthur Haas had no such concerns because of his financial independence.

> By the end of October 1912, I began my academic teaching career. I announced a two-hour lecture on the "History of Physics" that would span two semesters and I had the satisfaction of having an audience of about ten participants. Among them was retired section chief of a ministry (a senior civil servant) Schreiber, and a promising young physicist who later became a professor at Purdue University, Karl Lark-Horovitz.
>
> I made the mistake of working out the lectures far too carefully; the preparation of the course took almost a whole year and I had far too little time to follow the latest developments in physics. A lot of time was also claimed by the continuation of my involvement in outreach and education activities. I often gave popular lectures at the Viennese People's Education House "Urania," both individual lectures and lecture series [such as a series together with Philipp Frank] . I also worked on a so-called repertoire of photo lectures for the Urania, as well as being a member of their technical advisory board. One lecture I gave at the Urania

M. Wiescher, *Arthur E. Haas—The Hidden Pioneer of Quantum Mechanics*, Springer Biographies, https://doi.org/10.1007/978-3-030-80606-4_6

was titled, "Is the world infinite in space and time?" and this was subsequently printed in the *Archive for Systematic Philosophy*.

Indeed, the first lectures always present a special challenge for a young teacher. Nervousness, uncertainty and also inexperience in dealing with students cause every young lecturer to prepare and elaborate in great detail, while sacrificing necessary time for research. The careful preparation has its advantages if one wants to later write a book or other publication on the topic, since everything is meticulously recorded, but it also has the great disadvantage of introducing a lack of flexibility in dealing with the topic during the actual lecture. The inexperienced lecturer tends to rigidly follow the prepared script and thus will be less able to respond to spontaneously arising questions.

With ten students, it was a relatively small, special lecture. Such a lecture is not necessarily attended by students specializing in the field; participants either have a special interest in the topic or intend the lecture for further education. This is demonstrated by the two named participants of Haas' first lecture. Dr. Karl Schneider was the head of the section at the Imperial and Royal Austrian State Railways and, now retired, was hoping to expand his knowledge by taking classes at the university. He was presumably relieved of his college fees. The second mentioned student Karl Horovitz[1] had come, like Haas himself had, from a humanistic gymnasium where students learned classical languages, literature and history. He enrolled at the University of Vienna in 1911, where he studied chemistry, physiology and physics. His minor subject was pre-Socratic philosophy. For this discipline, he probably took the lecture offered by Haas who was known for his interest in the ancient philosophy of science.

But even if Haas had over-prepared his first lecture, he was a talented speaker. Joining his friend Philipp Frank, Haas honed his lecture skills at the Urania Public Education Center[iv] where he continued to offer talks after Frank had moved to Prague. Named after the muse of astronomy, the Urania had been founded in 1888 in Berlin, as a people's education association for expanding public knowledge. The Urania in Vienna followed the same model when it was founded in 1897. Designed in Art Nouveau style and equipped with an observatory, it started providing learning opportunities in 1910.

Viewed from across the canal in Fig. 6.1, the Urania's observatory was opened on May 20, 1910, on the occasion of the return of Halley's Comet, but courses had already been offered during the first months of the year. Several large lecture halls would provide opportunities for renowned scientists, artists, and intellectuals such as Albert Einstein, Thomas Mann, Thomas G. Masaryk and Max Planck to present their works to the public over the following years.

[1] Karl Lark-Horovitz (1892–1958) later linked the surname of his wife Betty Lark with his own. Due to interruption for military service during the First World War, he was not awarded a doctorate in physics until 1919. After a teaching assignment at the University of Vienna, Lark-Horovitz was one of the first who received a scholarship from the International Education Board (Rockefeller Foundation) to lecture in Canada and the United States in 1925; in 1927 he joined the physics department at Purdue University. There he took up a permanent position in 1928 and built a first-class research facility for modern solid-state physics.

Fig. 6.1 Urania buildings with observatory as seen from the opposite side of the canal

Young lecturers at the Viennese universities also considered this an excellent opportunity to present their views and theories to the public in popular scientific lectures. In particular, members of the Vienna Circle gave numerous lectures on a wide range of scientific, historical and philosophical topics. Philipp Frank had probably introduced young Haas to the Urania in spring 1910 when presenting a short two-month series of lectures on the development of modern physics.

During this first year, there would be a remarkable 129,010 people visiting the Urania. In 1912 more than 1,800 events would be organized for more than 300,000 visitors. This was a huge audience and underscores the great popularity of the sciences among the general population in this time before the First World War.[v] Especially the lectures of Ludwig Boltzmann were very popular with the public and the lecture halls were packed full.[vi] Philipp Frank and Haas continued this tradition, offering lectures and lecture series on scientific topics of the time. This was also true in Berlin's Urania where Erwin Schrödinger lectured as Planck's successor. The exact topics of Arthur Haas' lectures are not listed but would have been taken from his respective interests in classical as well as modern physics. He developed a successful style in these lectures, which he successfully used in many of his later popular-science lectures and writings.

Haas mentioned his lecture on February 16, 1911 in which he returned to his earlier topic, the finite universe; this lecture aroused so much interest, as he later noted, that it was published in a philosophical journal under the title "Is the World Infinite in Space and Time?"[2] He formulates:

> the question of the infinity of the universe, … actually consists of two interrelated problems, namely, whether the number of celestial bodies is unlimited and whether world events are without beginning and without end.

This is a question of early cosmology that Haas was posing at a time when celestial observation and imagination were still limited to the Milky Way. The idea of an origin was only being represented in religious doctrine and when philosophy considered origins, the discussion was of infinity.

In his essay, Haas first summarizes the philosophical perspective on the infinity of the universe, from Greek mythical ideas to Giordano Bruno and Immanuel Kant, and then rejects these arguments point by point with physical arguments and examples from mechanics to thermodynamics. He bases his arguments primarily on entropy and the second law of thermodynamics. He also discusses the potential case of a yet undiscovered law, a law that would exist parallel to the conservation law of energy and demonstrate the net sum of entropy as constant. Currently, this law posits that if entropy increases in one part of the universe, it must decrease in another part or alternately, if one takes into account time, it requires that the direction of entropy varies with time—all of these ideas have become concepts of modern cosmologies.

The second argument, the observation of radioactive decay, again points back to Haas' time as a student with Stark. Since the possibility of nuclear reactions was unknown at that time, it was assumed that all radioactive elements decay toward lighter elements by alpha emission, such as radioactive uranium and its yet undiscovered radioactive mother elements. If the universe is infinite then, going back in the past, the mass of the respective mother elements must increase toward infinity since each decay process requires a decay time. Further, each decay process is also associated with the release of decay heat, so one would expect a higher temperature in the present universe than had been previously observed. These are the two pillars of Haas' thought and in his arguments, he continues to disapprove of the infinite and favor a finite universe. He will not abandon this concept nor these arguments throughout the rest of his life.

Then, as Haas became increasingly successful in his teaching, worries began to mount at home:

> Everyone in my parents' household was extremely depressed during this time of my first lectures. There was no improvement in the condition of my father, although he was able to resume a large part of his usual professional activities. On the 17th of December we celebrated his 30th wedding anniversary. A few days later his health took a down-turn from which he would not recover. In addition, with the imminent danger of war, at the end of December my brother was called to the military at the Russian border.

---

[2] Arthur Erich Haas, "Ist die Welt in Raum und Zeit unendlich?" *Archiv für Systematische Philosophie* 18 (1912): 167–184.

Fig. 6.2 Otto Haas and Yella Haas at the time of the death of father and husband, Gustav Haas

After New Year, my father's condition seemed to improve; he started to read again and yes, he even played the so-called "Dame" game with me [something akin to chess]. So it happened on the evening of the 14th of January, a Tuesday. He won without me contributing anything; then I retired to my own work (I lived in my brother's room during his absence) when I heard a terrible, heartrending cry from my mother and at once hurried to see my father had collapsed from a stroke. It happened at 9:28 p.m.

The immediately summoned doctor could only determine his death. These were the most terrible hours. Of course, my uncle Fritz hurried over. All night my despairing mother kept watch with me on the stretcher. My mother's brother finally came from Hohenau. He managed to calm her down a bit. The next day my brother arrived from the Russian border. On the 17th of January we buried my father in the Döbling Cemetery, in the firm knowledge that there could be no better father than ours. My mother has since remained a completely broken woman [Fig. 6.2].

The year 1913 passed in quiet and deep mourning. When the university summer holidays began, I went for a month to Starnberg on Lake Starnberg near Munich; the weather was very unfavorable. I was assigned to the cabin of the royal family in the bathhouse, to honor my courage, being the only bather in icy weather. I spent the rainy days in Munich, mostly in the Deutsches Museum für Technik and Innovation.

The rest of the summer I spent with my mother and sister. I drove over Innsbruck to a lovely mountain lake near Trento, the St. Christoph Lake; from there we drove in the car to visit Görz, then to the famous Grotto of Adelsberg, finally arriving in the seaside resort Lovrana [later Laurana] on the Adriatic coast at Fiume, from where the car was sent home. There we spent about two weeks together with my mother's sister (Aunt Laura Brum). In June of 1913 another trip was made to visit the city of Budapest for a few days, which I liked very much.

The family had to arrange my father's estate. The estate was quite large, about four million kronen, and corresponded to the thoroughly Austrian attitude of my father, that not one penny was invested outside of Austro-Hungarian values. With the exception of two inferior

> Brünn houses, which my father had to take over so as not to lose any money, everything was invested in bonds. About one million kroner was ready in cash, as my father had decided to buy an agricultural farm as the safest investment. Sadly, my father had been out of the sugar industry for a few years and his income from this revenue source was almost completely eliminated by the time of his death.
>
> My father's office was taken over temporarily by my uncle Fritz, in order to continue it until my brother could take over. The farmland that my father had in mind, though not in a specific form, was never purchased. My inheritance was almost exclusively invested in so-called safe-for-sale bonds and Wiener Bank shares.
>
> There was no need to change the standard of living of my family. The apartment and automobile could be maintained even after the income was reduced through the loss of my father's earnings. I myself lived relatively modestly and of my income, which was several times the salary of a full professor. I did not even spend a third. The only luxury I allowed myself at that time was the design of my library and the improvement of the interior of my apartment.

The sudden and unexpected death of his father forced Arthur Haas to deal for the first time with tiresome financial matters. Until then he had been dependent on his father and his generous support, not giving any further thought to these matters. But things had suddenly changed. The assets were sufficient but his father's income from the firm no longer existed. Nevertheless, the existing fortune of four million kronen was a considerable inheritance, which in 1913 corresponded to the current value of about 20 million euros. These were firmly and supposedly safely invested in Austrian state securities. However, the Habsburg multi-ethnic monarchy was fragile; as early as 1914 the value of the crown had already fallen by 50%.

> At the end of September 1913, the annual gathering of German physicians and naturalists (traditionally held every four years in Austria) met in Vienna. Not only did I give a lecture in the historical section, but my student Lark-Horovitz and another student were also lecturers. I rose considerably in the esteem of Sudhoff and Siegmund Günther, both of whom had come to the congress. At this time, I also became very interested in the pure physics sessions.
>
> The second year of my academic teaching began at the end of October 1913. I had announced a two-hour lecture on the history of the principles of theoretical physics. But after only a few weeks, I had to give up this lecture in which, besides Lark-Horovitz, also Gerda Laski (later, a lecturer at the University of Berlin and a significant researcher in ultra-red radiation, who died young) and Hilda Geiringer (later a professor at the University of Brussels) had enrolled. The reason for this abrupt change was my sudden appointment as an associate professor at the University of Leipzig.

Haas names specifically Gerda Laski and Hilda Geiringer, two young physics students who may serve as examples demonstrating the attitude of the Vienna university system toward the education of women. Education of women had been guaranteed by law since 1896 but the idea behind this was the training of female teachers for the grammar school system, in order to remedy the general lack of teachers. The aim was not to produce independent women researchers with their own research programs and their own agendas. Despite the primacy of the male figure, who was solidly integrated into the societal structure of the time, both of the female students mentioned here had different ideas. They would develop their own academic goals, meeting demands as necessary to achieve their ambitions in the politically difficult times of the early twentieth century.[vii]

Gerda Laski graduated from high school in Vienna in 1913 and earned her doctorate in physics at the University of Vienna in 1917, attending Arthur Haas' lecture in her first semester. A forerunner for women in science, she had a successful career[3] but died early, in 1928.

Hilda Geiringer, also known as Hilda von Mises (from 1943), was also studying at the University of Vienna in her first semester. Her field was mathematics and she completed studies in 1917. In 1921, after four more years in Vienna, Geiringer became assistant to Richard von Mises (1883–1953) in Berlin. She had a successful but also extraordinarily eventful career in mathematics, where she developed new approaches in probability theory and in using statistical methods for tracking population genetics. Her career was marked by many difficulties and setbacks.[4]

Arthur Haas followed the careers of both women and especially that of Hilda Geiringer who, like him, was close to the Vienna Circle through her relationship with Richard von Mises.

Haas may have enjoyed the attendance of these two students in his lecture but his relationship with Sudhoff in Leipzig seems to have paid off. When Haas received an offer for a position through Sudhoff, he quickly abandoned his lectureship in Vienna.

> My relations in Leipzig actually reached back to the day when I met Sudhoff in 1908. A short time after I started arrangements for the Vienna habilitation, Sudhoff had made the suggestion to habilitate in Leipzig. My father was still alive then and it was my plan to go to Leipzig only as an associate professor, not a private lecturer.
>
> Surprisingly, I actually managed to achieve this goal. My new faculty position for the history of physics was supposed to be an exemption but, on behalf of the cartel of German academies (Leipzig, Berlin, Munich, Göttingen, Vienna and Heidelberg), it was coupled with the offer to serve as editor for the publication of a new volume of Poggendorff's *Biographical-Literary Hand Dictionary for the Exact Sciences*. For this position I was to receive an annual salary of 5,000 marks.

So Haas was successful in his negotiations with Leipzig, although they could only offer him a teaching position without pay. At least it was a step forward from an unpaid private lecturer to an unpaid extraordinary or associate professor, if one

[3] Gerda Laski (1893–1928) worked, from 1918 to 1919 after her doctorate, as an assistant at the University of Göttingen and in 1920 as an assistant at the Institute of Physics of the Technical University of Berlin. There she was introduced to the experimental technique of infrared research. This included the investigation of selected chemical substances using infrared radiation. From 1924, Laski was director of the infrared department at the Institute of Fiber Chemistry of the Kaiser Wilhelm Society, which was closed due to lack of funding. In 1927 she became a co-worker at the Physikalisch-Technische Reichsanstalt, to establish an infrared laboratory. The Kaiser Wilhelm Institute of Physics provided Laski with a monthly stipend until her death in 1928.

[4] Hilda Geiringer (1893–1973) was on the list of appointments as an associate professor at the University of Berlin in 1933 but was removed because of her Jewish background. She moved to Brussels but after only one year, she followed Richard von Mises to Istanbul; he had emigrated there because of his Jewish roots. He headed the newly founded Institute of Applied Mathematics where Geiringer also found a position. During this time Mises was in close contact with Philipp Frank who persuaded him to emigrate again, to the US in 1939. Mises, like Frank, was offered a professorship at Harvard University and Geiringer joined him at Cambridge. She married Richard von Mises in 1943 and, after some initial difficulties, was offered a professorship at Wheaton College in Massachusetts.

disregards the tuition fees of attendees. But this shortcoming was offset by the offer of a paid editorial role for Poggendorff's *Biographical-Literary Dictionary for the Exact Sciences*. The salary of 5,000 marks corresponds to a current annual salary of about 22,000 euros. This was certainly not a great financial incentive but Haas felt that he had sufficient assets at the time and preferred a quasi-secure professorship to the uncertainties associated with a private lecturing position. He accepted the offer. In addition, he was attracted by the task of being able to serve as editor of an important scientific series, which did not bring much income but garnered a certain reputation.

The *Biographical-Literary Dictionary for the History of the Exact Sciences* was founded in 1863 by Johann Christian Poggendorff (1796–1877), while not directly connected it complemented the journal *Poggendorff's Annalen der Chemie und Physik*. Poggendorff was the editor of this journal that came to play an important role in German scientific communication in the nineteenth century. The aim of the handbook was to compile the biographical data and the most important scientific achievements of the natural scientists themselves. It was conceived as a reference work but the almost exponential growth in the number of scientists required an enormous amount of reading and editorial work, in order to keep the handbook up to date. Published in several volumes, the first two were compiled by Poggendorff himself. His successor was his former student, the physicist, meteorologist and acoustician Arthur von Oettingen,[5] who held a professorship in Dorpat, Estonia until 1893. After his retirement, he accepted an honorary professorship in Leipzig where he published the third and fourth volumes of the handbook, essentially living on his Russian pension. Arthur von Oettingen wanted, however, to fully retire and not concern himself with the publication of the next volume. Arthur Haas, with his broad scientific and scientific-historical knowledge and interest, appeared as the ideal candidate for the succession[viii]:

> The details were discussed during a visit to Leipzig where I gave a guest lecture in the spring of 1913. At that time, I was cordially received by the two physics professors, Otto Wiener and Theodor des Coudres. Wiener in particular did everything imaginable to make the necessary arrangements not only with the University and the Saxon Academy but also with the Saxon Ministry of Education.

An invitation to present a guest lecture is always an important component of negotiations for a position in academia. The faculty can evaluate the candidate personally and vice versa; this gives a first impression and often serves as the initial basis for internal scientific cooperation. In his talks and negotiations with the representatives of the physics institutes in Leipzig, Haas seems to have developed a good personal

[5] Arthur Joachim von Oettingen (1836–1920) was of Baltic nobility and came from Ludenhof (today: Luua Manor) in Livonia. He studied astronomy and physics at the University of Dorpat and then worked in Paris in the laboratories of Antoine César Becquerel (1788–1878) and Henri Victor Régnault (1810–1878). Later in Berlin, he performed research in the laboratories of Heinrich Gustav Magnus (1802–1870), and collaborated with Johann Christian Poggendorff (1796–1877) and Heinrich Wilhelm Dove (1803–1879). In 1868 he became professor in Dorpat where he founded a meteorological observatory.

relationship with both the experimental physicist Otto Wiener[6] and the theoretical physicist Theodor des Coudres.[7] This probably provided Haas with a strong argument for moving from familiar Vienna to unloved Germany.

> In the early fall of 1913, the Saxon ministry asked me if I would accept an offer. I agreed but made a condition that I should be able to remain an Austrian citizen so that I would not be drafted in Saxony for military service. This was also granted. In the last week of November, at the Café Bürgertheater, I was as usual leafing through the newspapers and accidentally had a Saxonian journal in my hands, where I read, to my surprise, about my appointment as associate professor at the University of Leipzig. When I got home, I found a telegram from Sudhoff in which he congratulated me; it crossed with a telegram in which I congratulated Sudhoff on his 60th birthday.
>
> Then came the formal decree of appointment and, for reasons that I in retrospect do not quite understand, I hurried to Leipzig to immediately start of my lecturing, without losing a day.
>
> The Viennese lecture was, thus, interrupted and I had to resign from the faculty of the University of Vienna. But for those students who were interested, I unofficially finished the lectures during the Christmas and Easter holidays. I was very proud when almost all of the previously participating students actually joined these unofficial lectures.

This speaks to the teaching style that Haas had developed in his lectures because, with his official resignation from the Vienna university, the course would no longer have earned credit. Only personal motives and interest in the subject matter would have motivated his students to participate. None the less, Haas had decided on Leipzig and he had to leave this student following, including the women. He also had to abandon his daily visit to the Café Bürgertheater (Fig. 6.3) when he relocated to little-appreciated Germany, this time in Saxony.

Notes

i. Joseph Unger, *Die Universitätsfrage Österreichs* (Wien: Gerold, 1853), 73.
ii. Ibid., 74.
iii. Jan Jakub Surman, "Habsburg Universities 1848–1918. Biography of a Space," (dissertation, Wien, 2012), 482.
iv. Gudrun Wolfschmidt, "URANIA in aller Welt—Ausbreitung und Wirkung des Urania-Idee," *Communications in Astroseismology* 140 (2008): 92–103.

[6] Otto Heinrich Wiener (1862–1927) was a student of August Kundt at the University of Strasbourg, where Wiener received his doctorate in 1887 with a thesis on the phase change of light during reflection and methods for determining the thickness of thin layers. Wiener is known for the experimental proof of standing light waves. In 1890 he succeeded in determining the wavelength of light. He was professor at the University of Giessen from 1895. In 1899 Wiener became professor at the Institute of Physics at the University of Leipzig. Together with Theodor des Coudres, he built up an excellent physical institution there.

[7] Theodor des Coudres (1862—1926) began studying natural sciences and medicine at the University of Geneva in 1881, later in Leipzig and Munich. He successfully completed his studies at the Friedrich Wilhelm University in Berlin in 1887 with a thesis on the optical constants of mercury under Hermann von Helmholtz. In 1889, Coudres was assistant to Gustav Heinrich Wiedemann at the University of Leipzig and habilitated in 1891, from where he was appointed to the University of Göttingen in 1895. In 1901 he went to the University of Würzburg and took over the chair of theoretical physics as associate professor. In 1903 he succeeded Ludwig Boltzmann at the University of Leipzig, where Coudres worked until the end of his life.

Fig. 6.3 The coffee-house culture in Vienna was fully appreciated by Arthur Haas. This was considered a place for social meetings as well as offering an opportunity to read the newspaper for hours, with a cup of mélange and a glass of water. The Café Bürgertheater was the place Arthur Haas preferred for this activity, which he would miss in the following years

v. Klaus Taschwer, *Wissenschaft für viele—Zur Wissenschaftsvermittlung im Rahmen der Wiener Volksbildungsbewegung um 1900* (Wien: Promotionsarbeit, 2005) and Andreas Daum, *Wissenschaftspopularisierung im 19. Jahrhundert* (München: Oldenbourg Verlag, 2002).

vi. Ludwig Boltzmann, *Populäre Schriften* (Leipzig: J.H. Barth, 1905).

vii. Edith Prost, "Emigration und Exil österreichischer Wissenschaftlerinnem," in *Vertriebene Vernunft* I, ed. Friedrich Stadler (Münster: Mandelbaum Verlang, 2004), 444–470 and Ilse Korotin, ed., "10 Jahre 'Frauen sichtbar Machen' biografi A—datenbank und lexikon österreichischer frauen," *Mitteilungen des Instituts für Wissenschaft und Kunst* 63 NR. 1–2 (2008).

viii. Cf. Karl-Heinz Schlote, *Von geordneten Mengen bis zur Uranmaschine—Zu den Wechselbeziehungen zwischen Mathematik und Physik an der Universität Leipzig in der Zeit von 1905 bis 1945* (Frankfurt: Verlag Harri Deusch, 2008), 38–43.

Chapter 7
The Leipzig Professorship

> My mother was very proud when she escorted me to the train station to travel to Leipzig, although she was worried, like me, about the upcoming separation. We consoled ourselves with the fact that I still maintained the Vienna apartment and would spend at least six months of the year with her.
>
> After my arrival in Leipzig, I immediately began my two-hour lecture on the history of physics in the 19th century, since I had to make up for the lessons that I had missed at the beginning of the semester; it was a regular lecture in which I had about fifteen students. In the summer semester, I taught the same number of students about the historical development of the mechanical principles, which also appeared in print.

Arthur Haas gave this first lecture in the new Physics Institute which had just been completed in 1909, shown in Fig. 7.1. With a lecture hall for 350 students and approximately 2000 square meters of laboratory space, it was probably the largest physics institute in Germany at that time. The Institute of Theoretical Physics, which was established under the same roof, not only had its own lecture hall but also laboratories and a workshop. Both institutes together had four professors. Otto Wiener and Theodor des Coudres were the full professors and directors, while Erich Marxand George Jaffé[1] were the more junior associate professors, with an additional

[1] George (Cecil) Jaffé (1880–1965) studied mathematics, physics and chemistry at the Ludwig-Maximilians-University in Munich, specializing in physical chemistry. One of his academic teachers there was the chemist and 1905 Nobel Prize winner Adolf von Baeyer. Jaffé continued his studies at the University of Leipzig under the later Nobel Prize winner Wilhelm Ostwald, from whom he received his doctorate with a thesis on studies of supersaturated solutions. On Ludwig Boltzmann's recommendation, Jaffé moved to the University of Cambridge and from there, at the invitation of Nobel Prize winner Pierre Curie, to Marie and Pierre Curie's laboratory in Paris. In 1908, Jaffé habilitated in Leipzig and lectured as a private lecturer. In 1911 and 1912 he worked again in the laboratory of the Curies in Paris as a Carnegie scholar, before he was appointed extraordinary professor of the University of Leipzig in 1916. After the First World War, Jaffé was appointed as a regular full professor of theoretical physics in Leipzig in 1923. From 1926 to 1933, he taught theoretical physics as a full professor at the University of Giessen. Jaffé was only there for six years because, being of Jewish descent, he was dismissed in 1933. In 1939 he emigrated to the US where he was Visiting Lecturer until 1942, then Associate Professor, and from 1946, Full Professor at the Louisiana State University in Baton Rouge.

M. Wiescher, *Arthur E. Haas—The Hidden Pioneer of Quantum Mechanics*, Springer Biographies, https://doi.org/10.1007/978-3-030-80606-4_7

Fig. 7.1 The new Institute of Physics of the University of Leipzig was completed in 1909. Located at Linnéstraße 5, it offered both lecture and research facilities

six assistants and private lecturers. Haas would come to develop good contacts to this group.

Haas' first lecture, for fifteen students, would probably have taken place in a smaller seminar room. Unfortunately, there is no record of how Haas structured his lecture on the history of physics in the nineteenth century, but it can be assumed that he used this same material when he wrote his two-volume textbook on theoretical physics with special regard to its modern problems. This textbook was published in 1919 and 1921 and we will return to it later.

For his second semester in the summer of 1914, he offered a course dealing with not only the history but also the mathematical formalism of mechanics. He developed this topic further and published his first book in the same year.[2] His historical approach and description of the resulting mathematization of the field followed an interesting concept that was probably developed in his first lecture. However, he only partially pursued this historical approach in his later books.

> Of course, I also immediately devoted myself to working on Poggendorff's *Hand Dictionary*. To date, the dictionary covered, in various supplementary editions, all scientists up to the years 1902–1904. Now a new fifth volume was supposed to be published to include the increasing number of scientific researchers up to the time of 1913. I started the work on a carefully thought-out plan, first having my assistants process the *Physikalischen Berichte* and the corresponding scientific and reference journals in mathematics and chemistry.

[2] Arthur Erich Haas, *Die Grundgleichungen der Mechanik—Dargestellt auf Grund der geschichtlichen Entwicklung. Vorlesungen zur Einführung in die theoretische Physik, gehalten im Sommersemester 1914 an der Universität Leipzig* (Leipzig: Veit, 1914).

The *Physikalischen Berichte*[3] was an addendum to the *Annalen der Physik* which was at that time the most important journal for the German-speaking physics community, the place to publish experimental and theoretical results and thoughts. Haas' approach therefore demonstrates a systematic method for including all important writings and their authors. His predecessor von Oettingen had already summarized the publications of the last decades of the nineteenth century but in view of the increasing number of physicists and authors, Haas wanted to limit himself essentially to the decade prior to his arrival in Leipzig. This was, however, an optimistic plan as he was also seeking to rejoin the scientific community as an active researcher.

Haas enjoyed life in Leipzig during this first year. He talks about the friendliness of the faculty who immediately welcomed him as one of their own; he mentions several invitations, receptions and dinners, and does not forget to mention the beautiful daughters—Hilda, Elsa, Susi and Heidi—of the various hosts.

> I was received with the greatest cordiality by my colleagues, even though, of course, outside of Austria human warmth is a bit stiffer than in Vienna itself. My real boss was Otto Wiener, then fifty-seven years old. He was an outstanding experimental physicist who had gained world-renown status by providing photographic proof of standing light waves and, at that time, he dealt mainly with problems of color photography. He was quite pleasant and relaxed as a Southern German, although he was, otherwise, extremely German-national in his thinking. He often invited me to his house, where I was greeted warmly by his wife. He had two unmarried daughters, Hilda and Elsa; the latter ended later in a suicidal tragedy.
>
> The second physics professor was the theorist Theodor des Coudres, who was well-disposed toward me and whom I venerated. He was a typical representative of the classic Helmholtz school. As such, he had astonishing knowledge and admirable critical talent but only little familiarity with modern physics. He was of the same age as Wiener, a native of Hesse, and a member of a respected Huguenot family. He was a bachelor and he greatly neglected his appearance but used his great salary to travel almost every year to all parts of the world, for as long as possible.
>
> Physics was also represented by Honorary Professor Arthur von Oettingen, who had once been a professor at the University of Dorpat and carried the Russian title of Excellency. He lived with his daughter and invited me frequently as a guest. He was already almost eighty years old. Among the other physicists, I was especially friendly with Erich Marx and George Jaffé, as well as Schiller[4] and Rother.[5]

---

[3] The "Physikalischen Berichte," as Haas called them, should not be confused with the journal of identical name which was published in Germany after 1920.

[4] Ludwig Schiller (1882—1961) studied from 1901 to 1919 at the Technical University of Munich and the universities of Leipzig and Göttingen. In 1911 he received his doctorate as Dr. phil. in Leipzig. Schiller remained there for his habilitation in physics and airship aviation; it was during this period that Haas met him. From 1921 to 1925, Schiller was a private lecturer in Leipzig for physics and airship aviation, from 1925 to 1926 he was an untenured extraordinary professor, and from 1926 to 1945 he became a tenured extraordinary professor for applied mechanics and thermodynamics. In November 1933, he signed the declaration of commitment to Adolf Hitler, made by the professors at German universities and colleges. In 1945 he was evacuated to Weilburg/Lahn when the US Army withdrew from Leipzig. He lived there as a private scholar until 1958. In 1958, he was appointed associate professor of applied physics at the Technische Hochschule Darmstadt where he remained until his death.

[5] Franz Rother (1887—), a student of Otto Wiener, was awarded a doctorate in 1914 in Leipzig on a topic in the physics of gas discharges and electron emission. In 1923 he moved to work with

Also, professors of other disciplines invited me quite frequently to their homes, especially the astronomer Bruns,[6] the physical-chemist LeBlanc[7] (who had a very beautiful daughter Susi), the Vienna-born geologist Kossmat[8] who also came from Austria, the law professor

the manufacturer of medical technology, Reiniger, Gebbert & Schall AG in Erlangen, as technical director. There he worked on the development and production of X-ray tubes. He returned to Leipzig in 1926 and habilitated there in the same year. In 1928, he moved from Leipzig to Berlin, to the Institute for Radiology at the University of Berlin. He continued to publish scientific papers up to and including 1934. Nothing is known about his further curriculum vitae.

[6] Ernst Heinrich Bruns (1848—1919) studied mathematics, astronomy and physics at Berlin University from 1866 to 1871. From 1872 to 1873 he was employed as a geodesist and mathematician at the Pulkovo Observatory in Russia. In 1873 he became an observer at the observatory of Dorpat (now Tartu) in Estonia, where he remained until 1876. During this time, he also worked as a lecturer at the University of Dorpat. In 1876, Bruns was appointed associate professor of mathematics at the University of Berlin. He also worked at the Prussian Military Academy and the Geodetic Institute in Potsdam. In 1882 he went to Saxony as a full professor of astronomy at the University of Leipzig and director of the Leipzig Observatory.

[7] Max Julius Louis Le Blanc (1865—1943) studied chemistry at the universities of Tübingen, Munich and Berlin from 1883 to 1886. He received his doctorate in Berlin and later worked as an assistant to Wilhelm Ostwald in Leipzig, habilitating there in 1891. In 1895, Le Blanc was appointed associate professor but moved in 1896 to the newly founded electrochemical research department of Farbwerke Höchst in Frankfurt am Main. In 1901 he was appointed to the new chair of Physical and Electrochemical Chemistry at the Technical University of Karlsruhe. In 1906, Le Blanc moved to the University of Leipzig where he worked until 1933 and was its rector from 1925 to 1926. In November 1933, he signed the commitment of professors at German universities and colleges to Adolf Hitler and the National Socialist state.

[8] Franz Kossmat (1871—1938) was professor of mineralogy and geology at the Graz University of Technology. From 1913 to 1934, he was director of the Geological State Office of Saxony and director of the Geological-Paleontological Institute of the University of Leipzig. In November 1933, he was one of the signatories of the commitment of professors at German universities and colleges to Adolf Hitler and the National Socialist state.

Koschaker,[9] and the Germanist Sievers[10] who had a pretty daughter Heidi. In their house, I joined a very funny evening in which all visitors had to be dressed as characters from the comic poems of Wilhelm Busch. A venerable personality with whom I also had a lot to do, in connection with the *Poggendorff*, was the private scholar Wilhelm Feddersen.[11] He was of Danish descent, at that time over eighty years old, and much and universally honored as the true discoverer of electric waves [1859]. For decades he lived with his wife in Leipzig and often invited me to lunch on Sundays.

In weekly meetings of a small group to which I was also invited, I enjoyed the get-together with the great philosopher Wilhelm Wundt,[12] who was already over eighty years old and often talked to me about the history and philosophy of the mechanical principle. Very positively inclined toward me was the mineralogist Rinne,[13] in whose house I was invited quite often.

---

[9] Paul Koschaker (1879—1951) came from Klagenfurt and first studied mathematics at the University of Graz before deciding to switch to law school. In 1903 he received his doctorate in law from Graz. His interest in Roman law led Koschaker to the University of Leipzig but he habilitated at the University of Graz in 1905. In 1908 he was appointed associate professor of Roman law at the University of Innsbruck and, in 1909, full professor at the German Karl-Ferdinand University in Prague. In 1915, after Haas' appointment, Koschaker was appointed to the chair of Roman and German civil law at the University of Leipzig. He remained there until his appointment to the University of Berlin in 1936, where he particularly promoted the research of cuneiform writing. In 1941 he went to Tübingen where he retired in 1946.

[10] Eduard Sievers (1850—1932) was a philologist of classical and Germanic languages. He studied German language and literature in Leipzig and Berlin and in 1871 became associate professor of Germanic and Romance philology in Jena, where he received a full professorship five years later. In 1883 he went to Tübingen and in 1887 to Halle, from where he was appointed to Leipzig in 1892.

[11] Berend Wilhelm Feddersen (1832—1918) came from Schleswig and was awarded his doctorate in Kiel in 1854 with a thesis on spark discharge. From 1858, he lived as a private scholar in Leipzig. In 1895 he recognized, from experiments, that individual electrical spark discharges consist of (damped) oscillations. Feddersen was co-editor of the *Poggendorff Biographical Encyclopedia*.

[12] Wilhelm Maximilian Wundt (1832—1920) is one of the founding figures of modern psychology. Wundt studied at the universities of Tübingen, Heidelberg and Berlin from 1851 to 1856. After graduating with a doctorate in medicine, he became assistant to the physicist and physiologist Hermann von Helmholtz. In 1864, Wundt became associate professor of anthropology and medical psychology, and he published a textbook on human physiology. His main interest according to his lectures and classes was, however, more in psychology and related subjects. He wrote the first textbook in the field of experimental psychology. In 1875, Wundt was appointed professor of philosophy at the University of Leipzig where he opened the first institute devoted exclusively to psychological research in 1879.

[13] Friedrich Wilhelm Berthold Rinne (1863—1933) is considered the founder of salt petrography. He studied at the University of Göttingen where he received his doctorate in 1883. From 1883 to 1886, he was an assistant at the Mineralogical-Petrographical Institute of the University of Göttingen. In 1885 he became a private lecturer for mineralogy and petrography. From 1887 to 1894, Rinne was an assistant at the Mineralogical-Petrographical Institute of the University of Berlin. In 1894 he was appointed professor at the Technical University of Hanover. From there he moved to the University of Giessen in 1904 and further on, in 1908, to the University of Kiel. In the same year,

> I was quite intimate with the Professor of Psychology Wilhelm Wirth[14] and I was often invited to visit with him and his young wife and their children; they both came from Bavaria. Sometimes I also visited the young philosopher Bergmann[15]; he was the son of a pastor and married to a Jewess. He later became, as a true buffoon, a furious champion of German paganism.
>
> Through Dr. Thesing[16] and his wife (he was the owner of the publishing house Veit), I met the antiquarian Jolowicz,[17] owner of the largest German bookstore. I was bound in friendship, for decades, with Jolowicz and was often invited to his house (and later, with my wife). He enjoyed a very high reputation and he later received an honorary doctorate from the Technical University in Karlsruhe. In his house I met among others Ostwald (the great chemist) and Svante Arrhenius from Stockholm, both Nobel laureates.

These invitations and his growing number of friends and acquaintances show that young Professor Haas had integrated himself quickly and successfully into life in Leipzig. His connections were not only contacts with physicists but also representatives of other academic disciplines, ranging from philosophy to psychology, from German studies to mathematics and from linguistics to law. This diversity also reflects his broad scientific interest.

---

however, he switched to the University of Königsberg and in 1909 he finally established himself at the University of Leipzig.

[14] Wilhelm Wirth (1876—1952) was a student of Wilhelm Wundt. He had first studied law in Munich in 1894 but then changed to philosophy as his main subject, while also studying mathematics and physics for the higher teaching profession. In 1897 he received his doctorate with a thesis on experimental psychology. After a guest visit in Leipzig, Wilhelm Wundt offered Wirth an assistant position and he habilitated there in 1900. Together with other researchers, he founded the Society for Experimental Psychology in 1902. In 1908, Wirth was appointed professor in Leipzig. In 1933 he signed the commitment of professors at German universities and colleges to Adolf Hitler. In 1935 he began developing apparatus for target practice on behalf of the Wehrmacht. In 1940 he became a member of the NSDAP. In 1943, Wirth's seminary and private apartment were totally destroyed in an air raid; Wirth applied for retirement at the age of sixty-eight and moved to Bavaria with his family.

[15] Ernst Bergmann (1881—1945) studied philosophy and German literature at the University of Leipzig and received his doctorate in 1905. He then continued his studies in Berlin. He later returned to Leipzig where he was granted the status of a private lecturer at the university in 1911. In 1916 he received the position of extraordinary professor. There he developed a religious philosophy with mystical aspects. He soon adopted the ideas of the NSDAP and became one of its prominent academic propagandists. He officially joined the party in 1930. In Bergmann's theological work *The 25 Theses of the German Religion*, he declared the Old Testament and parts of the New Testament of the Bible to be useless in Germany. He claimed that Jesus was not a Jew but an Aryan and described Adolf Hitler as the new Messiah. Bergmann committed suicide in 1945, following the occupation of Leipzig by the Allies.

[16] After studying zoology, Curt Egon Thesing (1879–1956) became editor of the popular science journal *Kosmos* and in 1913, with Otto val Halem as a partner, founded the publishing house Veit & Comp. In this publishing house, Arthur Haas published his first books during the war. The Veit & Comp. publishing house was incorporated into the Walter de Gruyter publishing house in 1919—despite some political reservations on the part of Thesing, who was politically left-wing. In May 1920, he withdrew from the company as a partner for health reasons. He then lived as an independent writer and translator. After the National Socialists came to power, he was briefly arrested and his books were banned.

[17] Leo Jolowicz (1868–1940) took over a bookstore in Leipzig in 1898 and developed it into the largest and best-known scientific antiquarian bookshop in Germany, Gorch Fock. In 1906 he founded

However, the two publishers Curt Egon Thesing—who had founded the Veit publishing house in Leipzig and who produced *Kosmos* as one of the first popular science journals—and Leo Jolowicz, founder of the Akademische Verlagsgesellschaft (Academic Publishing Company), became particularly important for Haas personally. It was here that he made his first connections to the publishing industry, enabling him to become one of the most successful authors of popular scientific writings over the following two decades. In an interview, Austrian American physicist Victor Weisskopf (1908—2002), when asked what motivated him to study physics, recalled these popular scientific writings of his youth:

> There was also a magazine on a somewhat lower level than *Scientific American,* which is called *Kosmos* and which still exists in Germany. Quarterly they issued a popular book in the field of science and I devoured those. There were good and bad books in physics—not too many popular books. Haas, a theoretical physicist in Vienna, wrote a few smaller books of medium difficulty.[i]

Haas showed less appreciation for other aspects of social life, especially when it came to personal contact with the population; he could be harsh and full of prejudices.

> Leipzig had excellent, indeed world-famous concerts in the historic Gewandhaus. These were conducted by Nikisch[18]; I attended regularly and they gave me a lot of pleasure. The theaters of Leipzig were less good and often showed horribly tasteless trash. The actual entertainment centers in Leipzig could make a Viennese feel nauseous.
>
> The tastelessness was manifested in everything and everyone looked terrible to me. This was true from the clothing of the people to the super-gauche kitsch of the Leipzig Battle Memorial, which was evidently intended to retain the memory that, in the decisive battle, a large portion of Germans fought other Germans who were receiving foreign pay, while the Saxons themselves, with more national consciousness, took French money, but then switched sides during the fight.[19]
>
> The most painful thing for me was to listen to the local people, where the South-East German flair of dialect was very strong. The language, an indescribable stultification of German carried forward with singing, hissing and gargling sounds, provoked severe gastric complaints in a foreign listener. And, the population outside the upper classes was almost

---

Akademische Verlagsgesellschaft, which became one of the most famous scientific publishing houses. Publications included: the *Zeitschrift für physikalische Chemie*, the *Handbuch der Experimentalphysik*, the *Handbuch der Radiologie*, *Rabenhorst's Kryptogamen-Flora*, *Bronn's Klassen und Ordnungen des Tierreichs, the results of enzyme research and the results of vitamin and hormone research.* Among the authors were the Leipzig luminary of physical chemistry: Wilhelm Ostwald, Svante Arrhenius, Pierre Curie and Marie Curie. When the National Socialists came to power, the publishing house was "aryanized" and Jolowicz was gradually pushed out, due to his Jewish ancestry. He left the publishing house in 1937. He applied for an exit visa in 1939 but could not leave Germany. Jolowicz died in 1940, possibly by suicide. The book inventory of his Gorch Fock store burned to the ground in a bombing raid on Leipzig in 1943.

[18] Arthur Nikisch (1855–1922) was a Hungarian conductor who performed internationally, holding posts in Boston, London, Leipzig and also in Berlin.

[19] Here Haas refers to the decisive battle of the combined Austrian, Prussian, and Russian armies against Napoleon's Grande Armée, which contained Italian, Polish, Rhenish, and Saxonian troops in October 1813. Due to the participation of multiple European troops and nations on both sides and with shifting alliances, it is called the Battle of the Nations and is counted as the largest battle in Europe prior to World War I.

Fig. 7.2 The Café Merkur in Leipzig, which emanated homey familiarity for Arthur Haas

> misshapen. Everything showed these horrible effects, even at a young age—a terribly degenerate offspring of Kashubians and Obotrites, or whatever all the wild Slavic tribes were once called.[20] Regardless, these very people, twenty years later, would be immensely proud in their belief of belonging to the noblest race in the world, while all others will only represent some kind of animals.

In order to separate himself from the descendants of the Kashubs and Obotrites, Haas looked for places that reminded him of his familiar Vienna environment, such as Café Merkur on Dittrich Ring in Fig. 7.2. The coffee house was run in the manner of Viennese reading cafés, offering Haas shelter and tranquility as well as opportunities for exchange with countless artists and writers.

> Under such circumstances I was quite missing the opportunities for entertainment in Leipzig. I regularly went to the famous Café Merkur with its countless foreign newspapers, made trips as often as possible to Dresden and Berlin, and at the beginning of every holiday, without losing an hour, I went to Vienna, and from there I returned as late as possible to Leipzig. I had maintained my apartment in Vienna, whereas in Leipzig I lived during the first years in a Pension Peters near the central station on the Ring.

His escape to Vienna—seeking the shelter of family or the protective company of family members—took him quite often from Leipzig. This is why he initially stayed in a Pension (guesthouse) near the railway station.

[20] Indeed, Saxony had a mixed German-Slavic population. The Slavic tribes were the Sorbs and the Wends. They lived in the region called Lausitz (Luzice) along the rivers Neisse (Nysa) and Spree. Today, their region includes the south-east part of the Brandenburg state and eastern Saxony. The Obotrites and Kashubians were Slavic tribes that lived in the area along the Baltic Sea, today part of Poland. Haas uses their names in a somewhat derogatory manner.

> I spent the Christmas holidays of 1913 with my still completely heartbroken mother. She then came, accompanied by Uncle Felix, to my official inaugural lecture, "The Spirit of Hellenism in Modern Physics," held on January 17, 1914. To this celebration in the festival hall of the university, the so-called Aula (which adorned a famous painting by Max Klinger, *Hellas*), my cousin Dr. Ernst Grünfeld also came; he was with the University of Halle at that time, as a lecturer in economics (later he was professor and even dean).
>
> On the day after the inaugural lecture, which was a great success, I drove with my mother to Berlin, where I showed her the city. Unfortunately, in the evening she had to return to Vienna with Uncle Felix. The inaugural lecture was published, by the publisher Veit, in booklet form. Later, to my delight, it was translated into Greek and published by A. Chalas in Constantinople.

This introductory lecture by Arthur Haas, as newly appointed professor at the university, was thus devoted to his favorite classical topic, "The Spirit of Hellenism in Modern Physics" which he submitted for publication shortly afterwards.[21] In this lecture Haas claimed for Thales the idea of constancy and unity in physics and for Pythagoras the idea of using mathematics to describe physical events. This argument leads to Democritus as the father of the A-Thomos idea, atoms being the smallest indivisible particles. According to Boltzmann, this concept of atomism had become the foundation of modern physics; yet today, atomism can only be expressed mathematically in the formalism of quantum mechanics.

Haas' lecture was certainly given in the spirit of his contract as professor for the history of physics but subsequently, his interests turned away from historical concepts. He came to focus more and more in the direction of modern physics and its questions and possibilities, shifting with the times:

> I spent Easter and the Pentecost holidays in Vienna. In Vienna, as the year of mourning [for his father's death] was over, I again took part in numerous societies, such as a traditional Viennese party at the horse race concourse, in the Botanical Garden, during which I saw for the last time Emperor Franz Joseph up close. Then I took my leave of the cheerful, happy life that gave Vienna, until 1914, its fairytale splendor.

This is Haas' reflection about the end of an era, an ending that began with the outbreak of war in 1914 and concluded four years later in the catastrophic collapse of the Habsburg multi-ethnic empire. The war was a catastrophe for Vienna; the cultural center of a wide-flung Habsburg Empire was turned into the hydrocephalus of the small state of Austria. It was also a catastrophe for the Viennese Bourgeoisie and especially a catastrophe for Jews, who suddenly found themselves exposed to the various nationalities in central Europe without imperial protection.[ii] And it may even have been a catastrophe for the multiethnic people of the empire; while they had entered independence, they were left to pay the price with endless wars, civil wars, and ethnic cleansing that would last for the remainder of the century.[iii] And so, Haas carried on professionally as he realized the personal impact:

> In the summer of 1914, I held, with particular success, the aforementioned lecture on the historical development of the mechanical principle and also organized a literature class on

---

[21] Arthur Erich Haas, "Der Geist des Hellenentums in der modernen Physik," Antrittsvorlesung am 17. Januar 1914 in der Aula der Universität Leipzig (Leipzig, 1914).

classical works of physics (as I had done in Vienna). Within the library of the Physical Institute, I set up a rich historical subdivision.

My famulus [assistant] Karl Jörissen was especially devoted to me. The following year Jörissen lost his young life on the battlefields, as a child of a widow. She consoled herself with the phrase she had been given, that it was necessary "for the sake of the freedom of the seas." He had also helped me in publishing my lecture notes as a book.

Just before the end of the semester, the book on which I had been working since the autumn of 1913 was published, so that I was able to present copies to all participants in the lecture. The book was published by Veit in Leipzig and was reviewed very favorably by the critics; thus, *Physikalische Zeitschrift* wrote that the underlying idea of the book must be described as very intriguing and carried out with exemplary diligence.

Karl Jörissen was a young student of physics and history and he was assigned to help Haas as teaching assistant for the mechanics lecture. He probably also helped in compiling the data for the *Poggendorff Handbook*. Jörissen came from the small village of Niederlößnitz near Dresden and had received the draft order during the summer holidays of 1914. He exchanged several letters with Haas to provide him with practical instructions and to aid with maintaining the organization of the fall course. Jörissen wrote about himself: "I myself now say physics and history, valet [farewell] hopefully not forever."[22] Jörissen was transferred to the western front.[23] During the first two years of the war in the trenches, he was promoted to sergeant. Haas provided him with physics and history literature to overcome the boredom of daily life in the trenches. Jörissen was grateful: "Avenarius gave me as much pleasure as quantum theory and the historical writings have already been passed on to the officers. Because here, everyone is bored, and yet you don't have the necessary peace and quiet and contemplation to start something real."[24]

Karl Jörissen was killed on September 6, 1916 in the Battle of the Somme, which cost the lives of more than a million young men. It had started with the assassination of a single man, the Archduke Franz Ferdinand of Austria and his wife on June 28, 1914 in the city of Sarajevo in Bosnia and Herzegovina, by members of the Serbian terrorist group the *Black Hand*. Again, Haas was left with the impact and its strong impressions:

On June 28, 1914, I was deeply shaken by news about the assassination of the Austrian heir to the throne and his wife, which spread like wildfire. I had no sympathy for the victim—he was rather disliked in general. I had only once observed him closely at a solemn meeting of the Academy of Sciences, of which he was curator. He was throwing, like a little boy from his heightened seat, little paper balls, in a most undignified manner, at his family members who were sitting in the front row.

However, the manner in which all of the German newspapers were pushing Austria into war with Serbia, apparently obeying higher orders, filled me with the gravest concern and foreboding. Although I had learned to despise humanity, I did not consider it capable of doing what it did in the next four years. On August 2, 1914, the "grosse Zeit" [big time] began. For the brave Leipzig, it began with the classic announcement made by the emperor in front of

[22] Karl Jörissen to Haas, September 28, 1914.

[23] Grenadier K. Jörissen, XII Armee Korps, 23. Inf. Division, Leibgarde Regiment 100, 4. Kompanie.

[24] Karl Jörissen to Haas, February 9, 1916.

the Reichstag, "Nu aber wollen wa se dreschen" [Now let's kick butt] and ended four years later with the equally classic saying of the King of Saxony, presumably after appropriate consumption of liquor, that the people "seinen Dreck alleene machen solle" [should take care of their own junk].

I just could not understand the development of world events. In my heart I have always been a sincere admirer of French culture and taste, in my mind a no-less sincere admirer of Anglo-Saxon might and above all, strongly influenced by Wilson's writings on American liberty. The rabid radicalism of the new political leaders, of the Bohemian and Moravian Germans (Schönerer[25] and Wolf[26]), had awakened in me strong sympathies for the Czech Republic.

That Austria did not release Trento[27] after the national unification of Italy has been wrong in my opinion. Yet, I swore by the Viennese culture and I respected Austria's historical mission. I also felt a certain traditional veneration for the aged emperor of Austria, who survived my father even though already regent at his birth. But I was convinced that in a war, for Austria to fight as an ally of its mortal enemy Prussia, a defeat would be less fatal than a victory.

Apart from certain sentimentalities, for many years I had been thoroughly republican and democratically minded but above all pacifist with anti-militarist sentiments. I knew only one enemy of humanity, of my fatherland and of the German people—Prussian militarism, whose annihilation seemed to me a prerequisite for a necessary cleansing of Europe. My inner attitude about the dreadful event of World War resulted naturally from this opinion.

One may argue on the question of responsibility for the war, and in fact this question has been debated for the last century with shifting results. While this discussion will undoubtedly continue unmasking the shortsightedness and lack of comprehension of all political leaders,[iv] it is clear that in the early days of the war, Europe was flooded with enthusiasm. Many people, especially the younger generation, hoped for a purification of encrusted structures within not only the European power structure but also the internal rigidity of the states. Only a few foresaw that the war would not end with purification in victory. Rather, the collapse of all political and economic structures led to imperfect replacement by new, nation-based states. This,

[25] Georg Heinrich Ritter von Schönerer (1842—1921), a wealthy landowner in Lower Austria, was leader of the German Nationalists and later of the All-German Unification, from 1879 until the turn of the century. He was a fierce opponent of political Catholicism and a radical anti-Semite who exerted a strong influence on the young Adolf Hitler, who regarded Ritter as one of his role models. On the other hand, Ritter was a strong supporter of social programs, which he also personally enforced on his properties. From his estate administrators he demanded: "Give the people what is right!" Andrew G. Whiteside, *Georg Ritter von Schönerer. All Germany and its Prophet* (Graz/Vienna: Styria, 1981).

[26] Karl Hermann Wolf (1862—1941) was a radical German editor and writer. He represented the national or even Greater German goals of the All-German Unification and later also of the Radical German Party. At times a follower of Schönerer, he later represented his own, more radical political goals.

[27] The city of Trento was, before Napoleon, part of the Holy Roman Empire. In 1814 after the defeat of Napoleon, Trento became part of the Habsburg Empire and was officially administered by the province of South Tyrol, which was primarily German populated. Italy, after its unification in 1866, sought to include the Italian populated areas south of the Alps, including the city of Trento. This was one of the reasons why Italy, nominally allied with Germany and Austria in the Triple Alliance, joined the Entente of France, Great Britain, and Russia against its former allies after the beginning of World War I.

in turn, led to the rise of radical right-wing forces and thus to the second war, whose consequences still haunt us today, four generations later.

In the beginning, German academics and cultural representatives stood in large numbers behind the Kaiser and the military. Ninety-three signed the appeal "An die Kulturwelt!" (To the Cultural World!),[v] denying the allegations of war crimes raised by the Entente and seeking to make the German position clear. Among the ninety-three were leading scientists, including Philipp Lenard, Walter Hermann Nernst, and Max Planck. Many of the more pacifist scientists, like Albert Einstein, did not sign and many of the signatories withdrew their signatures later.[vi]

Haas was undoubtedly one of the more liberal representatives of the physics community at the time. However, the passage of time and his personal experiences during the post-war years, with radical right-wing forces emerging in Germany and Austria, influenced and exacerbated his views. He was opposed to the dominance of the military and particularly the militarism emanating from neighboring Wilhelmine Germany, and he strongly and repeatedly criticized these developments. He also did not stand with the disintegration of the Habsburg Monarchy into a series of individual states. Rather, he supported internal democratization with a greater equality of rights and voices for the various ethnic groups within the Monarchy. Haas saw himself as an Austrian of German origin and not as a Czech, as his personal decisions after the collapse of the Empire testified. Nor did he see himself as a Jew. However, he was increasingly told, during the post-war years in his own country, that that was exactly what he was.

Notes

i. "Oral History Interviews, Victor Weisskopf," accessed January 31, 2021, https://www.aip.org/history-programs/niels-bohr-library/oral-histories/4944.
ii. William O. McCagg, *A History of Habsburg Jews*, 201–222.
iii. Timothy Snyder, *Bloodlands—Europe between Hitler and Stalin* (New York: Basic Books, 2010) and Mark Mazover, *Dark Continent—Europe's Twentieth Century* (New York: Vintage Books, 2000) and John Lampe and Mark Mazover (eds), *Ideologies and National Identities: The Case of Twentieth-Century Southeastern Europe* (Budapest: Central European University Press, 2004).
iv. See Christopher Clark, *The Sleepwalkers—How Europe went to War in 1914* (London: Harper, 2012).
v. Juergen Ungern-Sternberg and Wolfgang Ungern-Sternberg, *Der Aufruf An die Kulturwelt!: Das Manifest der 93 und die Anfange der Kriegspropaganda im Ersten Weltkrieg* (Stuttgart: Franz Steiner Verlag, 1996).
vi. Stefan L. Wolff, "Physiker im 'Krieg der Geister'—die 'Aufforderung' von Wilhelm Wien," *Acta Historica Leopoldina* 48 (2007): 41–62.

Chapter 8
War and Post-War

The war did not begin in the Prussia that Haas so hated. War started when Austria moved against Serbia in order to humiliate her annoying neighboring state. Conflict between the two had again and again put the south-eastern regions of the Habsburg Empire, such as Bosnia and Herzegovina, into unrest and discord. Austria wanted to use the murder of the crown prince, by a Serbian terrorist movement, as an excuse for punishment—with German backing. But this set into motion chain reactions that led to the general outbreak of war, the circumstances of which are now well known and often commented upon. The number of publications and books on this issue are as varied as the number of interpretations, evaluations, and recriminations but, regardless, it ended in disaster. The impact for the Haas family began with Otto:

> When the Austro-Serbian war began and my brother was summoned to the military, my mother, with my sister, immediately rushed back, from their vacation in the Salzkammergut, to Vienna in great dismay. I joined them at once, especially as the semester ended coincidentally with the outbreak of the war.
>
> My brother was in the Bohemian town of Dobrzan near Pilsen [today Plzeň], in the replacement squadron of the second Dragoon Regiment to which he belonged. After spending a few days with my mother, I traveled to Dobrzan to bid my brother farewell before he went to the front.
>
> His departure was delayed a bit, so I was with him for about a week. Then I traveled back to my mother in Vienna. Our faithful Miss Barnett was unfortunately just then on vacation in England and could not come back to us. (She returned to Vienna only after the war.) In spite of the summer, we did not leave Vienna and we were utterly unable to think of anything but war.

But the war affected the military status of Arthur Haas as well. As a one-year volunteer, he was a member of the military reserve and so he was drafted. Even his original exemption from regular service did not help him during these first days of crisis when everyone was brought forward. Haas willingly followed. In spite of his official subordinate position, he did not have a difficult or dangerous time during his military service. This was probably due to his professorial title, which has always brought special prestige in Austria. But the boredom of the never-ending service was eating away at him.

M. Wiescher, *Arthur E. Haas—The Hidden Pioneer of Quantum Mechanics*, Springer Biographies, https://doi.org/10.1007/978-3-030-80606-4_8

At the end of October 1914, my military service at the home front began, which lasted until the end of May 1917. Since this had nothing whatsoever to do with my actual self, details are not worth mentioning. I was drafted, as a former one-year volunteer, to the cavalry, by the Brünn home-front command. I was assigned to a so-called horse-stable hospital, which was initially located for a month in Traiskirchen near Vienna and then moved to Brünn for the rest of the war.

In this stable, I spent more than two years lazily working, completing the officers' correspondence with various authorities and dealing with other administrative matters. My superiors, a nearly seventy-year-old retired Hungarian Lieutenant Colonel v. Török (still feeling a little boyish) and a sixty-year-old retired hussar rider, Count Sizzo-Nori, treated me with extraordinary tact and sensitivity; when I came to a restaurant where they were dining, they never failed to ask me to join them at their table.

In contrast to the troops in the field, he was not doing badly; he was even allowed to have his own apartment and drink his daily mélange in the café of the Deutschen Haus (German House). He made acquaintances with local lecturers, such as Arthur Szarvassy[1] and Alfred Lechner,[2] and also cultivated other pleasures, including the annual summer retreat.

For two years I was completely incapable of focusing on any physics-related matters. In order to distract myself somewhat from the awkwardness of the present situation, I studied, in a curious and almost infantile manner, grammar-school textbooks in geography and history, as well as higher mathematics from university textbooks. Obviously, I knew my native city of Brünn very well and I was happy to hang out in the Augarten and I spent, especially on full moon, many nights on the Spielberg. In my free time I hiked the surroundings of Brünn. I was often, frequently twice a day, in coffee houses or in the Deutschen Haus eagerly reading newspapers.

I lived in the immediate vicinity of my birthplace, such that I could look down into the beautiful arcaded courtyard. I had rented a room on Schlossergasse 1, a side street of the Great Square. Of course, I had very few visitors. I was often with my uncle Dr. Louis Gomperz, and sometimes with the very brilliant associate professor of physics at the Brünn

---

[1] Arthur Szarvassy (1873—1919) studied at the University of Vienna and was awarded his doctorate in 1898 with an experimental dissertation under the supervision of Franz-Serafin Exner, before turning to theoretical physics. This was probably very much to Szarvassy's liking, for Lise Meitner learned the beginnings of theory from him as a very young student and described him as her first "true" teacher and as an extraordinarily stimulating tutor, when he was an assistant at the Institute of Theoretical Physics. In 1903 he went to the German Technical University in Brünn and habilitated there after two years with a theoretical thesis. In 1905 he was appointed extraordinary professor of physics at the Brünn University of Technology. He died in 1919, shortly before his intended appointment as full professor.

[2] Alfred Lechner (1884—1944) studied mechanical engineering at the Vienna University of Technology and attended lectures in mathematical physics and astronomy at the University of Vienna. In 1910 he was awarded a doctorate in technical sciences and became an assistant in mechanics at the German Technical University in Brünn. Until 1918 he lectured there on mechanics and material strength—an area that linked him to Arthur Haas. After the First World War, Lechner was appointed assistant lecturer in the Department of General and Analytical Mechanics at the Vienna University of Technology. In 1931 he became full professor of general mechanics. Lechner worked on the expansion of his institute's experimental equipment and on the restoration of the rather old-fashioned facilities he had inherited at the Technical University of Vienna. In addition, like Haas, he gave scientific and popular-science lectures at the Vienna Urania, among others, and was appointed to its advisory board in 1926.

Technical University, Szarvassy who was married but had no children. I also had pleasant interactions with a young private lecturer at the technical university, Alfred Lechner, who later became a professor of mechanics at the Technical University of Vienna. In 1916, I suddenly became very interested in photography and soon took many beautiful portrait and landscape shots. In 1916, I took a three-day holiday to Budapest (only for pleasure) and then a ten-day trip to the Wolfgangsee in Strohl (in the Salzkammergut).

Family life seems to have normalized despite the increasing hardships of the war. While Arthur was in the Homeland Service in Brünn, Otto served at the front with promotion to the general staff and their young sister was successful in her first semester at university; only the war and the generally more difficult life situation seemed to burden the family. The war dragged on and no one believed in the fast victory nor in the peace that had been promised and hoped for in the first year. Haas considered it his duty to visit the family in Vienna as much as possible and his Brünn superiors did not put any obstacles in his way:

Of course, I saw my mother as much as possible. I often visited her on my free days in Vienna and she also came several times to Brünn. In 1915 my sister passed the Gymnasium-Matura with a certificate that, like the ones of her two brothers, had the best grade in all subjects. In the fall of 1915, she enrolled at the University of Vienna in order to study history.

As a soldier, my brother was in great danger during the first period of the war but less so later, after he was assigned to command troops as an orderly officer. Now and then he came, on leave, to Vienna. He was promoted to lieutenant and later to captain, and also received various war medals.

My mother worked for the Red Cross and also received the "Red Cross with War Decoration" as an award. She seemed to be slowly overcoming the dreadful pain of my father's death when, in the autumn of 1916, only four years later, she fell seriously ill. Already in the spring of 1916, my mother had begun to feel unwell; she did not pay much attention to it and did not complain, but it worsened considerably during the summer. She decided to go to a sanatorium in Sulz-Stangau near Vienna for a special cure. There I spent some days with her and then visited her more often. However, no one took the weakening of her health seriously, generally thinking that she was suffering from the issues that often affect women in their fifties. But at last the cause was found and at the end of September our family doctor, Dr. Siegfried Lederer who had also treated my father, made the devastating diagnosis that my mother was suffering from a rare but very dangerous adrenal gland disease, the so-called Addison's disease. Of course, we tried everything possible, consulting famous doctors, but it did not come to anything. Some improvements were made but it was hopelessly over. My mother became bedridden, and she was always growing weaker without suffering any actual pain.

For Christmas 1916, she distributed very carefully and lovingly, as every year, the Christmas gifts to family members and domestic workers. My brother, whom she had seen a few weeks before, was missing, of course, because he was on the battlefield. She gave me a small but excellent camera, with which I still photographed my children more than two decades later. I returned to Brünn but, on December 30th, a telegram reached me that my mother had fallen into a coma-like state, which I knew meant the end. I immediately rushed to Vienna. My mother hardly recognized me and fell into unconsciousness. On the afternoon of December 31, 1916, a Sunday, she breathed her last breath (at 5:21 p.m.). My sister, I, and several of our closest relatives were present; my brother was on his way from the front to our parental home, where we had hastily summoned him.

It had been the greatest wish of my mother, to recover for the anniversary of my father's death, Jan. 14, in order to be able to decorate his tomb on that day. She had constantly

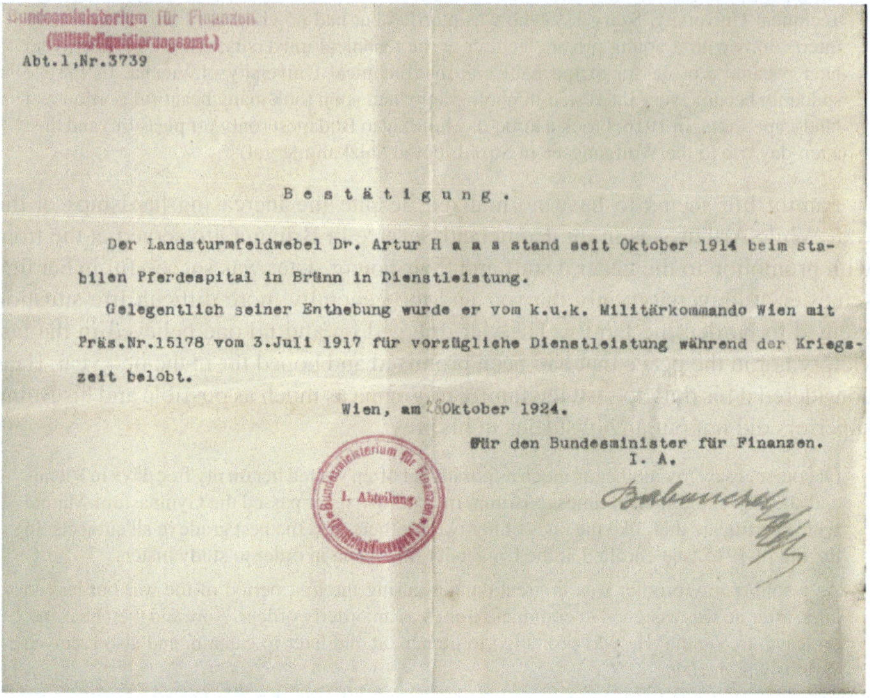

Bundesministerium für Finanzen
(Militärliquidierungsamt.)
Abt.1,Nr.3739

Bestätigung.

Der Landsturmfeldwebel Dr. Artur H a a s stand seit Oktober 1914 beim stabilen Pferdespital in Brünn in Dienstleistung.

Gelegentlich seiner Enthebung wurde er vom k.u.k. Militärkommando Wien mit Präs.Nr.15178 vom 3.Juli 1917 für vorzügliche Dienstleistung während der Kriegszeit belobt.

Wien, am 28 Oktober 1924.

Für den Bundesminister für Finanzen.
I. A.

Bundesministerium für Finanzen (Militärliquidierungsamt) I. Abteilung

Babanek

Fig. 8.1 Confirmation by the military authorities, verifying the service of National guard Sergeant Dr. Arthur Haas at the Horse Hospital in Brünfign

> visited my father's grave and was very unhappy that she had not been able to do so in the last few weeks of her illness. Before the sad day of remembrance came, she was buried on January 3 at my father's side. We lost her before she was fifty-five, and her heart's desire to see grandchildren remained unfulfilled. I was hurt by the death of my mother, whom I had cared for with great love, especially in her bitter last years. It grieved me much more than my father's death, as much as I had been shaken by that. My mother had literally died of a broken heart, in grief over the loss of her husband. Since she had lost her will to live, she had not been able to offer sufficient resistance to the attacking illnesses.

After this fateful blow, Haas had had enough of military service. In Brünn, he felt himself cut off from his scientific research. His brother Otto was at the front and could offer neither comfort nor conversation. His sister Margarethe, who was many years younger than him, was old enough and independent enough to attend university under the care of Uncle Fritz, and Uncle Fritz himself was guiding the firm through the increasingly gloomy war years. Arthur no longer saw any need to continue his visits to Vienna. The casual nature of his service gave him the opportunity to leave it completely (Fig. 8.1).

> After I recovered from the unexpected death of my poor mother, I strove above all to return to Leipzig, to continue my scientific work. I did not want to return as long as my mother lived because visits from Leipzig to Vienna during wartime would hardly have been possible.

Now, because of the censorship, an exchange of letters between Austria and Leipzig did not seem appropriate. I decided to travel to Leipzig on a vacation trip, to inquire about my return.

In March, 1917, I spent a week in Leipzig in order to, as I said, check on the progress of the Poggendorff dictionary. There, I asked some influential acquaintances to persuade the Saxon Academy, called the Academy Cartel, to appeal to the Austrian Ministry of Defense with a request for my dismissal from military service so that I could continue in Leipzig on the *Poggendorff* project, which allegedly was important for war purposes. The request was indeed granted; at the end of May 1917, I was dismissed from military service in order to take up my position at the Saxon Academy.

A few months later, to my great astonishment, I received a diploma. The Vienna Military Command had awarded me with a diploma recognizing my service during the war [Fig. 8.1]. It was touching that during the last days of my stay, de Coudres came personally to Brünn to see me before I returned to Leipzig. He had undertaken the long and difficult journey in wartime only to advise me before my arrival in Leipzig, suggesting how I should deal with certain difficulties that had arisen in connection with the *Poggendorff* edition. We spent a day for two in Brünn and after that, I returned to Leipzig with great eagerness.

After burning my uniform in the oven of my Brünn apartment, I went first to Vienna. As I had told the military authorities, I spent fourteen days there collecting material for my Leipzig mission. In Vienna, I met not only my sister but also my brother, who was on leave from the front. Then I traveled to Leipzig to resume my work on the *Poggendorff* as well as my lecturing at the university.

The death of his mother had freed Haas from the lethargy of his military posting and his coffee-house existence in Brünn. A further motivation for returning to normal academic life was the unexpectedly easy reprieve by the military authorities who released him and sent him to Saxony on a mission, supposedly important to the war effort. To his satisfaction, Haas was able to return in time for the beginning of the summer semester.

But Haas had been warned by Theodor des Coudres during his unforeseen visit from Leipzig, which had certainly been a costly undertaking in wartime; he had gone to prepare Haas for things that were not as they had been before, since his salary from the *Poggendorff's* budget had been allocated elsewhere during his absence.

Significant changes had occurred with the *Poggendorff* editorial process since the beginning of the war. Professor von Oettingen, who had published the fourth volume about ten years previously, had been living in Leipzig from his Russian pension. This pension was lost with the outbreak of war and Oettingen was deprived of his sole source of income. When I was called to military service, he had approached me to ask if I would show the generosity of leaving him my salary and position for the time of the war, so that he would be protected from emergency. I agreed but must say that the behavior of v. Oettingen upon my return showed little gratitude for my noble gesture. Although he remained well-disposed toward me and often invited me to private visits, he had rigid ideas of his own and caused me constant difficulties in the commission of my work. In the end, I finally left the project essentially to him alone.

Haas gradually withdrew from the project. He was still under the impression that, after the death of both parents, he had sufficient assets to meet his personal needs. In addition, he had grown tired of the project as well as of history of science, and he now wanted to devote himself entirely to actual research in theoretical physics. Therefore, he even stayed away from the editorial office after Arthur von Oettingen died in September 1920. Thus the fifth volume of Poggendorff's *Biographical-Literary Dictionary*, which covered the period between 1904 and 1922, was not completed until 1926 under the editorial leadership of the mathematician and historian Paul Weinmeister (1856—1927).[i]

After Haas withdrew from the project, he returned to his lecturing activities. These mainly focused on developments in modern physics, from quantum physics to the new theories of relativity. He had completely abandoned his earlier activities in science history, while connections with Karl Sudhoff had already become dormant since the time of Haas' departure for military service. Haas' writings never again mention his former mentor.

It is also significant that Haas did not comment on the war nor on developments at the front, although events and shifts on the eastern front affected his brother personally and were significant for Austria. The initial success of the Russian Brusilov Offensive in the summer of 1916 had led to the temporary collapse of the Austrian front, which and could only be stopped by the deployment of German troops from the Western Front. Romania's entry into the war and the offensive in Transylvania, which also could only be successfully repulsed by the deployment of German Army units, had stabilized the Eastern Front by the summer of 1917 but politically, these maneuvers had clearly illustrated Austria-Hungary's great dependence on the military might of Germany. At that juncture, events had reached the exact point that worried Haas the most.

The collapse of Russia after the October Revolution of 1917, the armistice of December 1917, and the peace dictate of Brest-Litovsk, in which Austria-Hungary also had high expectations, are not mentioned by Haas in his autobiographical notes. Is this omission because he was totally immersed in his lecture and writing activities or is the explanation that for him the subsequent events surrounding the collapse of the Central Powers made any discussion irrelevant?

> In Leipzig, while I continued to give science-history lectures, I gradually switched to theoretical physics. These lectures, in which I treated the latest developments in physical theories in an easy-to-understand way, almost became a societal event at the university. I had several professors, including Meteorologist Wenger, as permanent listeners in these courses. At the colloquiums, I also frequently lectured on problems related to atomic, quantum and relativity theory. Some smaller essays, including one about an atomic nucleus model, appeared at that time in the *Physikalische Zeitschrift*.

Besides Einstein in Berlin, at the time Haas was indeed the only lecturer in either Germany or Austria to offer classes on the theory of relativity.[ii] This explains the great popularity of his lectures and the participation of academic colleagues like Robert Wenger.[3] In addition to this particular lecture, Haas also gave a whole series of other lectures in theoretical physics. In winter semester 1917/18, he taught the theory of electricity and light; in the following summer semester, the last before the

[3] Robert Wenger (1886—1922) succeeded the Norwegian Vilhelm Bjerknes (1862—1951) as director of the newly founded Leipzig Geophysical Institute in 1917. Bjerkness had accepted a call from Bergen in Norway and so, in turn, obtained a war release for Wenger, his former assistant, to succeed him as head of the institute. Unfortunately, Wenger died very early and only had a few years to manage the institute.

end of the war, he presented the modern problems of theoretical physics, introducing the concepts of electron, gas, radiation, quantum, and relativity theory. A second lecture in this semester dealt with the history of physics from its beginnings to Galileo. This was followed in the winter semester of 1918/19 by the continuation of the history of physics, covering the time from Galileo to the end of the eighteenth century. Haas also again offered his lecture on the theory of relativity—unaffected by the German collapse, the November Revolution and the short but turbulent Leipzig Soviet-Republic, as well as the occupation of Leipzig by German army units in May 1919.[iii] In the summer semester of 1919, Haas offered a lecture on modern quantum theory and in the following winter semester a lecture on atomic physics and molecular phenomena.

In addition to his teaching during the last year of the war and the first post-war period, Haas immersed himself doggedly in his scientific work, turning a blind eye to military and political developments. In 1917 he published a scientific article on the atomic nucleus,[4] in which he tried to unite the Rutherford model of the atomic nucleus with his earlier model of equilibrium positions of groups of electrons within a homogeneous positive nucleus. While this latter approach was physically obsolete, Haas' publication was a provocation to the German physics community, which had followed the call made by Wilhelm Wien, in early 1915, to avoid quoting English science and its publications in order to strengthen the German physics community.[iv] Haas' continued interest in Rutherford's work clearly held precedence over the call of his colleagues and he may have been able to place this article through his personal association with Theodor Simon, his former supervisor in Göttingen and editor of the journal. On the other hand, the united front of German-speaking physicists against the "perfidious Albion" had already largely dwindled and the work may not have raised any concerns for *Physikalische Zeitschrift's* editorial staff.

One year later in 1918, Haas published a second article on the relationship between the gravitational constant and the fundamental constants in electron theory. In this article he described the gravitational constant as a function of the specific charge of the electron and the mass ratio between the electron mass and the mass of the hydrogen atom.[5] In so doing, he took up the approach that he had used in his habilitation to demonstrate the correlation between Planck's constant and the Rydberg constant. This short article is a first example of Haas' lifelong fascination with fundamental constants in physics and his search for a deeper meaning and connection between the seemingly so different disciplines of gravitational physics and electrodynamics. This fascination with field constants would unfortunately lead him down a false path in his scientific thoughts and ideas. He was not the only one; in fact, fascination with the comparison of the fields themselves has dictated the development of modern physics in the twentieth and twenty-first centuries. Yet, Haas was also busy with other endeavors:

---

[4] Arthur Haas, "Der positive Kern der Atome," *Physikalische Zeitschrift* 18 (1917): 400–402.

[5] Arthur Haas, "Über eine Beziehung der Gravitationskonstante zu den Grundgrößen der Elektrizitätstheorie," *Physikalische Zeitschrift* 19 (1918): 330.

> My main activity in those years was, however, the work on my legacy, writing a textbook of theoretical physics. I had already developed the idea for such a work in the spring months of 1914. The outbreak of the war pushed the notion into the background until I took it up again with full intensity in spring 1916. By that time, I had amassed enough material and the military authority that employed me at the time, provided me with an appropriate place to work in a rather spacious mansion. This enabled me to remain undisturbed for much of my service and to work on the manuscript. The project continued to progress in Leipzig. First the mechanical, then the electrical, and finally the optical part was finished. A few weeks after the end of the war, I experienced the immense pleasure of seeing the first volume of the *Introduction to Theoretical Physics* published and available in the bookstore.

It was indeed an achievement to complete an introductory textbook on "Theoretical Physics," [6]especially in addition to his regular scientific work and a full teaching load. In doing so, Haas had used his lecture notes and materials to prepare the textbook. The first volume, published in 1919, looked at the problems of classical mechanics and hydrodynamics, the theory of electromagnetism and electrodynamics, and the physics of vibrations and light as well as electron theory. The latter covered more modern observations, such as cathode rays, X-rays and the phenomenon of radioactivity. Apart from heat theory, the textbook summarized nineteenth century physics. Heat theory and the resulting quantum physics, as well as the physics of relativity and gravitation would be discussed in a second volume, two years later.

Haas' two-volume work is considered the first modern textbook of theoretical physics and became a scientific bestseller almost immediately. It was published in Germany in several editions, the second edition being published by Walter de Gruyter & Co. in Leipzig in 1921. In 1923, a completely revised and expanded edition was published. It was a great success throughout the decade and provided its author with a regular income until the book was banned in Germany in 1934. By 1924, the book had been translated into English, followed by several English editions in Great Britain and the United States. A review in *Nature* commented on the book in 1930, "… this book, which is in the clear and precise style characteristic of all Prof. Haas' publications, is invaluable." This textbook laid the foundation for his success as a scientific author and he was approached by several publishers for reprints and translations.

Immersing himself in this work, Haas had emotionally turned away from the war and its propaganda, both of which he hated. However, he did not completely turn away from social life; visitors from Vienna such as Felix Ehrenhaft[7] and his former student

[6] Arthur Erich Haas, *Einführung in die Theoretische Physik* (Leipzig: Veit Verlag, 1919).

[7] Felix Ehrenhaft (1879—1952) studied physics in Vienna and received his doctorate in 1903 at the Institute of Physics under Franz-Serafin Exner's supervision. From 1904, Ehrenhaft was an assistant at the I. Physikalisches Institut under Exner and, in 1905, received his habilitation. He remained in Vienna as a private lecturer. During these years, Haas probably met him, although Haas was not mentioned further in Ehrenhaft's notes. In 1912, Ehrenhaft became Associate Professor, in 1920 Full Professor and Chairman of the newly founded III Physics Institute of the University of Vienna. As a Jew, Ehrenhaft had to leave Austria in 1938. He continued his work first in England and then in the USA. In 1947 he returned to Vienna as director of the I. Institute of Physics. However, he only had the status of a visiting professor. He died shortly before reaching retirement age. Cf. Joseph Braunbeck, *Der andere Physiker – Das Leben von Felix Ehrenhaft* (Wien: Techisches Museum Wien, 2003) and Wolfgang L. Reiter, *Aufbruch und Zerstörung* (Wien: LIT Verlag, 2017), 421–423.

Gerda Laski, who had in the meantime received his doctorate, were a particular joy to Haas. On the other hand, he gradually alienated himself from his Leipzig colleagues, although he worked closely with one of his students and assistants Harry Schmidt.[8]

> Relationships with my colleagues remained well, although of course socializing in the old form had ceased. I still took part in the professors' afternoon excursions and attended other regular meetings. Closest to me was the already mentioned meteorology professor Wenger, a contemporary of mine from Strassburg; he had much of the more refined French culture, although German was his mother tongue.
>
> In the spring of 1918, we were visited by Ehrenhaft and Dr. Gerda Laski from Vienna, on their way to Göttingen; I gave a big "evening" in their honor. My favorite student and famulus [assistant] was Harry Schmidt from Hamburg. He later became a professor at the Polytechnic in Cöthen and was, at the same time, a lecturer and later a professor at the University of Leipzig. With him, I went for long walks several times a week and he helped me with loving zeal in reviewing my book corrections.

Yet even in these last years of war and work, and despite the misery growing around him, Haas could not and would not give up his beloved summer vacations; the coast had become a stranger to him and he used every opportunity to return to Austria. He felt increasingly foreign in Leipzig and wanted to return to his homeland, although it had changed dramatically in the meantime.

> In the summer of 1917, I spent about a month in Strobl at the Wolfgangsee with my sister. We had a great time together with the beautiful Susi Devrient, who was the daughter of famous actors at the Burgtheater in Vienna[9]; she later found a terribly tragic ending as the wife of the chemist Ernst Hauser.[10] For Christmas 1917, I was back in Vienna and together

[8] Harry Schmidt (1884—1951) began, in 1913, to study physics, chemistry, mathematics and philosophy at the University of Leipzig. There he received his doctorate in 1919 with a thesis in theoretical physics, "On the possibility and stability of equilibrium states of resting and rotating electron groups within a generally non-equivalent sphere of homogeneous positive electricity." This topic corresponded to the interests of Arthur Haas at the time. Schmidt habilitated at the University of Leipzig in 1926. From 1937 to 1945 he was a research assistant at the German Experimental Institute for Aviation in Berlin. In 1945, Schmidt was appointed full professor of applied mathematics at the Martin Luther University Halle-Wittenberg and later became director of the Institute of Applied Mathematics there. After two years of serious illness, Schmidt died of pulmonary tuberculosis in September 1951.

[9] Max Devrient (1857—1929) and his wife, the actress Babette Devrient-Reinhold (1863—1940). Their marriage was divorced shortly after the birth of their daughter Susanne.

[10] Ernst Hauser (1896—1956) was a well-known speleologist and chemist who led various cave expeditions in Austria in the 1920s. He married to Susanne Devrient (1900—1920) in 1918. The latter died on December 17, 1920 in Vienna under unknown circumstances. Ernst Hauser was declared a prime suspect on the basis of the charges brought forward by his mother-in-law; he was accused of poisoning his wife. No traces of poison were found but Hauser's statements contained numerous contradictions. After the authorities closed the investigation, the case was pursued mainly in the anti-Semitic tabloid press. Cf. Dr. Walther Rode, *Motion to open the preliminary investigation against Dr. Ernst Hauserand comrades for the crime of murder* (Self-published in Vienna, 1927). Hauser left Austria in 1921 after his doctorate and went to Göttingen. Since Ernst Hauser was of Jewish origin, he lost his position in Göttingen and emigrated to the US when Hitler came to power. A summary of the Ernst Hauser case can be found in, Peter Danner, "Ernst Alfred Hauser—a speleologist in the crossfire of anti-Semites, mother-in-law and the press," comment in *German Speleologists and Karst Explorers* 63 (3), (2017): 86–93.

> with my siblings (my brother had also come, from the front). I visited my mother's grave at the anniversary of her death.
>
> During 1918, I did not return to Vienna because crossing at the border was too difficult. I spent a month of summer vacation, in 1918, in Tutzing on Lake Starnberg near Munich. In September 1917, I had visited Hamburg for a few days just for fun. In May 1918, I drove to Göttingen for a series of lectures that Max Planckgave on the problems of light quantum theory. I visited Berlin a few times, and on the way back from Tutzing, I also stayed in Nuremberg.
>
> In Vienna, my sister continued her university studies; she had been with a governess since the death of my mother, a Miss Bernhard. After the death of my mother, my sister moved from the Stubenring home to an apartment in Landstrasser Hauptstrasse 1; there she lived with her governess in a mezzanine apartment of five large rooms in a giant modern building, only a few minutes' walk from the old apartment.[11] The law firm remained at the old site and for the time being it was run by Uncle Fritz, who managed the estate after my mother's death by simply investing everything into war bonds.

Margaretha's move from the long-familiar apartment on Stubenring can be explained by her urge for independence, albeit a bit dampened by the employment of a governess who did, however, quickly become a friend. But there were other concerns about the stability of the firm; Friedrich Haas, as asset and estate manager, had converted the inheritance from Yella Strakosch into war bonds that threatened to become worthless in view of the rapidly deteriorating situation. Nevertheless, Margaretha lived comfortably with five rooms; it was a luxurious apartment in an impressive Secessionist house built in 1905/1906, on the site of an older inn. Undoubtedly, this was not a cheap residential area.

But despite his fixation on lectures and writing, together with his travels and family concerns, Haas could not blindly ignore developments in Leipzig:

> In Leipzig, the suffering of the population was getting worse and worse. Almost all were emaciated to skeletons. The misery of the children was terrible but even those who had enough money were unable to really feed themselves. Worse than the constant hunger, however, were the incessant poisonings that one suffered by the "enjoyment" of the "substitute food" (mineral oil instead of actual oil, etc.).
>
> The population had been becoming restless since 1917. As soon as I returned to Brünn from my visit to Leipzig in March 1917, I answered my uncle's question about the impressions I had gained in Germany, saying that in five years' time Germany would definitely be a republic. By the summer of 1917, one could hear critical comments everywhere, quite frequently in railways and at pubs and inns. In other words, an unabashed opinion was shared by simpler people that "the soldiers should rather shoot at their superiors than at the French."
>
> But even in 1918, the women of the "better class" complained with deep regret into their empty coffee jar that the war was not strong or cruel enough, while in the evening their husbands, in the tavern sipping their replacement beer, took the view that in order to secure the possession of Belgium and the French mining regions for Germany, other parts of France must necessarily be annexed for strategic reasons. In faculty circles, Professor v. Öttingen suggested that the Austrians would have to occupy Rome, whereupon the Americans could easily be forced to "pay Germany until they turn black." My children certainly had more political maturity at the age of five than the entire Leipzig University combined.

---

[11] After Margaretha Haas' marriage and departure from Vienna, her brother Otto and his young wife took over her apartment.

It ended as it had to. At the beginning of November 1918, the German empire ended in the well-known way, in unprecedented personal cowardice. In Leipzig, the soldiers tore the epaulettes off of the shoulders of the officers on the street and spit in their faces. At the same time, the dictates of the peace conditions were made public. The Leipzig professors did not, at first, pay attention, since they were outraged by the fact that a huge red flag had been hoisted on the city tower to a height that no one could reach. When the rector of the university, the ultra-hyper-national mathematician Otto Hölder,[12] was politely asked by a delegation of Red soldiers to put a red flag on the university as well, he quietly agreed. This was a man who had been shouting, only a few months before, that he was ready to sacrifice the last drop of blood for his royal house.

The war was over. The students who had not fallen in the war returned. The chief representative of the Ultra-German Nationals, the psychologist Krüger[13] who had not ventured away from his war-mad desk throughout the war years, was not embarrassed to address them, in his homily, as his "comrades in arms." When I happened to talk about how deeply sad it was that so many young German scholars had died, I received the answer, "but that does not matter, since many English scholars died as well."

The sudden and ill-prepared collapse, caused by the decision of the German military to call a ceasefire, came as a surprise to a population who had been psychologically trained by military propaganda over the preceding four years for the certain final victory. The war had boosted national sentiment and it is not surprising that the unexpected collapse had extreme internal consequences such as the strengthening of radical nationalist movements. However, it was not the German generals who were accused of broken promises and blamed for the failure of war. Blame was instead placed on the mutinous sailors who, in the last days of the war, refused to sail out

[12] Otto Hölder (1859—1937) first studied engineering at the Stuttgart Polytechnic and from 1877 at the University of Berlin where he switched to mathematics. In 1882 he received his doctorate at the University of Tübingen, went to the University of Leipzig and in 1884 became a private lecturer in Göttingen. In 1889 he followed a call to Tübingen where he was offered a professorship in 1890. From 1896 to 1899 he was at the University of Königsberg, 1899—1927 at the University of Leipzig (in 1918 as rector). At the end of the war in 1918, he was president of the German Mathematicians' Association. In November 1933 he signed the declaration of commitment to Adolf Hitler by the professors at German universities and colleges.

[13] Felix Krüger (1874—1948) habilitated in 1903 at the Institute for Experimental Psychology of Wilhelm Wundt at the University of Leipzig. In 1906 he was called to the University of Buenos Aires and headed the Laboratory for Applied Psychology there. In 1910 he was called to Halle and appointed head of the psycho-physical collection. In 1912/13, Krüger taught at Columbia University in New York. As far as his behavior during the war was concerned, Haas did him an injustice: Krüger volunteered for the war and took part in the battles at Verdun and Cambrai, the summer battle and the Romanian campaign. In 1917 he was appointed to succeed Wundt in Leipzig, where he very successfully headed the institute for the next twenty years. Despite or perhaps because of his war experiences, he retained his German-national sentiments and was an opponent of the Treaty of Versailles. In 1919/20, during the Kapp Putsch, he took part in the battles of the Leipzig volunteer regiment. During the Weimar Republic, he gave lectures in German ethnic and right-wing extremist circles and associations. Krüger welcomed the National Socialists' seizure of power as he hoped that the spiritual renewal of the German people was imminent. Yet, he did not share the anti-Semitism of the National Socialists and continued to supervise his Jewish doctoral students. From 1936 onwards, the Saxon Ministry of Education prohibited him from "holding lectures and exercises for the duration of the current semester." It was recommended that he give up the office of rector, for health reasons, which he did on January 8, 1936. In April 1938 his application for early retirement was granted.

with a fleet that had been blockaded for years; they were simply unwilling to go into final heroic battle against the superior British naval power and meet certain death. The plan had been that the *Nibelungentreue*, the loyal march to war on the side of the Habsburg Dual Monarchy, was to be followed by the *Nibelungentod*, the death of the loyal with the sinking of their fleet. It is not surprising that the sailors refused the fate intended for them. And yet, the refusal caused an uprising that triggered a general revolution in November 1918. However, this uprising was defined as a *Dolchstoß*, a stab in the back of the victorious army[14]—a legend that was spread by the very same generals, above all by Field Marshal Ludendorff who had just described the military situation as hopeless and had urged a ceasefire.

It is no wonder that the people felt they had been betrayed, that all of the victims and the suffering had been in vain; naturally someone had to take the blame. The war psychosis was widespread and protracted, and it was certainly not limited to the Leipzig professors. So, it was almost inevitable that no peace could be achieved. The war was still dragging on, especially in the East; Freikorps (militias recruited from former military troops) and other irregular units[v] fought for their ideals and conceptions, contesting borders that had been decided in the Versailles peace treaty. The new states of the former Habsburg Empire fought as well, for their borders and their reorganization, claiming their national interests. Minorities and minority populations within the former Dual Monarchy, especially the Jewish population, suffered and still suffer from the consequences of this war and the subsequent peace treaty, which was seen by many as a dictate rather than a treaty. It was "The Pity of War"[vi]that ended in a peace that meant the end of peace.[vii] And Arthur Haas' personal world was not exempt:

> War psychosis possessed the entire Leipzig faculty and no cure seemed to help. I could not possibly stay any longer in Leipzig but I was too committed to just abandon everything and walk away. I requested and received permission to finish the lecture of the winter semester at the end of January, and I left for Vienna. I went with the firm decision to teach in Leipzig for the summer semester of 1919 but then to not return, except for shorter visits. Financially, it was no loss since I did not receive any salary. I saw no chance that that would change in the foreseeable future.

A citizen of the Habsburg Monarchy, Arthur Haas was one of those victims of the war but he did not yet know it. First, he had to return home. It was then that the disintegration of the Habsburg Empire into various national states caused him to realize that he was suddenly no longer an Austrian. He had become a Czechoslovakian and thus a minority, not only as a part of the German-speaking population in Brünn but also as a Jew in a city now called Brno. He did, however, still believe himself to be Austrian and his first goal was to become so again. He would later recall:

> With the founding of the Czechoslovak Republic, I became a Czechoslovakian citizen because I was registered as having been born in Brünn. Yet, I welcomed heartily the proclamation of the German-Austrian Republic, on November 12, 1918, although, on the other hand, I felt sympathy for the tragedy of the last Austrian Emperor Karl. Both then and later,

[14] The notion of the *Dolchstoß* (stab in the back) also invoked the Nibelungen legend with its cowardly murder of Siegfried by Hagen von Tronje.

I have shared the view of a French historian, that Emperor Karl was the only honest man among the many leading statesmen of Europe at war.

But during the collapse of the old Europe, in a time that seemed to be without hope, Haas first considered leaving Europe and building a better future somewhere else.

My friend Harry Schmidt thought I was joking when I stated, in November 1918, that I saw only one way out for me, emigration to America. Just a few months later, my sister also thought I was joking when I told her that we were completely impoverished.

It may have been just an idea when he shared the idea of emigration with Harry Schmidt in Leipzig. But this idea increasingly solidified, especially when it became clear that the former, now independent states of the Habsburg Empire could not agree on a common policy and particularly not on a common fiscal policy, which would result in inflation and the collapse of monetary values. The fortune earned by Gustav Haas disappeared within a few months and the family seemed impoverished.

With this statement Arthur Haas' autobiographical notes, which I have followed thus far, come to an end. He did return to Vienna as planned, to start his career anew. This development will be followed in the next section.

Notes

i. See Martin Henke, *Flinke Funken im schnellen Spiegel* (Hamburg: Book on Demand, 2000), 152–153.

ii. Karl-Heinz Schlote, *Von geordneten Mengen bis zur Uranmaschine—Zu den Wechselbeziehungen zwischen Mathematik und Physik an der Universität Leipzig in der Zeit von 1905 bis 1945* (Frankfurt: Harri Deutsch, 2008), 107.

iii. Werner Bramke, Silvio Reisinger, *Leipzig in der Revolution von 1918/19* (Leipzig: University of Leipzig, 2009).

iv. Stefan L. Wolff, "Physiker im 'Krieg der Geister,'" 43.

v. Cf. Matthias Sprenger, *Landsknechte auf dem Weg ins Dritte Reich – Zu Genese und Wandlung des Freikorpsmythos* (Paderborn: Schöningh Ferdinand GmbH, 2008).

vi. Cf. Niall Ferguson, *The Pity of War* (London: Basic Books, 1999).

vii. Cf. Margaret MacMillan, *The War that ended Peace* (New York: Penguin Canada, 2013) and *Paris 1919* (New York: Random House, 2002).

Part II
Post-War Struggles 1920–1924

Chapter 9
Vienna Misery

Haas had already decided to leave Germany in January 1919, to resign from his professorship at the University of Leipzig and return to Vienna. He was tired of the increasing nationalism of his German colleagues who could not accept defeat. And he was homesick.

Going back to his homeland was, however, no longer an option. With the end of the war had come the disintegration of the Habsburg Empire. The end had been initiated in October 1918 with the founding of Czechoslovakia, the South Slav state (later Yugoslavia), and the state of German Austria. This was followed by the fall of Galicia to Poland and Hungary's withdrawal from the Union. The separation that resulted in individual states was sealed in 1919 with the Treaty of Saint-Germain. Further provisions were made with the Treaty of Trianon in 1920. These events and treaties marked the end of the Austro-Hungarian Empire as Haas had known it. With these developments, his personal situation became unclear because he had lost his Austrian citizenship and counted as a citizen of the newly founded state of Czechoslovakia. His hometown of Brünn was now called Brno, in which the German population became a minority. His new Czech passport is shown in Fig. 9.1.

Now a Czech citizen registered in Brno as notarized by the new administration (Fig. 9.2), Haas traveled under this new citizenship to Vienna where he needed to explore possibilities for being accepted as Austrian. It was 1919 and his vacation trips to the summer resorts in Carinthia had become difficult and only possible under limited conditions as the entries in his passport show. Europe was divided, her lands separated from each other. Where before the war there had been free movement and exchange, the borders were now nearly impenetrable. Even between what had formerly been nearby communities, considerable administrative effort was required to bridge newly established boundaries.

Although Haas now belonged to the Slovakian part of Czechoslovakia, he wanted to establish himself in Vienna where he saw his hopes. He had no affinity for the newly founded republic with its uncertainties about how and where it would develop. While he had supported greater independence for the Czech and Slovak populations, he had seen possibilities only within the framework of the Bohemian state as part

M. Wiescher, *Arthur E. Haas—The Hidden Pioneer of Quantum Mechanics*, Springer Biographies, https://doi.org/10.1007/978-3-030-80606-4_9

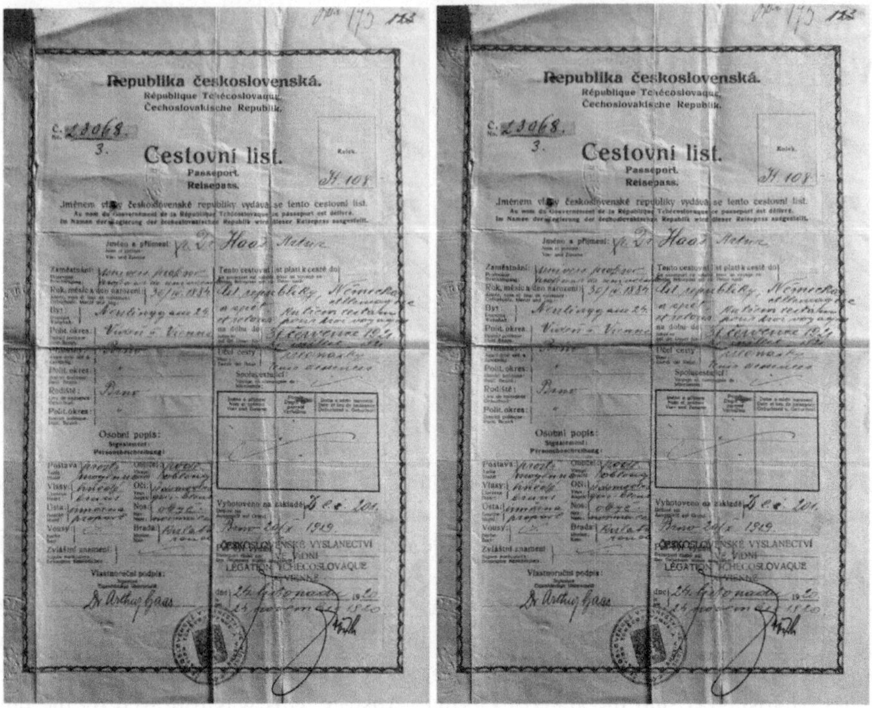
Republika československá.
République Tchécoslovaque.
Čechoslovakische Republik.
Cestovní list.
Passeport.
Reisepass.
Jménem vlády československé republiky vydává se tento cestovní list.
Osobní popis:
ČESKOSLOVENSKÉ VYSLANECTVÍ VE VÍDNI
LÉGATION TCHÉCOSLOVAQUE À VIENNE

Fig. 9.1 Czech passport issued in 1919 for Arthur Haas who was now a citizen of Czechoslovakia since he had been born in Brünn, which had been renamed as Brno

of a reformed empire. In addition, although he spoke a little Czech from his youth, Haas saw himself as a German speaking Austrian.

Beside these personal concerns, there were also political considerations that surfaced with the collapse of the empire. In the first censuses of the new republic, in 1921 and 1923, separate criteria were established for different ethnic and minority groups. In addition to Czechs and Slovaks who were in the majority, Germans, Poles, and Hungarians had to identify themselves as such through language affiliation. Jews also had to identify themselves, although not on the basis of language but on racial grounds. This segregation resulted from the shortcomings of the Treaty of Saint-Germain, which did not grant Jews their own nationality and made only vague statements about protection of minorities without the support of firm guarantees from the Entente powers. Despite these uncertainties, the Czech state under Masaryk, unlike other successor states of the Habsburg Empire, offered the Jewish population protection and concessions to a large extent.[i]

Whether concern about his Jewish origins played a role in Haas' considerations is unclear. If so, then probably only a minor one since the extent of the anti-Jewish movement was not yet fully foreseeable in 1919/1920. In addition, Haas was assimilated to such an extent that at the time he hardly considered himself as belonging

Land Mähren. Stadt Brünn.

Nr. 201

Heimat-Schein,

womit vom Stadtrate der Landeshauptstadt Brünn bestätigt wird, daß

Arthur Haas

Charakter und Beschäftigung Universitätsprofessor

Alter geb. am 30. 4. 1884 in Brünn

Stand ledig

in Brünn das Heimatrecht besitzt.

Eigenhändige Unterschrift der Partei: Dr Arthur Haas

Brünn, am 20. 10. 1919

Für die Gemeinde: I. A.

Ober-Magistratsrat

STADTRAT DER LANDESHAUPTSTADT BRÜNN

Fig. 9.2 This Brünn certificate of origin, issued in October 1919, was the legal basis for Haas' new citizenship

to this minority. He wavered. His world had disappeared and he began to think seriously about leaving Europe and looking for a future elsewhere, possibly in the United States. Nevertheless, he decided to return to what was then German Austria.

In the spring of 1919, with the permission of the Leipzig university authorities, he travelled to Vienna, Brno and back to Vienna to settle the question of his citizenship before the official end of the semester. He collected his certificate of origin and his new passport identifying him as a Czech citizen. With these papers Haas had the necessary documents to apply for a change of citizenship in Vienna. Many citizens of the former Habsburg Empire were making these applications, either to return to their homeland or to be allowed to remain as citizens in Vienna or other centers.

General conditions were in chaos. The army had disbanded; soldiers and even deserters only wanted to go back to their often-distant homeland. There was hunger; the old established trade and supply structures had collapsed due to the closing of the borders and people flocked to the countryside to buy from farmers. Trains were overloaded; long-distance connections were disrupted, and local railways were overcrowded and poorly coordinated as many railway officials sought to return to their homes, in their now new countries.[ii] All offices and especially the registries were understaffed and overwhelmed. So, Arthur Haas had to wait.

Uncertainty was waiting for him not only with regard to his nationality but also his future career plans. He hoped to be accepted again as a lecturer in Vienna in order to start his academic career anew. If this did not work out, Haas thought he could put his technical and physical knowledge to industrial use. The sugar industry of the Strakosch brothers was not an option at the time. The sugar refinery in Mödriz was lost and the factory in Hohenau in Lower Austria remained, from 1918 to 1922, under state management along with other companies.

Haas therefore worked together with the Viennese chemist Paul Tschelnitz[1] on a project for improving the electrical conductivity of metals and also less conductive materials such as carbon. Increasing electrical conductivity reduces electrical resistance, resulting in improved current transmission as well as lower power usage and heat loss. Haas saw a wide range of industrial possibilities such as reducing losses in high-current lines and electrical machines, improving submarine cable connections, introducing galvanic separation processes and, for domestic use, improving the light bulb's performance and lifetime.

The idea seemed simple; he wanted to chemically enrich the conductor material with small amounts of radioactive substances. As he states in his patent application:

> These substances emit α and β radiation that generates, by scattering and other energy-transfer interactions, free electrons in the conductor, thus increasing its conductivity.[2]

The energies released during α and β decay are indeed more than sufficient to excite electrons from the atoms or from the bound configuration of the valence band of a solid material, transposing them into the so-called conduction band. The problem that Haas encountered was in low-cost production of such materials. The patent specification explains this:

> In metallic conductors, the incorporation of the radioactive substances is purposefully achieved by alloying such materials with the metal. In order to prevent the radioactive substances from releasing energy into the surrounding air, a layer of non-activated metal of

[1] Paul Adolf Tschelnitz (1896—1963) was the son of Bela Tschelnitz (1863—1938), senior director of the S. Tschelnitz company, an oil refinery and grease factory. Many members of the Tschelnitz family were sent to concentration camps after 1938 because of their Jewish origin. Dr. Paul Tschelnitz survived and immigrated to the United States in 1950, where he took the name Paul Adolf Telsen.

[2] Patent application filed on October 9, 1919 by Arthur Haas and Paul Tschelnitz, submitted to patent lawyer Dr. Richard Reik in Wien, Franzensring 18.

such a strength is deposited on the metallic conductor, e.g., by electroplating, such that the radioactive radiation emitted into the surrounding air is suppressed.

Depending on the type of radiation and the material used, these would be micrometer to millimeter thin layers for absorbing radiation. More details are unfortunately not given in the application. It is an interesting idea but intensive radiation from high-energy decay processes does not only release electrons into the conduction band; there are also negative consequences. As was demonstrated in 1946 by the Hungarian physicist Eugene Wigner[3] in the Manhattan Project,too much radiation causes a disruption of the lattice structure of metals. This leads to impurities and thus to a decrease and not the hoped-for increase in the conductivity of metals.[4] These results were also confirmed in a later summary work by Wigner's student Frederik Seitz (1911—2008), regarding the influence of radiation on the thermal and electrical conductivity of various solid state materials.[iii] It is doubtful that Haas and Tschelnitz ever made preliminary experiments to support the predictions of the patent application but the incident demonstrates a further change in Haas' thinking—away from pure, value-free theoretical physics and toward the arena of physical applications. It would, however, take some time before this shift was complete.

While Haas was thinking about these problems, he returned from Vienna to Leipzig to give the lecture on modern quantum theory in the summer semester of 1919, as can be seen from university records.[5] In the following winter semester he offered a lecture on atomic physics with the topic, "Theory of Molecular Phenomena," but Haas also travelled to Vienna several times during these months to reestablish old connections and look for opportunities to resume his lecturing activities at the University of Vienna. In doing so, he also rebuilt old connections with the Philosophical Society and gave lectures on the topics of modern physics that had his attention during these years.

On May 5, 1919 he spoke on "Physics as a Geometrical Necessity," a topic occupying him in connection with his book on vector calculus that was published in

---

[3] Eugene Paul Wigner (1902—1995) was born in Budapest; the family was of Jewish origin and moved to Austria in 1919, in the wake of anti-Jewish developments. They did, however, return to Hungary after the political situation stabilized. Wigner studied at the Technische Hochschule in Berlin and later worked at the Kaiser Wilhelm Institute before accepting a position at Princeton in 1930. He later became one of the leading theorists on the Manhattan Project and worked in the subsequent non-military or semi-military nuclear reactor development programs. Eugene P. Wigner, *Survival and the Bomb* (Bloomington, Indiana University Press, 1969).

[4] This type of radiation effect, now known as the Wigner effect, has been studied extensively for a variety of materials. Since the 1950s, the effect in graphite has been part of reactor applications; the neutron flux in the reactor creates internal radioactivity by trapping neutrons in the surrounding materials. See E. P. Wigner, *Symmetries and Reflections - Scientific Essays* (Bloomington: Indiana University Press, 1967).

[5] Haas himself writes that he taught the theory of relativity in the summer semester of 1919, but university records show that he gave the lecture on relativity the previous winter semester and devoted himself to quantum theory in the summer, an understandable mistake given the confusion of the times. See Karl-Heinz Schlote, *Von geordneten Mengen bis zur Uranmaschine - Zu den Wechselbeziehungen zwischen Mathematik und Physik an der Universität Leipzig in der Zeit von 1905 bis 1945* (Frankfurt: Verlag Harri Deutsch, 2008), 109.

revised form two years later.[6] In early 1920, Haas was again in Vienna and gave a lecture on "Atomic Harmony of the Spheres of the Atom" at the Philosophical Society on May 11th. This topic was based on an analogy between the firmly defined orbits of the electrons in Bohr's hydrogen model and the orbits of the planets, which according to ancient ideas were linked to fixed harmonies:

> Once the Pythagoreans spoke of a harmony of the spheres, of a supernatural music that is said to be governed by simple, integer relations in space; today we know that the dream of the Pythagoreans is, in fact, that world in miniature; it is realized in the atom, whose chords we do not hear but recognize in the photographs of the spectra.[7]

This theme guided the new direction of Haas' scientific thinking, which was moving away from the history of physics and classical mechanics and toward the new atomic physics. He was also increasingly interested in this connection between atomic physics and cosmological questions, such as the parallels between the microcosm and the macrocosm. On December 20, 1921, Haas was back in Vienna and gave a lecture on "The Problem of Primordial Matter in Modern Physics." This is a topic that he had already addressed earlier when he argued against the infinity of the universe, based on limitations in the radioactive decay chains.

In the philosophy of the pre-Socratics, primordial matter is the "One" or the "Being" from which everything arises. Haas conceived of the atom as this primordial matter since, according to the latest findings of the Rutherford scattering experiment, the atom has its mass concentrated in its center, the atomic nucleus. Based on the charge equivalence of the positive atomic nucleus and the surrounding shell of negative electrons, Haas considered the concept of electricity or the electric field as a possible primordial substance. As he states:

> In electricity, the new physics can see the uniform primordial matter that researchers have been searching for thousands of years and from which all things that can be perceived by the senses are formed.[8]

This concept of primordial matter giving birth to particles and waves in our universe could be interpreted as a first glance into the modern concept of string theory as an ultimate way to express the nature of matter and all its interactions.

In Vienna, the city had transformed from the capital of a huge empire to the giant hydrocephalus of a small republic; a large part of the intellectual life remained but instead of flourishing and glittering, it had become dreary and grey. New structures, ideas and initiatives had to be developed and Haas wanted to join in this rebuilding. He resumed his old practice of giving public lectures on modern aspects of science at Urania in the winter of 1919/1920. He decided to publish his lectures in a book for laymen, *Das Naturbild der Neuen Physik* (The Natural Image of the New Physics), which went into print in Leipzig in 1920. The publishing house he had previously used, Veit & Comp., had been purchased by a new publisher, De Gruyter, in 1919

[6] Arthur Haas, "Die Physik als geometrische Notwendigkeit," *Naturwissenschaften* 8, (1920): 121–127.

[7] Arthur Haas, *Das Naturbild der neuen Physik* (Bern-Leipzig: De Gruyter, 1920), 91.

[8] Ibid., 35.

along with all of the publishing rights. De Gruyter thus became the main German publisher of Arthur Haas' books for the next twenty-five years.

His new book covered all aspects of modern physics without mathematical formalism and in easily understandable language. The electromagnetic theory of light was followed by molecular statistics, thermodynamics in the Boltzmann sense and electron theory—the current understanding of atoms and radioactivity for that time—relativity, and finally quantum theory. He writes on the last two topics:

> Like the theory of relativity, quantum theory, which emerged almost simultaneously with it, is not a branch of physics but physics itself, treated from a completely new angle.[9]

The book became a great success demonstrating that despite the difficult post-war years, the public's interest in scientific questions had not only been maintained but had even increased. The second edition was published by De Gruyter as early as 1924 and the third, revised edition in 1932. The book was translated into English as early as 1923,[10] with a second edition published in 1930.

Haas had to return to Leipzig for the summer semester of 1920—even if it was only to keep the Venia Legendi—and he gave a lecture on the same topic "Das Naturbild der Neuen Physik." He requested leave for the following winter semester, a time that he again spent in Vienna, but gave a final lecture in Leipzig on Einstein's theory of gravity in the following summer semester of 1921.

Haas made a list of lectures that he intended to prepare for the following five years, in the event that he stayed in Leipzig. He had chosen the topics: "Electrons, Atoms, Molecules"; "The World View of Modern Physics," in which atomic theory, relativity and the physics of stars were to be covered; "The Foundations of Modern Atomic Theory"; and "The Periodic System of Elements from the Point of View of Modern Atomic Theory."[iv] This wide range of scientific topics was at the forefront of physics, again underlining his broad interest in modern science.

However, it did not come to this. In July 1921, toward the end of the summer semester of 1921, he received official permission from Vienna to become an Austrian citizen. With the end of his scheduled lectures, Haas also ended his lecturing activities at the University of Leipzig and immediately returned to Vienna. He did, however, keep his contract as an unpaid lecturer and returned to Leipzig in the winter semester of 1922 to give at least one of the promised lectures. He maintained his connections with the University of Leipzig until 1927, providing some advice and counsel to the new editors of the *Poggendorff Handbook*. Only in 1927, after Haas had received his permanent position at the Vienna Academy of Science, did he officially resign.

With his Austrian citizenship (Fig. 9.3), Haas again had the opportunity to give lectures in Vienna as a private lecturer (Fig. 9.4), an opportunity he used in the winter semester 1921/22. He gave this lecture on vector analysis and its application in physics, selecting specific examples from the theories of classical mechanics and electrodynamics as well as the modern theory of relativity.

---

[9] Ibid., 69.

[10] Arthur Haas, *The New Physics—Lectures for Laymen and Others* (London: Methuen & Co, 1923).

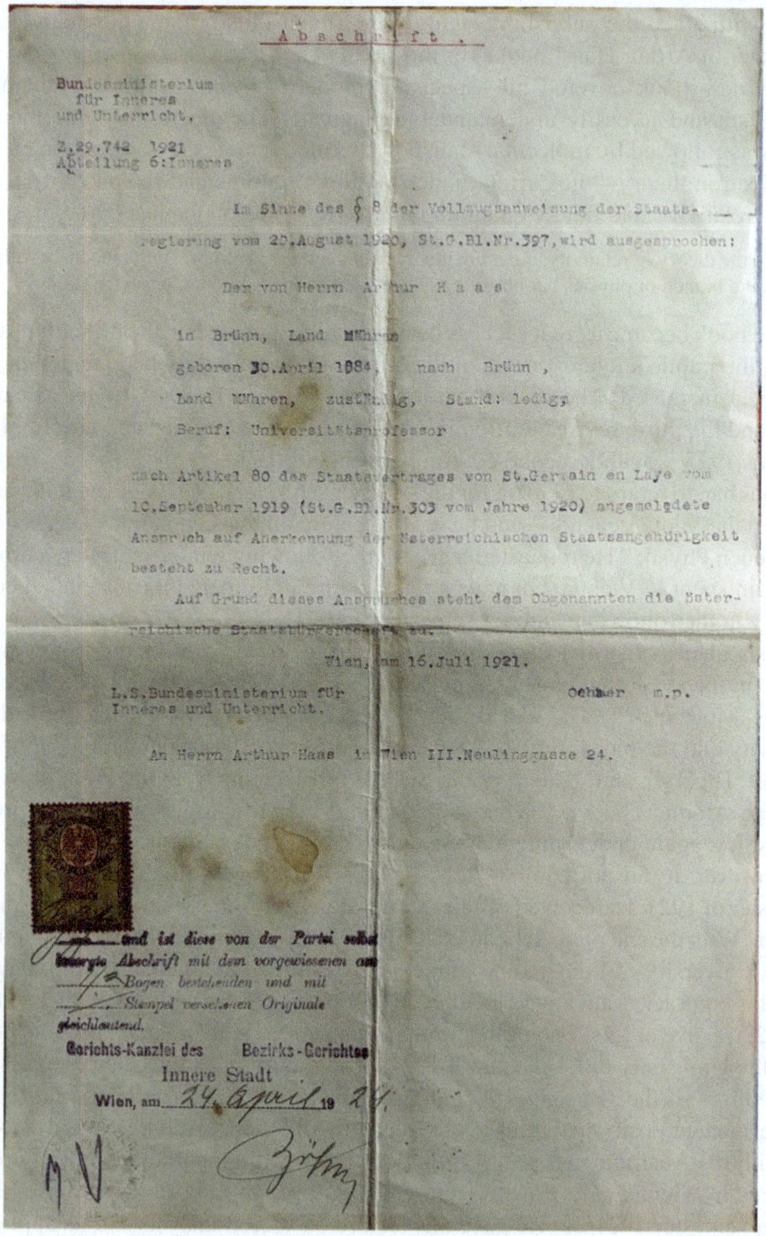
Abschrift.

Bundesministerium
für Inneres
und Unterricht.

Z.29.742 1921
Abteilung 6:Inneres

Im Sinne des § 8 der Vollzugsanweisung der Staatsregierung vom 20.August 1920, St.G.Bl.Nr.397,wird ausgesprochen:

Der von Herrn Arthur Haas
in Brünn, Land Mähren
geboren 30.April 1884, nach Brünn,
Land Mähren, zuständig, Stand: ledig,
Beruf: Universitätsprofessor

nach Artikel 80 des Staatsvertrages von St.Germain en Laye vom 10.September 1919 (St.G.Bl.Nr.303 vom Jahre 1920) angemeldete Anspruch auf Anerkennung der österreichischen Staatsangehörigkeit besteht zu Recht.

Auf Grund dieses Anspruches steht dem Obgenannten die österreichische Staatsbürgerschaft zu.

Wien, am 16.Juli 1921.

L.S.Bundesministerium für Inneres und Unterricht. Oehmer m.p.

An Herrn Arthur Haas in Wien III.Neulinggasse 24.

... und ist diese von der Partei selbst erzeugte Abschrift mit dem vorgewiesenen aus 1/2 Bogen bestehenden und mit Stempel versehenen Originale gleichlautend.

Gerichts-Kanzlei des Bezirks-Gerichtes
Innere Stadt
Wien, am 24. April 1924.

Fig. 9.3 Declaration of Austrian citizenship for Arthur Haas in July 1921

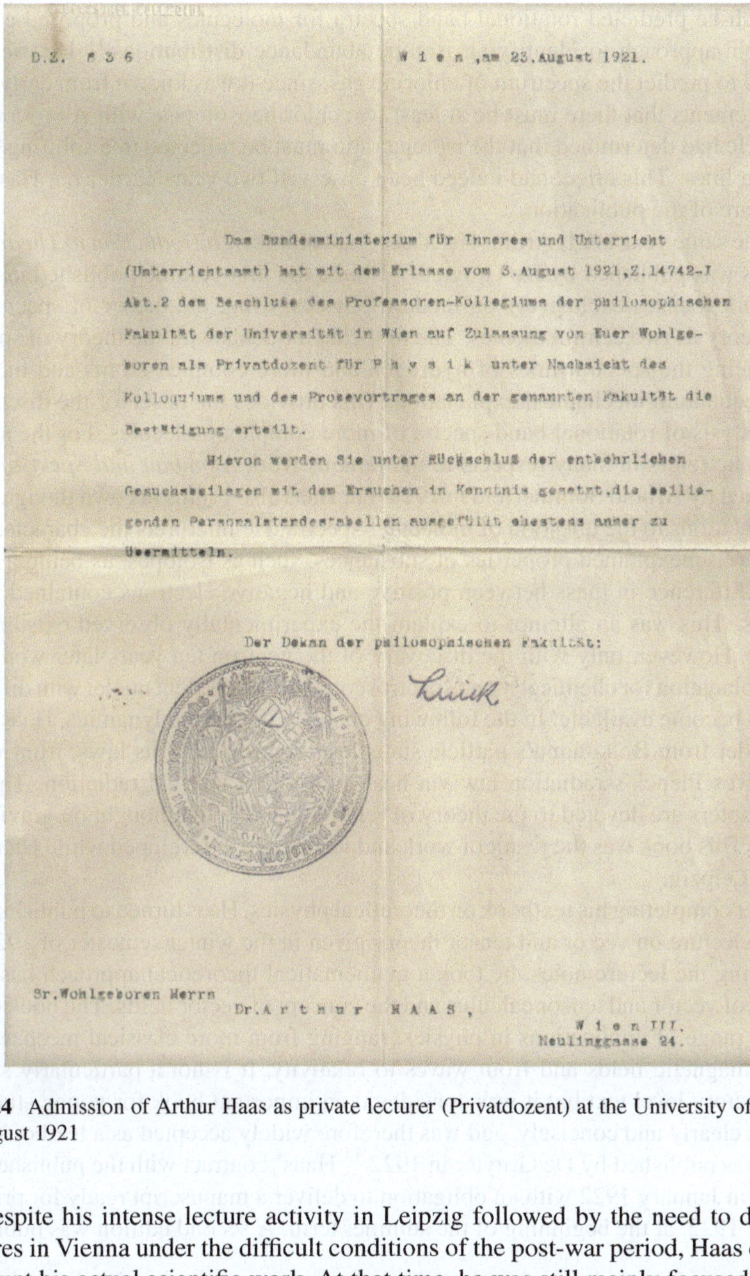

D.Z. # 3 6 W i e n ,am 23.August 1921.

Das Bundesministerium für Inneres und Unterricht (Unterrichtsamt) hat mit dem Erlasse vom 3.August 1921,Z.14742-I Abt.2 dem Beschluße des Professoren-Kollegiums der philosophischen Fakultät der Universität in Wien auf Zulassung von Euer Wohlgeboren als Privatdozent für P h y s i k unter Nachsicht des Kolloquiums und des Probevortrages an der genannten Fakultät die Bestätigung erteilt.

Hievon werden Sie unter Rückschluß der entbehrlichen Gesuchsbeilagen mit dem Ersuchen in Kenntnis gesetzt,die beiliegenden Personalstandestabellen ausgefüllt ehestens anher zu Ueermitteln.

Der Dekan der philosophischen Fakultät:

Luick

Sr.Wohlgeboren Herrn

Dr.A r t h u r H A A S ,

W i e n III.
Neulinggasse 24.

Fig. 9.4 Admission of Arthur Haas as private lecturer (Privatdozent) at the University of Vienna in August 1921

Despite his intense lecture activity in Leipzig followed by the need to develop lectures in Vienna under the difficult conditions of the post-war period, Haas did not interrupt his actual scientific work. At that time, he was still mainly focused on the connection between the quantum nature of the atom and the results of experimental spectroscopy. Haas imagined ever greater precision in his theoretical predictions, for example regarding the spectroscopy of molecules. These topics were included in his lectures and he also sought to publish his ideas. In 1921 he wrote a paper in which he predicted rotational band spectra for molecules and proposed a spectroscopic approach to identifying isotope abundance distributions.[11] He used this method to predict the spectrum of chlorine gas, since it was known from early mass measurements that there must be at least two chlorine isotopes, with A = 35 and A = 37. He had determined that the isotope ratio must be reflected in a splitting of the rotation lines. This effect had indeed been observed two years earlier but Haas was not aware of the publication.[v]

In the same year 1921, Arthur Haas' second volume of *Introduction to Theoretical Physics* was published by De Gruyter.[12] Unlike the first volume published in 1919, this work focuses exclusively on modern physics, using the theory of spectra and heat theory as two central themes. The first chapter deals with the theory of spectra, introducing the spectral lines of hydrogen and more complex atoms and includes Bohr's quantum mechanical explanation. This provides the basis for the discussion and analysis of rotational band spectra of more complex molecules. For the atomic spectra he refers in many respects to the famous work *Atombau und Spektrallinien* published by Arnold Sommerfeld in 1919, but Haas also brings his own thoughts and interpretations to the question of molecular spectra. He interprets the characteristics and as yet unexplained properties of substances, such as isotopes, as being a result of the difference in mass between positive and negative electrons contained in the nucleus. This was an attempt to explain the experimentally observed existence of isotope. However, only with the discovery of the neutron ten years later would the real explanation for chemically and spectroscopically equivalent nuclei with different masses become available. In the following chapters on thermodynamics, Haas leads the reader from Boltzmann's particle statistics to entropy and its laws, from which he derives Planck's radiation law via heat theory and thermal radiation. The last two chapters are devoted to the theory of relativity and a first thought on gravitation theory. This book was the result of work and lectures that developed while Haas was still in Leipzig.

After completing his textbook on theoretical physics, Haas turned to publishing his Vienna lecture on vector and tensor theory given in the winter semester of 1921/22. Following the lecture notes, he took a mathematical theoretical approach based on the use of vector and tensor calculus and the concept of vector fields. The book offers a wide range of applications in physics, ranging from more classical mechanics to electromagnetic fields and from waves to relativity. It is not a particularly simple introductory level text but it was considered an important book for young students, written clearly and concisely, and was therefore widely accepted as a textbook. The book was published by De Gruyter in 1922.[13] Haas' contract with the publisher was signed in January 1922 with an obligation to deliver a manuscript ready for printing in May 1922, at the beginning of the summer term. A second edition was published

[11] Arthur Haas, "Rotationsspektrum und Isotopie," *Zeitschrift für Physik* 4, (1921): 68–72.

[12] Arthur Haas, *Einführung in die Theoretische Physik II* (Berlin-Leipzig: De Gruyter, 1921).

[13] Arthur Haas, *Vektoranalysis – in ihren Grundzügen und wichtuigsten physikalischen Anwendungen* (Berlin-Leipzig: De Gruyter, 1922).

Fig. 9.5 "German Publisher Scum," Arthur Haas' opinion of his German publishers after he received less than 1,000 marks in royalties for his first two books and the new edition of his third book

by De Gruyter in 1929. It was undoubtedly an outstanding achievement to get the book ready for printing within a few months.

But time was pressing. Concerning the financial arrangement that Haas had made with the publisher, the latter in a letter dated November 1922, agreed to pay the author a total of 1,000,000 marks in outstanding payments for 3,000 volumes of *Introduction to Theoretical Physics* and 1,500 copies of *Vektoranalysis*. This included the obligation to submit a revised version of the first volume of *Introduction to Theoretical Physics* by March 1923.

The apparently enormous sum of one million marks is explained by rapidly growing inflation in the German Reich. Inflation was partly caused by economic misery in the wake of the lost war but also by having to pay war reparations to the victorious powers as agreed to in the Treaty of Versailles of 1919. According to the treaty, the reparations were to be covered in gold currency and foreign exchange but the state obtained the fees internally by printing additional paper money. The additional paper money was also thought to help in quickly settling the enormous debt burden of the Reich, 150 billion marks. All of this led to inflation.

In January 1920, the Mark was only a tenth of its pre-war value, in October 1921 only a hundredth, and in the autumn of 1922 only a thousandth. Thus, the amount that De Gruyter paid to Haas was equivalent to a purchase value of 1,000 marks. Even this income quickly evaporated when hyperinflation broke out the following year in 1923, reducing the amount to mere pennies. It is no wonder that Haas described the corresponding folder in his files as "German Publisher Scum," as seen in Fig. 9.5. He had no choice but to sell his work at these prices. Rapid inflation and the increasing devaluation of his assets, invested in government securities and shares, forced him to continue writing and publishing in order to be able to live on the subsequent royalties, which he calculated in a letter of complaint to the publishers as 1½% of the sales income.[14] It was of some help that in 1923 the second edition of his very successful book on theoretical physics was published and became a widely used textbook by many universities in Austria and Germany.

[14] De Gruyter charged a retail price of M 17.50 for the paperback and M 20.50 for the hardbound version of the second volume of Haas' *Introduction to Theoretical Physics*.

Loss was widespread and he was not alone; the bourgeois class had placed a large part of its wealth in supposedly safe war investments with hopes of receiving large returns after victory. So had Uncle Fritz who managed the family's assets after the death of Yella Haas, investing primarily in war bonds. These bonds were useless after the lost war and a large part of the family fortune evaporated. In 1924, Haas commented on the situation in a letter: "It cannot be denied that the catastrophe has robbed me of two-thirds of my fortune and one has to cut back accordingly!"

Even worse than speculation losses, with the collapse of the Habsburg Empire and its economic structure an economic crisis developed. The newly founded individual states with their various economic interests, as well as the fragmentation of currencies and bank assets, formed the reason and background for the crisis. In the peace treaties, Austria and Hungary were held responsible as the sole legal successors of the Monarchy and had to bear the costs of reparations. In addition, all other successor states made their own claims on the remaining assets of the former Dual Monarchy. Thus, there was not much left for the claims made by the victorious powers, claims payable in gold currency. The announcement of the peace terms led to a collapse in the value of the crown, with inflation reaching almost 100% per month. The Austrian government increased the production of paper money to finance its budget, which led to hyperinflation as in the German Reich.

In the following years, the devaluation of money increased at an incredible rate. At the beginning of 1920, there were 12 billion crowns in circulation and by August there were already more than one trillion, a 100-fold amount. Political missteps, such as the unofficial referenda on unification with the German Empire in the west of Austria, caused additional trouble. The referenda ended with a clear vote to join Germany, which angered the victorious powers who then stopped granting credit. Consumer prices doubled every month and, in the summer of 1922, reached 5,000 times their pre-war level.[vi] All in all, the level of consumer prices rose more than 5,000 times between 1914 and 1922; within the following two years, they rose almost 14,000 times. While hyperinflation reduced the burden of government debt, it also impoverished the holders of war bonds and thus the bourgeoisie, eliminating money that the Haas family had counted on.

Haas was forced to take a closer look at his own assets and those of his younger sister. Margarethe Haas, commonly known as Gretl, had married the civil servant Edi Gerstenberger and she lived with her husband and daughter in Sattendorf, a small village in Carinthia near Klagenfurt. Arthur and Otto Haas discussed among themselves and with their sister how best to invest and preserve the various shares of their assets, as documented in numerous letters. But these steps did not help; the sibling's financial resources were in decline. Escalating difficulties were expressed in a growing number of pointed remarks between the siblings. Thus Arthur wrote to his sister in November 1922, in big brother manner:

> But your letter of November 10th only confirms to me that you are very careless in your financial affairs, and I hope that at least from now on you will take more care of them.

Edi Gerstenberger tried at that time to take advantage of his wife's fortune, investing in a small company for the production of cyclecars[15] as well as in the corresponding construction of a warehouse in Sattendorf. The company failed with heavy losses and Arthur commented in an extraordinarily strong letter:

> Yesterday's family council, which included Uncle Fritz, Uncle Felix, Otto and myself, unanimously agreed that it would not make sense for EDI to use any of your assets or our capital to run his business... and the financial sacrifices you are making for the realization of EDI's "plans" must necessarily come to an end, while sacrifices will probably have to be made in liquidation.[16]

The liquidation costs amounted to 57 million crowns. Reminders for fiscal responsibility followed previous reminders and this correspondence continued throughout the inflationary period and beyond. Clearly Arthur was devoting a considerable amount of his time not only to physics but also to buying and selling stocks, bonds, and other securities in order to save his remaining assets as well as those of his sister. The individual buying and selling transactions are too complex to be pursued further in this framework. But with the eventual stabilization of currency through the introduction of the shilling in December 1924, their personal loss though considerable was not as great as initially feared. Arthur Haas, together with his brother Otto, invested their remaining money largely in material assets, securing more for their livelihood than they had thought probable.

To this end, Arthur and Otto carried out a rigorous exchange of business letters during these critical years. In the meantime, Otto had married in 1919 and taken over his father's law firm at Stubenring 14. As a practicing lawyer, Otto joined in partnership with Uncle Fritz and after his retirement, with a Dr. Beck. Here too, the main issue was the state of the firm's and family's assets. These were to be preserved by investing in fixed values, such as houses and land, and by speculating on shares. Arthur made some quick short-term purchases and sales but with little or no profit. While Arthur was reproaching his sister Margarethe for her bad investments, Otto reproached Arthur for his not fully thought through financial transactions that brought losses instead of profits. Yet, Arthur still seemed largely optimistic about the future while, as the lawyer, Otto prepared the monthly accounts. Thus, on August 22nd, Otto sneered sarcastically at Arthur for his balance-sheet estimates:

> I think your financial figures are very optimistic... I arrive at a significantly lower final sum than you. It may be that, as a result of your greater "chutzpah" and persistent application of the principle of not losing sleep over money, you have achieved a significantly greater increase in assets in recent months than I have.[17]

---

[15] A cyclecar was a small, light and inexpensive car produced in Europe and the United States between 1910 and the early 1920s. The purpose of cyclecars was to fill a gap in the market between the motorcycle and the car. Edi Gerstenberger wanted to establish his own Austrian company and also manufacture for the tourism industry in Carinthia.

[16] Arthur Haas to Margarethe Gerstenberger, letter from October 4, 1923.

[17] This is an interesting statement that reflects Arthur in his younger years as very relaxed in matters of money, a habit that definitively changed only two years later.

Only on a few occasions did a personal word find its way into their numerous letters about buying and selling securities. The brothers planned several fixed investments because, with prices and shares falling, fixed investments such as houses were the remaining cheap option. However, the market was flooded with speculators who had preceded them and so in 1922, the only successful purchases made were two apartment buildings in Brno and several apartment buildings in Berlin.

The brothers hoped to balance the cost of financing with rental income but the onset of inflation in Germany quickly frustrated this plan; rents were paid regularly but the money had lost its value, while maintenance and repair costs rose rapidly. On top of this came costs for a legal representative in Berlin as well as costs for local administrators. The burden could not be sustained for long. Constant concern and struggle followed as the Haas brothers tried to sell these tenements for at least some profit.

In December 1923, Otto Haas had the real-estate department of the Berlin department store Wertheim appraise the houses in US dollars, in order to have a reliable yardstick in this time of strongly fluctuating prices. The Berlin property on Seesenerstraße was valued at $8,000, on Tauroggenerstraße also $8,000, on Wielandstraße the value was $6,000 and the apartment building on Landsbergerstraße $2,500. The house in Kriemhildstraße and the two Brno houses were not valued due to local uncertainties. Plans, especially on Arthur Haas' side, to sell the houses in order to invest the proceeds in safe securities with gold backing failed because in such times the shortage of money was too great to carry out such transactions.

The winter semester of 1922/23 took Arthur back to Leipzig where he was desperately looking for a permanent position. Now, with his rapidly shrinking fortune, he preferred a bad offer in Leipzig to the completely unpaid lectureship in Vienna where the hoped-for promotion to an extraordinary professorship was not forthcoming. Over the years, considerable resistance had developed to the appointment of Jewish professors. This was the case not only in Vienna but at all universities in Eastern Europe and especially in the successor states of the Habsburg Empire. Legally discriminatory steps against Jews were demanded, above all by radical right-wing student fraternities and anti-Semitic professors and lecturers.

This pressure was partly a consequence of political and social changes that had accompanied the collapse of the Empire. In addition, great uncertainty in the academic job market was coupled with the prevalent assumption, often reinforced by propaganda, that Jewish academics dominated and blocked employment opportunities.[vii] As a result, many of the students and young academics of Jewish origin went - until 1933 - to the much more liberal Weimar Republic of Germany, seeking better conditions and better career paths.[18] Such resistance to the appointment of new Jewish professors was common in Vienna, as will be discussed in more detail later, and it is possible that this is also why Haas considered the step of returning

[18] These often highly talented scientists—including Eugene Wigner and Viktor Weißkopf, to mention only two examples—moved to the United States after 1933, where they played an essential role in the development of the scientific structure and also in the subsequent development of nuclear weapons and the nuclear industry. See Gordon Fraser, *The Quantum Exodus—Jewish Fugitives, the Atomic Bomb, and the Holocaust* (Oxford: Oxford University Press, 2012).

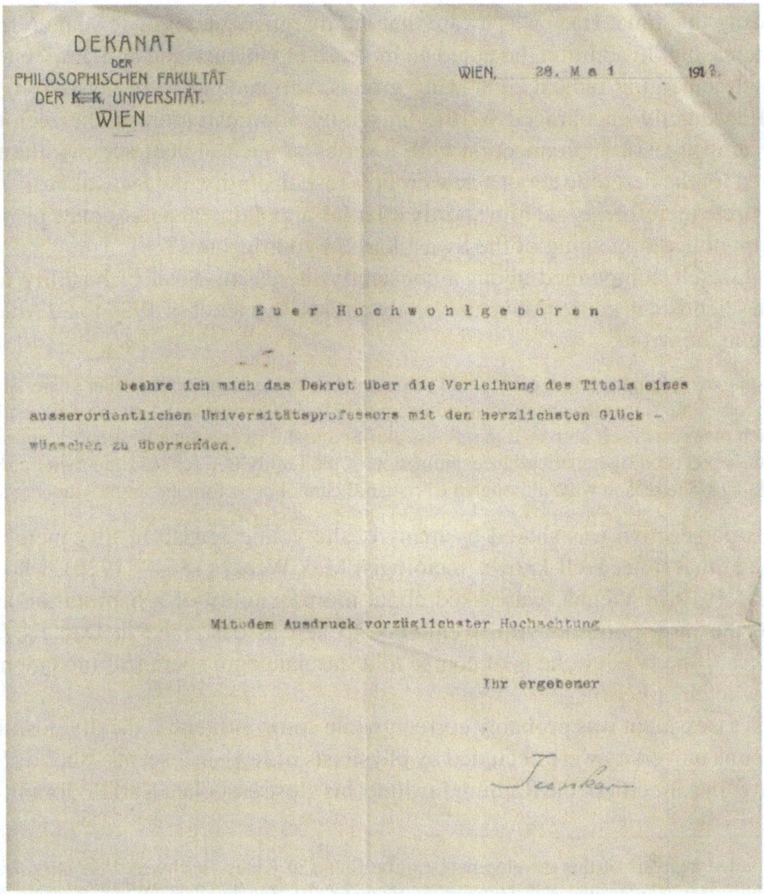
DEKANAT
DER
PHILOSOPHISCHEN FAKULTÄT
DER K. K. UNIVERSITÄT
WIEN

WIEN, 28. Mai 1923.

Euer Hochwohlgeboren

beehre ich mich das Dekret über die Verleihung des Titels eines ausserordentlichen Universitätsprofessors mit den herzlichsten Glück - wünschen zu übersenden.

Mit dem Ausdruck vorzüglichster Hochachtung

Ihr ergebener

Fig. 9.6 Letter from the Dean's Office, University of Vienna dated the beginning of the summer semester 1923, informing Haas about receiving the title Associate University Professor

to Leipzig. For, despite his academic successes, he saw his hopes of a permanent appointment dwindle.

Perhaps as a result of this attempt, the University of Vienna took action; Haas' lectures were considered successful and his possible emigration to Leipzig caused the university concern. In 1923 with the beginning of the summer semester, Haas was appointed Associate Professor at the University with the letter shown in Fig. 9.6.

This was not, however, a satisfactory arrangement for Haas; as in Leipzig, it was not a budgeted post. He had already achieved academic rank a decade earlier in 1913, at the University of Leipzig. This lack of progress did not reflect the pace that the young Haas had once imagined for himself in his academic career. In addition, he was forced to earn money by other means since his inheritance was gone and, while the position involved teaching, the post still came without salary.

During this time, Haas was aware that the marginalization of Jewish colleagues was increasing. In addition, he noted an increase in violence and anti-Jewish actions by newly emerging radical right-wing groups, actions that were not only ignored but often quietly encouraged by the university administration or the rectorate.[19] In December 1922, in connection with a series of such violent actions, the media reported for the first time about a new group's responsibility: the Swastikans.[20] Haas' sister Grete therefore asked him shortly after his appointment as associate professor: "Do you notice something of the Swastikans in your lectures?".[21]

He himself complained about a constantly increasing level of hostility toward Jewish scientists in conservative university circles. In a letter of 1924 to an American colleague, he wrote:

> Despite everything, I am denied a paid professorship at German and Austrian universities. The reason for this is the chauvinism now raging in these German and Austrian universities, which manifests itself above all in ruthless anti-Semitism. Every consideration for appointment is preceded by a meticulous examination of the family tree for absolute racial purity, and that is the reason why, although a Christian, I cannot hope for paid employment.[22]

This perspective was shared by many of the young Jewish faculty members in Vienna at that time. Well-known sociologist Max Weber (1864—1920), who spent the year 1919 in Vienna, was asked about the possibility of a habilitation and he quoted the inscription above the entrance to hell as described in Dante's *Divine Comedy*: "If he is a Jew, he is of course told 'lasciate ogni speranza' (abandon hope all ye who enter here)."

This assessment was probably correct; while some influential faculty positions at the Vienna university were occupied by physicists of Jewish descent—such as Stefan Meyer,[23] the director of the Radium Institute, his close associate Karl Przibram,[24] and

---

[19] A detailed analysis of this development can be found in Klaus Taschwer, *Hochburg des Antisemitismus Der Niedergang der Universität Wien im 20. Jahrhundert* (Wien: Czernin Verlag, 2015).

[20] See "Pack schlägt sich …," *Arbeiter-Zeitung,* 14.12.1922 (mit antisemitischen Untertönen: 'Hakenkreuzler - gegen»Hakennäsler', d.A.), zitiert nach Klaus Taschwer, "Nachrichten aus der antisemitischen Kampfzone - Die Universität Wien im Spiegel und unter dem Einfluss der Tageszeitungen,1920–1933," in *Reichweiten und Außensichten,* eds. Margarete Grandner and Thomas König (Göttingen: V&R Academic, 2015), 110.

[21] Margarethe Haas to Arthur Haas, March 24, 1923.

[22] Arthur Haas, "Letter to Willis R. Whitney—first director of the GE Research Lab, Schenectady, NY, USA," from July 9, 1924.

[23] Stefan Meyer (1872—1949) was one of the pioneers in radioactivity research. He studied at the University of Vienna as a student of Ludwig Boltzmann and Franz Exner. With the discovery of radioactivity, Meyer concentrated on this field of work and in 1910 became director of the newly established Radium Institute in Vienna. In 1939 he was retired because of his Jewish origins but survived the persecution of the Jews, hidden in Bad Ischl where he died in 1949. See Wolfgang L. Reiter, "Stefan Meyer: Pioneer of Radioactivity," *Physics in Perspective* 3 (2001): 106–127.

[24] Karl Przibram (1878—1973) is one of the pioneers of early solid-state physics. As a student of Ludwig Boltzmann and Franz Exner, he dedicated his early work to the condensation and conductivity phenomena in gases and to Brownian motion. Under the influence of Stefan Meyer, his lifelong interest in mineralogy and the influence of radioactive radiation on minerals began. Because of his

Fig. 9.7 Left: Professor of Paleontology and Paleobiology Dr. Othenio Abel, as rector in 1932/33. Right: The seminar room of the Institute of Paleontology where the Bear Cave Faculty met regularly to discuss the latest personnel developments at the University and to agree on their collective coordination tactics

Felix Ehrenhaft,[25] professor of physics and director of the 3rd Institute of Physics—there was a growing resentment against the "Jewish element." This was tirelessly pursued and propagated by a network of eighteen conservative anti-Semitic faculty members, the so-called Bear Cave of the Faculty of Philosophy. From the early 1920s, the interventions and agreements of this group made it almost impossible for Jewish or left-wing academics to habilitate or obtain a habilitation in philosophy or science at the University of Vienna.

The Bear Cave Group was named after their meeting place in the Institute of Paleontology, which was decorated with several stuffed bears.[viii] The dominant member of the group was the paleontologist Othenio Abel (1875—1946), known as the founder of paleobiology and pictured with the Bear Cave in Fig. 9.7.[ix] Abel played an important role as Dean of Philosophy from 1927—1928 and as rector of the University from 1932—1933, and he had a far-reaching influence on all personnel decisions for the entire university.

Jewish descent, Przibram had to leave Austria in 1938; he found refuge in Belgium and returned to Austria in 1946 as professor of experimental physics at the University of Vienna. See Wolfgang L. Reiter, "Karl Przibram: Radioactivity, Crystals, and Colors," *Physics in Perspective* 21 (2019): 163–193.

[25] Felix Ehrenhaft (1879—1952) was an Austrian physicist who also studied in Vienna under Exner and received his doctorate in 1903 with a thesis on metal colloids. In 1905 he completed his habilitation, in 1912 he became associate professor, and in 1920 full professor and chairman of the newly founded III physics institute of the University of Vienna. He carried out experiments on Brownian molecular motion and, complementary to Millikan, on Caltech's charge measurements on the smallest particles. Whenever Einstein came to Vienna, he was a guest of Ehrenhaft's. In 1938, Ehrenhaft emigrated as a Jew, first to England and then to the United States. In 1947 he returned to Vienna as a visiting professor. Joseph Braunbeck, *Der andere Physiker – Das Leben von Felix Ehrenhaft* (Wien, Leykam Verlag, 2003).

In physics, the group successfully prevented the habilitation of Karl Horowitz (1892—1958) who later emigrated to the United States and became professor at Purdue University. Another case was Otto Halpern (1899—1982) who was rejected in his attempt to habilitate in theoretical physics.[x] In 1928 he was awarded a Rockefeller Fellowship, which enabled him to work with Werner Heisenberg (1901—1976) before emigrating to the United States. Halpern held professorships at New York University, then MIT and, after the war, at Berkeley.

In Vienna, there were actually no longer any opportunities for Jewish scientists to have an academic career and in 1928, Hans Thirring[26] advised his aforementioned student Victor Weisskopf to move to Göttingen and work with Max Born in the less hostile environment of Germany. The limited number of Jewish or liberal faculty members—including Meyer, Ehrenhaft, Mark, and Thirring—were in no position to oppose the coordinated majority vote of the conservative Bear Cave Group.

Haas became increasingly desperate; his academic career seemed to be blocked despite his scientific and literary successes. He saw no possibility of obtaining a regular paid teaching position either in Austria or in Germany.

Due to the success of his latest books, he decided to concentrate on writing about scientific topics for the public, in order to make a living without having to deplete his remaining financial reserves. As early as 1924, he published a general presentation of the latest results in atomic physics[27] but this presentation was soon outdated by new research results of an experimental and theoretical nature. Therefore, already in 1929, a completely updated and extended new edition of the book was published with a total circulation of 3,000 volumes. Approximately two-thirds of this book was new and the other third was substantially reformulated. While the first edition was mostly based on a largely empirical description of newly observed phenomena, from X-rays to radioactive decay with electron theory as an attempt to explain them, the second edition was already completely based on a quantum mechanical description of the observed phenomena, including the new wave theory of his friend Erwin Schrödinger. In addition, substantial deletions were made from the old text so that one can speak of a largely new book.

Also, in 1924, the second edition of Haas' book, *Naturbild der Neuen Physik* was published and this too appeared in a considerably expanded form. While the first version was printed in 1,500 copies, the second edition arrived with 3,000 copies.

He was successful as a scientific writer but his hopes still lay in the possibility of getting a fully paid professorship; without this he saw no prospects for a professional future (Fig. 9.8). Therefore, Haas tried once again to leave Vienna where living and

[26] Hans Thirring (1888—1976) was professor of theoretical physics in Vienna and worked in particular on the formulation and improvement of general relativity. He was a student and successor of Friedrich Hasenöhrl as director of the Institute of Theoretical Physics until 1938, when he was retired early because of his pacifist attitude and his open opposition to the annexation of Austria to the German Reich. After the war he was recalled to his previous positions. "Oral History Interviews, Hans Thirring," accessed January 31, 2021, https://www.aip.org/history-programs/niels-bohr-library/oral-histories/4912.

[27] Arthur Haas, *Atomtheorie—in elementarer Darstellung* (Berlin-Leipzig: De Gruyter, 1924/1929).

Fig. 9.8 Arthur Haas in the early 1920s, in Leipzig and Vienna

working were an increasing source of frustration for him. With his new economic difficulties and in view of the anti-Jewish sentiments at the university, which were obvious to him, he began to apply for positions outside of Austria and Germany. He waited impatiently for answers from the universities and professors he had contacted and when they arrived, they were mostly negative. Yet some of the letters promised hope, such as an invitation from the University College in London to give a colloquium there in May 1924. These efforts, almost driven by desperation, are described in more detail in one of the following chapters.

On March 17, 1924, Haas noticed an advertisement for a lecture on the American education system to be given by a young woman named Emma Huber. He was obviously interested in the topic, as he was constantly thinking about emigration and change. Haas was particularly interested in North America and its universities, as he had heard enthusiastic reports from friends and acquaintances about their facilities and also about the teaching conditions prevailing there. He had probably also read Boltzmann's account of his trip to California with the assertion that "it is best at home," but that home offered Haas neither a job nor a future. He decided to attend the five o'clock event, to get more information.

This decision marked a turning point for the young bon vivant Arthur Haas who had, despite the mounting difficulties, sailed rather successfully through life without worrying too much about income and obligations. Despite the financial losses of the post-war years, he had maintained sufficient fiscal resources through financial speculation, affording him a regular income from his stock market investments. He was able to live in the way that he was accustomed to and still follow his interests in physics and his ambitions in science without monetary handicaps. He desired a regular faculty position, yet that was more for the social prestige than for the salary.

Haas' ambition to America but so far, he had not seriously inquired about the possibility. Meeting Emma Beatrice Huber would change all that. Being with her became his goal and this forced him to rethink his entire attitude toward life. He could not continue as he had been under the "persistent application of the principle of not losing sleep over money," as his brother had noted. With meeting Emma Beatrice Huber, Haas discovered the need for a regular home, profession, and income. This was the turning point in the carefree life of Arthur Haas; Emma had her own ideas and goals, which also influenced him. We therefore step back and learn about her origins and background in order to fully consider later developments.

Notes

i. Cf. Kateřina Čapková, *Czechs, Germans, Jews – National Identity & the Jews of Bohemia* (New York: Berghahn, 2012), 26–41.

ii. Peter Eigner, Günter Müller, eds, *Hungern – Hamstern – Heinkehren – Erinnerungen an die Jahre 1918 bis 1921* (Wien: Böhlau, 2017).

iii. Frederick Seitz and J. S. Koehler, "Displacement of Atoms during Irradiation," *Solid State Physics* 2 (1956): 305–448.

iv. See Karl-Heinz Schlote, *Von geordneten Mengen bis zur Uranmaschine*, 108–117.

v. E. S. Imes, "Measurement in the near infrared absorption of some diatomic gases," *Astrophysical Journal*, 50 (1919): 251–276. For a detailed discussion of the isotope effect and its interpretation, Helge Kragh, "The isotope effect: Prediction, discussion, and discovery," *Studies in History and Philosophy of Science Part B: Studies in History and Philosophy of Modern Physics* 43 (3), (2012):176–183.

vi. See Erik Eybl, *Von der Eule zum Euro - Nicht nur eine österreichische Geldgeschichte* (Klagenfurt: Mohorjeva - Hermagoras, 2005).

vii. See Regina Fritz Grzegorz Rossoliński-Liebe and Jana Starek, eds, *Alma Mater Antisemitica - Akademisches Milieu, Juden und Antisemitismus an den Universitäten Europas zwischen 1918 und 1939* (Wien: New Academic Press, 2016).

viii. Karl Taschwer, "Geheimsache Bärenhöhle," in *Alma Mater Antisemitica—Akademisches Milieu, Juden und Antisemitismus an den Universitäten Europas zwischen 1918 und 1939*, eds. Regina Fritz, Grzegorz Rossolinski-Liebe, Jane Starek (Wien: New Academic Press, 2016), 221–242.

ix. Karl Taschwer, "Othenio Abel. Antisemitischer Fakultäts- und Universitätspolitiker," in *Universität – Politik – Gesellschaft, Göttingen*, eds. Mitchell Ash and Josef Ehmer (Wien: Vienna University Press bei V & R press, 2015), 287–292.

x. Klaus Taschwer, "Der verlorene Schlüssel des Otto Halpern - Zur Karriere eines theoretischen Physikers, den man in den 1920er-Jahren auf infame Weise aus Österreich vertrieb und nie mehr rehabilitierte," *Der Standard*, October 31, 2012.

Chapter 10
The Origins of Emma Beatrice Huber

Emma Beatrice Huber was born on July 11, 1896 in Munich as the daughter of the building contractor Michael Huber (1861—1928) and his wife Theresa née Raith (1862—1899). Michael Huber was a successful businessman who traded in building materials and ran a successful paving and asphalt business Huber & Meiler GmbH. In Wilhelminian Germany at the beginning of motorization and increasing road construction, this was undoubtedly a profitable and prosperous business. Emma Huber had a two-years-older brother, Georg Huber, who was destined to take over his father's company.

Michael Huber was born in Landshut and, as the youngest of five children, he was intended for the priesthood after the early death of his father. Although strictly catholic throughout his life, after military service he changed direction and opened his own construction business in Munich, the rapidly growing capital of Bavaria. He married Theresa Raith from Munich but she died early, in 1899, so that he was left to raise his two children (Fig. 10.1). As a businessman and single parent, he had considerable help from his cook Marie and several maids. This was undoubtedly difficult for the children, who developed a close bond with each other over this time. The family was part of the parish of St. Anthony's Church at the Capuchin monastery in the Isarvorstadt, a newly built district in 1895. Catholic care played a major role in the upbringing and education of the children from a young age.

Georg Huber went to the local Realgymnasium after attending the parish's Catholic elementary school, to prepare himself for taking over his father's business. Emma was accepted in 1902, at the age of six, into the Catholic elementary school at Tumblingerstraße near Kapuzinerstraße and she spent eight school years there, where day care was provided; her favorite subjects were drawing and religion. According to the first entries in her poetry album of 1909, after father, brother and cook Marie, the drawing teacher Ella von Zlatarovic and the religion teacher Josef Wallner were the people closest to the young girl. Although no other teacher received this honor, her closest friends and classmates wrote little sayings and wisdoms in her booklet.

M. Wiescher, *Arthur E. Haas—The Hidden Pioneer of Quantum Mechanics*, Springer Biographies, https://doi.org/10.1007/978-3-030-80606-4_10

Fig. 10.1 On the left are the recently married Michael and Theresa Huber; on the right, Michael Huber with his two children Georg and Emma after the death of their mother at the turn of the century

The class had thirty pupils and education was structured according to the necessary domestic subjects. Beside the basic subjects of religion, reading, spelling, penmanship and basic arithmetic, the curriculum focused on domestic matters, kitchen instruction, handicraft, knitting and mending. With the additions of drawing, singing, gymnastics and French language, Emma Huber finished in 1910 with good grades. She was then allowed to spend another year in the same building to complete the compulsory year of girls' further education, where practical household subjects were taught in depth.

Emma had good grades and so when she had turned just fifteen years old, her father decided to have her trained as a teacher. He sent her to the Maria Ward School in the town of Kempten in the Allgäu, in the autumn of 1911. A contemporary photo of the school is shown in Fig. 10.2.

The school had been founded in 1861 by the Sisters of the Blessed Virgin, an order founded by the English nun Maria Ward (1585—1645) and closely associated with the Jesuits in their pursuits for education and training. In Germany the order was called "Die englischen Fräulein" (The English Misses), in keeping with the origins of the founder. The class photo from 1912 shows Emma Huber with her classmates in Fig. 10.3, many of whom became long-time friends.

In Kempten, Emma was to be prepared for teacher training. She spent two years there with her schoolmates in the close fellowship of boarding-school students. The school was to provide middle-class daughters with a secondary education, focusing especially on French and also English as modern languages.

The educational goal of the school was to provide "a solid all-round education in a patriotic spirit on a religious basis, which would enable the pupils to fulfil their future profession in order to keep up with the demands of the times." The affiliated boarding school where Emma lived,

Fig. 10.2 The Maria Ward School in Kempten where Emma Huber lived as a boarding-school student for two years, from 1911 to 1913

Fig. 10.3 Emma Huber's class picture at the Maria Ward School in Kempten. Emma is sitting in the front row, the third student from the right

was intended for such schoolgirls whose parents did not want to send their daughters to an educational institution as early as the fifth year of school and who wanted to pursue more practical teaching goals, and for schoolgirls who wanted to devote themselves to commercial subjects or whose effectiveness was likely to be limited to the domestic circle. The Institute School offered the following subjects: religion, German, civics, arithmetic, accounting, chemistry, home economics, health, drawing, needlework, gymnastics, as well as French language, shorthand and singing.[i]

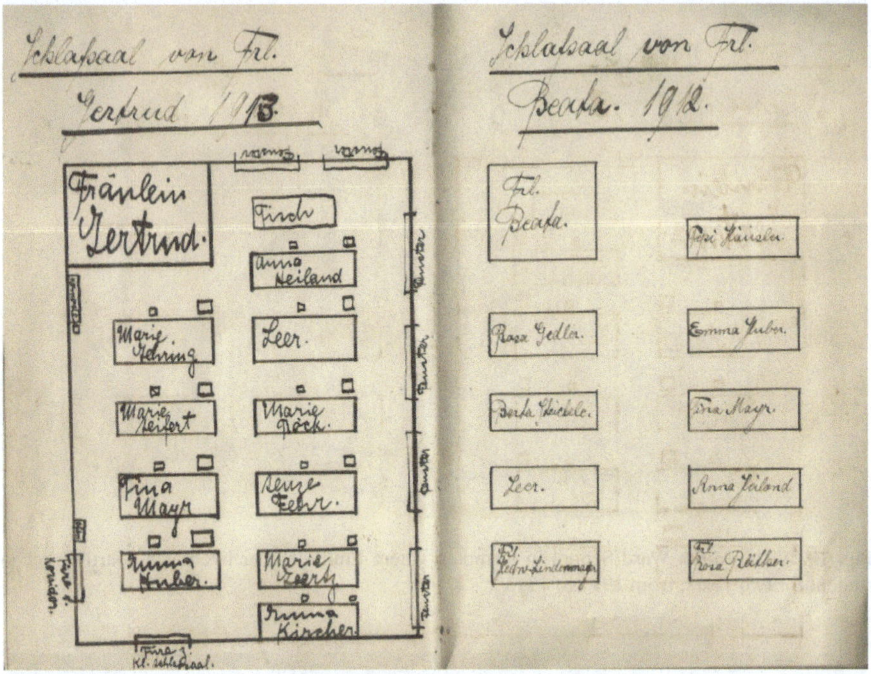

Fig. 10.4 Layout of Emma Huber's dorm rooms in Kempten during her stay in 1912 and 1913

Michael Huber obviously had practical teacher training in mind when he sent his daughter to boarding school in Kempten.

The students shared a common dormitory for about ten students, each of whom was supervised by one of the English Fräuleins (Ms. Beata and Ms. Gertrud) who also slept there. Emma Huber carefully recorded the dimensions of the dormitories and the distribution of the beds in her poetry album, depicted in Fig. 10.4.

In 1913 the school was visited by two American Catholic nuns, Sister Mary Isabel (Fig. 10.5) and Sister Mary Beatrice from Xavier College in Chicago, who wanted to learn about the educational methods of the school. Sister Mary Beatrice Sheehan was the head of the Art Department at Xavier and was particularly interested in European art. Since Emma's English was already considered very good, she was asked to be the guide and escort the American nuns through the museums of Munich, especially through the collections of the old Pinakothek (The main museum in Munich with an extensive collection of pictures and paintings from the thirteenth to the nineteenth century). However, Sister Mary Beatrice died unexpectedly on July 10, 1914 in Munich; she was only forty-nine years old. Emma helped Sister Mary Isabel with her German, assisting with arrangements for transfer of the body to Chicago. In gratitude, Emma was invited to Chicago for the following school year, to study at Xavier and perfect her English.

Fig. 10.5 Sister Mary Isabel in a photo from Emma Huber's poetry album

Then in August 1914, World War I broke out. Emma had only just turned eighteen and her father was naturally hesitant to send his daughter overseas. Nonetheless, in view of German successes on the various fronts, Michael Huber gave in and allowed his daughter to travel on a neutral ship to neutral America where she would eventually obtain an American college degree and then become a language teacher. This was undoubtedly an unusual and remarkable decision to send his daughter on a dangerous Atlantic crossing during the war, especially since studying in America at that time was something uncommon and not necessarily seen as a career-enhancing step.

On August 22, 1915, Emma Huber set out alone to travel by train for several days via Stuttgart, Heidelberg, Frankfurt, Mainz, Cologne and Emmerich to Rotterdam, where she arrived on September 2nd. It was her first great journey and she describes it with graceful words, observing it with big eyes:

> "When the angels travel, heaven weeps," says a proverb and indeed it does not lie, for when I finally wanted to leave Munich on August 22nd, 8:22 a.m., heaven was inconsolable about it; I thought it would dissolve completely.[1]

It was a slow journey with frequent train changes and she used the time to stroll through the different towns and cities where she had to stop. She enjoyed the journey, visiting churches and synagogues, and she reflected on the beauty of the river valleys of the Neckar, Main and Rhine. From Mainz to Cologne, she took the steamboat gliding down the river without any hurry:

---

[1] Emma Huber, *Travel Diary* (unpublished, 1914).

> Once again, the river storms against a jagged cliff bank, only to flow quietly in a majestic course between St. Goar and St. Goarshausen. Here Rheinfels Castle rises steeply from the river opposite the Loreley Rock, where waters make its shadow appear mysteriously deep and dark. And the Loreley Rock, how deeply did it affect me? It was an indescribable feeling that captured me as I sailed past it. The others on the ship felt the same way, because they enthusiastically started to sing the touching song by Heine: "I don't know what it means." [Ich weiß nicht was soll es bedeuten.][2]

Emma's entries are full of young, romantic descriptions of her surroundings and experiences. Not a word, not a thought about the approaching war, which she had to face for the first time when crossing the border into the neutral Netherlands; but the experience was barely realized:

> I had arrived at the border station in Emmerich. What did I care about Emmerich, if it had not been a border station? But so, I had to take my small suitcase and wander with the others into a small, dirty room where I had to deliver all my books, stationery, maps, and whatever else I could remember, so that they eventually could hand it back to me. I don't want to think about all the trouble and difficulties I had, otherwise I'll get angry.
>
> At every second step I had to pass soldiers who examined my passport very carefully, then turned it around a few times, then read it again and finally handed it over to me with a stern face. And that happened every time and every time it pricked my feet and hands more. But "Patience brings roses" and "Thy fate is the common fate of all; into each life some rain must fall; some days must be dark and dreary," Longfellow once said and this was a small consolation to me. No, but I never forgot that trial.

Late at night, she finally reached Rotterdam where she stayed in a guesthouse run by Mrs. Bredée, who also helped her during her stay:

> We took a cab and drove to the Holland-America line. Here I saw my steamer for the first time, which was to take me to the other side. When I saw it, it impacted me very powerfully. Again and again I had to look at this enormous colossus of a ship. After I had taken possession of my cabin and everything was put in order, Mrs. Bredée and I went by boat across the Meuse and then by tram home.

The ship was the flagship of the Holland-America Line, the Dutch steamship *D.D. Rotterdam* that had been put into service in 1908. In two days, Emma was to set sail under its neutral flag—hopefully not to the afterlife but rather, to New York as Fig. 10.6 shows.

On her last day before embarking, Emma visited the museums of The Hague and then drove to nearby Scheveningen to view the sea: "My first course was, of course, the sea—which I was to see for the first time. The impression was really great—I could only stare in amazement without end."

Here Emma Huber's notes end; only empty pages follow. She didn't tell anything about her journey on the almost empty ship, about disembarking by boat to Ellis Island, or the immigration procedures that were probably even more painful than those she had been required to follow in Emmerich. She also wrote nothing about

[2] This is a song written by the Jewish-German writer and poet Heinrich Heine (1797—1856) who lived and died in self-imposed exile in Paris. The song about the treacherous Loreley, combing her beautiful long blonde hair while luring shippers to their deaths, was so loved in Germany that the Nazis did not dare to forbid it and so just declared it a poem written by a German writer.

Fig. 10.6 The *Rotterdam* of the Holland-America line with Emma's 2nd class ticket to America in September 1915. This was one of its last trips during the war, due to the increasing dangers of unlimited submarine warfare. Details about the ship can be found at: accessed December 10, 2020, https://en.wikipedia.org/wiki/SS_Rotterdam (1908)

her first impressions of New York, where she visited the Statue of Liberty and St. Patrick's Cathedral, and still nothing about the long rail journey that followed, from New York to Chicago on the Lake Shore Limited or the New York Central Limited of the New York Central (NYC) Railway Company. This experience must have been so different from her rail journey in Germany because even years later she was talking about the fast Pullman cars that made travelling a luxury. Maybe there were too many impressions. Maybe the records have been lost.

This journey to America marks the beginning of a fascinating new life in a dynamic society, so different from the controlled life at home in ponderous Munich or in Kempten. This was a new life whose speed and dynamism she enjoyed and carried within her. She would transfer its impressions years later to Arthur Haas through her stories and descriptions, so that it became for him the land of longing, the land where milk and honey flowed, the land where freedom and science were valued.

However, back to the young Emma Huber who at first had an unbiased attitude toward this new country but then sought protection and security in the Catholic environment that was so familiar to her, in Saint Francis Xavier College. Saint Xavier College grew out of a small Catholic school for girls, which had been founded in 1846 by the Irish order Sisters of Mercy at the request of Bishop William Quarter.[3] The city of Chicago was less than ten years old when the sisters, under the direction of Mother Mary Francis Xavier Warde, R.S.M., responded to the bishop's call and founded Saint Francis Xavier Female Academy, which later became Saint Xavier College (now Saint Xavier University). The institution is considered the oldest state-recognized college in the city of Chicago and is, since 1901, located about twelve miles outside of downtown Chicago in a park-like area of over 5,057 acres, shown in Fig. 10.7.

In 1912, the school received a certificate for the development of a higher education program, "a university in which may be taught all branches of higher learning." This is why the two nuns were sent to Germany, to establish training centers, gain experience, and perhaps recruit students. In 1915, Saint Francis Xavier College was officially opened for women; the school offered a classical education program in religion, classical and modern languages, mathematics and history. Five students were enrolled this first year,[4] one of whom was Emma Beatrice Huber. Four years later, in 1919, two of these students left the college with a successful degree, Emma Beatrice Huber was not among them.

During her college career, Emma Huber became a victim of the turbulence of the times.

Xavier had a tri-semester system, autumn, winter and spring, with twelve to thirteen weeks of classes each. During her first autumn at the college, Emma attended

[3] Founding the school was part of a major educational campaign by the Catholic Church in the then predominantly Protestant United States. The Church was responding to the policy of private universities, from Harvard to Princeton, who, out of their Protestant tradition, did not want to admit either Jews or Catholics. However, the growing number of Catholic immigrants, from Irish to German, required education and training opportunities for their descendants. Within a decade, the United States saw the emergence of a multitude of Catholic colleges and universities.

[4] Today, over one hundred years later, the number of students is 3,500.

Fig. 10.7 St. Xavier College in Chicago as it appeared on postcards at the time of Emma Huber's attendance

lectures in English, French and Latin. During the following winter and spring semesters, she deepened her knowledge by devoting herself mainly to English and French literature. In the following trimester of 1916, she took drama and theatre classes as well as other literature courses. Mathematics lectures are not recorded, although this was part of the teacher training plan.

In America, Emma was also receiving an education from the city itself since Chicago was a rapidly growing center of the Midwest. The big city, so different from Munich, was not burdened by history and tradition but driven forward by the dynamics of the new age and its technologies. The first skyscrapers rose to the clouds and cars roared on the crowded streets. She was met with department stores and countless people moving about the city; double-decker buses and trams traveled in close succession and above her head rattled the "El," the elevated railway network of the city center that had been completed shortly before the turn of the century. There were huge stations from where never-ending trains, pulled by powerful locomotives, rolled into the sheer endless expanses of the West, to new shores with magical names that evoked dreams of departure and freedom. And then there were the museums where, thanks to the new wealth created by industry and growth, the art of the "old continent" was being collected and exhibited. Emma spent long hours in these museums. It was a world of new beginnings and hopes where peoples came together as immigrants. It was the twentieth century. That's how Emma Huber saw it and that's what shaped her image of America (Fig. 10.8).

However, toward the end of the spring term, her situation dramatically changed when American president Woodrow Wilson declared war on Germany. On April 6, 1917, the many Germans who were living in the United States suddenly found themselves marked as enemies. On presidential order, German immigrants were declared "Enemy Aliens" and identified as an internal security threat. These warnings from the highest authorities, about possible ethnic-German traitors and spies who

Fig. 10.8 Contemporary postcards from the estate of Emma Haas née Huber. Above, two views of State Street as the main shopping district of the city. Below left, the Art Institute built in 1893, one of the largest museums in America. Below right, the mighty La Salle Street Station built in 1882, which served as the terminus for the NYC Railway Company and from which Emma Huber probably entered the city for the first time. (From the collection of Emma Haas née Huber)

were an internal threat, triggered hysteria and anti-German emotions among the population, which in some places led to lynching. Most Germans had to register in the following weeks and months and were interned in the four camps that had been established.

This was not what Emma Huber had imagined as possible but she escaped several years of internment by giving up her student status. On the advice of the nuns, she entered the order of the Sisters of Mercy. In the fall of 1917 after finishing a lecture on the French novella, Emma Huber disappeared within the monastery walls of the college as Sister Beatrice shown in a photograph, Fig. 10.9.

But she did not remain idle; besides following the rules of the order, she served as a teacher for children in the early grades. Sister Beatrice also took music lessons from the nuns, focusing primarily on harmony, and gave music lessons for the children.

One year later in November 1918, the war ended with the defeat of the German and Habsburg Empires. Since Germans in America were interned for another two years, Emma Huber remained a nun during that time. Yet although she was part of the order, there were still considerations before she would take final vows. Seven probationary years are required by the rules of the order, during which the sisters can change their minds and withdraw from final ordination. This is what Sister Beatrice did, although she did not do leave light-heartedly. Short notes and sentences scattered

Fig. 10.9 The only preserved picture of Sister Beatrice in the typical habit of the Sisters of Mercy

over her booklets and flyers show that she struggled with herself and her God over what decision she should take.

> I confess that I need a lot of grace to do this because without God's help I can't do anything, not the slightest thing. There is so little will in me that I fear that there is no one in this world with so little strength of character as me. [...] I have had only few Christian thoughts; I feel our exercises as a burden and many times it is torture to go to the reading or just open my mouth during the lectures. [...] I have seldom kept my resolutions and I have hardly ever made good reflections in the morning. I also have performed my prayers with the greatest distraction and have often been very careless in punctuality. On the whole, I am very moody, impatient, defiantly nervous and unloving, also malicious and childish.

With these thoughts she left the convent in the spring of 1922 but stayed at the school for a few more months in order to take the classical Greek courses necessary for graduation, in the winter and spring semesters. Her graduation certificate of August 1922, signed by the Dean of the College, Mother M. Irma, says: "Miss Huber has an average grade of A- in the above numbered courses. She was successful in her management of children in the class room and is hereby recommended as a teacher of elementary classes or as a special teacher of music." Primary school teacher or music teacher, is this what Emma dreamed of and what she expected from life?

Her lectures in music history, art and design had awakened new interests. Emma, seen in Fig. 10.10, followed her dreams which drew her back to the old world. But not to Germany, which was in turmoil and rapidly declining through unrest and revolution as she learned through letters sent by her father. For art, music and architecture there was only one destination and that was Paris, the artistic center of the old world. She spoke fluent English and had studied French for years. Paris was her goal. As the family chronicle notes, she cut her hair, said goodbye to the Sisters of Mercy and embarked on the long-delayed return journey to Europe in March 1923.

The family saga tells of unfriendly treatment by the unfriendly French; after all, the city had recently been German enemy territory. This is supposedly what drove Emma

Fig. 10.10 Emma Huber in 1923, after she had left the Order and returned to Europe

to Vienna but questions arise: Was she really in Paris, even for only a short time? Or, did Paris remain a dream that she could only fulfil later in life? The latter seems to be the case since nothing appears in the surviving documents, neither addresses nor notes indicating such a visit. For although Emma Huber kept extensive address books recording all the people she met during these years, no French or even Parisian names and addresses can be found among her notes.

Emma Huber instead seems to have returned to Germany and for the following months she seems to have been the guide for a group of American clergy and sisters led by a Father Seraphin from Augsburg. The destinations listed are Munich and the castles and lakes in the area, the cities of Karlsruhe and Baden-Baden in the west, and Vienna and Budapest in the east. There were mixed American and German groups of five to ten members; in such groups, of course Emma Huber's language skills would have been very useful, of course. It is possible that this work enabled and financed her trips back home and also allowed her to earn money for future study. In between these tours, Emma mostly stayed with her father in Munich where they went mountain hiking, an activity he especially loved. She also visited school friends in the Upper Palatinate and Bavarian Forest, although she preferred staying for weeks with her friend Eugenie Grau at the Ammersee in the southwest of Munich, depicted in Fig. 10.11.

Mr. Georg Dessauer, a civil servant from Vienna with his English wife, also took part in one of these guided trips. Mr. Dessauer had organized the English Club of Vienna at Johann Strauss Gasse 42, where the small English-speaking community of Vienna met regularly in a multi-story apartment building near the main railway

Fig. 10.11 The Ammersee in the evening on a postcard from Emma Haas née Huber's notes. She spent long summer weeks at the Villa Grau

station. When a group of clergies whom Emma was escorting later visited this club, Dessauer joined them on a steamer trip down the Danube to Budapest, in September 1924.

Since Emma Huber's goal was to study architecture and design, it is not clear why she chose Vienna. But with the Dessauer family she had found cheap accommodation at Johann Strauss Gasse 36, in the long line of tenement buildings that run along the street from the English Club. Then, based on her interest in design and her language skills, she was offered a small position as stage designer. At the Theater in der Josefstadt, which had been leased by Max Reinhard in 1923, she oversaw an elaborate redesign in the style of the Teatro La Fenice in Venice. She worked under a contract that expired with the completion of the reconstruction and the first performance on April 1st. Therefore, she tried to make some additional money in March by giving lectures about America and American life and culture. This always attracted a large number of listeners in the difficult economic times. It also attracted the young Professor Haas.

Note

i. Annemarie Engeßer, *Als die "Englischen" nach Kempten kamen – Die Geschichte der Maria-Ward-Schule in der Stadt Kempten* (Kempten: Privatdruck, 2012), 54.

Chapter 11
The Year 1924—Hope and Despair

Arthur Haas was not only taken with the subject and the way it was presented; he was also extremely impressed by the young woman who gave the lecture. He spontaneously approached her and invited her for dinner. Emma Beatrice Huber had only been in Vienna for a short time and knew no one in the city except the Dessauer couple and some of their elderly friends, so she accepted the invitation.

It must have been a pleasant evening. Since their common interest was America, Emma probably talked about her studies at Xavier College, about American cities like New York and Chicago, and about all of her impressions of the country. She was happy to do so because she missed American independence and freedom of movement, especially since she had neither a good job nor much money. Arthur Haas asked and asked, wanting to know everything about American life and her living conditions, so the evening passed as if by magic in lively conversation. Despite this cordial beginning, the relationship was still very formal toward the end of March when she refused him another invitation for Monday, March 24 and also on the following Wednesday, due to indisposition.

> Dear Professor
>
> Thank you very much for the kind invitation for tonight. I can't come - I'm terribly sorry. I have to cancel Wednesday as well.
>
> Since Friday I've been "good for nothing" as they say in English. Very reluctantly I even gave up yesterday's trip. - I just did not feel well enough. I am quite impatient with myself about it because I would quite like to spend a few hours in your company again.
>
> Next week I would be quite happy to see you again.
>
> Please don't be angry with me for my refusals,
>
> I will now deliver the letter to you personally, so that you will receive it safely this evening.
>
> Best regards
>
> Emma Beatrice Huber

Yet only a few days later on April 5th, the "Du" was introduced and from "Dear Herr Professor" the salutation quickly changed to "Dear Arthur". America remained their theme; the stories of her experiences painted a land where milk and honey flowed

M. Wiescher, *Arthur E. Haas—The Hidden Pioneer of Quantum Mechanics*, Springer Biographies, https://doi.org/10.1007/978-3-030-80606-4_11

Fig. 11.1 The open-air swimming beach on the Danube waters, which Arthur Haas had enjoyed since the early days of his studies, had been established in 1900 by the alternative healer Florian Berndl who sought to realize there his ideals of a natural way of life. The air- and sunbathing of the Gänsehäufel soon became an attraction for many Viennese, who were drawn by the often-unusual advertising for this unique open-air swimming pool

for Arthur. It quickly became his goal to leave Austria and move to this Promised Land. He was worried about his imperfect English but Emma agreed to help him in learning the language as well as with structuring and organizing his lectures.

The need for her assistance was reinforced when he received an invitation to travel to London. His book brought him recognition and he had become a well-known science writer also in the English-speaking scientific community. In the spring of 1924, Arthur took leave from Vienna to give a lecture on "Objective and Subjective Physics" at University College in London. There he was exposed to the less formal university life that characterizes the academic world in Anglo-Saxon countries.

In preparation for this trip, Emma worked diligently with him to improve his English and lecturing style, as he constantly emphasizes in his letters. But that was not all; meals together became visits to the silent movie theater on Schottenring and, due to the unusually warm spring weather, to the popular Viennese Freibad (outdoor pool), Gänsehäufel on the Danube in Vienna (Fig. 11.1).

It was a glorious spring for the young couple. The sun was shining and nature was enticing, as the poet wrote:

> The air is completely free of dust, the water surface like an endless lake. The island consists of willows and Danube sand. It is a labyrinth of willows, a jungle, a giant wickerwork protecting this island like a sanctuary of life, which brings life energies to the body of the poor city dweller! In the prater on the street, one is still affected by the poison of the big city, but a quarter of an hour later you can bathe yourself clean from all the poisoning! Out

of the tangle of cool pastures, nature looks at you lovingly, spreads her regenerative powers on you without forcing you! Like a holy island, it is in physiological truthfulness, a modern fountain of youth from the old fairy tale! The maltreated, half-embroidered skin now drinks here with its billions of pores all light and air, filling itself, busily seeking to compensate for all sins, while the soul, stimulated by God's peace, joins in and removes worry, creates inhibitions and inertia! The eye breathes in the landscape and the ear rests in the vast silence from the terrible noises of the city! Seagulls fly in the pure air. Water plants harken up from the water. May the people treat this island with respect, if not even with devotion! May not exuberance, that may of course be borne by the surplus of vital forces, disturb this paradise of originality that the city of Vienna has opened up to its tired children!

Water, pure Danube sand and pure air should bring us closer to God from our innermost being. Times are hard. Bodies and souls are, without them knowing it for sure, overtired and bitter, as it were unconsciously quarrelsome and capricious, open to injustice in their irritable weakness! So, may nature bring peace and order! May this Danube island "Gänsehäufel" be a respected place, a place of pilgrimage for sinful bodies. And who, who has not sinned here below?[i]

Who would not fall in love? The young couple shared in these pleasures that led to joint breakfasts, opera, and concerts in Vienna and also Salzburg. The latter was probably linked with Emma Huber's hope to design the stage set for a performance of Karl Vollmöller's play, *Das Mirakel*, planned by Max Reinhard for August in the Kollegienkirche. However, this did not happen. The performance failed due to internal and financial disputes, but Emma Huber had already left the theatre group at the beginning of April.

The engagement took place in Salzburg on April 13, 1924. It was a surprise for the bride's family in Munich but agreement came with a friendly letter from Michael Huber on the birthday of his future son-in-law.

I want to start right now by wishing you all the best for your birthday today; the date I found in my notes. May heaven grant you good and lasting health and may your marriage with my Emma be as joyful as I wish you with all my heart. Understanding and merging into each other in all thoughts and actions should be the foundation of a good marriage.

And now to business! As you know, due to the unfortunate aftermath of the war and the terrible inflation of the last year, the financial situation in German countries has fallen behind very badly, so that we here in Germany have to pull out all the stops in order to get back on track [...]

Emma is not a very poor girl, even though I am not able to equip my child appropriately—the engagement came as such a surprise and was so unexpected that I couldn't have stepped in with the best will in the world in such a short time.

I wish you all the best for your further development and hope that there will be no more obstacles to keep you from the appointed time for the wedding.[1]

The exact wedding date was not fixed but initially Arthur Haas was optimistic that it would be soon. His next task was to evaluate his financial resources. His quick calculations established that, due to losses in recent years, his monthly income was only sufficient for his own personal, comfortable life. This included taking his young fiancée out on the town once in a while but was certainly not sufficient for building a new household to the standards he believed were necessary.

---

[1] Letter from Michael Huber to Arthur Haas on April 30, 1924.

Haas drew up a detailed budget to see how much money was needed for a two-person household and the sum exceeded his current expenses by a considerable amount. First, he abandoned the idea of a big apartment that would have caused considerable additional cost; this was followed by dismissing the idea of a maid who would have had no place in the present apartment anyway. Further cuts were made by careful planning and by reducing entertainment and shopping plans. It was a detailed, written plan but Emma Huber did not seem to give it the necessary serious consideration. As her letters demonstrate, she seemed to favor a more flexible plan for housekeeping.

Haas had calculated a budget totaling eight million crowns[2] and without being able to finance these household costs through a monthly income, the marriage seemed irresponsible. Therefore, his next step was to do everything possible to develop a regular income that would enable him to marry.

His plan was to sell the Berlin houses that he had purchased with his brother Otto as a fixed investment in 1922. Doing this as quickly as possible, he could then invest the proceeds in safe foreign securities. He calculated that the fixed interest would allow him to live with his wife in reasonable circumstances despite inflation. In addition, he assured himself that he had a good chance of obtaining a full professorship soon. He thought that Vienna would be acceptable but outside of Austria with its economic crisis would be even better, preferably in the USA. However, the economic difficulties caused by hyperinflation caught up with the largely optimistic and naive young Haas who, for the first time, soon became fully aware of the impact of the economic crisis on his personal life.

Since the bride-to-be had no income and her father was in poor health, the young couple agreed that Emma should return to Munich while Arthur went to London and his lecture. He had high hopes for his further career, preferably in the Anglo-Saxon world of science, and was counting on the trip to answer this need. As soon as his financial affairs were settled by the sale of the Berlin houses or, alternately, by a permanent position that would guarantee him a regular income, the wedding should take place. Arthur Haas thought that he would have these issues resolved within a few weeks. He had no idea what a burden his decision had created for the young relationship.

Arthur Haas and his fiancée enjoyed the warm May days and spent their last weekend together in Salzburg, where they visited churches at Emma's suggestion and attended concerts on Arthur's advice. From Salzburg, Haas traveled back to Vienna to set out on his long train journey across Switzerland and along the Rhine to London, while Emma Huber took the train directly to Munich to visit her father. There she planned to spend what they hoped would be only a few weeks until the wedding. Contact was to be maintained, if possible, by daily—on Arthur's side—numbered letters. These letters would also inform her about the steps that he was taking in order to finance their extended household with a regular income. Emma's

[2] In these days of hyperinflation, conversion was difficult. So, Haas oriented himself to the gold mark and the dollar. In 1924, 8 million crowns would correspond to about 500 goldmark or $120, which today corresponds to a purchase sum of about $1,800.

letters were mostly just answers to Arthur's more regular correspondence but, as he often complained, her responses showed little interest in his sometimes too exhaustive descriptions and lamentations.[3]

The letters provide a detailed look into the personal life of Arthur Haas during the crisis in Austria and during the development of his relationship with Emma Huber as his future wife. These letters also illustrate the impact that this crisis had on the academic world and the Austrian middle class. The young couple's correspondence also draws attention to Arthur Haas' personal changes from someone for whom science was the only goal to someone who had found a new dimension in faith, which had not been noted before if one disregards the early influences of his friend Karl Jellinek. It is therefore significant that Haas chose Karl Jellinek, who had lived for years in distant Danzig (today Gdansk), as his best man.

The letters also show Arthur Haas as a deeply emotional person with strong mood swings under the real or imagined pressure of circumstances, moods that seem to have been countered by Emma Huber as a much more relaxed personality. These moods demonstrate Haas' need to plan all developments carefully; any deviation from these plans caused him extreme unease and even panic. These letters show the human being and not only the scientist that he presented in his memoirs.

Apart from these personal aspects, the letters also demonstrate the network of professional relationships that existed within the scientific community, which had been rebuilt after the war and re-established for mutual support. The letters record where and how personal sentiments, particularly anti-Jewish attitudes, influenced this support network beginning as early as the early 1920s. The correspondence will therefore be reproduced below, albeit in an abridged form and clarified by explanatory comments, recording Arthur Haas' transition from a scientific bon vivant to a family man within the science administration of the Vienna National Academy.

The first letters describe Arthur Haas' journey to England for his invited talk at the University College in London. He anticipated the lecture with a great deal of hope that this event would provide the perfect opportunity for making an impression in the Anglo-Saxon world, an impression that would be followed by an offer for a faculty position. He was correct in that his host was Professor Frederick George Donnan[4] who was known for attracting German scientists to England and the United States. Haas' letters reflect his hopes but also describe the many unexpected difficulties and obstacles that he faced, which frequently turned hope to despair.

Returning from life-filled London, roaring with traffic, trade, and change, to bleak, gray, hungry, and distressed post-war Vienna plunged Haas into never-ending despair, which he described to his more optimistic bride in long, plaintive letters. The man

[3] Usually there was a period of three days between the letter and the reply, except with express letters. This has been taken into account in the presentation of the letter sequences provided here.

[4] Frederick George Donnan (1870—1956) was Professor of Chemistry at University College London and one of the fathers of physical chemistry in Britain. He worked mainly in the field of colloid chemistry and osmosis. After 1918 he was very much involved in international scientific exchange and after 1933, he worked actively to enable German scientists to immigrate to England. F. A. Freeth, "Frederick George Donnan 1870—1956," *Biographical Memoirs of Fellows of the Royal Society* 3 (1957): 23–39.

who had been so carefree was confronted, during the few months of spring and summer in 1924, with economic reality. Haas was filled with hopelessness regarding Europe and his dream of an American paradise solidified.

The letters begin on the first day of Arthur's trip to England, one day after bidding farewell to his fiancé who he called by the endearing nickname, "Bice".

Arthur to Emma Innsbruck, May 27, 1924.

No. 1.

Dearest Bice

At the moment the train is stopped for half an hour in Innsbruck and I want to use the time to send you the first letter. I slept like never before, from midnight to five, in Vienna. Now I am awake again and for the first time I miss the expectation of seeing you on the same day. How sad I was on Sunday before our engagement because that was the only day that I did not see you and now so many such days are to come! I can't tell you how much my heart aches and even if I could, I probably wouldn't - to not bore you with it.

... Today I can see almost nothing of the landscape; there is thick fog and everything is covered in clouds; everything is also flooded but sometimes a snow-covered mountain top peeks down through the clouds and that is wonderfully beautiful.

When you receive these lines, you will have already spoken with your father. I hope that he, like you, has confidence in me and is convinced that I will work with all my strength to create a secure foundation for both our existences as soon as possible.

[…]

Just write me soon and very much. Give my best regards to your father and brother.

Yours sincerely, Arthur.

Letter number two was a postcard from Basel, which Haas wrote in the station concourse.

Arthur to Emma Brussels, May 28, 1924.

No. 3.

My dearest Bice

You must have received the letter from Innsbruck and the card from Basel. I will use the short stop in Brussels to tell you more.

The trip over the Arlberg was very nice - then came the Rhine and, for the first time in ten years, I was outside the dungeons of the "Central Powers." The impression Switzerland made was wonderful. Everything so nice, so clean. The trip went first past two magnificent lakes (Waller and Zurich), then there was the immense Zurich railway system with enormous wire lines because most locomotives run on electricity. In Zurich, Professor Schrödinger came on the train and brought me a huge box full of wonderful delicacies: salmon, sardines, ham, bananas, oranges, cakes, Lindt chocolate, wonderful wine. It took me many days to get through it, and everything was so neatly packed with paper napkins and labels. I was quite touched by so much kindness.

After a twenty-minute stay, I left Zurich at 2:15 and was in Basel at 4:45. Since the train didn't leave until 7pm, I had time to see the city and have my growing beard cut. The city is quite nice and if luck would have it that I would go there, we would certainly feel comfortable. However, as far as I was able to conclude from the attached restaurant menu cards, Basel seems to be more expensive than Vienna for a meal in a middle-class restaurant (e.g.) Menu: 4 courses, 3 Fr 40; Menu: 3 courses, 2 Fr 50.

At 7 o'clock in the evening I traveled on and after only three minutes I was in Alsace-Lorraine, i.e., according to the new geography in France. In the evening light, the view of the landscape was very beautiful. When we arrived in Strasbourg at 9pm, it was already dark. I slept very well and had a fluent French conversation with my coupé comrades, who were very kind to me as an Austrian.

In the morning sun, I saw the Belgian countryside and a glimpse of Brussels; now at 8 o'clock I will immediately continue and, after traveling through Ghent and Bruges, I will arrive at the coast in two hours.

I hope to receive a message from you soon. I don't need to write that I keep thinking of you and keep looking at your two pictures (the new one and the passport picture) - I showed them to Prof. Schrödinger, by the way.

[…]

Hail and kiss your Arthur

This is one of the few times that Arthur Haas mentions his much more successful fellow student Erwin Schrödinger who lived in Zurich at the time.[ii] Only with this brief episode does Haas portray Schrödinger as the warm-hearted person who came to the aid of his suffering fellow countryman.

Letter number four from Dover is a postcard in which Haas enthusiastically announces his arrival in England. Once in London, he was overwhelmed by his impressions and began to write in greater detail.

Arthur to Emma London, May 29, 1924, c/o Prof. Donnan 23 Woburn Square WC1

No. 5.

My dearest Bice

[…]

I crossed the canal in splendid weather and a completely calm sea. The view of Ostend, the sea with its ships and seagulls, the rising of the chalk cliffs of Dover were impressions I will never forget. The crossing lasted from 10:30 - 2:30; then it was a two-hour ride to London, where I arrived at 4:30. The first sight of English meadow landscapes densely populated with herds was marvelous, the car journey from the station to the apartment overwhelming. Woburn Square is surrounded by a beautiful park and is full of single-family houses, narrow and four-story. One of these is the house of Professor Donnan who lives there with two sisters. They are incredibly nice to me. They served me an opulent tea followed one hour later by dinner (omelet, bowl with chicken, peas, salad, strawberries with whipped cream); afterward, on a 2½ hour walk, Professor Donnan led me through the main streets of London, which seemed to be quite empty but can hardly be called beautiful.

On the way we also talked about America, about the possibility of being invited for "courses" there and receiving a permanent position. Prof. Donnan, who knows the circumstances in America very well and will be there again in July, explained that I literally only need to want to go there; every university there would be glad to have me. Since he judges everything very soberly, I believe that I can at least draw the conclusion that it would be possible for me to find a position at any of the American universities. And Prof. Donnan promised me his strong support. He wants to do everything he can to help me. In England there is probably nothing to hope for.

I slept very well in wonderfully pleasant guest rooms and was picked up at 10 o'clock by Mr. Emil Hatschek,[5] an outstanding chemist who took over for today's needs in a most amiable manner. We drove around a lot on bus roofs, saw the "City," London Bridge, St. Paul, the harbor, and had lunch together. Now I am writing this letter at 2 pm in his flat.

Tomorrow at 5:15 pm is, as you know, my lecture, preceded by official tea with the rector of the University College, to which about thirty to forty invitations have been sent. Dinner with Professor Donnan is later.

The English is excellent; I speak with many mistakes but fluently and confidently. With Prof. Donnan and his family, I speak only English, with Mr. Hatschek of course German.

I am very curious about your first letter and hope to have it in my hands soon. Warmest greetings and kisses from your Arthur

Arthur to Emma London, May 30, 1924, c/o Prof. Donnan 23 Woburn Square WC1

No. 6.

Bice darling

Today is the "big day" and unfortunately, I will hardly be able to write to you in more detail. I will write quickly before the tea, which the rector of the University College is holding in my honor just before the lecture.

Above all, I want to thank you very much for your so very kind lines from Salzburg, which I received this morning (the mail comes here about ten times a day). How comfortable I would feel in London if I wasn't longing so much for you, Bice, and if the uncertainty of the future didn't weigh so heavily on my soul!

After I wrote letter No. 5 to you yesterday, I was alone on Oxford Street and at the Selfridge department store. Then I met up again with Mr. Hatschek who took me through London's parks before we dined lavishly (as his guest) in a very elegant restaurant. Afterward, I went to his apartment due to heavy rain and after 10:00, I was still together with the Donnan family for more than an hour.

You cannot imagine how nice everyone is to me. This morning I was alone in Hyde Park, busy observing life on the famous Rotten-Row [Fig. 11.2]. I had lunch with Prof. Donnan (as his guest) in the dining hall of University College where I saw hundreds of male and female students, the latter mostly pretty and with boy's hair - quite different from our girl students. Afterward I was with Prof. Donnan in the big hall for professors of the university, who were sitting together over black coffee. I became acquainted with many of them and spoke fluent English with all of them about all kinds of problems.

[…]

The question that concerns me most is what will happen about America. I asked Mr. Hatschek what he thought of Prof. Donnan's statements about this. Mr. Hatschek is always a skeptic and pessimist, but he literally said: "If Prof. Donnan promises to do something for someone and says himself that he will achieve it, then he will certainly achieve it". God give it in this case! If I could only breathe a sigh of relief! Will today's lecture be my swan song, my farewell to science or the basis for a new ascent? That's a terrible question. - Greetings and kisses, dearest Bice, Arthur.

Many greetings also to your father!

---

[5] Emil Hatschek (1868—1943) was originally from Budapest and moved to England with his parents before 1900. In 1900, he became a naturalized Englishman. He was an expert in colloid chemistry and published several books on the subject.

Fig. 11.2 Around 1890 in Hyde Park, Rotten-Row was where riders, carriages and people strolled and enjoyed the mutual display

P.S. Sir Henry Strakosch[6] is unfortunately in Geneva; his brother invites me for Monday evening. After Monday, June 2nd, don't write back to me in London. I will probably leave on the morning of June 6th! Arrival in Vienna on June 8th.

Arthur to Emma London, June 1, 1924
No. 7.

Bice darling,

When you read how yesterday went, you will forgive me for not being able to write to you for one day.

After I wrote letter No. 6, I went at 4:30 on May 30th to have tea, which was given in my honor by the rector of the University College. I was very disappointed since it was a trivial formality. At 5:15 pm my lecture began. Attendance was "not bad" but disappointed me. A very large lecture hall, which could hold about 300 people, was only half full; the agonizing heat was probably to blame. [...] I had immediately grasped the situation to the extent that I knew, with regard to my future, that only one thing mattered - the impression the lecture would make on Prof. Donnan, both in terms of its content and my ability to lecture in English; afterward, he spoke very positively about the lecture, not only to me but also to third parties.

6 Sir Henry Strakosch (1871—1943) is the son of Eduard Strakosch, the brother of Arthur Haas' grandfather Bernhard Strakosch. Sir Henry built a successful banking business in England. He was a close friend of Winston Churchill and often supported him through financial crises. Strakosch played an important role as a financial advisor to British and Colonial government authorities, for which he was ennobled in 1921. See Marie-Theres Arnborn, *Friedmann, Gutmann, Lieben, Mandl, Strakosch* (Wien: Böhlau, 2002), 118.

Fig. 11.3 Stadium in the center of the exhibition at Wembley, presenting the achievements of the British Empire

The chair of the lecture session, Prof. Porter, Prof. of Physics at University College, introduced me to the auditorium in a two-minute speech, in mumbled English of which I understood nothing. Then I began. I had the feeling that it went very well. I spoke more slowly than at the rehearsal. I didn't stare at the manuscript at all and at the beginning of each sentence, I thought through its content. The attention was marvelous. If I had a stormy beginning, there were applause afterward, which caused me to take at least twelve bows at longer intervals. Then Prof. Porter took the floor for a kind of discussion about the lecture. He spoke for a few minutes. Then Prof. Donnan had his say. He described the part that I will play in the forming of modern atomic theory (great applause), then my "merits" through my books (good applause) and then praised the lecture today (good applause). After, we went to Prof. Donnan's for dinner. [...]

Prof. Donnan, who kindly dedicated his entire Saturday to me from 9 am to midnight, took me to Wembley Park for the big exhibition [Fig. 11.3]. We were there in ten minutes, since the trains were speeding along at breakneck speed. The impression was tremendous.

We visited the Palace of Industry, the palaces of New Zealand, Australia, Canada, India, Burma, the old London Bridge; one was more magnificent than the other. (If only I had you with me, my Bice!) Then Prof. D., who is a typical man of the world, took me to lunch in the restaurant of the exhibition "Lucullus." Grapefruit, a wonderful fish, roast and vegetables, asparagus, strawberry cake, bar drink, wonderful Moselle wine, café. (And already at 8:00, I had had an "English Breakfast.") Afterward, over a black coffee, came the highly significant conversation with Prof. D., which I will report on later. Afterward the Palace of Engineering, the most advanced one can think of, and then into the stadium, which holds almost 125,000 people. There English military music was played by more than 1,000 men, first performing military marches and then, just think, they played Wagner for one-and-a-half hours.

From there home, dinner at home (as always with lots of wine), then in a flash change to tuxedos and, as the guest of Prof. Donnan, to the Criterion Theatre where a comedy was excellently performed. I could understand everything! Afterward, just think, Prof. D. took me to a very distinguished restaurant with a concert, another soup and again bar drinks and punch. Home at 12:30. English hospitality is gigantic. Since I arrived in London, I have not spent a penny except for the hairdresser!

And now the most important thing, the conversation with Prof. Donnan over black coffee after lunch. Prof. Donnan spontaneously promised to do everything possible for me and suggested that I speak to him quite openly. You cannot imagine what a good person Prof. D. is. So, I talked candidly about everything, after I had told Ms. Donnan the day before about our engagement; they are very interested in you, noticed your picture. (Prof. Donnan showed no interest in this respect; I think because he is as a matter of principle against marriage.)

Then Prof. Donnan told me all of his plans. I would need to go to America, to the chemistry department of a university where I would teach atomic theory, because the chemists lack contact with modern physics. At two universities, both of the very first rank, Prof. D. has highly influential and blindly devoted former students, namely at Princeton University (three-hour train ride from New York) and the paradisiacal Pasadena in California. His students are professors at each. He has no doubt that I could be employed by one of these as a well-paid professor and he wants to make the most concerted effort in this direction.

When I asked him about a salary of $500-1000 a month, he said I should expect $700. This is enough for a family in a small town but not enough for a family in New York to live comfortably. But I could make a significant amount of extra income from industry contracts. When I remarked that I am a theoretical physicist, he said that especially in the USA, industry is mostly looking for this specialty. He knew this precisely because his own primary income was not from his professorship but from his work for English and American industry. I could also earn huge amounts of money if I would write popular-science books in English for an American publisher. Concerning the language difficulties, Prof. D. said that I already know enough English to be able to give unprepared scientific lectures but that I would in any case very quickly overcome the language deficiencies.

Prof. Donnan is so respected in America that he will receive, for example, $1,000 for a single lecture at the Franklin Institute in Philadelphia in September. I can see that he really appreciates me from the fact that he spontaneously wrote a foreword to the English edition of my large two-volume book, which will be published in London in a month.

I will of course continue all efforts in every direction with tenacity and will immediately seize every professional opportunity that presents itself to me wherever it arises; if I am then redeemed by employment with either Princeton or Pasadena, all the better.

But in any case, I have to tell you that yesterday's meeting with Prof. D. has really awakened my courage and self-confidence again.

[…]

P.S. By a happy coincidence, the physicist Prof. Millikan from Pasadena (formerly in Chicago, you know his name) will come to London in mid-June and Prof. Donnan will talk to him about me.

With this trip to London, Haas seems to have successfully taken the first step toward establishing a network with the Anglo-Saxon scientific community – a close community that emerged after many of the previously established connections between German and American scientists had been disrupted during the war. Frederick George Donnan was sincere in his efforts to rebuild these connections and considered it his life's work to offer German scientists an opportunity in England and the United States. This was in part because he had received his doctorate in 1896 in Leipzig from the leading chemist of his time, Wilhelm Ostwald. Afterward,

Donnan maintained close and personal relationships with German scientists, and he also developed great respect for the achievements of German science during the war.

During the war, Donnan worked on the extraction of nitrogen from the atmosphere by absorption in plant material, for the production of ammunition; this was a goal that Fritz Haber on the German side developed with success and more directly through the Haber-Bosch process. Besides the well recorded and widely applauded charity of Donnan,[iii] there was possibly also a strategic interest in these efforts to attract capable scientists to England and America after the first war. His colleague, Emil Hatschek who was born in Hungary, had set an early example to be followed by many others. Likewise, Donnan helped in particular many young Jewish scientists to escape after 1933 and to build new lives and new careers in England.

Arthur to Emma London, June 2, 1924
No. 8.

Bice darling,

Why don't you write to me about if and how you arrived in Munich and how you found your father? [...]

Now at least my future appears in a rosy light and my worries are only about the next few months. Yesterday, I confidentially told Mr. Hatschek about Saturday's conversation with Prof. Donnan regarding America and asked him for his opinion. He, who is very pessimistic in everything, literally told me: "If D. promises you something, you can absolutely rely on it". And on the other hand, Prof. D. has assured me that he will do so.

It would be wonderful if next spring we could settle down together in America and if we succeed, we could live there very well! Of course, "the dove on the roof will not let me spurn a sparrow in my hand". I will also eagerly continue efforts to achieve something in Vienna, at least as a provisional solution. On the issue of the houses in Berlin, I also have greater hope for better sales opportunities, as the economic situation in Germany is improving.

[...]

This morning I was alone; I went early through the city, to Guildhall, in the streets, to the British Museum, further through the streets. Now, at 7:30, my cousin Friedrich Wilhelm Strakosch will pick me up, with whom I will spend the evening.

[...]

Emma to Arthur Kapuzinerstraße 20, Munich, May 28, 1924.

My dear Professor

Your first letter leaves a lot to be desired. I will carefully keep it as important incriminating evidence for our later divorce because of your coldness.

Well - just don't imagine that I will ever send you a kiss or eat breakfast with you again. And when I think of how often and how much I keep thinking about you, and that I even cried because I was so homesick for you - and then I get such an icy letter from you - well, I am really a fool. But just wait. I'm going to phone somebody and have fun too, just heavenly. See, the phone is already ringing. It's for me - Goodbye!

May 29, 1924.

It's Friday but I've turned down all the invitations and I'm staying in. I just want to pack my suitcase and go to England. I also got your card from Basel and your letter from Brussels. Thank you very much. I'm glad you're enjoying the trip so much. Tomorrow is your lecture. I'll send a telegram early in the morning, so you'll receive it before 5:00.

[...]

Arthur to Emma London, June 3, 1924.
No. 9.

My dearest Bice,

This morning I received your dear four-day letter of May 28-31. Thank you so much for that. I was so happy because I had another bad dream for an hour. Your letters are so dear and graceful, and mine are so crude. That's thc way we both are.

Were you really annoyed that my letters were so cold? You know well enough how I feel and it is not a phrase to say; I do not make a move in London without having you by my side in my thoughts. When I look something up in the *Baedeker*, while standing in front of a famous building, I imagine that I am not reading it but that you are at my side, explaining everything to me as the much cleverer one.

And when I am sitting in a luxury restaurant as the guest of someone who is dear to me (as I was yesterday with Mr. Strakosch), I send him away in my mind so that I can be with you, Bice, in undisturbed tête-à-tête. And believe me, I really have no other interest here than building a bridge for our comfortable future, which I hope to be able to provide for you before a year passes, maybe half a year from now in America. And I'm such a child; I see us buying the best Californian fruit in the paradise of Pasadena and sending it to those who show me so much kindness now in London.

My faith in an American future is getting stronger every day. It is good that I will be Mr. Hatschek's guest again tomorrow from morning to afternoon and that I will be able to talk to him some more.

[…]

Dear Bice, you know that the most important question at the moment is the sale of the houses; the sooner we succeed, the sooner we can be united; and if only we could do it now, so that we can be together in the summer. Please have a look in the newspaper, maybe at agents who your father recommends (but without all the expenses!!). If you find someone, let him make a detailed offer to my brother (office: Stubenring 14) with exact details of the payment method. But there must be nothing to pay if nothing comes of it! Don't be angry that I am bothering you with this request but "tua res agitur" and perhaps something more can be achieved from Munich than from Vienna.

[…]

Last night was very nice with Mr. Frederick William Strakosch.[7] He took me to one of the finest restaurants (Piccadilly Hotel) and afterward to the Little Theatre where all sorts of funny little sketches were performed. When it was very funny, I heard you shout, "Dear me" and "sooo right," once also "Sixtes" [Munich dialect] and then you gave me half a candy. Then I felt so much longing for you and I could not laugh anymore.

Best wishes, dearest Bice, and many deep kisses (on mouth, eyes and hair) from your Arthur

Emma to Arthur Kapuzinerstraße 20, Munich, June 1, 1924.

Dear Arthur,

I just don't feel like writing letters. I'd like to do everything over the radio.[8] I have received all your letters and cards - thank you very much for that. But you know, you could just as well address them to some grandmother - they are so terribly "harmless." Well - if I was a little more in the mood, my letters could not be printed with "skin and hair" in a newspaper - yours could. It all gets saved for incriminating material.

---

[7] Friedrich Strakosch was a distant uncle of Arthur Haas, born in 1869. Like his brother Sir Henry Strakosch, he was also a banker in London but played a far less prominent role.

[8] She refers here to the telegraphic techniques used at the time.

[…]

Of course, I immediately talked to Papa about the houses - he will do his utmost to help you and so will I, as a cheap pastime; the main thing in Munich is that a pastime is cheap.

I would actually know a lot more to tell you but, first of all, I'm not in the mood today. Secondly, I hope that tomorrow I'll feel more like writing and thirdly, I've written enough today.

I wanted to send a telegram to your lecture. But it wouldn't have gotten there in time, so I didn't. I am inclosing it with the letter. I didn't think of the radio.

Please, next time you write, remember that I'm not a grandmother yet.

Arthur to Emma London, June 4, 1924.
No. 10.

Dearest Bice.

I have just found your letter No. 3 for which I thank you so much. I am so sorry that you are so unhappy with my letters. I hope that you at least find me "fierier" in conversation than in my letters. But I want to pull myself together!

Above all, I would like to make up for one more "thank you" that I neglected. Mr. Hatschek told me today that the style of my lecture was very good (except for one passage where the meaning was distorted by the translation and where the word "raise" was used with the wrong meaning). I know that the effect of the lecture depended very much on the style and I told him with great pride that you helped me prepare the lecture and contributed almost everything to its success. I must thank you, my dearest Bice, very much.

I was very touched that you wanted to telegraph me before the lecture. I really don't deserve your kindnesses. I was also very happy about your little picture. How were the others? I am going right now to the local photographer "honoris causa"!

[…]

This morning I met with Mr. Hatschek; he showed me around the enormous South Kensington Museum (different sections on zoology, arts and crafts, science). Afterward he took me to an, as always, corpulent (I only meant opulent) lunch and then I said goodbye to him. Of course, we also talked about the USA. I learned from him that Prof. Donnan wants to arrange things in person, not in writing, when he goes to New York, Princeton and Philadelphia in September. He believes that if all goes well, I will receive the appointment around Christmas 1924.

[…]

Friday (the day after tomorrow), I will go back to Vienna! How bleak it will seem to me without you, Bice, Bice!

Hopefully I will get a letter before I leave and with at least one kiss in it.

Just don't queue up too much in Munich! And may you get to Vienna as soon as possible. I'll do everything I can to make sure that it can happen as soon as possible. Then I will have more than just your suitcase. If only everything had already been overcome. Then a different atmosphere would lead to more amusing letters.

Greetings and affection, dearest Bice, kissed by your Arthur

Emma to Arthur Kapuzinerstraße 20, Munich June 4, 1924.

Dearest Arthur,

What's the matter with you now? I'm so scared for you. I don't want to think about you being sick or anything else that might make you uncomfortable. On Monday, I received the letter you wrote to me right after your lecture - but since then nothing more. I am so terribly nervous about it because I am eagerly awaiting the report about your lecture.

This morning I'm taking the 11 o'clock train to the Upper Palatinate to visit a friend - but I don't feel like leaving Munich in such a state as I am now. If I don't hear from you soon, I'll keep telegraphing to England or Vienna until I know how you are.

Now I have written the letter and don't know where to send it. It will be too late to England. You won't be in Vienna until next week. That's terrible. If it goes on like this, I'll lose a kilo a day.

A big kiss from your Bice.

Arthur to Emma London, June 5, 1924.
No. 11.

My dearest best Bice!

I wonder if you remembered that today is a small anniversary - just two months ago. I have just retired to my room to write to you at this hour.

Since that day, which was the happiest in my life, the sky has unfortunately been covered with black clouds. But now I am confident that it will not be long before we can rejoice together in the sunshine. If only we were already over the very near future.

I have just had a one-hour conversation with a very clever woman who now lives in London and is translating my book into Russian. She has a good friend at Illinois University in Chicago, whom she will immediately contact about me. She thinks my chances in America are very good; she liked the style of my lectures here (your work!) very much.

[…]

Here in London, I am overwhelmed by a profusion of impressions. But when I have returned to bleak Vienna, if I have to again beg for a position every day, how will I bear it without you?

[…]

Emma to Arthur Amberg-Bayreuth, June 5, 1924.

Lumpacivagabundus!

Well, well, well. I'm scared shitless for nothing. You just couldn't write to me because of all the dawdling and lamenting. Just wait until I catch you; revenge is sweet.

But the letter was interesting. It goes without saying that I have read it repeatedly since it arrived. But your writing is miserable. It's a pig's claw effort to decipher your hieroglyphs. It's a pity Prof. Donnan is such a misogynist. We will discuss the content of your letters in person. Of course, I am very pleased with you. But that's all the compliments I'm going to pay you.

Maybe I'll wrap myself around your neck one day for this [Fig. 11.4]. Tomorrow, you said, you were leaving England. "Goodbye England, parting is painful!"

Just come home soon. I want to get married!

Bice

I might go back to Munich tomorrow because it's already been raining for two days.

Arthur to Emma London, June 6, 1924.
No. 12.

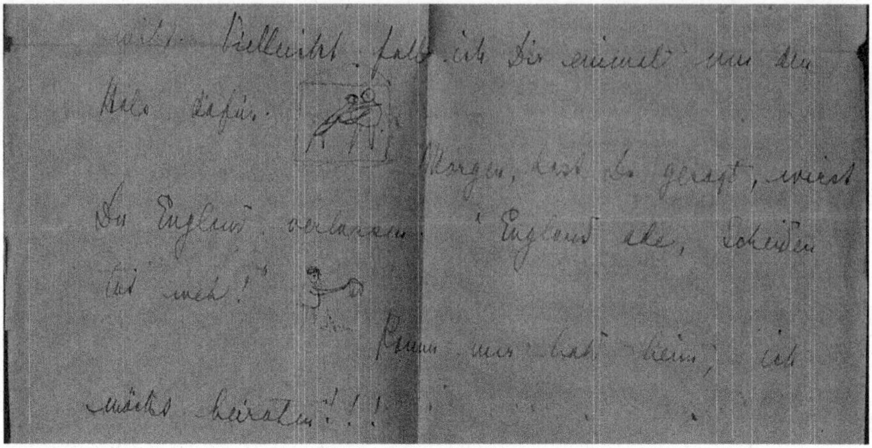
Vielleicht fall ich dir einmal um den Hals dafür.
Morgen, hast du gesagt, wirst du England verlassen. "England ade, scheiden tut weh!"
möchte heiraten!!!

Fig. 11.4 Handwriting and drawings by Emma Beatrice Huber

Bice darling,

After I wrote yesterday evening's letter, I went downstairs to Prof. Donnan's study and he then sat down with me in the "hall" for another hour.

I first asked him to allow me to dedicate the new book *Atomic Theory* to him, which will be published in four to six weeks. This made a big impression on him (it's not the book that I want to dedicate to you, but a finished and as yet unpublished one). Then he urged me to tell him everything about my situation - I did.

So, he has decided not to wait until he is in America in August-September but is going to write immediately to numerous American university professors who are his students, not just to Princeton and Pasadena.

He has no doubts about success.

[…]

Arthur to Emma On the Swiss Railroad, June 7, 1924.
No. 13.

My dearest, best Bice!

I am beginning this letter in Basel. I left London, as I told you, on Friday, June 6th at 2pm. In the morning I spoke again with Prof. D. for a long time. He said that it was a great pity that I had not told him everything two months ago. He was sure that he could have gotten me two summer courses in America for this summer (fee, $1,000 each). He can do it every year without any problem. For an immediate permanent appointment outside Vienna, it is a problem that no one knows me personally but he will certainly get past that.

Yesterday, Friday morning, I was still strolling around in London; at 4:00, I was at the sea; the crossing was not quite as quiet as the outward journey but still, for me, without any trace of sea sickness. In the evening at 9:00, I arrived in Ostend, which made a wonderful impression on landing, and then left by train. From Anglo-Saxon high civilization to Romanesque dung heaps; in Luxembourg our car did not reach the connection.

I was forced to stay in Luxembourg from 5 - 7:30 this morning. A splendidly picturesque city, beautifully situated on both sides of a very deep valley vaulted by beautiful bridges. The city is half French, half German. Then the train goes through Alsace Lorraine, which has become French. Magnificent, the blast furnaces of Lorraine are of picturesque effect. The Strasbourg cathedral is visible from afar. Since we missed the connection in Luxembourg, we went on in overcrowded excursion trains, with constant and torturous changes. Since the corridors were crowded, the French officers in Mulhouse helped me in the friendliest way by handing me my heavy suitcases through the window. […] When saying goodbye, they asked me in a friendly way if I am "de l'Allemange" [from Germany]!

[…]

Switzerland, which had appeared to me as a jewel box on the outward journey, gave me the impression of a lumber warehouse when I returned from England, and Austria gave me the impression of a dung heap.

[…]

Now the answer from Chicago will soon follow; I wonder how it will turn out! I do not have much hope. In any case, I will immediately refer to Professor F. G. Donnan, Professor of Chemistry at the University College in London as a reference, if the answer is not unfavorable.

Best wishes and many, many, many kisses! From your

Arthur

P.S. The Friday issue of the London *Nature* brought a report of my lecture! I will send you the issue as printed matter. But please send it back to me.

Many greetings also to your father.

Arthur to Emma Between Innsbruck and Salzburg, June 8, 1924.
No. 14.

Bice, honey, Bice!

A few hours ago, I posted a letter to you in Innsbruck but I must write to you again; I am longing for you so much. […]

The closer I get to Salzburg the sadder I get. How happy we both were here in Salzburg on April 13th, and how sad - at least for me - the last stay in Salzburg was. May we be together again in Salzburg very soon and be as happy as we were that day when we met your father.

2 PM And Hohensalzburg is Visible, O Bice!!

Arthur to Emma Near Wels, June 8, 1924.
No. 15.

Bice darling,

You will think me "Cuckoo"! What's wrong with him, that he writes me three letters within eight hours? Nothing. It's just that I'm so terribly, terribly longing for you.

I'm taking the Salzburg-Vienna route, which I've taken with you three times; after Salzburg I closed both eyes and thought you were sitting next to me in your traveling outfit with your red hat, which I love so much, leaning your head on my shoulder and stroking my hand a little bit, like you did for the first time at the opera the day before our engagement.

Now I am torn out of my reverie and have to think of the harsh reality. I wonder what kind of news I will find in Vienna! I am as pessimistic about an imminent position in Vienna as I am optimistic about an American position by Christmas. The most important thing at the moment seems to be the sale of the houses so that we can marry very soon.

[…]

After fourteen days, I now read Viennese newspapers again. The stock market situation is the same as it was fourteen days ago, no better and no worse. I found a new advertisement about Berlin houses. I want to investigate the matter tomorrow.

Emma to Arthur Villa Grau, Holzhausen, Ammersee, Pentecost Sunday Morning.[9]

Sixtus!

Now I'm back at Lake Ammersee. The day before yesterday I was still in Oberpfalz, yesterday I spent a night in Munich and today, that means right now, I am lying on a deck chair in the park and the bees and beetles are buzzing around me, some are crawling on top of me, of course the birds are chirping, through the trees I can see the sailing boats, today there is a regatta, and from far away I can hear a motorboat, then I will watch it as it goes by and if there's nothing more to see on the lake, then I'll watch the butterflies and mosquitoes again, and listen to the birds, and because I've done all that enough now, and perhaps I don't know what else I can do, I'll write you a letter because you are Arthur Haas. But it's not very comfortable. However, I just want to tell you that the Ammersee is the most beautiful place and all of Vienna, etc. cannot compare to it. Unfortunately, I have to go back to Munich in a few days because my girlfriend's parents suffered big losses this year. So, I don't want to live here for nothing for too long.

Afternoon

We just watched the regatta [Fig. 11.5]. I think there were at least twenty-five boats involved.

[…]

Today you will arrive in Vienna. Well, what then! Your Bice

This letter was written from Eugenie Grau's house. She was Emma's friend from her last days at school during the war in 1915, before traveling to Chicago. Emma visited Eugenie several times at the Ammersee and was drawn back there in times of personal stress. One can see that this letter is much calmer and friendlier than the ones written from her Munich home on Kapuzinerstraße.

Although now aware of its emptiness and post-war misery, Arthur was returning to Vienna where he had spent carefree days with Bice before his trip to London. This trip had opened his eyes to the world outside of Vienna, to the big-city life that he himself had once enjoyed and since forgotten, repressed through war and post-war years. Vienna now seemed all the emptier and more hopeless after his return.

Arthur to Emma Vienna, June 8, 1924.

No. 16.

Dearest, dearest Bice,

After arriving in Vienna, I found your dear letters of June 4th and June 5th (Amberg). I was so happy and read them again and again. Many thanks for that. Were you really frightened because you didn't hear anything for one day? I wouldn't have thought so! Anyway, in future no day shall pass without me writing to you. Hopefully it won't take too long at all until you can come to Vienna and then move into Neulingsgasse. […]

Arrival in Vienna was dismal. I found (at 9:30 in the evening) the table full of letters. Especially yours, which of course I read first; a letter from Prof. Lotmar[10] who said that

---

[9] June 8, 1924.

[10] This is probably the private lecturer Dr. Fritz Lotmar (1878—1964), lecturer in internal medicine at the University of Bern. He worked with Haas on improvements to the second volume of theoretical physics. Although a neurologist of great scientific reputation during his work in Bern and Munich, he was also interested in problems of physiological chemistry, mathematics and physics, which probably brought him into contact with Haas. In 1958, Lotmar even wrote about Einstein's theory

Fig. 11.5 Regatta on the Ammersee

Basel is absolutely hopeless; a letter from Prof. Ehrenfest[11] (Leiden) who took away all hope for Holland but assured me of his eager support for efforts to America; other letters of lesser importance. I cannot see my brother until Tuesday. He writes me that he has not

of relativity. Besides these broad scientific interests, Fritz Lotmar was very interested in modern art and is considered one of the closest friends of the expressionist painter Paul Klee.

[11] Paul Ehrenfest (1880—1933) from a Jewish family in Moravia, studied under Boltzmann and moved to Göttingen in 1903, at the same time that Haas arrived at the university. In 1911, Ehrenfest was appointed to succeed the Dutch physicist Hendrik A. Lorentz in Leiden, carrying out important fundamental work in the fields of thermodynamics and quantum physics. Despite common origins and education, the acquaintance was probably based more on professional than personal exchange.

yet been able to achieve anything positive about the houses. And not a single letter telling me that progress had been made on the question of a commercial position in Vienna. A Dr. Tüschel, whom I don't know, drove up in his car on Saturday and urgently asked for me; I should call him on Tuesday. (I think it's regarding a scientific film center!)

[…]

After lunch (during the morning I have been working on urgent book corrections) I went, thank God, to see an old childhood friend, engineer Hans Ellinger, with whom I have been close friends since I was four years old and who was absent from Vienna in early May. To my great pleasure he told me that he has a friend with a lot of cash who is desperately looking to buy Berlin houses. He called him immediately and made an appointment for this evening. He hopes for more than I have been offered so far. (He anticipates 1¼ to 1¾ billion crowns.) Moreover, he has a very good friend (no, a cousin) who is commercially active in the electro-technical field and earns a lot of money. He is looking for an employee and my friend wants to recommend me. I would, however, have to travel a lot. Finally, he wants to recommend me to another friend who has opened a successful radio shop.

[…]

After this letter to you, I will be writing a long one to Prof. Donnan and an even longer one to Prof. Ehrenfest who, as I already wrote to you, is also willing to support my move to America.

- Oh Bice, I'm so much longing for you, and yet we must be patient. If Ellinger succeeds in selling the house (and I would probably get all the money immediately) then we could get married right away, provided you know how to adapt to the given situation. If there is a corresponding professional income, we could certainly live comfortably. But let's not praise the day before the evening, and if it doesn't work out, let's continue to remain patient.

[…]

Also, many thanks for the fact that, as my brother tells me, you were so kind to write to him and Uncle Fritz.

Arthur to Emma Vienna, June 10, 1924.
No. 17.

My dearest Bice,

Today has brought progress in many ways. In the morning I worked out compilations of my scientific papers (a total of thirty) for Prof. Donnan, who needs them for America.

Then a visit to a friend, who was not at home, followed by conferences with my brother and two house agents; in the afternoon to my friend Ellinger. His friend is willing to pay 900,000,000 Kr for the houses immediately, but we hope to reach 1,000,000,000. The tremendous advantage would be that all the money would be paid immediately in Vienna in cash, whereas generally only down payments are available.

At present, the absolutely safe bonds of the League of Nations (in dollars) have an interest rate of more than 8%, and the Southern Railway Priorities, which are denominated in gold francs, have an interest rate of more than 10%. If I invest half of the proceeds from the 900 million crowns in League of Nations facilities, and half in southern runway priorities, I have secured a 6 ½ - 7 million monthly annuity. Since I just received from England 160 £ (approx. 50 million kroner) for the translation of my *Introduction* [*to Theoretical Physics*] and since another approx. 16 million kroner for my *Atomic Theory* are due, I can draw 3 million kroner per month for 18 months, whereby the income would increase to approx. 9 ½ million for the time being, money with which we would have to cover our living expenses until I have found a regular income. I can't imagine that I would not succeed in having this amount through autumn 1925 - not to mention my hopeful confidence in America.

Also, if nothing were to happen in America and unemployment were to continue in Vienna, I could at least earn a little more money through journalistic and scientific work and from lectures outside Austria.

Finally, Prof. Donnan said one thing was certain, that he could always get me summer lectures in the USA. Since I would get $2,000 in the summer and would then be able to save more than half, I could also create a certain income this way, but that would be very bitter since I wouldn't be able to take you with me. But let's hope that I will get a permanent position in America soon and we will go there together.

Of course, I still have certain reserves, as you know, in Brno houses, a share of other properties, pre-war claims, and my heavily burdened bank deposits are still active. If I can cash in on this, I would hope for about 300-500 million crowns. Further reserves are the later editions of my books. If the current 3rd and 4th editions of my two-volume work are followed by a 5th and 6th edition in a few years, this means for me 10% of 3,000 x 16 goldmark or 4,800 goldmark = approx. 80,000,000 crowns.

[…]

It should be kept in mind that while Haas was writing these letters of complaint, constantly evaluating his budget and calculating his earning potential, he was also steadily working on physics. He not only completed revisions for the new editions of his theory books—which entailed an intensive correspondence with specialist colleagues and helpers, such as Fritz Lotmar—but also created new books. These included the book on atomic theory already described in the last chapter, which was completed and published in 1924 during this extremely difficult year. These projects took an enormous amount of work during this period of deep personal crisis.

In addition, in the following weeks there was a rapid increase in his correspondence regarding the sale of his houses, which he saw as the only way out of financial misery. This all means that he must have been working to a strict schedule, in order to keep his life and workload moving at such a pass.

Arthur to Emma Vienna, June 11, 1924.
No. 18.

Dearest Bice,

Yesterday I was so hopeful. Today, my hope has evaporated. I'm all broken down. The gentleman who wants to buy the houses from me suddenly demands that I take any increase in value on myself, which is a difference of 250 million crowns. So we could not agree. We must continue to look for new ways forward to achieve the desired result. If only I could give you a final report. I haven't felt this desperate in a long time. My friend Ellinger knows of other possibilities and he'll investigate them right away. With haste and a heartfelt greeting and kiss from your Arthur

Thank you very much for your letter from Pentecost.

Emma to Arthur Villa Grau, Holzhausen, Ammersee, Wednesday after Penecost[12]

Before I begin, I want to tell you that I am again in the garden lying on an easy chair, so I am not responsible for any spots etc. or for bad writing, although I could never write as badly as you do even if I tried. It took me all morning to read your letters.

---

[12] June 11, 1924.

Neither am I responsible for anything that I am writing, for the birds are singing so beautifully that I love to close my eyes and listen to them. And you ought to see the flowers and trees around me - they are all so lovely. It's so still and restful here and yet there is so much life in this quietness. Arthur darling, I so wish you could sit here next to me and also admire the beautiful spot.

I am so sorry that I must leave this place either tomorrow or the day after, for I really cannot stay here any longer. The parents of my girlfriend are so nice to me but just the same, I must go back to Munich. My father not doing very well either; he has to repay so much from the last four months. This winter brought a dreadful loss for him, which he tries to make up for now. I hope matters go alright with you. And - I want to come to Vienna soon. I don't like to wait.

I thank you ever so much for your letters 11, 12, 13, 14 and 15. Of course I enjoy them. Papa is trying his best to help you sell the houses but until now without success.

I shall write again soon. Hugs and kisses from your Bice

P.S. Dear me, I forgot to begin this letter. Excuse me!

Arthur to Emma Vienna, June 12, 1924.
No. 19.

Bice, Bice dear.

I want the letter I posted last night to be followed by another one early in the morning; perhaps they will arrive at the same time.

I couldn't get any sleep last night. Even though nothing special happened, I had been so sure that the sale of the houses would be completed in the next few days and we would be able to get married - now the matter is in question again and that was like a blow to my head. But I still hope to sell the houses through my friend, who looks very earnestly for a buyer, so that a secure permanent monthly pension of 6 - 7 million crowns will result.

But now is the worst time to sell. No one has money to buy. And as a result of the terrible crisis that is now prevailing in Germany, houses are being offered at ever lower prices without finding a buyer. Everyone tells me: Wait until September, it is madness to sell now. But I can't get married; I can't even lift a finger if I haven't sold the houses, unless I find a well-paying job. In this respect I have some hope, although the pay is bad at first and the work (travelling often) will be very hard. But I will leave no stone unturned so that I can finally bring you home. This terrible situation, who could have guessed on April 5th!

Bice, dearest Bice, just write to me about what to do. Should I rush everything or wait a little? [...] It is expected that the dreadful crisis that so frustrates both the sale of a house and finding a profession, as well as the smallest amount of money, will soon be over. And I won't be able to get any news about America for at least a month.

[...]

Wouldn't you like to write again to the dean in Chicago and tell him that I was successful in London and that Prof. F.G. Donnan, Professor of Chemistry at University College, London WC 1, can give more detailed information about me?

Finally, one more thing. I read an announcement in the *Münchener Neuste Nachrichten* about a real estate agency Lennig & Cuhl, Munich, Zweigstraße 10. If you would be so good as to inquire there about the houses, but of course there must not be any expense, only a commission in case of a completed sale. (The essential thing is that the buyer takes over the value-increase tax!)

If only everything was already over! Today will be another day of excitement for me!

Should I send you kisses, something you so rarely do? I don't know if you want them but I kiss your picture so often now and deeply. Yours, Arthur.

This is one of Haas' first major attacks of desperation, caused by the stress of his work and confrontation with the rapidly developing economic crisis. The hyperinflation of 1923 in Germany had been halted by the introduction of the rentenmark in November. However, as in neighboring Austria, this came at the cost of a drastic redistribution of wealth. The bourgeois class lost its savings, people were faced with high debt, and companies lost their capital. Countless people chose to emigrate. Arthur Haas was not the only one driven to America.

At first, homeowners benefited from hyperinflation. The real value of their loans fell accordingly, while the houses retained their value; these are the circumstances that had driven Arthur and Otto Haas to invest in the Berlin housing market. However, in 1924 a tax on interest was levied in order to draw money from real estate assets. Numerous owners were unable to bear the new tax burdens and had to sell their properties, causing property prices to plummet by up to 50%. This was the situation in which Arthur and Otto found themselves, being forced to sell their houses quickly.

There was not enough money to revive the housing market; while the introduction of a temporary currency, the rentenmark, had ensured temporary stability, it was only considered a domestic currency while the stock exchanges still traded in the old mark. It was not until the introduction of the reichsmark in August 1924 that an effective increase in financial activities occurred and thus a stabilization of the housing market. However in June of 1924, the situation that Arthur Haas suddenly found himself in was rather dire with regard to his investments in the Berlin housing market. Therefore, it was his most pressing goal to convert the real estate values into money-generating share values. This was to plague him for the next two months.

Arthur to Emma Vienna, June 13, 1924.
No. 20.

Bice darling

I wonder if the post office will bring me another letter from you today.

The day yesterday, at the early beginning of which I wrote you the last letter, was again terrible. In the morning I had a long meeting with a house agent, then the most pleasant, long meeting with the local university's Professor Schaffer[13] who has been to America several times and will go back again in autumn for six months. He said that approximately three weeks ago, he had already sent out brochures about my work. He emphasized again that it is very easy to get an American professorship and if a respected English scholar is involved, I can be sure of it. He was skeptical about the $700 monthly salary. This might be the case at the most prestigious universities, otherwise $500 is common, sometimes even only $400. On the other hand, it is very easy to earn considerable extra income. This is quite different from Europe (for a lecture outside the university, often only $50).

Afternoon again, long conference with another crooked house agent. Next to this, a lot of work and in the evening, long talks with my sister-in-law and my brother. My brother

[13] Franz Xaver Schaffer (1876—1953) received his doctorate from the University of Vienna in 1899; his fields of study were geology and paleontology. He worked at the Natural History Museum in Vienna from 1900 to 1936 and also taught as a university professor at the university. He was mainly engaged in the analysis of geological strata while his somewhat older colleague Othenio Abel was more involved in the field of paleozoology. Schaffer did not apply for membership in the NSDAP until 1940, where he was admitted in 1944 as of January 1, 1941. At that time, he was already regarded by the Vienna NSDAP as an "enthusiastic National Socialist.".

continues to preach that I should not destroy my health by trying to achieve everything overnight; this cannot be achieved in twenty-four hours. He preaches easily. If he could feel how anxious I am for you. It seems so much stronger than reason. My sister-in-law and my brother both speak of you with sincere esteem. That's right, my sister-in-law's birthday is on June 18th, if you'd like to send her a postcard.

Today I have worked only for myself so that my new edition of the big book is finally finished and now I can get the big English fee. This afternoon I plan to see Professor Lampe[14] in the Urania, then practice my English for America in Max Carl Smyth's Progress Club. Afterward, I will be in the other club to see Mrs. and Mr. Dessauer; they will certainly not be well. It's horrible!

[…]

Greetings and kisses from your Arthur.

Arthur to Emma Vienna, June 15, 1924.
No. 21.

Dearest, dearest Bice.

[…]

In spite of desperate efforts to sell the houses there is no progress to see. News from America cannot be expected before the end of July. - And it is terribly difficult to find a job now. If you had patience, I would have it too. But your last letter from Wednesday after Pentecost, for which I thank you so much, contains the warning that you do not want to wait and want come to Vienna soon. Yes, if I could, but I can't until the houses are sold, not even without marriage. Since I can't get any money out of the bank, and my brother is also in trouble because he can't get a single one of his bills paid, I can only live as meagerly as you did in March, and therefore you're better off at home.

I can only continue to work: 1) to sell the houses and thus create a pension (about 5 - 7 million crowns per month), which would provide a secure existence, for which, even without taking up a practical profession, I would still have to pay about two million crowns per month from author's fees, which are secured for 24 months; 2) work with all means to get the American professorship, which seems promising but requires time; 3) work with all means to obtain a good Viennese profession, which may well succeed after the end of the current unprecedented crisis, but is extremely difficult at present.

[…] It's terrible in Vienna now. Dessauer, with whom I was together on Friday in the English club, is doing much worse than before Pentecost. All places of entertainment, all restaurants and coffee houses completely extinct!

Bice, if you like me just a hundredth of the way I like you, encourage me, give me strength with a few meaningful words. You're hurting me already. It's terrible that you had to experience this disappointment in me.

Bice, I long for news of you, if only I had you as my wife!

Be kissed by your Arthur

Emma to Arthur Kapuzinerstraße 20, Munich, June 15, 1924.

Dear Arthur,

---

[14] Prof. Felix Lampe from Berlin was an expert in the growing film industry, specializing in the documentary nature films that were of particular interest to the film and theatre department of the URANIA. Arthur Haas was especially interested in this new medium, as reflected in later lectures for the university on the physics of sound film.

Thank you very much for your letters. I have received them all in full but unfortunately, I cannot answer them in detail because I think I would write my fingers crooked; you only do what you want to do. I will save my energy.

Unfortunately, I haven't been able to go to the agent in the Zweigstraße yet, maybe tomorrow, because I did a little stupid thing recently and have had to stay in bed ever since. Daddy is sick in bed too. But I don't want to torture you with my worries. The expressions in your last letters - broken down, hope gone, desperate as it has been for a long time, not finding sleep - just teach me that you are busy enough with yourself now.

If I didn't love you so much, I'd simply advise you to stop worrying about it and wait for things to take care of themselves.

But I long for you and Munich makes me sick and nervous. I can't stand it like this for long - something will have to happen, at least on my part. So, that was just a very fleeting glance into my inner being. It's not meant to be a threat - but it's not just idle gossip either.

I am, as always, your Bice

Arthur to Emma Vienna, June 17, 1924.
No. 22.

My dearest Bice!

I just got your letter from June 15th. I didn't write you yesterday because I couldn't bring myself to write about the way things are. I can't sell the houses, and until I can, weeks will go by. And there are no job prospects at the moment. And with America, it will go on for so long.

Bice, for God's sake, what's gonna happen? I don't mean to despair despite everything, but unfortunately there is so little hope that I can give you the kind of pleasant life I originally hoped for, and I don't dare to expect you to live much more modestly. When I think about it, do you always believe that a five-room apartment is necessary? How will I pay for it? And that's why I can never sleep and why I long for a redeeming word, which would give me courage again but which I miss in your letters, although I beg you.

[…]

I want to go to Dessauer's tonight and talk to them, perhaps that will give me some courage.

What's wrong, you being in bed? Hopefully it's just a trifle! I hope your father isn't seriously ill either. I am so worried about you.

Bice, please answer my letter from the day before yesterday so I know where I stand. It just doesn't help that the catastrophe has robbed me of two-thirds of my possessions. Everything has to get easier!

[…]

Maybe tomorrow I'll write to you more clearly, today I'm close to madness.

Tender kisses from your Arthur

P.S. A little more hopeful because of America, also a little more hopeful because of house sales, but desperate because of job opportunities in Vienna. And would you be able to resign yourself to a simpler life? If only we could live like a full professor in Vienna!

Arthur to Emma Vienna, June 19, 1924.
No. 23.

Dearest Bice.

After I sent you the last letter the day before yesterday, I went to Dessauers' to talk to them a bit. But it felt strange when I went to the Johann-Strauß-Gasse for the first time, without any hope of meeting you there. By chance, Dessauers had invited some people (namely Miss Weigel, an Australian doctor and Mr. Kunz) for a cup of tea, nothing more, also tea biscuits. I stayed there until midnight; I couldn't talk about you but it did me good to be together once again - for the first time in ten days - with people in another capacity than that of a petitioner or as property for agents.

Yesterday was my sister-in-law's birthday and I was there at noon; Lenerl made herself very important. Unfortunately, my brother is also plagued by great worries (about the financial situation); he cannot get one of his big outstanding bills and expense paid. (Still, how much I would like to change places with him!!) Every evening I go to Mr. Carl Smyth's Progress Club to practice English (in case there would be an offer!). Today, on Friday I went to a lonely place on the Danube beach in Vienna and several hours in the open air and water (the first pleasure in ten days) did me good.

I don't want to torture you again by again telling you about the dire thoughts that dominate my mind. There are two small points of hope. In Berlin, the stock market is getting firmer again, which could also make selling houses a little easier. (I could have 700 million crowns on my hand immediately, but that is too little for me!) The second point of hope is what I learned from a June 13th letter written by a Dutch physicist who says that he received a letter from Prof. Donnan "a few days before" in which this man, who has great weight, speaks warmly about me going to America and asks the Dutch physicist to help. On June 5th, Prof. Donnan promised me that he would write a lot of letters about my case immediately and I am glad to know that Prof. D. kept his promise. Because the Dutch physicist surely only took his turn after many others.

[…]

You know how I long to have you in Vienna, but look (there's no shame in saying it) I can't get any money at the moment, as long as the houses are not sold, and I have to get by on 2 million a month. Because I saved a few pounds in London, I have those. What grief you would have to face if you came to Vienna now! In your father's house your life really only costs as much as the extra food that your cook buys every day and that is nothing compared to the costs of your stay in Vienna, which cannot be procured at the moment. If you would hear the whining of all the people in Vienna, it is heart-rending. And always news about dismissals of family fathers in all professions!

Bice, dear, what will you say to my last letters? Believe me, it's not just a saying, "the more I suffer for your sake, the more my love for you grows". If only you will be patient, it will be over. Don't be angry with me, but don't forget me. "Hail and kiss me," your Arthur.

Did it affect your acquaintances as terribly in material terms as it did me, or am I just an unlucky fellow? At least by mid-July I will already know what the situation in America is like and what my prospects are.

Arthur to Emma Vienna, June 24, 1924.
No. 24.

Dearest Bice.

Returning to my apartment a quarter of an hour ago, I found a letter from Prof. Donnan [Fig. 11.6], the copy of which (strictly confidential) I am sending to you first of all. In the bleak present, he shows me our future together in a rosy light. I hope that he also awakens in you the same strong hopes that he does in me; I believe, not wrongly, that Prof. Millikan, who was formerly at the University of Chicago and has now given three lectures in London, is indeed the premier American physicist and Nobel Prize winner. If he wants to help me, I am sure he will succeed.

If only he were here already, "the day"! Greetings and kisses from your Arthur

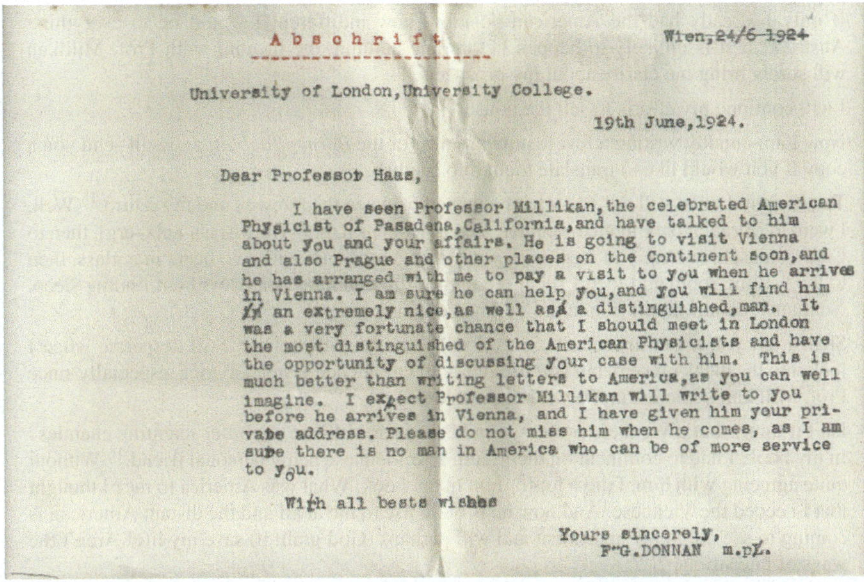
Abschrift Wien,24/6 1924

University of London,University College.

19th June,1924.

Dear Professor Haas,

I have seen Professor Millikan,the celebrated American Physicist of Pasadena,California,and have talked to him about you and your affairs. He is going to visit Vienna and also Prague and other places on the Continent soon,and he has arranged with me to pay a visit to you when he arrives in Vienna. I am sure he can help you,and you will find him an extremely nice,as well as a distinguished,man. It was a very fortunate chance that I should meet in London the most distinguished of the American Physicists and have the opportunity of discussing your case with him. This is much better than writing letters to America,as you can well imagine. I expect Professor Millikan will write to you before he arrives in Vienna, and I have given him your private address. Please do not miss him when he comes, as I am sure there is no man in America who can be of more service to you.

With all bests wishes

Yours sincerely,
F.G.DONNAN m.p.

Fig. 11.6 Letter of Professor Donnan from the Imperial College announcing the imminent arrival of Millikan in Vienna

Robert A. Millikan (1868—1953) was indeed one of the leading physicists in America at that time. He was considered an excellent teacher but was also an experimenter whose oil-droplet experiment led to the determination of the elementary charge. For this he received the Nobel Prize in the year 1923. Millikan gave his Nobel Prize lecture on May 23, 1924 on the topic "The Electron and the Light-Quant from the Experimental Point of View,"[iv] in which he describes his groundbreaking experiment. (Haas' visit to London must have taken place after the Stockholm lecture.) So, it is no wonder that Haas was waiting in suspense. Millikan's itinerary in the summer of 1924 was, however, still not clear and it seems to have been full of contradictions. In November 1924, Millikan also published an article in a somewhat unusual journal, *School Science and Mathematics*, on the subject of "The Astronomy of the Atom," a parallel to Haas' own interests. This is a concept that we will return to later.

Arthur to Emma Vienna, June 25, 1924.
No. 25.

Bice darling!

Yesterday morning after I received the very pleasant letter announcing Prof. Millikan's visit (a copy of which you have received), the midday post brought me another pleasant letter.

Prof. Lotmar in Bern has succeeded in getting the *Berner Tagblatt* to publish a monthly contribution of mine and to pay me 50 Swiss francs (approx. 620,000 östr, crowns).

If only I could find such good people as Prof. Donnan or Prof. Lotmar in Vienna!

In Vienna today, another catastrophe took place. One of the biggest banks in Vienna stopped its payments. Of course, this further aggravates the already terrible crisis.

If only I already had the American position, how indifferent I would be to everything Austrian! If it is unlikely to happen before late autumn, the meeting with Prof. Millikan will surely bring me clarity about my prospects.

I will continue my efforts to sell the house.

Now I am quickly writing a few feature articles for the *Berner Tagblatt* and will send you a copy if you would like to translate them into English.

That's right, I must still tell you what happened between the farewell and my return.[15] Well, I went to Mirabell park in nice weather, then to Hohensalzberg [fortress in Salzberg], then to Café Tomaselli (where you were with your father) for hours with two beers in a glass, then to the waiting room and then I spent the night riding on hard boards without finding sleep. No matter, soon we will ride in the Pullman car in America!

Shall I tell you about my mood? Since I talked to you, it's calmer. Still desperate when I look into the bleak present, but still full of joyful confidence about America, especially since Prof. Millikan's interest in me has been awakened.

Isn't it funny that a Viennese professor and Professor Millikan are bitter scientific enemies? In my book, I had to comment on their feud. The Viennese is my personal friend.[16] Without quite agreeing with him, I did a lot for him in my book. What was America to me? I thought that I needed the Viennese. And now he is of no use to me at all and the distant American is coming to see me in my apartment and will perhaps (God grant it) save my life! Aren't the ways of fate strange?

[…]

Arthur to Emma Vienna, June 26, 1924.
No. 26.

Dearest Bice.

Hopefully I'll soon get a few lines from you, which I'm really looking forward to. I am curious if the report of Prof. Millikan's upcoming visit has raised as much hope for you as it has for me.

In Vienna, the crisis is getting worse and worse every day. Everything is collapsing and it is horrible.

Today I worked out two newspaper articles for *Bern*, "Entertaining Topics in Modern Physics" and "The Age of the Solar System." I hope they'll be accepted. No matter how little it brings, it is very rewarding to at least make money once again.

Otherwise my life is very monotonous. Work, work, running around, lunch in the community kitchen, dinner in the form of two eggs in the café house and reading the newspaper. Insomnia has stopped, I sleep like a log.

I often look at the map of America in the very large atlas I have. Pasadena (where Prof. Donnan wants to take me and where Prof. Millikan is working) is just outside Los Angeles, 34° latitude, i.e., the same latitude as Cairo in Egypt and therefore probably so warm that you can bathe outside all year round and you don't need fur. It is 50 km to the sea, so that one surely spends free time there. Princeton is about 150 km away from New York, between New York and Philadelphia but closer to New York. Will we be eating grapefruit there soon?

Instead of complaining, it's probably better to hope and work.

[…]

[15] Following their last day in Salzburg on June 17, 1924.

[16] This is the debate between Millikan and the Viennese professor Felix Ehrenhaft about fractional charges, which the latter believed that he had found in a similar experiment.

Many, many, many warm kisses and greetings from your Arthur

Arthur to Emma Vienna, June 27, 1924.
No. 27.

My dearest Bice!

Unfortunately, no progress at all, although I think the chances of selling a house are better because of the improvement in the strength of the Berlin stock exchange, but it improved only by a nuance. Please, again, go to a Munich broker (without admitting the minimum expenses and purely informative) so that I know how to assess the chances in Munich and then report to me.

I hope you are well. Our lot is terrible, but hopefully patience will help us to reach our goal. If only I had already spoken to Prof. Millikan.

Now I'm going to an English lecture, a Professor Shepard from Columbia University in New York; it will be held today and tomorrow in the university's ballroom. Theme, "America and the Americans." I'm very curious.

Many, many kisses and best wishes from your Arthur

Arthur to Emma Vienna, June 28, 1924.
No. 28.

Darling, best Bice!

Still not the letter from you that I long for. At least again, thank God, we are a week down the road. Would that you could sleep until one day a telegram wakes you up with the calling from America!

[…] Yesterday I attended the very interesting lecture of Prof. Shepard on "America and the Americans." It was very well attended and will now be continued at 6 pm. It was very interesting and I understood almost everything. My sister-in-law is going to Gottersdorf with Lenerl tomorrow for four weeks. Unfortunately, things are also looking bad for my brother at the moment.

Since the last time, some new, very praising reviews of my books have been published. If you are interested I will send you a very small excerpt of them.

Many hearty greetings and kisses from your Arthur

Arthur to Emma Vienna, June 29, 1924.
No. 29.

My dearest Bice

After I wrote you yesterday I went to the second lecture of Prof. Shepard, which was again very interesting, and afterward to a meeting about the sale of a house; the lack of results filled me with deep depression. The situation is getting worse every day in every respect and unfortunately it is such that I can only put my hopes on America - unless there is a complete turnaround. My dejection is also increased by the fact that I saw in Salzburg that your economic sense is not sufficiently developed to allow you to lead a more modest life. (I do not blame God for that!) You can imagine with what terrible anxiety I am looking forward to meeting Prof. Millikan under these circumstances.

During the lecture by Prof. Shephard, I also met a cousin who was in America for more than a year. He claims that a modest life there is not much more expensive than a modest life in Vienna. By the way, I get a similar opinion from American newspaper advertisements regarding clothing. I spent today outside, in the morning in the Prater, in the afternoon (my second pleasure since returning from England) in the Gänsehäufel. I was alone and read continuously in an English-German conversation book, so that the English I will speak to Prof. Millikan will not make an all too bad impression.

Hopefully tomorrow will bring me a letter from you

Best wishes and kisses from your Arthur

Arthur to Emma Vienna, June 30, 1924. In a hurry in the park on the way to the Physics Institute.

No. 30.

Dearest Bice!

For the last quarter of an hour, I am more confident again and I want to quickly tell you about it and the reason.

I was with a gentleman who had advertised that he was looking to buy houses in Berlin. He made a very dignified first impression, received me in his elegant apartment and claimed to buy for Austrian aristocrats who wanted to invest their money. To my great amazement, he immediately made the offer to immediately pay me double the peacetime price in cash, whereby he would take over the increase in value. That would be 1,800,000,000 kr, which would correspond to a monthly pension of 11-12 million kr. But he will first look at the houses in Berlin and let me know on July 7th or 8th. Then, if need be, the deal could be closed immediately, and in a few weeks after the transfer has been made, the money could be paid out to me.

[…]

I received your first lovely letter since the Salzburg concert this morning.[17] I thank you very much for it. By the way, a few hours ago my friend Ellinger told me that a Mr. Ax, who wants to distribute radium stuff, would like to hire me starting in September. I would do scientific writing without needing an agent. According to Ellinger, 5-6 million a month or at least that's how it looks.

The possibility of employment with Mr. Ax is an interesting side remark. Radium was considered a miracle cure in the 1920s and was sold for an enormous range of applications, from the manufacture of luminous paints to applications in medicine and cosmetics. Physicists, especially the staff of the Radium Institute in Vienna, were had special knowledge and connections to industry. Arthur Haas certainly had in mind his former fellow student Viktor Hess. The later Nobel Prize winner of 1936 had gone to the United States in 1921 to become technical director of a radium factory, exactly the same company that employed young females to paint the dials on clocks and watches with fine brushes. They later became known as the Radium Girls.[v]

Arthur to Emma Vienna, July 1, 1924.

No. 31.

Dearest Bice.

Meanwhile, I live on the glimmer of hope that I told you about yesterday. I just received a telegraphed offer from a Berlin company offering to buy one of the Berlin houses. I did not accept it because it was too low, but at least I can see that there is a desire to buy houses again. I am also reassured by the fact that the situation on the Vienna Stock Exchange has improved somewhat today. The fact that despite my repeated urgent requests, you still haven't been

[17] This letter is lost.

informed (about achievable prices) by a Munich real estate dealer has actually disappointed me somewhat. So much for the main concerns, which unfortunately are not purely financial but on which our fate unfortunately depends.

Now the first news from America will soon arrive. Prof. Millikan will probably visit other parts of Europe before he comes to Vienna, but I will probably see him soon. The fact that your "Dean" hasn't been heard from is actually strange to me, although I didn't expect much from his support.

I am now very busy, finally editing and correcting the two books that are to be completed. The work is quite mechanical. Looking through the annotated sheets, making compilations and the like. At least at the moment my nerves are a bit calmer, so I am already working more efficiently.

I hope dearest Bice that these developments are not going too fast. I was very interested in the fact that you have a radio. I am glad that it gives you some diversion. I wish I had one too.

[…] Yours sincerely, Arthur

Arthur to Emma Vienna, July 2, 1924.
No. 32.

Dearest Bice.

Time goes by so slowly when it is unbearable and furiously fast when it is as happy as the time from April 5th to my birthday. I try to numb myself as best I can with intensive work. What do you actually do all day long? Watch and learn from your housekeeper so that if a button is torn off in America or elsewhere, you can sew it on? Also learn how to make ham and egg, which you love so much?

You are actually an even bigger child than I am but also, unlike me, a sweet one; I can see that in all your letters. We fit well together but at least one of us should have more sense for the harsh realities. I think that perhaps I am getting too much reality now - unfortunately, too late.

All of my acquaintances greet me with the words: "But you've lost weight." There must be something true about it and if it continues like this, I will be even slimmer than you one day. So at least all sorrow has a good thing; it's just a pity that, unfortunately, the body is also accompanied by an emaciated soul and spirit. But the cure is cheaper than Marienbad[18] (or infinitely more expensive, as you take it).

[…]

I got fifty marks for you today. Half of which I am sending you, the other half will follow in fourteen days. I'm very sorry that it can't be more. You know how miserable I am at the moment, having nothing more than you did in March (apart from the days of eating rice!). Hopefully I can send you the same thing again next month, hopefully, but I don't know for sure. "We look good," say the Viennese in such cases.

Emma to Arthur Kapuzinerstraße 20, Munich, July 1, 1924.

Hello Sweetheart!

How art thou? Did you get the eyebrow brush?

Actually, I should be very mad at you. I just called the office and got the answer that there was no mail for me there either. Your last letter (28) was written on Saturday. On Sunday

[18] Expensive spa town in Bohemia at the time, where Austrian, Russian, and German royalty gathered.

you have of course no time. Only this is the second or third time now. Really, I am seriously angry with you. I sit at home all Sunday, refuse all invitations and only go for a walk with Dad. But next Sunday, I won't have time to think about you either.

I expect an invitation from Spain at the end of the week. I must give a final answer. As I feel right now, I can already guess how it will turn out.

You know, I am not asking you to always stay at home and not have fun. But you don't need to forget me. You have promised me three times now to write every day - but you don't seem to be able to keep your promises.

Really, Arthur, I'm so sad because you've disappointed me again. I was in such a good mood when I started the letter but since the 5 o'clock post did not bring me anything from you either, I have to keep thinking that you don't take the saying, "One man, one word" very seriously. Bice

Emma to Arthur Kapuzinerstraße 20, Munich, July 2, 1924.

Dear Arthur,

I have just received two letters from you in the morning mail. One was written on Sunday evening, the other on Monday afternoon. The Sunday letter was posted on Monday at 11 o'clock, which explains the fact that I received it only today.

Many thanks for both.

I still feel very sad because your last letters did not really help to put me in a better mood.

You write that your depression is increased by the fact that you saw in Salzburg that my economic sense is not sufficiently developed to let me live a financially limited life. You didn't tell me this in Salzburg - but of all things, were these your contemplations, while you were reading "in an English-German conversation book," at the Gänsehäufel? When I go to the Gänsehäufel, I have fun.

But what saddens me so much is this: You know very well that I can fit into a modest life much more easily than you can and that my economic sense is completely developed in such a way that it can adapt to any way of life. You know all of this as well as I do but it is much easier to blame me than to say "mea culpa".

My opinion of you is this: A life of 7 million a month is very comfortable for you alone; it's the way you've been used to. You don't need to take up a profession and can continue to live carefree. If you were to marry now, you would have to share and limit yourself a little. For a man who has never done this, however, this may seem hard. You don't think so - maybe. I'm afraid you haven't reached self-knowledge. So, you tell yourself that I am the one who cannot submit but is still demanding.

Now I tell you once again what my demands are: I demand that you always act toward me in such a way that I can have complete trust in you - without many words. That is all I ask. If I have a husband who is worthy of all my trust, then I consider myself luckier than all other women. That I will be faithful to my husband in every way of life, I have enough character for that.

No more. It's amazing how sober we can be with each other. I always imagined the bridal time to be quite different.

Regarding my economic sense, I ask you again to be completely unconcerned because, to reassure you, I now cook every day with our cook - should your love go through your stomach.
Greetings from your Bice

Emma to Arthur Kapuzinerstraße 20, Munich, July 2, 1924.

Dear Arthur,

This is now the third letter from you today and the second that I am writing to you. I have just received this last one with the 5 o'clock post (written on July 1st – No. 31). You made a few remarks in it that make me extremely unhappy and put me in a worse mood than I have been for two days. I do not want to discuss them because it hurts me too much to talk about them.

I am afraid that those around me already suspect that something is going on inside me. You seem to enjoy torturing me. I hope the invitation from Spain arrives this week. Bice

Arthur to Emma Vienna, July 3, 1924.
No. 33.

Dear sweet child!

Thank you very much for your dear letter from the day before yesterday. I was very touched that you care so much to receive a letter from me every day. But you do me wrong and you are a big child. You have been in Vienna long enough to know that on Saturday in Vienna letters are neither delivered nor picked up. The natural consequence of this is that even if I write to you every day, you do not receive a letter on Tuesday, but two on Wednesday, and that is how it was this time, Sixtes! But it was sweet of you to telephone to see if there was a letter from me; I would never have believed you capable of that.

But think, Bice, today I'm much more confident, not yet optimistic, but not so terribly pessimistic. I have received a new offer for the house in Seesener Straße, extremely urgent, by telephone and telegram from Berlin. The day before yesterday they had offered me 18,000 marks immediately by telegram for the house (which is not the most valuable of my houses) + 60,000 marks in one year, but with a 30% increase in value from my side, which means about 16,000 marks net. Yesterday they already offered 33,000 marks immediately net, without me having to pay the increase in value tax. Considering that I need about 70,000 marks to create our subsistence level (which, in League of Nations bonds, amounts to 6-7 million crowns per month, to which must be added income from work and, without a change of profession, at least 2 million per month), you can understand why I am no longer so pessimistic. I replied to Berlin that they can have the house if they send my brother 6,000 dollars by July 10th. That would be a good start (I own half of Seesenerstraße, half of Tauroggener Straße, Landsbergerstraße and Kriemhildenstraße).

The second reason why I am a bit more optimistic is that today the long-awaited change of mood on the Vienna Stock Exchange has finally taken place and all stocks have improved.

[…]

Many, many kisses and greetings from your Arthur

Arthur to Emma Vienna, July 4, 1924.
No. 34.

My dearest, dearest Bice!

The post office brought me both your letters this morning, dated July 2nd. Thank you again for that. I can't tell you how touched I was by the content of your letters and how much I feel when you suffer.

[…]

I wrote you yesterday that one of the houses already has a firm offer (22,000 marks immediately and absolutely net), i.e., the amount that seems to me to be the minimum for this house, so that a permanent subsistence level is generally assured by the sale of the house. By analogy, I conclude that hopefully in the foreseeable future, through the happy sale of all the houses, there will be the possibility of marriage, even if nothing were to happen in

America. This is a hope that I did not have a week ago. All the indications are that there is a favorable turnaround on the Berlin housing market, just as there was a complete change of mood on the Vienna Stock Exchange yesterday.

The fact that you want to remain loyal to me even in the absence of comfort, Bice this moves me deeply. I trust that you are not only completely sincere but that you are also aware of the consequences of such a decision, because that means, without grumbling, renouncing everything that seems desirable when the financial situation does not allow it and refraining from any attempt to attack the substance of the capital on whose income we have to live.

As far as I am concerned, Bice, what you write about me is partly correct, of course. I am no stranger to the idea that I will have to lower my standard of living in the event of marriage, but I am determined to make this so-called sacrifice and to give up my scientific profession for your sake, if another can be found.

But it would be quite wrong if you believed (I hope you don't believe it) that I was looking for excuses, to "get out of the way," after the financial situation had completely changed. You must not doubt my complete sincerity.

I'm still working relentlessly toward the goal of securing our common minimal existence (which is still not secured today). Whether this will succeed, I don't know; but that it will succeed in the next few months seems more likely than the opposite; but that's just my subjective opinion. You also know that I am trying in three directions, independent of one another: Selling houses, the Viennese profession, America. In the first respect I'm more hopeful than I was a while ago, in the second one too, in the third point nothing new has come up yet. I don't expect an answer before mid-July; unfortunately, I don't know when Prof. Millikan will come to Vienna.

Hopefully, some of my increased confidence will be transferred to you.

Many hearty, deep kisses and greetings from your Arthur

I am very touched that you are really learning to cook, dear Bice!

You are much braver than it seemed at first!

Emma to Arthur Kapuzinerstraße 20, Munich, July 3, 1924, 5:00, Nachm.

Arthur,

You're a darling. You know I can't stay angry with you for long. Look, I love you so much. That's why I'm sometimes getting really mad at you.

I think that since being engaged to you, I've only just awakened to the fact that I'm just as much a woman as any other woman. Before, I thought I was an exception.

But you are naughty. You have no reason to think that I'm a bigger fool than you. And your economic sense is not so far off, either. You shouldn't want to pull out a splinter in someone else's eye when you have a beam in your own - or something like that, as the saying goes.

I'm giving you back the 25 marks. Not that I won't need it, but I want you to use it to go to Salzburg next Thursday July 10th and then go back to Vienna on July 11th. So, no more excuses. I'm definitely going to Salzburg.

But thanks very much for the money. You're a nice guy, but sometimes so...

I'm imagining that you're with me and I'm kissing you really hard, Bice

Emma to Arthur Kapuzinerstraße 20, Munich, July 4, 1924.

Dear Arthurlein,

Tell me, does your love go through your stomach, through your clothes or through your head? I'm about to become the universal cook. I can come up with anything. You have no idea how educated I am.

This morning I made a mushroom soup, mutton and beans with potatoes, and I baked a delicious cake. And almost all by myself; our cook was rummaging around. I cook every morning - in case your love should pass through your stomach.

In the afternoon I sew, or I study your physics book in case your love is different. - But if that is too boring for me, then I manicure my fingernails or read in a very nice entertaining book like *Casanova*, *Don Juan*, or other tales of morality. Harmless, isn't it?

But this afternoon, two of my girlfriends will pick me up for a swim - it will be the first time since I've been in Munich. I don't really feel like swimming since my Ammersee experience. Hopefully you got back the 15 marks. You know what you have to use it for. I have to think about who I'm going to pump this time for my travel expenses. But my real reason for sending the money back was another: I didn't want you to think that I was only making amends because you sent money. No, I'm not that bad, and I sent it right back to you so you wouldn't think that. Servus, Bice

In fact, Emma Huber did learn how to cook from the cook Maria, as shown by the recipes that she carefully wrote down for meat soup and meat loaf, goulash, beef fillet and luncheon roast, pies, pasta and noodle dishes, cake, and nut balls. The couple's private communication crisis seemed to be over.

Those parties involved in the broader international situation also seemed to agree on an approach involving greater diplomacy and new negotiations. As a result, the markets and stock exchange gradually stabilized in Berlin and, with a slight delay, in Vienna as well. A new conference was planned for mid-July in London, at which the question of reparation costs and Germany's increasing indebtedness were to be debated once again. For the first time since the war, Germany had been admitted as a full partner to an international conference and this was already seen as a positive event on the stock exchanges.

Arthur to Emma Vienna, July 5, 1924.
No. 35.

My dearest Bice!
How can I thank you for the two letters I received, this morning and at noon (the express letter and the one from yesterday)? I am so touched and so moved by your devotion: I would never have thought that you would want to spend your birthday with me. I will arrive in Salzburg on Thursday, July 10th at 3 o'clock in the afternoon and, as your train will arrive earlier, I ask you to wait for me like last time.

Only in two cases would I have to refrain from travelling to our second Salzburg meeting: 1) If Professor Millikan announces his visit for the 10th or 11th; 2) If the sale of the house were to be finalized on one of these days. Because in both cases, it is about our common existence and that is much too important. My love for you is not limited to words in letters, but we must finally reach our goal. But if I do not have to answer the call, for one of these reasons, I will be in Salzburg on Thursday.

Today, I have even greater optimism for two reasons. One of the houses is already as good as sold and has achieved the required price. This morning I received a telegram about Seesenerstraße, offering 24,000 marks (¾ immediately; ¼ deferred until December 10th, against an interest-bearing fine gold mortgage on the house). All charges are borne by the buyer. I have accepted. But even if it is paid, nothing changes in my catastrophic shortage of money. Because, only half of this house is mine (I have, as you know, two whole and two half). The 150 million kroner I will get as 3/8 of the purchase price will absolutely have to be placed into my bank account to reduce the so-called debit balance. But I am hopeful because the other houses are not worth less than Seesener Straße and 70,000 mark is the amount I need to secure our minimum existence.

The second reason I am in a better mood is that I hope to be able to write a feuilleton [serial] for the *Frankfurter Zeitung*, which, as I learned from a reliable source, pays 100 goldmark (more than 1,600,000 K) for a featured serial. So I hope to earn, working for both Bern and Sweden, 2-3 million monthly by writing serials. But it is still a big "?". After all, there is also hope that I will be able to work making educational films at a cinema company. A friend who is influential in this respect will make an effort. - I also received a letter from Prof. Donnan in London; a Dutch physicist of world renown and with excellent American connections is very much committed to me.

Tonight is a quarter of a year, Bice, since that moment when I first felt that this happiness was too great for me to bear. May God give it (ask and it shall be given to you) that when another quarter is up, we are united forever. And until then my dearest, best Bice, whom I love and love more and more from day to day, through all the suffering and all the sorrow and all the torment and all the worry, most heartfelt and hearty kisses from Your Arthur

Arthur to Emma Vienna, July 6, 1924.
No. 36.

Dear, good Bice,

Think to yourself, "I have always had a feeling that the day of the quarter anniversary just wanted to bring a lot of good things," and indeed, it brought me three bits of good, promising news. Early, the contract of sale of the Seesenerstraße house, noon, […], a pneumatic card with the request to call a man who buys Berlin houses immediately. I did it. He's desperate to get Landsberger Strasse. A few days ago he had offered me absolutely 12,000M, on the phone he already went to 20,000 goldmark immediately payable. Since I had seen at Seesenerstraße that I get more if I don't say, "yes" in the moment, I said that I will visit him on Tuesday. Perhaps, it is easily possible, I will have a better offer from another side by then, especially from the gentleman whose conversation on Monday, the June 30th, gave me good confidence.

So, 10 o'clock this evening then, and as you may suppose, I gave my entire thought to you in an hour of remembrance of what happened a quarter of a year ago, so I was now allowed to firmly believe that the next quarter will already find us united forever. Everything seems to be going well, and I hope that you do not doubt my sincerity in this matter.

On Thursday, the day before your birthday, I arrive in Salzburg at 3 o'clock in the afternoon like the first time, to meet you at the station.

[…] I kiss you my dearest Bice, many, many times. Yours, Arthur.

Emma to Arthur Kapuzinerstraße 20, Munich, July 6, 1924.

Arthur dear,

Thank you very much for your letters. Good luck with Dr. Millikan visiting. I just got the 5 o'clock post. A letter from you (26) was also there and one from Gretl and Otto. But Damnation - no - and no –your hieroglyphs are getting more and more hieroglyphic – don't you want to try to write decently? A high school student scribbles better.

I also have a very big atlas; I will take it out and see if you are right. Since my visit to Ammersee, swimming in the open air doesn't attract me anymore - on the contrary, when I think about swimming, I get a shiver. But the main thing is of course that I don't need to wear fur. But maybe I do need one. - You see I have your letters in front of me and criticize every line. –

I have no appetite for grapefruit at the moment either. I don't care what I eat. I'd like to stop it altogether - stop it all altogether. I wish I was Sleeping Beauty. But I would take care that you don't wake me up, but rather a beautiful, handsome, dear, good, clever, rich prince. But as long as I can no longer be Sleeping Beauty and no prince comes to wake me up - you are still my dearest.

Yes - yes - love - what foolish things one does - just because of love.

We got a radio set and yesterday we heard our first program.

Arthur to Emma Vienna, July 7, 1924.
No. 37.

Dear Bicelein!

So, as we agreed, I will meet you at the Salzburg train station at 3 o'clock in the afternoon on Thursday, July 10th with everything like the last time. I hope nothing will happen that will force me to telegraph. But if it's a matter of an exit strategy, that would come first.

My good mood is only temporarily affected at the moment. An "Overseas Radio Telegram," 15 minutes ago - beside myself with happiness, already believing in an American position, I tore it open in the greatest excitement; it was from Berlin. The Seesenerstraße buyer is resigning because my brother has made a reservation in the agreement that is too complicated. I am terribly annoyed about this (Oh, those lawyers!) but nevertheless the pleasant fact remains that housing prices have risen sharply. I am full of confidence that the houses will be sold so that we can soon marry. God grant it.

I heartily greet you, happy to see you again soon. Yours, Arthur.

Arthur to Emma Vienna, July 9, 1924.
No. 38.

My dearest Bice!

Half an hour ago I received a telegram informing me that your trip to Salzburg has been cancelled. I am very sad about this because I was looking forward to seeing you again, my dearest Bice, and now nothing will come of it. But also, the fact that you are ill worries me very much, although I try to console myself with the fact that it is only part of a more frequent malaise and nothing more. In any case, I hope for reassuring news in return.

[...]

My best Bice, what can I wish you for your birthday? If your most important wish is the same as mine - day and night, awake and asleep, in all my actions and guiding all my actions - then trust me that I will do everything, everything for this wish to come true. As powerless as man is against fate, I still hope that, give God, it will be and very soon. [...]

And many, many, many more intimate birthday kisses from your Arthur

I wanted to bring you a small lasting gift in Salzburg. Unfortunately, it can't be sent by post. At our hopefully not so far away next get-together!

Arthur to Emma Vienna, July 10, 1924.
No. 39.

Dearest, best Bice!

I think I know that you are fine, but I am still very restless and beg you to send me an urgent message about your condition.

Last evening, I received an express letter from the Berlin company. I saw from it that I would actually already be in possession of the money for the house in Seesenerstraße if the exaggerated (or perhaps necessary) precautionary clause in my telegram, which Otto advised me to take, had not thwarted the matter. As annoying as it is, it is gratifying. I would have got a higher price for it than what seemed to me to be the minimum price (to get a 6-million-month pension) necessary (I would have got 24,000M, but 20,000M would have been enough). On the other hand, the company immediately informs me that a buyer for the house Landsbergerstraße has come forward and wants to know my final price. (Because of the house the negotiations also include other interested groups.) Furthermore, this morning an announcement appeared in the *Neue Freie Presse* in which Berlin houses are being sought. Of course, I wrote immediately and carried the letter immediately to the Post Office. You see, I do not miss a minute. I hope that in the next few weeks the sale of the houses will succeed, because there are buyers again and the prices now seem to be attainable at the level buyers seem to be looking for. But everything is terribly complicated and requires calm and patience. (If Germany would not be in a terrible shortage of money, the house sale would be child's play. But a misfortune does not come alone.)

[…]

But now I'm very nervous about America, because the time is approaching when the first answers can be known.

Many hearty, deep kisses and many, many greetings from your Arthur

Arthur to Emma Vienna, July 11, 1924.
No. 40.

Dear, sweet Bice!

I cannot tell you how much my heart hurts today, missing you and you being so far from me on your birthday. How happy had I imagined this day to be, and now everything has turned out this way. Today I lost my good spirits again, although nothing bad happened. But look, I'm such a child and I firmly imagined that today's early mail would bring me some very good news, which I could then telegraph to Munich to put you in a happy mood, such as a cable call to America. But nothing came, nothing at all and that's why I'm so terribly depressed, as an old child.

[…]

As far as the status of the house affair is concerned, Mr. Winkler (who promised to pay me double the price of pre-war times on June 30th) will not return from Berlin until Monday (July 14th). Whether I will achieve anything with him is still uncertain, but everything looks better again. How these uncertainties torture my soul! And in negotiations you always have to act as if these were mere business problems and nothing more.

It's so strange. I used to ask myself so often, philosophizing on the divan, whether the whole world was real or just in my imagination. That was as long as I had my way. But since fate has been hammering away at me with such blows, philosophy has been thoroughly lost on me.

[…]

Yes, something very banal. Today I had myself weighed in a pharmacy; I have lost four kilos since the beginning of May and weigh 72 kilos without clothes and before lunch, which is just the weight corresponding to my body length. Are you satisfied? But if it continues like this, I will soon be a slim man. Sixtes, sixtes!!

Bice, Bice, I think about you all the time, but today. ... At 6:30 am I will meet the Dessauers and their friends at the Liebenberg monument for a small hike. I love being together with the Dessauers so much because I always think you will be there, too.

Many many, many heartfelt kisses and greetings from your Arthur

Arthur to Emma Vienna, July 12, 1924.
No. 41.

Fig. 11.7 A postcard of the Krapfenwald, a wooded hill near Vienna as it looked in 1910

My dearest Bice,

For God's sake, what's wrong with you? I'm so worried because there's been no mail from you today. You can't really still be sick? Why don't you let me hear from you?

Yesterday I was with the Dessauers at the English club in the Krapfenwaldl, which belongs to the Kahlenberg [hill in Vienna]. I didn't go for pleasure but to practice English in view of Millikan's visit. Mrs. Dessauer is going to London with the children on Wednesday for an indefinite period. That is not a sign that the Dessauers are doing better but a sign of the opposite, I believe. Things are going so badly that they can no longer live in Vienna.

Today brought some small progress; a "serious buyer" was found for the house in Tauroggenerstraße, which belongs to my brother and me in equal parts. Of course, I wrote immediately. I think that once a house is sold, the others will quickly follow and then everything will be over.

... In the last few weeks, the trade journals have published particularly favorable reviews of my books. My book *Atomic Theory* (dedicated to Prof. Donnan) is now completely finished and should appear in bookstores in about three weeks. Also, the big two-volume work on theoretical physics should be published in August (i.e., the second volume in third or fourth edition).

[…]

Best wishes from your Arthur.

It was a hot summer; despite work and complaints, Haas and the Dessauers decided to find some relaxation in nearby Krapfenwaldl on the Kahlenberg as depicted in Fig. 11.7. The forest offered an open-air swimming pool, the Krapfenwaldbad opened in 1923, which also attracted Haas. It was easy to reach the Krapfenwaldl by cogwheel railway, from where one has a wonderful view over the whole city of Vienna.

Even so, in the following days Haas seemed to still be looking for local summer retreats; he missed his annual summer excursions and wanted to get out of the hot and sticky city. He accomplished this with frequent visits to Kritzendorf on the Danube. Kritzendorf was easy to reach by taking the Franz-Josephs train and had become a favorite excursion destination and summer resort for many Viennese, after a river bath was created there by Danube regulation. Especially in the 1920s, this new swimming opportunity attracted thousands of visitors. In addition, there was a Viennese weekend colony with over 600 "stilt houses," which were used extensively on weekends by the better-off Viennese bourgeoisie. Here, too, Haas used his connections with influential persons and dignitaries of the city.

Arthur to Emma Vienna, July 13, 1924.
No. 42.

Dearest best Bice

The concern for your health is tormenting me more and more; hopefully tomorrow's early mail will bring me a message that will dispel all doubts.

Today I was a guest of Oberbaurat [architect's title in the city's bureaucracy] Adam, whom you also know as Vice President of the Geta, in his hut in Kritzendorf on the Danube. I behaved well and hardly spoke to any ladies. […]

Tomorrow, the desperate efforts to sell the house will begin again, if only they had succeeded!

[…].

You know, I would now much rather look at the two little pictures than at your big picture. Particularly, the one I took in Laxenburg on Easter Sunday.[19] It seems to me (is it just my imagination or like the picture of Dorian Gray?) that you look so sad in the big picture, and so funny and happy in the Laxenburg one.

My brother will leave in a while for his summer retreat and I have to stay here, completely abandoned and helpless. It will be terrible if a house doesn't sell soon. If I don't get the payment for *Atomic Theory* in August, I could ask my publisher for an advance. But, if this dog were to recognize that I need money, I suspect the next contract would be much less favorable. I prefer to bear the lesser rather than the greater evil. I have not yet received payment from the Swiss newspaper either. The minimal chance of getting a badly paid job seems to be improving a bit.

If only we were already out of Europe today!

Be dearest Bice, kissed dearly by your Arthur

Arthur to Emma Vienna, July 14, 1924.
No. 43.

Dearest dear Bice,

The absence of news from you fills me with the greatest concern and anxiety. Are you really sick or do you have unkind intentions because of Spain? If I do not receive a letter from you tomorrow, I don't know what I will do.

[…]

Today there is another small progress to report, as an agent from Innsbruck (big real estate office) spontaneously asked to sell my houses. It seems that the housing market continues

[19] Laxenburg is a small romantic town, about 20 km south of Vienna, famous for its imperial castle where Emperor Franz Joseph and Empress Sisi spent their honeymoon.

to revive. Mr. Winkler will not be available until tomorrow; the answers from all the others are still pending but I hope!

[…]

Today I write to you only briefly. I am too anxious because of the absence of all news from Munich!

Emma to Arthur Kapuzinerstraße 20, Munich, July 13, 1924.

My dearest, smallest Arthurchen,

You're touching!

Your letters are almost starting to be kind. Until now they have always been piano but the last one already showed a tendency toward forte. Really! I am very touched.

But my dearest Arthur, in all seriousness, I am very grateful to you for the letters, for they have been a great comfort to me. I was in bed from Tuesday until yesterday, Saturday, celebrating my birthday this way. I think I'm like I was back at the Ammersee, lying in the sun a little too long; too much sun seems to hurt me. But the doctor said it was not very serious; my people at home were terribly worried. Now I feel quite well again and will definitely be in Salzburg on Thursday the 17th. Listen! Comprendez vous! Understand! So, I'll see you again on Thursday at 3 o'clock.

I have a lot to tell you - I've heard about America.

So now I'm going to call it a day. I'll save all my news for Salzburg.

Servus and a kiss Bice

Dad sends his best.

Salzburg, which is easily accessible by train from Vienna and Munich, has been a convenient meeting place since they were there for the first concerts in April and May 1924. These meetings were of course time-consuming, which is why Haas was always worried about missing a letter or a meeting with a potential buyer for his Berlin properties. However, these meetings in Salzburg were also necessary for bridging and clarifying tensions and misunderstandings that arose. They traveled from Munich and from Vienna, met, strolled, visited churches or other sights, sometimes a concert, talked in a café, explained themselves to each other and traveled back to Munich or Vienna on their respective trains. The Vienna train left later, which is why Haas had to spend more hours with his thoughts in Salzburg.

Arthur to Emma Vienna, July 19, 1924.
No. 44.

Dearest dear Bice!

How unspeakably sad the half hour was yesterday when I could see you leaving Salzburg only from a distance, thick smoke from the locomotive covering my view of you. And when the train left the station hall...

Today has been a disappointment so far. Neither the day before yesterday, nor yesterday, nor so far today has the post office brought me a letter that has any meaning at all. No sign of life from Prof. Millikan, nothing from England, Holland, America, Switzerland, nothing on the newspaper articles. The gentleman I wanted to see on Thursday morning (about the houses) was still not back from Berlin this morning but he is supposedly "rolling in," so I will go again at 5pm.

It was so happy and beautiful the day before yesterday and yesterday! ... And now again the terrible uncertain hunt for the ultimate goal. Now that we have been to so many churches in

Salzburg and I have never forgotten the "three wishes," now it has to happen! Otherwise one would really have to despair of all justice being lost in the world.

I received a postcard from Mr. Dessauer saying that he is already a straw widower and that I should visit him. I will do it.

Then I received a letter from a sixty-year-old old lady in Leipzig to whom I had lent 180 marks in December so that she could get her confectionery business going again (as long as I was doing well, I was always a good guy). She wants to send me 60 marks _ 13 dollars, but doesn't know how. I took the liberty of asking her to send it to your father. Please take it into safekeeping for me and bring it to me when the time comes. The old lady's name is Clara Reuther, Leipzig, Plagwitzerstraße 27. Tell your father right away. Otherwise, he won't know what it means.

[...]

Many fervent kisses, dear Bicelein, and warm greetings also to your dear father, dear Arthur

Arthur to Emma Vienna, July 20, 1924.
No. 45.

Bice Darling!

Yesterday's visit to the agent Winkler, in whom I had had such high hopes, was a complete disappointment. It was a hard blow for me! Today, the Sunday edition of the *Neue Freie Presse* published two advertisements in which Berlin houses are being sought; for one, I have to write and for the other, I only have to visit. Of course, I will do everything immediately.

It is so terribly horrible that the best and purest things in the world depend so much on material dirt. If I hadn't been a dreamy idealist, I would have grasped this terrible insight in earlier years!

Today I was again in Kritzendorf, where I again caused a sensation because of my shyness around people. I was mostly alone with my English conversation book, at times also with Dr. Adam and a few minutes also with ladies to whom I was completely indifferent.

Hopefully tomorrow's mail will bring me something good after all. I have so many answers outstanding because of house sales! And America, the big mystery? This week Donnan is going to Canada and the USA. Once he's in place, things will probably move quickly.

If only the London Conference, which is now in session and whose outcome is optimistically assessed, would bring Germany the foreign shares![20] Then everything would pick up immediately and the housing market would also pick up.

Isn't it so strange? Two people, in Munich and in Vienna, their fates depending on world politics. And, somewhere in the USA there are already a few people sitting together at a university discussing what they should do; their solution will determine the future of the people in Munich and Vienna. How happy are the innocent locals! They sit on the same ground from childhood to old age, take a wife when grown and a son soon follows. Also hard work but there is peace and quiet, none of the terrible torments of the soul. We poor unfortunate people with brains!!

Many fervent kisses and greetings from your Arthur

Please send me a copy of the lectures that you heard. I would be so interested!

[20] The London Conference on reparations held from July 16 to August 16, 1924, dealt with Germany's payments to the victorious powers. For the first time since the First World War, Germany was accepted as an equal member of an international conference. The conference culminated in the contractually agreed upon Dawes Plan in which, in addition to the annual reparation installments, the foreign contractual partners granted Germany a loan of 800 million gold marks to stabilize the Weimar Republic, which brought the economic stabilization Haas had hoped for.

Emma to Arthur Kapuzinerstraße 20, Munich, July 19, 1924.

My dearest Arthur,

When I started the letter yesterday, there were so many things I wanted to tell you. I didn't know where to start. So I got lost in thought and sat there quietly for a long time, thinking and thinking. And today I feel the same again. There is too much going on in my head to be able to think clearly.

When I'm at home, it's as if I've become a completely different girl. I don't want to do anything. I just want to sit all alone and think. I think that lately I've become even more nervous than I used to be. Every little unusual thing upsets me terribly. Dearest Arthur, the waiting is terrible. Sometimes I feel a completely powerless rage toward you, my pride and joy. My whole self is rebelling against this unbearable waiting - but when I'm calmer, I'll trust you again. You would be the worst person if you betrayed my trust. I know you're doing your best to shorten this terrible time.

Dad is upset too but I defended you. Dad's not in a good mood lately - I suspect he's worried about business - and he's not feeling very well.

And another thing, dearest Arthur. I wasn't really so nice to you in Salzburg, I know. I was angry with myself. I had been looking forward to seeing you again - I was counting the hours, already days before. But when I arrived in Salzburg, I was suddenly overcome by a stupid feeling, that is, I had no feeling at all. I didn't care whether you came or not, and this indifference remained until once again, I didn't see you. Then the longing for you came back, a great longing of mine. God bless you, darling. Bice

Arthur to Emma Vienna, July 21, 1924.
No. 46.

Dearest Best Bice!

Today after returning home at noon, I found the enclosed letter to you, which I will send immediately. Hopefully there is nothing in it that will take you away from Munich! I noticed that the letter is slightly damaged. Of course, I don't know if someone has opened it or tried to open it, and I think it is also your wish that I don't follow up on that suspicion.

This morning I went to the bank that is interested in buying the Berlin houses. They will give me an answer in eight to ten days but I don't have much hope. By the way, I have seen all sorts of things saying that houses are again being eagerly bought. Also, today the real estate department of the huge Berlin department store Wertheim asked me to entrust them with the sale of my houses - one more sign that the housing market is getting better.

Today I wrote two very long letters, one to a "job agency for postdoctoral physicists in Berlin, set up by the Fachgemeinschaft Deutsche Hochschullehrer der Physik" and a second to Prof. Gans at the University of La Plata in Argentina. I informed him that I would accept a professorship at a South American university if I was paid enough and asked him for his support. I am sure of your approval. This does not mean that my hopes of North America have somehow diminished; this doubles the chances. If an offer comes from North America and another one comes from South America, the latter will be rejected.

How we span the whole world with our thoughts when even the distance between Munich and Vienna is much, much too great.

Be many, many, many times kissed and greeted fervently by your Arthur

This letter is an indication that Haas is gradually giving up hope of an academic career. Marriage seems urgent to him, but marriage and subsequent married life require a regular income, an income that does not seem to be available in academia. Sale of the houses could have provided the necessary income but had not worked out.

The hurdles and resistance involved in securing a permanent position at the university seemed insurmountable and America seemed to remain a distant dream. Thus, Haas was gradually making himself comfortable with the idea of alternative opportunities in industry. For a forty-year-old physicist with a purely theoretical background, this was undoubtedly not an easy decision to make and, as further developments show, not an easy decision to put into practice. Haas staggered between hope and despair, as expressed in each of these letters. The resulting monologue of failures and setbacks was not something that Emma Beatrice wanted to contribute much to.

Arthur to Emma Vienna, July 22, 1924.
No. 47.

Dearest Best Bice!

Thank you very much for your dear letter from Sunday. It moved me so much - and made me so sad. Dearest Bice, it is as you write it. You can absolutely trust me, in the sense that I do and will do everything, everything, to achieve the goal. I am completely honest with you. But it's so terrible. We're in the thickest fog and can't see a hundred yards. Since I returned from Salzburg, not a single message has come to me that gave me new confidence; not a single letter, not a single letter about America, not a single word from Millikan. It may be a coincidence here, but it depresses me terribly. And I could bear everything if I wasn't tormented by the idea that you are suffering. Work is the only remedy that helps, just don't brood on and on.

I have had several long meetings, all with the same refrain: There has never been a more terrible crisis during the lifetime of anyone living today; it would be madness to think of marriage at such a time. It may all be logical but if only such a detached person could look into his heart! And the idea that Europe is so miserable while America, with its immense treasures, has no idea what to do! If only the London Conference could turn things around.

I wrote very, very long letters to a German physicists' agency in Berlin and to Prof. Gans in La Plata, Argentina, who holds me in high esteem and is very influential. I explained to him that I would immediately accept a sufficiently paid appointment to South America, unless I had received another one earlier. I certainly acted in your interest. The congress in Toronto, to which Prof. Donnan is going and for which I have great hopes, will take place from August 6th to 13th. Prof. Donnan will then travel first through Canada, then in September to Philadelphia, Princeton, New York, and so on. I couldn't wish for a better friend, if only he were able to help!

[…]

Today my sister-in-law came from Gottsdorf to Vienna for two days to do some shopping [Fig. 11.8]. She is lucky to have time for that!

You write that you think you were not dear enough in Salzburg. No, that's not true. You were very, very dear and good. But unfortunately, in many respects, you still have a perspective (not toward me but toward things) that does not match the seriousness of the situation. But this is not a reproach! And lest you rack your brain over what I mean, I mean your brusque indignant refusal to help somehow - to find support in Munich for my desperate efforts, which would make unification possible by selling the house. But I don't want you to do anything about it right now!

Be kissed fondly and heartily and greeted by your Arthur

Arthur to Emma Vienna, July 23, 1924.
No. 48.

Dearest Best Bice!

Fig. 11.8 Arthur Haas' sister-in-law Margarethe Haas née Hirsch

In the fog in which we have to grope around, a light has been shining again for two hours. I don't know whether it is a real one or a will-o'-the-wisp. My brother's lawyer, Dr. Beck, urgently called me and told me that a friend of one of his clients is ready to put a billion crowns (minus 5% - 50 million commission), absolutely net, on the table for my three houses. Then everything would be saved!! 900 million results in a 6 million monthly pension at 8% interest in League of Nations shares. Then we don't have to wait much longer (maybe only a few weeks). But he is currently living in a sanatorium outside Vienna and I have already experienced too many bitter disappointments to rejoice too soon. Dr. Beck has written in my name that I agree. Now we have to wait and see. But in any case, I have something to hope for again!

A professor from Graz who has been a technical director in a radium factory near New York[21] for 2 ½ years, since the war, and to whom I turned for advice, which he also gave me, wrote at the end of the letter, "By the way, I do not doubt that the recommendation of Prof. Donnan will be more than sufficient to pave the way for you". He may be right.

By the way, when I leafed sleeplessly through a university almanac last night, I discovered that there are about seventy universities in the United States, many of them quite insignificant. If only I could get into one! At least I would not be bored. Most of them are on the East Coast, but there is one in Salt Lake City and a few are in Texas and Nevada!

Now I'm in a better mood again. Will this with Dr. Beck be the main hit or a failure?

Greetings and kisses from your Arthur

Arthur to Emma Vienna, July 24, 1924.
No. 49(Fig. 11.9).

[21] The Graz professor is Victor F. Hess (1883—1964) who received the Nobel Prize in 1936 for his measurement of cosmic radiation. Victor Hess worked from 1910 to 1920 with Franz Exner and Stefan Meyer at the University and Radium Institute in Vienna. Hess knew Haas from the years before the war. In 1921, Victor Hess became technical director of the US Radium Corporation where he remained until 1926 before returning to Austria as a professor at the University of Graz. The US Radium Corporation in New Jersey produced luminous paints, using a mixture of glue, zinc sulfide and radium powder. In addition to clock faces, luminous paints were applied with fine brushes to light switches, doll's eyes, fishing lures, compasses, aeronautical instruments, house numbers and the numbers on theater seats. The company later gained regrettable fame due to the increase in radiation-related illnesses that were reported by its workers, the so-called Radium Girls. See Roger F. Robinson, *Mining and Selling Radium and Uranium* (Cham: Springer, 2015), 152–153.

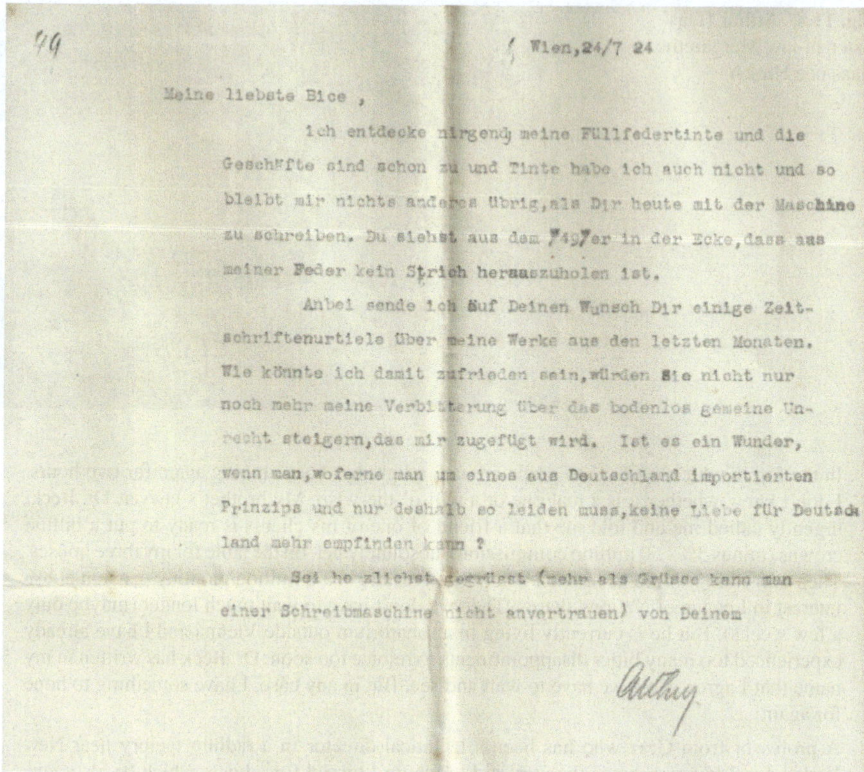

49

Wien, 24/7 24

Meine liebste Bice,

ich entdecke nirgends meine Füllfedertinte und die Geschäfte sind schon zu und Tinte habe ich auch nicht und so bleibt mir nichts anderes übrig, als Dir heute mit der Maschine zu schreiben. Du siehst aus dem 49er in der Ecke, dass aus meiner Feder kein Strich herauszuholen ist.

Anbei sende ich auf Deinen Wunsch Dir einige Zeitschriftenurtiele über meine Werke aus den letzten Monaten. Wie könnte ich damit zufrieden sein, würden Sie nicht nur noch mehr meine Verbitterung über das bodenlos gemeine Unrecht steigern, das mir zugefügt wird. Ist es ein Wunder, wenn man, woferne man um eines aus Deutschland importierten Prinzips und nur deshalb so leiden muss, keine Liebe für Deutschland mehr empfinden kann ?

Sei herzlichst gegrüsst (mehr als Grüsse kann man einer Schreibmaschine nicht anvertrauen) von Deinem

Arthur

Fig. 11.9 Letter No. 49 from Arthur to Emma

Arthur to Emma Vienna, July 25, 1924.
No. 50.

Dear good Bice! Darling!

Number 50! Horrible! The fiftieth letter, when will there be an end, a release from this time of torment? All hope now clings to the fact that through Dr. Beck the house sale may succeed. I am terribly excited about this.

No other letter about the houses. - Today I received a short letter from Prof. Donnan in London, which depressed me very much because of its lack of content. He writes:

> I hope you received my last letter in which I informed you that I had spoken to Professor Millikan and that he would call upon you in Vienna, and I hope you had a talk with him, as I feel certain that he would be able to render you very valuable help. As regards your new book on "Atomic Theory," in which you are going to pay me the very high honor of dedication, please send it to my address at 23 Woburn Square WC1. I fear I shall have to defer the pleasure of seeing it until I come back from America at the end of September, as I am due to sail from Liverpool on the 25th of July. With all best wishes and many thanks...

I am in such dismay. I am afraid that the time for a possible visit by Prof. Millikan might have passed. Has he already been to Vienna without having informed me, or will he not come to Vienna at all due to a change in his travel plans or will he come after all? - I also feel bitter that Prof. Donnan does not explicitly assure me that he will support my cause in America, but I think that does not mean anything. He will certainly make an effort.

From Prof. Ehrenfest - Leiden [Holland], who has excellent American connections and who, together with Prof. Donnan, is eagerly supporting me, the answer is still pending, although he did promise, in a letter dated June 2nd, "in four weeks." In any case, it is unfortunately, unfortunately certain that four days ago Prof. Donnan was not yet in possession of a favorable American offer.

Oh, Bice, dear, if only it were already over. I would carry everything, everything, if I could already have you by my side. But this terrible uncertainty. …

[…]

Be kissed many, many times and greeted by your Arthur

Arthur to Emma Vienna, July 25, 1924.
No. 51.

Dear dear Bice!

I wrote to you three hours ago but I write again because the afternoon post brought me three good letters.

A Viennese gentleman contacted me, whom I had written regarding a newspaper ad, and he wants to buy my houses without an agent, but only after he has visited them in Berlin. I immediately reached an agreement with him by telephone; on Monday we will have a meeting.

Then the owner of the real estate office in Innsbruck, to whom I had quoted a very high price, wrote to me that he has no doubt that he will be able to obtain a deal because of his good connections.

Then Professor Pirani,[22] chairman of the employment center for physicists of the German Physical Society, informed me in a very long letter that he has immediately taken several steps. However, his letter begins with the words: "It is of course very difficult for a gentleman with a literary and scientific reputation such as yours to find a position that is even remotely satisfactory".

His suggestions include the Siemens group, the A.E.G [Allgemeine Elektrizitäts-Gesellschaft] group, a film company that creates scientific films and "certainly (as Professor Pirani writes) would have used both hands a year ago."

He also named three gentlemen in America I should contact (which I will).

Dr. Whitney of the General Electric Company in Schenectady (whereby I, Dr. Pirani writes, can take knowledge of my name and my activity for granted).

Dr. K. G. Frank, New York, 114 Liberty Street, representative of the Siemens & Halske Company.

"Electrical World" New York, 10th Avenue 36th Street.

[…]

[22] Marcello Pirani (1880—1968) is one of the pioneers of vacuum technology. He worked at Siemens and Osram on the development of incandescent lamps and moved to the Technical University of Berlin in 1928. He left Berlin in 1936 due to political developments and emigrated to England where he worked on industrial applications of vacuum technology. In 1953, he returned to Munich and then to Berlin as a consultant to Osram. He is well known to every experimental physicist as the inventor of the Pirani tube for measuring vacuum.

As we have already seen in earlier letters, Haas applied for jobs outside of Austria and in Germany in 1924. In the latter he did not see much hope, while his desire to leave his homeland grew. He saw faraway places, especially North America, in an ever-more glorified light. First there were academic positions: Basel in Switzerland, Leiden in Holland, in Egypt at the Ecole de Medicine in Cairo, and also universities in South America. He did not even exclude Russia as a possibility; in fact, many colleagues in similar situations - mostly with great social hopes for the success of the revolution - followed the call to the newly founded top universities of the Soviet Empire.[vi] Haas was, however, fixated on the great hope of North American universities, a hope that he could not let go.

Through an exchange of letters with Marcello Pirani, Haas also again considered the idea of working in industry, either in German companies such as Siemens or in North America at the GE Research Laboratory in Schenectady, New York. Possible work in the film industry or the radium business, as mentioned in an earlier letter, seems to have vanished, never mentioned in further letters. However, with all of these applications for long-term positions, Haas always kept the short-term goal of selling the houses in mind.

Arthur to Emma Vienna, July 26, 1924.
No. 52.

Best Dearest Bice!

Today I find it very difficult to write to you because I'm afraid I have to tell you about yet another disappointment. The friend of Dr. Beck's client has changed his mind again and the sale of the house will not happen. I didn't have much hope because the matter didn't seem to be very serious to me. But you know how much I (like you) like to dream and so I had already seen us both in a few weeks on our honeymoon in a summer resort - and now this hope has been destroyed again. I feel as though I have again been hit over the head with a hammer.

Now my hope is on the meeting Monday morning with a Mr. Mandl who is acting as an independent buyer for the houses, next to which I'm also hoping for a notice from half-a-dozen agencies. After all, there are more and more advertisements in German newspapers in which people are looking to buy houses, while advertisements offering houses are becoming rarer. The signs of a turnaround are unmistakable on the German housing market and that gives me courage.

Today I wrote a very long letter to Dr. Frank[23] in New York and another to Russia for permanent appointment. Professor Pirani - Berlin pointed out this possibility in a second letter that I received this morning.

I am unlikely to receive a new message from Professor Donnan before the end of August; that is bitter. From Professor Ehrenfest, I expect news any day. What has happened to Millikan?

What would you say if we finally landed in Berlin? Today I wrote a long letter about this possibility to Professor Pirani.

[…]

[23] In 1908, the German physicist Karl Georg Frank set up an information office in New York to represent the interests of Siemens & Halske, alongside those of other German companies. After World War II and the resumption of business relations, Frank mainly sold scientific measuring instruments, laboratory equipment and arc lamps on the American market, insisting on a simplified design that seemed to him to be more useful for the American end user.

Farewell for today, may God grant that I may soon send you some definite good news. Many, many kisses and fond greetings from your Arthur

Arthur to Emma Vienna, July 27, 1924.
No. 53.

Dearest Best Bice!

[...]

Perhaps something will come out of the intensive efforts I am making to get a job in the scientific research institutes of big industry.

Today the newspapers are once again publishing the immediate appointment of some brainless private lecturer as a full professor! But he probably wears a swastika as an ornament on his pocket-watch chain; that's all right, it's more important than a brain.[24]

Today it is quite cold here, so I canceled the planned bathing excursion to the Danube. In the morning I was in the Prater, now in the afternoon I work at home on newspaper articles, of which I am preparing a small supply. Then I want to get this letter and others (job applications) to the main post office, so that they leave today.

Who will be the redeemer from my distress, who will allow me to bring you home? Will it be Millikan or Donnan or Prof. Pirani in Berlin or Prof. Wien in Munich, to whom I am writing today because he has excellent relations with South German industry, or will it be Mr. Mandl, who wants to buy my houses from me, or the Agent Kanalgitterbestandteil or something like that, or Professor Gans in Argentina? It would be funny if it weren't so sad. Only one thing I know for sure. It won't be one of my relatives, although there are several among them who could rescue me without feeling it themselves.[25] Oh, how bad people are!! Are they really all created in God's image? And yet, if they had lost all faith and would then only think of you, they would win it back.

[...]

Arthur to Emma Vienna, July 28, 1924.
No. 54.

Dearest dear Bice!

[...]

Oh Bice, for the time being, writing to you is so painful for me because I can't tell you anything about progress. This morning I went to see Mr. Mandl, a fine man who wants to buy 1-2 houses in Berlin and was interested in mine. But who knows if he plans to buy mine right now! He will not come back from Berlin until the 11th of August. An offer came from

---

[24] An interesting remark showing that Haas was well aware of the reasons for his difficulties in obtaining a permanent position at the university. The candidate for the chair, whom Haas calls a "brainless private lecturer," is Silesian Otto Reche (1879—1966). Reche succeeded the late Rudolf Pöch to the chair of ethnography in Vienna. Haas was right, Reche had already carried out racially motivated experiments with Pöch on prisoners of war. In Vienna and after 1927 in Leipzig, Reche developed into one of the most prominent protagonists for the integration of "ethnology" into university study, under the primacy of a racist anthropology as accepted by Nazi ideology. See H. Glenn Penny, *Matti Bunzl, Worldly Provincialism: German Anthropology in the Age of Empire* (Ann Arbor: University of Michigan Press, 2003).

[25] Here one senses a certain criticism of the extraordinarily wealthy Strakosch family of his mother, as well as his successful cousins in England and Austria. However, it is not clear whether and to what extent Arthur Haas approached the family for help, or whether he only expected spontaneous assistance.

Berlin to buy the houses but too low, would only give a monthly pension of about 4½ - 5 million kroner.

I wrote long letters to Leipzig and Vienna, to professors with good American connections. Every evening I have been talking with a professor who is very sure that America will take me. I wonder if he is right.

Anyway, a whole new factor has now been added: German and American industry. Now there are more opportunities. - And it is good that the stock market situation in Berlin is constantly improving, and improving considerably, so that the development of the housing market again improves, but still - the terrible uncertainty ... and the heartlessness of my relatives, who could help me so easily, but who don't dream of it.

[…] I have been thinking, it has been ten years since the day that caused the catastrophe,[26] the consequences of which I still feel so terribly. I may have survived the war without any physical damage but I would happily take on a lot of physical pain if only I could be spared the suffering of the soul that has tormented me for a quarter of a year. And in my family, not a compassionate heart!

Today I will write many more letters. I've already sent four letters and four or six will follow. A long one back to New York, to the head of scientific research at the General Electric Company in Schenectady (NY) and also answers to house-purchase advertisements in yesterday's Berlin daily newspaper.

When I think about how good I felt every year at the end of July in the summer resort - and yet, despite everything, I am so much better off this year; I have at least one hope that I didn't have before – to soon have won you forever!

[…]

Be kissed by your Arthur

Arthur to Emma Vienna, July 29, 1924.
No. 55.

Dearest Bicelein!

What is it about you not writing me anything? I am so depressed, more than a week since your last letter, and especially on your name day!

Today it looks much better again. For a start, when I came home a few minutes ago there was a telegram: Buyer offers an increase on yesterday's express-letter offer (express letter has not yet arrived) 2,1000 marks net cash carries increase in value. That is what I hope for, so I am satisfied. Only I do not know yet, which house? I will accept as soon as the express letter has arrived. Thank God it goes forward! (And there are strong increases in the Berlin stock market, which is automatically transferred to the housing market.) At first, I was terrified by the telegram because I thought you were ill.

In some respects, I'm becoming more confident again. I hope to get a job in the electricity industry with a big company. Efforts are being made both in Berlin and Vienna and, unlike all other industries, the electricity industry is not in decline but, on the contrary, is taking in new people. Starting salary is not high, maybe 4-5 million a month for me, but combined with the houses and the income from my writing it could be enough, even if the income does not increase and America comes to nothing. I am not as despondent as I was a few days ago!

[…]

Emma to Arthur Kapuzinerstraße 20, Munich, July 28, 1924.

[26] On June 28, 1914, the heir to the Austrian throne Franz Ferdinand and his wife were shot dead in Sarajevo, an assassination that within a few weeks triggered World War.

Dear Arthur,

First of all, thank you for your letters.

The day before yesterday I sent you the English book *The Missing Million.*[27] You must have received it already.

Today Papa received the money from the lady in Leipzig. She sent 30 marks and 17 dollars, not the amount you expected.

I don't understand what you meant in your last letter: I should send you 10 marks, 10 pfennigs! Please explain yourself.

In this letter I am also enclosing a copy of my college transcript. I have not made a copy of the certificates from the universities (Chicago University and Loyola University) because they are in the name of Sister Mary Beatrice.

I am also including 5 marks in this letter, if you would be so kind as to pay the photographer for my photographs and have them sent to me. I thank you in advance for your trouble.

[…]

Arthur to Emma Vienna, July 30, 1924.
No. 57.[28]

Dearest Bice.

Although I have been very angry with you since this morning, I would like to tell you about the progress that has been made today.
The telegraphed offer to buy the house refers to Landsbergerstraße 12, as a new telegram tells me. I accepted without reservation so unless something extraordinary happens, the house can be considered sold. The proceeds represent a monthly pension of 2.2 million crowns (at 8% interest). But this is only one of the three houses.
Sale of the others will probably happen soon. Of course, it is still difficult to convert the rentenmark in Berlin to Austrian crowns in Vienna, and that may take some (though not long) time.
Furthermore: The head of the Vienna Siemens-Schuckert-Werke has asked me to introduce myself to him tomorrow morning. However, I would still need the approval of the Berlin head office to be hired; I have approached them several times. If it came to employment (Factory in the III district not far from my apartment), I would probably start on October 1st, with decision having been made in September. Expected salary 4 - 5 million monthly and 13 times a year + 12% wife allowance.
Tomorrow I will tell you more. Best wishes Arthur

Arthur to Emma Vienna, July 31, 1924.
No 58.

Dearest Bice.

The most meaningful thing about the above quote [Fig. 11.10] is that one of the most important American physicists explains that there is a great lack of theoretical physicists in America. So even if Prof. Dempster does nothing, I can hope that Prof. Donnan and Ehrenfest will succeed if they make a serious effort to help me, as I am convinced that they will.

[27] This mystery novel was published in London in 1923. Written by Edgar Wallace, it is the tale of a millionaire who mysteriously disappears on the eve of his wedding. This probably served the purpose of improving Arthur Haas' English.

[28] Letter No. 56, probably the cause of the following argument, is missing in the collection. It may have been destroyed in anger.

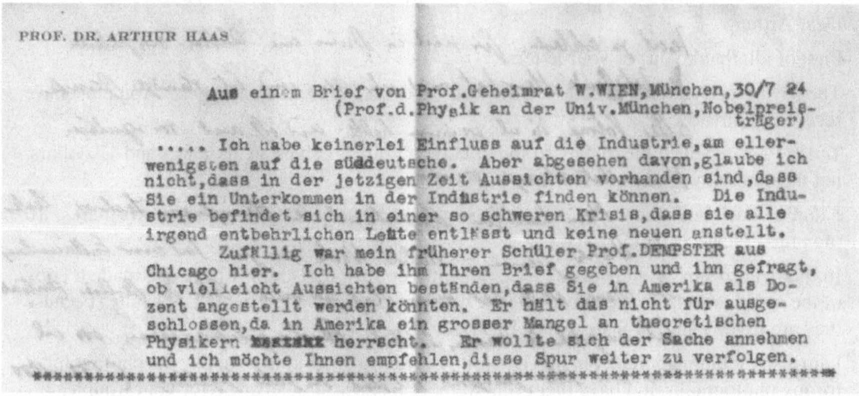

PROF. DR. ARTHUR HAAS

Aus einem Brief von Prof.Geheimrat W.WIEN,München,30/7 24
(Prof.d.Physik an der Univ.München,Nobelpreisträger)

..... Ich habe keinerlei Einfluss auf die Industrie,am allerwenigsten auf die süddeutsche. Aber abgesehen davon,glaube ich nicht,dass in der jetzigen Zeit Aussichten vorhanden sind,dass Sie ein Unterkommen in der Industrie finden können. Die Industrie befindet sich in einer so schweren Krisis,dass sie alle irgend entbehrlichen Leute entlässt und keine neuen anstellt.

Zufällig war mein früherer Schüler Prof.DEMPSTER aus Chicago hier. Ich habe ihm Ihren Brief gegeben und ihn gefragt, ob vielleicht Aussichten beständen,dass Sie in Amerika als Dozent angestellt werden könnten. Er hält das nicht für ausgeschlossen,da in Amerika ein grosser Mangel an theoretischen Physikern herrscht. Er wollte sich der Sache annehmen und ich möchte Ihnen empfehlen,diese Spur weiter zu verfolgen.

****

Fig. 11.10 Exerpt from a letter from the Nobelprize winner Wilhelm Wien to Haas, promising new opportunities in America

A new project also emerges. The professor of physics at the German Technical University in my hometown Brno has died.[29] A very distinguished elderly gentleman, who has excellent relations with the Brno University of Technology, visited my brother yesterday to suggest me as the successor in Brno. The gentleman in question is very fond of me and is a constant visitor to all the lectures I have given; he even comes frequently to my regular university classes ...

My visit this morning with Chief Engineer Lechner, head of the Siemens Halske Works in Vienna, was partly a disappointment. The Viennese plant has no place for me. But the Berlin headquarters has extensive scientific-research laboratories where I could find a job that suits me very well and with good pay (500 – 1,000 marks per month). From Vienna, no work will be done in this direction. That would certainly mean exchanging dear Vienna for the hideous Berlin.

Nothing new on the house question! I am waiting since we have concluded by telegram that I will be, in Vienna in a fortnight at the latest, in possession of the equivalent in Austrian crowns. I also hope that the other two houses will be sold soon (because the number of buyers is increasing rapidly, with numerous new advertisements every day) and once the three houses have been sold - then you shall, if you wish, be welcome in Vienna as Dr. A. H.! I am much more confident than a week ago.

[…]

Arthur to Emma Vienna, August 1, 1924.
No. 59.

Dearest Bice

Also today, thank God, further progress.

An express letter has come about Landsbergerstraße, which can probably already be considered sold (unfortunately, much too cheaply), and also a telegraphed offer to buy Seesenerstraße for 7,000 dollars,[30] although it is not clear whether the buyer will pay the capital gains

[29] This was Gustav Jaumann (1863—1924) who had strongly supported Arthur Haas' application for habilitation in 1911.

[30] Because of the unreliability of German currency, people tended to use the US dollar as currency.

tax; otherwise I would have to pay 2,000 dollars but, since I would give Seesenerstraße away for a little less than 5,000 dollars, I would in any case have reached the price I need.

I confidently hope that I will sell the three houses quite soon [...]

Best wishes and kisses from your Arthur

I was very interested in your college transcripts. I read *Missing Million* with great interest. I myself have a "missing million" (gold crowns), which the state's crooks, knowingly or unknowingly, have stolen from me! If I still had it, I think I could very cleverly spend and consume the interest while at your side! But instead of me having the money, who would use it nobly, the missing million is now with the crooks who eat with a knife. Isn't that unfair?

Arthur to Emma Vienna, August 2, 1924.
No. 60.

Dearest Bice

I'm very bitter that you don't write anything to me. One unkind letter in twelve days: That was all!

I picked up your pictures today (by the way, he charged 180,000 Kr for them).

Since my letter yesterday, the following further progress has been made

1) Privy Councilor Wien (Munich) invited me to write two chapters for a handbook of experimental physics to be published by him, work totaling 1,200 - 1,400 pages which would be 70 - 80 sheets. For each sheet, the publisher offers me 180 goldmark. That would be a total of 12,600 - 14,400 marks or about 200 - 220 million crowns. The work would take about three years (according to my estimation) so that for these three years a monthly income of 6 million crowns would result. I still have to wait for a firm commitment and by then I will have a better idea about my situation.

2) On America, another way has opened up for me. On Sunday, I wrote to Privy Councilor Prof. Sudhoff in Leipzig asking him to intercede for me with Americans belonging to his circle of friends. In particular, I mentioned David Eugene Smith,[31] Professor of the History of Mathematics at Columbia University in New York. Prof. Sudhoff now informs me that by happy coincidence Prof. Smith is in Switzerland as the guest of a good friend of Sudhoff's; he wrote immediately and hopes for success. I had originally been a professor for the history of physics and thus a colleague of Prof. Smith's, while Sudhoff is a professor for the history of medicine. He arranged for my appointment in Leipzig in 1913. We became acquainted in Salzburg while attending a "Salzburg Concert" (1909). This avenue does not seem hopeless to me.

3) Regarding the houses, a real estate dealer in Munich has met each Friday with an interested party in Berlin. The result is not known to me yet.

[...]

Yesterday and today, I have had long conferences about the Brno professorship (quite hopeless) and then regarding a project involving a consortium that wishes to take over my Berlin houses (not serious enough).

---

[31] David Eugene Smith (1860—1944) is considered one of the fathers of mathematic history and education in the United States. He studied law and mathematics at Syracuse University but then spent a year in Göttingen in 1891 under the tutelage of Felix Klein, who promoted and stimulated his interest in mathematical history – as did the young Arthur Haas, fifteen years later. After Smith's return to the US, he spent several years at various schools and smaller colleges in New York State before moving to Columbia University in New York in 1901 where he remained until his death.

I am still in a good mood but it is very much affected by the fact that you show yourself so unlovingly toward me.

Best wishes from your Arthur

Emma to Arthur Kapuzinerstraße 20, Munich, August 1, 1924.

Dear Arthur,

I won't keep a letter like that for an extra minute. You can pick it up and salt it if you want.

I don't at all see that I have done anything wrong and if a justification were necessary, I would give it completely.

I won't read such a letter a second time, since I am already indignant enough after reading it once.

If your next letter is written in a different tone, then I will answer it differently.

Regards Bice

P.S. I enclose the desired stamps. But don't try to justify yourself again with an excuse about your hieroglyphics. They are not responsible.

II. P.S. (Enclosed again 5 marks for my photographs - I will do it until I get them).

This is probably Emma's outraged answer to letter number 56 of July 29, 1924. That letter is missing but probably contained more sharply formulated complaints and accusations from Arthur about the lack of letters from her side. Emma Huber's indignation was so great that on the same day she wrote two letters following this one, all of which arrived in Vienna with the morning mail on August 4th. The mail between Munich and Vienna took on average two to three days, depending on when the letter was posted. In the case of express letters, transport time was shortened to about a day.

Emma to Arthur Kapuzinerstraße 20, Munich, August 1, 1924.

Dear Arthur,

As one shouts into the forest, so it echoes back. I won't start to become friendly - doesn't occur to me at all. You have started to be nasty - so, start in a different way when you want to repent.

If you think that you annoy me when you don't send me kisses, then you're completely wrong - personally I get as many as I want and they don't really have any value in writing. So! –

I was infected by an epidemic that is currently raging in Munich. I only weigh 96 pounds, but this week I'm feeling fine again - because a lovely flower like me doesn't spoil. That was one reason why I didn't write, another is that: Since I doubt very much if you'll ever find a job for yourself, I've again been looking for one for myself. I immediately got a favorable answer and was accepted as an English teacher. I am enclosing the envelope here; the letter itself I have to keep as identification because the director is coming to Munich on August 6th to talk to me in person. I would have to commit myself to at least one year and if I do, I will not break my promise under any circumstances. Should I accept or not? Let me know the answer before August 6th (by telegram if necessary). The acceptance or rejection of this position depends on your answer because you know best what the future looks like. If I have no answer from you before Aug. 6th, I will accept the position.

Regards Bice

Emma to Arthur Kapuzinerstraße 20, Munich, August 1, 1924.

Dear A.,

I just threw today's second letter into the mailbox but maybe you are unnecessarily concerned about my illness last week. In general, it wasn't a dangerous disease. Although many thousands of people in Munich fell ill, most of them were only very slightly ill but I was quite badly affected. The most unpleasant thing was that I lost a lot of weight. But this week, I have gained weight again.

The doctor has made the following observation about us: I am suffering a great deal mentally, either because I have great longing for my distant bridegroom or for some other reason. Due to these inner sufferings, which become worse and worse because I don't talk about them with anyone, my body has become very weak and very susceptible to any illness. Organically I am missing absolutely nothing, I am quite marriageable, just a little anemic. And at the moment very "heart sick" - no wonder!

You can cure my heart disease - the doctor knows that too.

So - what do you think was wrong with me?

Greetings Bice

Arthur to Emma Vienna, August 3, 1924.
No. 61.

Dearest Bice

How much longer do I have to be mad at you? When will a nice letter come from you?

Early this morning I received a telegram from the Innsbruck agent, "awaits today's express letter, failure to conclude otherwise". I told him the ultimate price for each of the three houses is $7,000 and I'm curious to know what the express letter will say. I confidently hope that the sale of the three houses will succeed, now that I have satisfactory offers from four sides. The only thing to do is keep a cool head, be patient and don't rush things.

Of course, it is conceivable that such negotiations could drag on for weeks. Of course, I am eager for us to marry before the end of August but I don't know if it will be possible to conclude the negotiations that fast. Perhaps in a week, beginning today, all of the houses will be sold and the money will be in my hands - but one cannot make any predictions. Perhaps it will take another eight weeks. Only one thing seems certain to me: the prices on the Berlin housing market are solid; now sales for the minimum prices I have in mind are feasible. Whether it will succeed and how quickly it will be done depend on my skill and on that of the agents.

If it is successful, I will always have 6 million crowns a month. Even if I do not find a job and the American appointment does not come about, my writing work will ensure a monthly subsidy of at least another 2 million crowns for about ten years.

Show yourself to be so kind that I dare to send you more than just warm greetings. Yours Arthur

I can't go to Salzburg for the time being because I can't leave. On the other hand, I may have to travel to Berlin (at the expense of a potential house buyer!).

Arthur to Emma Vienna, August 4, 1924.
No. 62.

My dearest best Bice!

The Post Office just brought me your three dearest letters from August 1st. I am extremely worried that you were sick again. What on earth is wrong with you? Now that I know why you haven't written for so long, I want to be nice to you again.

I read the position advertisement to which you have responded, and thought of you at the time.

You have now seen from my letters, which you have received since August 1st, that a complete change in the situation has occurred. Today the situation is such that I am firmly convinced that we will be able to marry within six months of our engagement.

I have just now received an express letter and, especially since I am receiving 21,000 marks for Landbergerstrasse this week, I now have no doubt that the other two houses will soon be sold at a favorable price. But I do not know in advance how long it will take to settle.

It is not out of the question that we could perhaps marry before September 1st. But dear Bice, unfortunately we don't know if another unexpected political event will cause a sudden sharp setback before the houses are sold. Hopefully not, since the political situation is getting better and better,

As soon as I have the proceeds from the sale of the three houses, I am immediately ready to get married!

I've been in an excellent confident mood for a few days now (although I'm very concerned about you).

Let me know immediately how you are doing because I am very worried about you.

Warmest greetings from your Arthur

And stop pouting!

Arthur to Emma Vienna, August 5, 1924.
No. 63.

My dearest Bice!

I am very concerned about your health. Are you really healthy again, are you putting on a lot of weight again, and what actually was wrong? Some kind of sore throat or something?

I hope you're in as good a mood as I am. To know that you are so far away is of course horrible for me but now I finally see that my goal is within reach and I don't have to wait in uncertainty. If nothing unexpected comes, we will succeed!

Today I am compiling all the notarial deeds for the house on Landsbergerstraße. I hope to receive the money this week. As far as Seesenerstraße is concerned, negotiations are still pending. As far as the house on Kriemhildstraße is concerned, no buyer has contacted me yet but he will surely come along soon.

Once the houses are sold and we are united, then we can wait to see what will become of our America dream and whether my efforts to find employment in Vienna or Berlin will be successful. As far as Vienna is concerned, it looks completely hopeless at the moment.

In any case, Professor Wien's offer, of which I wrote you and which will bring me a total of approximately 220 million crowns over a period of approximately 3 years, is very significant and can also be considered a highly favorable turn of events.

[…]

Hopefully, I can tell you about new progress tomorrow. I am writing several letters a day about the houses, answering the increasing number of buyers' notices.

Many, many, many warm kisses and greetings from your Arthur

I am very busy and also busy with the preparation of the apartment.

This offer by Privy Councilor Professor Dr. Wilhelm Wien to the already not-so-young Privatdozent Haas is indeed important help in the latter's career. It was only through such indirect support, which was generally recognized and valued as a great honor, that advancement within the field of science was possible. Wien's support can be seen as an attempt to create a professorship in the German-speaking world after all. This is especially underlined by the fact that in a second letter, Wien repeats the offer but also mentions the possibility of other candidates for this honorable and lucrative task. A subtle hint, which is immediately recognized as such by Arthur Haas.

Arthur to Emma Vienna, August 6, 1924.
No. 64.

Dearest Bicelein!

There is nothing new to report today. Tomorrow is a critical day of the first order and hopefully a lucky day. Tomorrow the sale of the house on Landsbergerstr. in Berlin will probably be completed and I will become the owner of 21,000-mark proceeds, which my Berlin notary will take possession of tomorrow. - I can't think that everything is again being just a bluff. If only it doesn't fail at the last moment because of a swindle; I am as excited as I was before my high school exam!

Why do you write so little? Since your last letter, I've been so worried about your health. Please, if you can, allay my concerns. Are you putting on weight again?

Privy Councilor Wien from Munich wrote to me again, asking if I would like to accept his proposal (which would mean about 220 million crowns for three years of work). I have already accepted in a letter that crossed with his. This is of utmost importance because even if I do not find a job and employment in America, I will have secured an income of 6 million crowns per month for three years.

Today, on the advice of Privy Councilor Wien, I wrote directly to Professor Dempster[32] of the University of Chicago, who is currently in Göttingen. Just think, in one year we could be on a summer retreat at Lake Geneva![33] ...

A Viennese lady, whose daughter is married in Chicago, told my sister-in-law yesterday that her daughter is desperate despite very favorable income conditions. I wouldn't mind though; the worst place with you is better than Vienna without you.

Greetings and kisses from your Arthur

Emma to Arthur Kapuzinerstraße 20, Munich, August 5, 1924.

Dear Arthur,

So - I'm not taking the job. I hope fate will not let us down again. If I were forced to look for a job for the third time, I would accept it immediately - in any case. But I put trust in your efforts.

---

[32] Arthur Jeffrey Dempster (1886—1950) was a Canadian-American physicist best known for his work in the field of mass spectrometry and his discovery of the uranium isotope 235U in 1935. He studied at the University of Toronto from 1909 to 1910 but moved to Germany to do his doctorate near Vienna. At the beginning of the First World War, he returned to the United States where he became a professor at the University of Chicago. During the Second World War, Dempster worked as head of the Metallurgical Laboratory or "Met Lab" at the University of Chicago where he was part of the Manhattan Project, studying the materials necessary for making atomic bombs. Through his connection with Haas, some of these investigations were also carried out at the University of Notre Dame. In 1946, Dempster accepted a position as department head at the Argonne National Laboratory. He died on March 11, 1950 in Stuart, Florida, at the age of sixty-three.

[33] Here Haas is not referring to Lake Geneva in Switzerland but to Lake Geneva in Wisconsin, which in the 1920s was a preferred holiday destination for vacationers from Chicago and the neighboring city of Milwaukee.

Of course, I follow the sale of the house with the greatest interest, and I hope and wish nothing more than that you will soon have happy success.

With my health I am doing well again - I think I have already gained weight again, which I lost during my illness.

I try to relax more now; the doctor says it is absolutely necessary. I prefer to sit on our veranda - it is so beautiful outside, so quiet and the flowers bloom so magnificently. But the wild vine already has red leaves and they fall off with the slightest breeze. One just flew toward my table. I enclose it. It is also a reminder that autumn is coming again.

Best wishes and thanks for your dear letters. Your Bice

P.S. Thank you very much for the pictures. I needed them very much and I am happy that I can send some more to America. But I am surprised that you had to pay 180,000 Kr. They only cost me 65,000 Kr.

I haven't yet received the linen that Mr. Dessauer sent me.

Arthur to Emma Vienna, August 7, 1924.
No. 65.

Dearest Best Bice!

Today I'll be late for my letter-writing! I can't report anything new yet; no news about the houses, neither good nor bad, has arrived by this hour.

In any case, everything looks good because the London Conference will undoubtedly end successfully and as a result, all securities and housing prices are already steadily rising. Also, the shortage of money (which has been the main obstacle for selling houses so far) is gradually ending in Germany.

[…] I have already achieved quite considerable success: Professor Wien has assured me a monthly income of about 5-6 million crowns for about three years.

[…]

Today only the most heartfelt and intimate kisses from your Arthur

P.S. Perhaps the enclosed letter from my students[34] might interest you, so that you can see that my efforts have fallen on fertile ground after all. I am now more optimistic in this respect and cannot imagine that I will not also achieve material benefits from this. As far as America is concerned, I still have the feeling that all will succeed there but how long I will have to wait is still uncertain. In any case, the waiting is made much easier by Prof. Wien's proposal.

Emma to Arthur Kapuzinerstraße 20, Munich, August 5, 1924.
My dear Arthur,

[…]

Healthwise I feel excellent. I should have been sick for a day or two this week but I felt so good that I could even go into town and do some shopping. That hasn't happened for years. I can eat and sleep again like I haven't in months. Since the worries about our future have become clearer, my health has also improved.

Yesterday I spoke to the director of the reform school.

Best wishes and a kiss from your Bice

Arthur to Emma Vienna, August 8, 1924.
No. 66.

[34] This is a letter from a Viennese student on vacation; he praises Haas' new textbook unreservedly in comparison to other current textbooks in physics.

My dearest Bice!

Thank you so much for your dear letter from yesterday. I am so glad that your health has recovered and that my confidence has been transferred to you and is increasing your well-being.

Unfortunately, I still have no news about the Landsbergerstraße house. The notarial power of attorney that I had sent arrived at the Berlin notary early yesterday (Thursday). I would be worried if not for, on the other hand, the undeniable improvement of the Berlin housing market, which convinces me that the sale of the houses will probably succeed in the near future.

[…]

I have read *Missing Million*; the book reads very easily and is very interesting. Now I'm reading another book by the same author, called *Chick*. I'm sure that I'm going to learn a lot of English this way.

[…]

If only everything were already over. The situation is getting better every day but if only I could hold you in my arms forever!

Many fondest kisses and greetings from your Arthur

Arthur to Emma Vienna, August 9, 1924.
No. 67.

My dearest, dearest, best Bice!

Still no news from Berlin about Landsbergerstraße! I don't know what's going on. I'm terribly nervous but, on the other hand, I'm telling myself that even if some kind of trickery were to make it fail, I don't have to despair.

It seems to me that the situation on the housing market is such that the market prices have already reached the level I need. But of course, it takes luck to find the buyers. It is difficult now in August because only people with money buy, and some of them are on their summer vacation. Our permanent representative is also currently away from Berlin, which makes things more difficult. And the lady who wanted to buy the house in Landsbergerstraße is about to leave.

So it's all about patience again. I've written a huge number of letters again today about the houses. ... Oh Bice, I am so afraid for you and I still can't have you with me, ... but I want to have courage and I hope it all will be over soon. If only the first house were sold.

I wrote to Professor Dempster in Chicago where he's staying. - The congress in Toronto started on August 6th; if only Prof. Donnan keeps his promise to talk on my behalf. I think I can be sure of that.

That my brother left this morning makes me sadder. There was at least one person with whom I could sometimes talk a little - I still miss him now.

[…]

I've become so pious and devout; does soul trouble make you pray? If only my prayers could be fulfilled! All night long and over and over, I ponder on why I have no news about Landsbergerstraße – and I tell myself that if it had failed, I would have been informed. So I have hope from the lack of any news. Maybe it's just sloppiness or small difficulties that have delayed things a bit? - And tomorrow, Sunday, again no mail that would bring good news.

I hope your health continues to be good; don't take the matter of the delayed house sale too much to heart. - Of course, I'm going to get rid of the worst pangs of conscience - have I sinned? I don't know, only God knows - but why I have to suffer like this, I don't know.

Be kissed a thousand times by your Arthur

One can see from this letter that Arthur Haas was living with a very high level of stress. This is one of the few times that he makes a clear religious statement, giving a hint of his belief in God. He prays in part to the punishing God of his Jewish forefathers but also to the Christian God who should have been available to help in times of need and to whose church Haas has belonged since the Göttingen days. This stressful situation is also reflected in the following letters, in which he tries to follow the principle of reason instead of relying on his God.

Arthur to Emma Vienna, August 10, 1924.
No. 68.

Dear good little Bicelein!

I want to send this letter as an express letter so that it arrives at the same time as the last one. I wrote the last one in a state of deepest depression; I am sending this one in a mood of returned confidence.

[…]

Now the buying chances outweigh the selling chances more and more. I have a lot of new contacts and I am therefore convinced that new offers will now be more frequent and easier than before.

The previous technique of selling was misguided. It failed the first time (before our last meeting in Salzburg) because I answered too arrogantly at the behest of my brother. My reply failed to engage the limited patience of the prospective buyer because he had to leave Berlin. This time it failed with Landsbergerstraße (probably it failed, although I haven't given up all hope yet) because I was one day too late. My power of attorney should have left on Monday but Otto, who was actually completely exhausted, declared that he had no time; he did not draft it for me until Tuesday and so, instead of Wednesday, the power of attorney arrived on Thursday. However, on Tuesday evening I received a telegram saying that the sale had to be made on Wednesday. - If it failed, that would be the reason. My brother had, of course, no idea what terrible emotional suffering the delay of one day would cause me.

And finally, with respect to the new offer for Seesenerstraße ($7,000 but, probably, I would have to carry the increase in value of $2,000), my immediate presence in Berlin was urgently requested by wire but I could not get myself to go due to the unclear nature of the offer.

As one who lives so far away from Berlin, competition with people who are living in Berlin, who can be on the spot immediately, is terribly difficult to overcome - this difficulty must be addressed. Although it costs money, I took the necessary steps yesterday. A representative of my brother and myself is now authorized to conclude the purchase contract in my name within twenty-four hours and to receive the money on our behalf. My brother is of the opinion that one can have blind trust in this man.

I am convinced that Seesener- and Landsbergerstr. would surely have been sold already if I had gone to Berlin on the spot in both cases, so I confidently hope that the new method will soon achieve the goal.

[…]

What gives me the most confidence is the prevalence of buying over selling notices in Berlin newspapers, the increasing number of buying notices for Berlin houses in the Viennese newspapers, and above all the constant, very significant increase of all securities on the Berlin stock exchange.

Heartfelt kisses best regards from your Arthur

Emma to Arthur Kapuzinerstraße 20, Munich, August 9, 1924.

Dear Arthur,

Your letters give me more and more courage - I thank you very much for that. I must confess that only two weeks ago I considered our cause to be completely lost. Now I am thinking more calmly about our future and that is why my health is also doing well again.

Do you know what I like to dream now? - Of your apartment. I don't really have any clues because you never told me about it - but I know that it is beautiful. I have several nice things at home, which I will have sent to me later if necessary. I am waiting for your letters every day with an impatience that is understandable.

We have great pleasure with our radio. If the weather allows it somehow, we all sit outside on the veranda and listen because we also have connection there. Or we sit in the living room where we have connection too. Our cook also has an apparatus in the kitchen, so she can listen in her free hours and in the evening. Her pleasure is of course boundless. In general, it is a pleasure just to look at our old cook. She looks so happy and satisfied and she gets so roundish that we are very proud of her. In five or six years, she will celebrate her silver-service anniversary with us.

My dad and brother are terribly busy - but up to now nobody has had money to pay the bills. Daddy can sometimes be happy when he brings in enough money to pay the workers. Love
Bice

P.S. I already wrote to your sister-in-law in Gottersdorf at the beginning of this week. Hopefully she will get the forwarded letter. It's not very interesting as I remember, but what can one write? - I know so little about your sister-in-law.

Arthur to Emma Vienna, August 11, 1924.
No. 69.

My dearest Bice,

Thank you very much for your letter from the other day. You are so kind and good and well-behaved. How sad I was when I thought that your well-being might again be affected by the destroyed hope of selling the Landsberg house. [...] But apart from that, it doesn't look bad. I already calmed down quite a bit during the night between Saturday and Sunday and even more in the course of yesterday.

I have gone through Sunday's *Berliner Tageblatts* again: more than twenty house-purchase ads (I'm writing to all those who offer a viewing, that's about 8!); this amount used to be a rarity. On top of this, there are no meaningful sales advertisements. We are seeing the emergence of a new house-buying boom and so on, and I hope to see more good news next week; I have already made every effort to change my selling technique so that the buyer can close within twenty-four hours; I hope that will work. - The Berlin stock exchange is constantly on the rise, as is the Viennese exchange. On Friday, I managed to sell 20 million shares without losing any money; this is the amount by which my choking balance of debt to the bank is now reduced.

Oh, that I could write to you about other things than this horrible housing market and the stock exchange. Unfortunately, the key to my happiness lies in these disgusting things.

By the way, it's strange - even today not a single line about the Landsbergerstraße house from my Berlin notary or from the company. The matter is as mysterious as a detective novel. In the end, was it only delayed? Or did German censorship of letters break up the act and leave it lying around for a few days? In Germany, anything is possible and especially letters to lawyers and notaries are often "officially opened".

Today I received a letter from Privy Councilor Wien. He expressed his satisfaction about me taking over the work on his manual. I wrote that I would be finished in two years. He said that that would be a bit too long and so he wants to give me only the "mechanics" at first, maybe the "optics" to someone else, and something "modern" for me later. I answered immediately by express mail that I have realized in the meantime that the task is not as difficult as I had thought; I assured him that I would be finished more quickly and I urgently asked him to therefore also give me the optics. I hope that he agrees to this.

Of course, I will not be able to work it out with the same care as my previous books; I will need to write about 1,100 – 1,400 pages in 1½ years! But if I do it, I could earn over ten million crowns a month in these 1½ years. You can imagine the satisfaction it gives me to realize that my "intelligence" is also financially exploitable. I also wrote to Privy Councilor Wien (he lives in Munich on Leopoldstraße) that I want to start work as soon as possible and that I would like to have regular payments from the publisher. Hopefully this will not fail too!

Today the post office brought me a letter that touched me very much. In 1918, I had a famulus[35] in Leipzig who always took a walk with me in the evening. One day he (the twenty-three-year-old) surprised me by telling me that he had become engaged and asked whether he could continue the walks with his bride, as a threesome. So, the three of us became good friends. He is now a (badly paid) teacher at the Polytechnic in Cöthen (Anhalt), where I visited them last December. I am in correspondence with him and two months ago I informed him of my engagement. I had asked him to wait with congratulations until my situation was more pleasant. He has respected that so far. But today I received a touching letter from Mrs. Schmidt and a lovingly handmade embroidery was enclosed for you with the wish that we may use it soon. I was so touched, may her wish come true. Now, Bice, be confident. I think you can be - I've always been a pessimist. But a little patience. The more patient you are, the quicker you get there.

Most heartfelt kisses from your Arthur

Arthur to Emma Vienna, August 11, 1924.
No. 70.

Dearest Bice,

Just as I was about to take letter 69 to the post office, together with the six letters that I had written in response to yesterday's ads in the Berlin daily newspaper, *Tageblatt* a telegram messenger brought me a telegram: "Cash buyer 5,000 dollars, all costs, also buyer bears increase in value, Bannet"!

I am telegraphing the acceptance (it is Seesenerstraße) and at the same time sending express letters to Bannet and my representative so that the deal can be closed on Wednesday.

One should not praise the day before the evening, but you see that my confidence is well-founded. If only it will be so! Most heartfelt, Arthur...

Arthur to Emma Vienna, August 12, 1924.
No. 71.

Dearest Best Bice!

You already know from my urgent letter posted yesterday at the main post office shortly before midnight (with nine other letters) that I received a new telegraphed offer for the house

[35] Harry Schmidt (1894—1951) married Johanna Fernholz in 1921. From 1923 to 1937, he was a professor of mathematics at the Polytechnic Köthen (Anhalt). He habilitated and in 1933 he became a non-permanent extraordinary professor of mathematics at the University of Leipzig. During this time Harry Schmidt published, *Einführung in die Vektor- und Tensorrechnung* (Leipzig: Max Jänecke, 1935) which built on his early work with Arthur Haas, following his book on vector analysis in idea and basic concepts of application to physical problems. Schmidt's wife died in 1941. A new edition of his *Introduction to Vector and Tensor Calculus* was published posthumously in 1953 by VEB Verlag Technik der DDR.

on Seesener-Straße at 9 [o'clock] yesterday evening. I have given both the company making the offer and my realtor such precise instructions, taking all eventualities into account, that one would think that the purchase would have to be completed within the next few days. But one should not praise the day before the evening. I am in a very good mood again. - Yesterday, even on the Vienna Stock Exchange, where everything is soaring, my securities rose again by twenty million crowns. Oh, why do we live in a world where I can only tell you about such disgusting things!

I had written to Prof. Dempster of the University of Chicago on the advice of Prof. Wien, but the letter came back with a note that the addressee had already left, unknown forwarding address.

My Aunt Jetti wrote me today from Baden that she and her husband have been very ill. I feel sorry for her but not for him, because he does not feel sorry for me either.

Now I want to go swimming again in Dianabad.[36] The mail is unlikely to bring me anything today; mail from Germany posted on Sunday would have already arrived and would not, of course, have included any business transactions.

So hopefully the week will bring the first real success! I am sure that the second and third houses will follow very quickly from the first. If only I had you in my little apartment. Once the first success is achieved, I will choose a different topic in the letters: My apartment and its transformation for you.

Be kissed a thousand times from the heart by your Arthur

Arthur to Emma Vienna, August 12, 1924.
No. 72.

Dearest Bicelein!

We just received a telegram from the Dreher-Roske company saying that the house on Landsbergerstraße has already been sold.

I wanted to telegraph you immediately, but I will wait until a telegram from my notary informs me that the full amount of money has been received (21,000 marks).

Only then will I cheer and telegraph you. I'm still too suspicious.

Kisses from your Arthur.

Arthur to Emma Vienna, August 13, 1924.
No. 73.

My sweet little Bice!

It is like a miracle that I still cannot understand - suddenly everything turns out for the better. I still cannot believe it. Unless something quite unexpected happens, the full proceeds from the sale of two houses will be in my hands by the end of this week, and if I can telegraph you one day (hopefully soon) that the third house has happily also been sold, then you know that the marriage will take place within a fortnight at the latest. In addition, I am very happy with the publication contract with Privy Council Wien, which gives me a well-paid job over a longer period of time. So, I can put the question of my profession on hold until I know what will become of America.

---

[36] It must have been too cold for the open-air pool. Opened in the middle of the nineteenth century, the Dianabad was one of the first European indoor pools. Haas would have used the luxurious new building completed during the war, which offered two swimming halls (a sports pool for men and for women there was a wave pool), steam and tub baths, sunbathing, spa, shops, hairdressing salon, pedicures, a laundry, a dog bath with a veterinarian and other attractions.

[…]

Today I also received a letter from the wife of Prof. Ehrenfest from Leiden-Holland. She informs me that her husband (who is very influential in America) has gone to Russia and I should not expect any news from him before the beginning of October.

As far as America is concerned, I say to myself now, on calm reflection, that I can hardly hear anything before the end of September and that an offer will hardly be made before January 1925. But thank God, now that our future is assured elsewhere, it is not so urgent, and it is perhaps better that I learn some more English before we cross the ocean. I think it is certain that sooner or later I will be able to find a position in America. Now that my textbook is published in English, I will become better known in the USA.

Yesterday, a natural philosopher from a village in the Black Forest visited me. I was expecting a bearded hermit but was met by an elegant young man. He owns a property there, is worldly wise and was in New York last year. So I went for a walk with him and had him tell me about America and its colleges. He said that apart from the apartment and any extras, life there was not much more expensive than in Germany.

Hopefully I can write you a second time today with new messages. What a joy it is for me to write to you now, in this mood of "sky-high jubilation". If only I could chat with you but I hope that you are reasonable enough to understand that I cannot go to Salzburg at the moment. I am only leaving my apartment for short periods of time because of possible telegrams. I hope you are feeling very well now too. Be tenderly, tenderly kissed by your Arthur

Please also greet your father from me.

Arthur to Emma Vienna, August 14, 1924.
No. 74.

My favorite little Bicella!

I am very happy today too. I confidently hope that Mr. Schirrmacher in Berlin will sell the Seesenerstraße house today. As far as the Landsbergerstraße house is concerned, the property manager has already received notification, from the mediating company Dreher & Roske on August 8th or 9th, that the house has passed into the possession of a Miss Rosa Markus. Accordingly, the purchase contract must have been concluded on the 7th or 8th, which would have only been possible if the 21,000 goldmark had already been documented at the notary. The notary was to immediately inform me of this by telegram and transfer the money to my Viennese bank. But I have neither received a telegram nor any other message from the notary, nor has my bank been notified from Berlin. Somewhere there is a mess, which I am very upset about (but of course the money is not lost in any case).

Today is terribly busy. I am completing the last corrections to my new edition of the big two-part textbook on theoretical physics. I received the pages this morning and will return them tomorrow.

When I am sure, when my notary tells me that the money has been collected for me so that the matter is settled, I will telegraph you immediately.

Isn't it strange that the buyer of the first house has the name of a saint, although she will surely be the opposite of that!

I hope you are in a very good mood now and that it will only be a short time until our separation is over.

Be kissed many times by your Arthur

Arthur to Emma Vienna, August 15, 1924.
No. 75.

My dearest little Bice!

I am waiting for you with such longing, but the waiting becomes easier for me because now everything is turning out well, thank God!

At the same time as this letter, I am sending you a telegram informing you that the house in the Landsbergerstraße has been sold by the notary. Everything is in order. The contract was concluded on August 7th and the money was received by the notary. I don't know why the money has not yet been posted in Vienna nor why the notary, that vile swine, has not yet sent me the requested wire-message or letter. I learned the facts of the matter from an express letter that has just arrived from Mr. Schirrmacher (my confidant) who, at my request, made an inquiry at the notary (Baron Lüdinghausen).

Furthermore, Mr. Schirrmacher informs me, regarding the Seesenerstraße house, that the notary came to an agreement with the intermediary company Bannet & Fischer immediately after receiving my express letter on August 13th at 9 [o'clock] in the morning. They told Mr. Schirrmacher that the buyer was a Swede and they had immediately telegraphed him in reply to my letter (which also apparently arrived August 13th at 9 [o'clock] in the morning). B. and F. will call Mr. Schirrmacher and say that "the matter will be settled tomorrow or in the next few days" (tomorrow would mean August 14th). Finally, an express letter from Dreher & Roske arrived in which they announced that they also had a "first interested person" for Kriemhildstr. I have already told him to go ahead, the day before yesterday.

Today I still have a lot to do to complete the final part of the new edition; I am working on the fourth part of my *Introduction to Theoretical Physics*, mainly indexing names and subjects. - From tomorrow on, I want to dedicate several hours a day to my apartment.

I am so unspeakably happy, Bicelein. You can't imagine what a stone fell from my heart when I sold my first house. It has always been a terribly tormenting question of whether or not the houses could be sold at all. And if it did not succeed, what could I do with you?

The sight of the red leaf you sent me caused agonizing despair for me just a week ago. Now it is a talisman and tells me that hopefully I will be unspeakably happy when the leaves on the trees are red.

Be kissed a thousand times by your Arthur

Arthur to Emma Vienna, August 15, 1924. No. 76.

Dearest Bicelein!

Today, after lunch, I suddenly had a crazy fear that something was wrong with the sale of Landsbergerstraße, particularly (and maybe I was right) that the money had not yet been paid. I was attacked by a terrible agitation and immediately rushed to Uncle Fritz, taking all the files with me. I was with him until 8 [o'clock]. I was terribly excited because the notary public had still not confirmed receipt of the money. A few minutes after I had come home rather depressed, the bell rang and the postman brought the telegram, a copy of which is above, which makes me overjoyed and relieved me of all worries. - I hope that Seesener and Kriemhild will follow soon after.

Many fond kisses, my little Bice! Arthur

I will immediately take your letter to the main post office and post it as an express letter so that you will have the good news tomorrow. How happy I am!

Emma to Arthur Kapuzinerstraße 20, Munich, August 14, 1924.

My dear Professor,

So you're really doing your thing. Hope it won't rain again. Sunshine is such a beautiful thing.

[…]

The day before yesterday I was invited for coffee with Miss Pladioko (the American lady from New York) and we agreed that she, Miss Weigl and I would fly to Vienna together in an Aeroplane. - Maybe!

If they fly earlier, then I will fly alone to Vienna. But I do fly. Fly toward me? It would be wonderful if we would meet in the air.

I still like dreaming of my new apartment. Write me about it soon. I'm so curious. Good-bye.

Bice

[…]

I read this letter once again and concluded that I am actually demanding. Excuse me.

Arthur to Emma Vienna, August 16, 1924.
No. 77.

Dearest Bice.

I was very curious to see how the news would resonate with you given the complete turnaround that has - thank God - occurred in our situation. A terrible, heavy stone fell from my heart and instead of despairing uncertainty, the decisive success that has been achieved has established firm confidence that the goal will soon be reached; I had hoped that your letters would also reveal something of this. But the letter I received from you this morning was so indifferent and cold that I do not understand it at all - the most important thing you have to tell me is that you will fly from Munich to Vienna. That you are still a child! And that you still don't understand the harsh reality of life.

Although Landsberger Str. has been happily settled, unfortunately the settlement is delayed a bit, by how long I don't know. The notary Baron v. Lüdinghausen writes to me: "The agreed purchase price of 21,000 rentenmark was paid to me immediately in cash by Miss Markus. I have taken this amount into safekeeping for you for the time being, because the payment of this purchase price is made by the buyer with the usual and, in view of the existing legal provisions, self-evident condition that the purchase price may only be handed over to you after approval of the contract has been granted by a competent municipal authority. You are probably aware that since February 1923, every purchase contract for a property requires the approval of the authorities in order to be legally effective. In the present case, there should be no objections to the approval, and it is now to be expected that it will be granted as soon as possible. I will then immediately transfer the amount after deduction of my expenses (165 M)".

I don't know how long the "shortest time" lasts, whether it is only a few days or two to four weeks.

You have again cooled my joy a great deal with your letter that shows so little warm feeling. If you absolutely want to fly from Munich to Vienna, I will have to wait in any case until I have sold all the houses and secured a sufficient working income. My concern about whether you will really be able to find your way into a simple life without grumbling and sulking and murmuring has become very intense again, at least since I received your letter.

With warmest regards from your Arthur

[…]

Why did you again choose "My dear Professor" as the title of this letter from August 14th? What have I done now that you don't like?

Undoubtedly a rather condescending, patronizing, teacher-like response to a fresh, joking letter from his fiancée. Haas' failure to recognize her dream, of him meeting her in the air, can only be explained by the considerable stress and relief he had experienced in recent months. He had written to her about this in recurring letters and she had only replied on occasion. In her last letter she had expressed her real feelings, a light-hearted cheerfulness about the fact that the wait was finally over. This is an American tendency that, despite convent school, Emma had picked up in Chicago. For Arthur's part, he insists on Central European seriousness and does not recognize the light, exuberant joyful tone. He continued to brood for nights on end as he had done in the case of delayed letters and replies from realtors and physics colleagues.

Haas' perceptions can be understood by considering the perspective of men in his time with regard to women, a way of thinking that he had already demonstrated concerning his fellow students. As with Lise Meitner, Haas approached females with a certain arrogance. In the following years of male friendship and bachelorhood, he did not learn enough to change or expand this view. From his understanding, with which he had grown up, the woman should follow the man, since women cannot be trusted in their own independent decision-making abilities—they always remain children and need guidance. While Emma Huber had already demonstrated on several occasions that she could make her own decisions very well, he had not yet realized this over the course of their short relationship. For him, a happy marriage could only be achieved on his terms and within the framework of his financial plan; Haas simply dismissed Emma's ideas as pure capriciousness. Misunderstanding turns into misinterpretation and accusation, which brings the relationship to the brink of failure.

Arthur to Emma Vienna, August 17, 1924.
No. 78.

Dearest Bicelein!

Your letter, dated August 14th that I received yesterday, has indeed given me a sleepless night. I am very much afraid that your assurance, that you want to fit into a very modest life, is a meaningless phrase and that you don't actually think anything of the kind. You are still indulging in illusions that are unfulfillable and will disappoint you. It is absolutely essential that we have an in-depth debate on economic issues. You know that everything pleasant in life costs money. Therefore, a person can only be satisfied and happy if he has no desire from the outset for things that do not suit his economic situation.

In particular, a marriage can only be happy if the wife is sensitive enough so as to not express wishes that cannot be in any way fulfilled within the budget at hand. For, if the husband really loves his wife, he will suffer anyway - from not being able to do many things for his wife that he would like to do. If, however, he still constantly hears his wife's unmet wishes expressed, everything will be spoiled.

A thorough debate on economic issues is absolutely essential between us. You know that (as long as I am not in America, and I don't know whether I will even go there) I can probably only arrange our life on a budget of 8 million crowns - a sum that is higher, incidentally, than that of Dessauer who, for example, has to feed two children.

How I imagine such a budget, you can see from the supplement to these letters. I ask you to study it thoroughly and to comment on it point by point in a very detailed answer.

If you do not return the letter because you find it too boring or too time-consuming, or because you consider the imposition of such a discussion a personal insult, then you force me (I am serious!!) to postpone our marriage until I am able to generate a disposable income of 20 million crowns a month. And that means waiting forever, if I don't get out of Europe, or until I am called to America.

I would also like you to bear in mind that even if I had a total monthly income of 12 million, the budget guidelines that I am sending to you could not be exceeded. This is because I would then have to calculate an additional 2 million for a third family member and 2 million more for the rent of a slightly larger apartment. And even if I did have an additional 13 or 14 million monthly, I would put it aside as savings (in the form of a life insurance policy, for example).

Dearest Bicelein, I swear to you that I offer no insidious intention and no unkindness with my letter. But I want to know that, if I'm going to waste my houses for a ridiculous price to be able to bring you home, at least the basis for a happy marriage is being created and that I can rely on you to not express any wishes that do not fit into the budget. - It is also essential that you have clarity beforehand about what life in marriage will be like for you so that you will not be disappointed after we are wed.

Anyway, you know how much I suffer (with respect to you) from the fact that the May disaster destroyed most of my assets.

Be deeply kissed and greeted by your eternally faithful Arthur

Don't be angry about this letter out of whimsy. I wrote it out of love and necessity.

But he could not stop and needed to continue with his reproaches. These frustrations must have been accumulating in his mind during the past weeks. By presenting himself as an example, he intended to guide his woman so that she would surely follow. He wanted to be praised for his work, for his successes, for everything he had done for her in his imagination but without making any allowance for her personality, which he simply could not do from his patriarchal way of thinking. For Emma's part, her answer was clear and definite; and so, he gave in and perhaps was even prompted to think about things differently. However, given the shock of his inflationary losses, he could not completely abandon his economic concerns and his budgetary way of thinking; he had drawn up a budget with great care and detail, as seen in Fig. 11.11, but without an understanding of the situation.

Arthur to Emma Vienna, August 18, 1924.
No. 79.

Dearest Bice.

From your letters, from those you write (only one for a week) and from those you do not write, I am afraid that the turnaround, praise God, in our situation makes very little impression on you. When I receive good news at 10 o'clock in the evening, I immediately write a letter to you and carry it to the main post office that night - but rarely does a letter come from you and when it does, it is as indifferent as the one I received on Saturday the 14th of August.

Today I received a report from my representative, Mr. Schirrmacher. He informs me that a new prospective buyer, Arthur Caesar Behrendt (one of the many dozens to whom I had written based on advertisements), is considering the purchase of the Seesenerstr. and Kriemhildstraße houses. Schirrmacher would like to close with him unless Benett & Fider pays the $5,000 for Seesenerstraße today, Monday.

I'm quite confident that a second house will be sold this week - but I'm not really happy about all this because I don't get any really warm letters from you.

Today I finished work on my *Introduction to Theoretical Physics*. In the next few weeks, no less than four of my books will be published: *Atomic Theory* (about 200 pages) in green cover, the second volume of the 3rd & 4th editions of *Introduction to Theoretical Physics*

Richtlinien für ein Monatsbudget von acht Millionen Kronen monatlich

Wohnungsauslagen	1,500.000
Ernährung	2,500.000
Bekleidung	2,000.000
Arzt	500.000
Taschengeld	1,500.000
	8,000.000.

Zu "Wohnungsauslagen: Gerechnet sind ca.450.000 K für Bedienung, ca.100000 K Mietzins und Steuer,* Rest für elektr.Licht,Gas und Beheizung sowie kleine Wohnungsreparaturen. Was im Sommer an Heizung erspart wird, dient zur Bestreitung der Mehrauslagen einer Sommerfrische gegenüber dem Normalverbrauch.

* und Wäsch

zu"Ernährung". Pro Person und Tag stehen für Ernährung leider nur 40.000 K zur Verfügung. Wie kann man damit auskommen? Es ist z.B.im Kriegsministerium ein vorzügliches Restaurant ohne Tringeld, wo man für 10500 K eine Riesenportion Rindfleich bekommt oder für 15000 K ein sogenanntes "grosses Menu" sehr reichhaltig, dort bekommt man auch für 12000 K ein gutes Abendmenu. Die Räume sind sehr sauber, aber während der normalen Essenszeit sehr überfüllt. Kommt man hingegen vor 12 Uhr, so ist es fast leer, ebenso vor 7 Uhr. Ein sehr gutes Menu bekommt man auch billig im Restaurant Gilly in der Nähe des physikal.Instituts. In den meisten "Restaurants" macht man *mit Menus um 15000 K keine guten Erfahrungen. Des Abends könnte man auch zuhaus Schinken mit Eiern oder dgl.essen. Bei vernünftiger Einteilung muss es möglich sein, mit 40.000 K täglich derart auszukommen (=80.000 K für beide zusammen), dass für die Frau zum Frühstück und zur Jause und zum Gabelfrüstück auch Eier, Butter und Marmelade sowie Speck u.dgl. zur Verfügung stehen. -

* sonst

zu "Bekleidung". Es ist sowohl für Frau als auch für Mann mit einem Jahresverbrauch von je zwölf Millionen (= 1 Million monatlich) gerechnet. Darin sind inbegriffen: Kleideranschaffungen, Wäscheanschaffungen, Anschaffungen von Hüten und Schuhen, Krawatten usw. sowie alle Reparaturen von Kleidungsstücken aller Art.

zu "Arzt". Es ist sowohl für Frau als auch für Mann mit je drei Millionen jährlich gerechnet, wobei auch die Auslagen für den Zahnarzt mit einbegriffen sind.

zu "Taschengeld". Es ist gerechnet mit monatlich je 500.000 K für die Frau, für den Mann und für gemeinsame Auslagen (in der Hand des Mannes). Persönliche der Frau wären : Elektrische, wenn allein, Friseur, Konditorei, Briefpapier, Briefmarken, Kodak, ev.Theater allein, Geschenke kleiner Art u.dgl. persönliche des Mannes: Elektrische, wenn allein, Friseur, Briefpapier, Briefmarken, und ähnliches wie bei der Frau usw. Gemeinsame Auslagen: Gemeinsame Fahrten in der Elektrischen, gemeinsamer Besuch von Kinos, Vorträgen, Theater, Ausflüge, Abonnement einer Zeitung u.dgl.

Zu Ernährung wäre noch nachzutragen, dass z.B. eine angenehme und billige Möglichkeit eines Nachtmahls auch wäre, im Kaffehaus ein Paar Frankfurter (=9000 K) pro Person mit Zeitunglesen und überdies ev. zu Hause je zwei Eier, kommt pro Person auch nur auf 15000 K.

Für ein Stubenmädchen ist in der jetzigen Wohnung unmöglich Platz zu beschaffen. Die Wohnungsfrage muss solange vertagt werden, als sich das Amerikaproblem geklärt hat.

Fig. 11.11 The monthly budget envisioned by Arthur Haas for the new household after the marriage. Here he carefully estimates the running costs for maintaining the apartment, for food and clothes as well as the monthly allowance for his young wife (about 400 pages) in sky-blue cover, the first volume of *Introduction to Theoretical Physics* (at Constable & Co, London) and the Swedish translation of *Nature Picture of New Physics* (Norstedt & Söner, Stockholm).

I'll wait to send the description of my apartment until I finally receive a friendlier letter from you. I have bombard you with joyful telegrams and express letters - and the echo?

Best wishes from your Arthur

In any case, I want to see you soon. If there are still delays with the complete execution of the project, I would be in favor of us meeting again in Salzburg. But you must leave the choice of the time to me because only I know if and when I can leave Vienna without danger. So please wait for new news.

Many kisses from your A.

Emma to Arthur Kapuzinerstraße 20, Munich, August 18, 1924.

Dear Arthur,

I am still so indignant about your last letter that I am not able to send you the desired love letter in the way you want it, with exuberant love affirmations.

When a man thinks so badly of his bride then I do not understand why he wants to marry her at all. I do not deserve the bad opinion you have of me. I mean, if a man does his utmost and leaves nothing undone in order to offer his future wife a comfortable life, and he shows himself worthy of a great trust but his bride appears to be cold and disinterested, then this girl is, in my opinion, the worst creature among God's creatures. So I would actually have pronounced my own condemnation at this point for the cold, disinterested appraisal of your efforts that you have very clearly reproached me for in your last letter.

My dear Arthur, if you require me to express my love and interest in you with selected, rehearsed expressions every time I write, then I am sorry that I do not possess such writing talent. Until now, I thought you knew that I prefer to keep my feelings deep in my heart because they are mostly too sacred to be seen by anyone, especially a feeling like my love for you. Nobody knows how deep it is and nobody should know it - not even you. And today, I think least of all you because you don't understand me at all.

You do not think I am capable of making even the smallest sacrifice without having a selfish motive. You don't believe in my love but you believe in any selfish reason that makes me want to become your wife. Didn't I tell you earlier that you are the first man in my life to whom I have confessed my love? Even if you didn't feel my heart rejoicing, this confession was the most holy thing I had ever made to a human being.

I could pass away in shame when I think that you might compare me to those women who professed their love to you in the past. If your knowledge of women only comes from these women, then you will never understand me. I have spoken to you repeatedly about my love for you and my interest in your efforts - which, as I told you earlier, is actually against my nature - but you still don't believe it, don't believe that there can be something sacred in a woman's heart, but you still do believe in a cursed selfish reason.

I am at a loss about how to behave toward you. I always thought you were sensitive enough, I mean sensitive enough to be able to read between the lines and know my jubilation and my joy in your successes, to know with what longing I await your letters and with what desire I wish the most urgent achievement of my goal. But you have not noticed any of this. Therefore I am at a loss. Bice

Emma to Arthur August 18, 1924, afternoon.

Dear Arthur,

While I was trying to put my first letter into the envelope, I received your letter of August 17th.

I am so sad about it that it breaks my heart. If you don't trust me enough, then it would be better to break up.

I have assured you before in a letter that I would like to share every way of life with you, be it ever so humble and as long as I have confidence in you - but if you don't believe my first promise, then no second promise is of use.

To make a slave out of me, one who mechanically renews her promise every week - I am inwardly too proud for that. I have already promised you once quite sincerely what you again demand from me today. If the promise given at that time is not enough for you, then it is the end between us. I will not marry a tyrant even if I love him. Bice

Arthur to Emma Vienna, August 19, 1924.
No. 80.

My dearest best Bice!

I just found your double letter of the 18th when I came back home. I am very sad that you are so upset - I certainly did not want to hurt you - so be good as before. It was certainly nonsense when I attached an exaggerated symptomatic significance to your statement about absolutely having to take the plane from Munich to Vienna. And it is nonsense that I made so much of it that your last letter contained nothing at all about greetings and the like - I have complete confidence in you in every respect, that goes without saying and I just wanted you to get a clear picture of the economic situation. I would therefore be very grateful if you would give me your views in detail on whether you approve of the outlined budget that I sent to you, in terms of its distribution between the various items, or what you would like to see changed. Look, Bice, if, unlike before, I am too pedantic in the future, you must tell yourself that it is for your own sake.

You must be provided for (albeit, unfortunately, only modestly) for the rest of your life, whatever my fate and career may be: that is why from now on, the property as such must not suffer the slightest reduction, only gaining in its normal income. My income from work may be used up, but possibly not entirely. Surely you must understand me!

This morning I entirely devoted myself to tidying up the apartment; I discarded almost a hundredweight of paper of all kinds to make room. - First of all, I want to describe this one room. It looks like this [Fig. 11.12]:

The room has two windows and two doors. The bookcases are built of American walnut, the lower part with wooden walls and four rows with a glass front; so altogether, twenty rows almost all with very nice leather-bound volumes are visible. The wallpaper is red, the ottoman red, the table has a green velvet cover, above it a chandelier. At the desk a green lamp. The club fauteuil [armchair] brown leather. Between them I will put a bigger and nicer table (which is in the bedroom for the time being) and then with this table they will form our "dining room" (for breakfast and cold supper and snacks). Meanwhile there is a small copper table between the club chairs; it could be put in the corner where I made a "?" on the sketch.

The bedroom now looks like this [Fig. 11.13]. Between the windows there is a cupboard with horizontal compartments for linen and a very large mirror (matt nut).

The toilet table is very nice with a large mirror and eighteen drawers. The cupboard next to it is for hanging clothes. To the right of the door is a small cabinet for hats with four big compartments. The table next to the bed should be placed in the living room between the club chairs? It is in beautiful baroque style. This then makes room for a second bed.

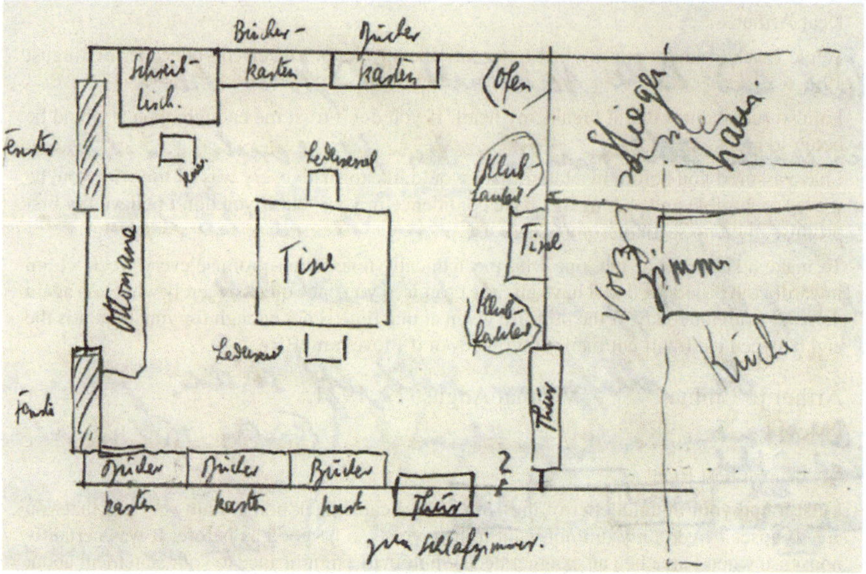

Fig. 11.12 Living room/library at Neulingsgasse 24, a drawing by Arthur Haas

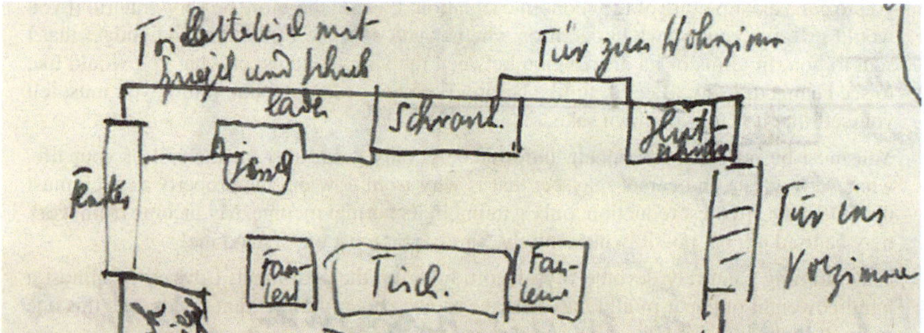

Fig. 11.13 Bedroom as drawn by Arthur Haas

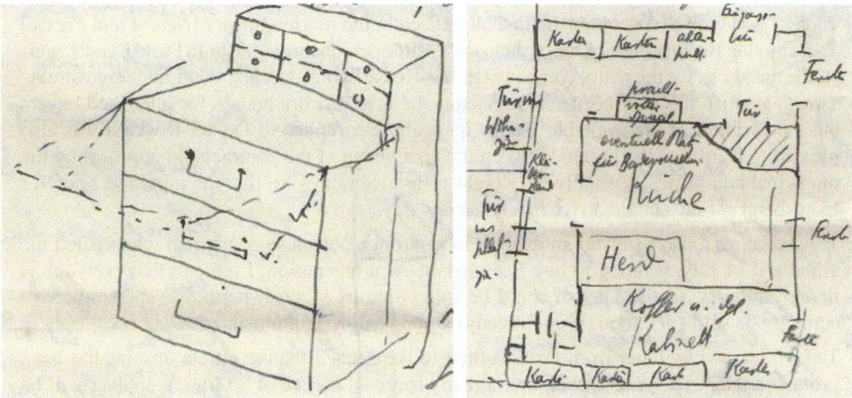

Fig. 11.14 Chest of drawers and kitchen in a drawing by Arthur Haas

The writing desk is beautiful, a lovely baroque piece with three large drawers. The upper part folds down and can be used as a writing surface and there are also compartments like this on top. Do you find your way around the miserable drawing? [Fig. 11.14] The dotted lines mean the unfolded surface, which otherwise closes the whole thing. You'll love your writing table. I want to put it where the toilet table is now, move the toilet table further away from the window and move the cupboard (that is now next to it) to the cabinet, to make room. Otherwise, the two beds would hardly have room. I think it's better to have your desk in the other room, apart from mine; then we'll disturb each other less. Of course, you can write on mine as much as you like. I think I'll be in the physics institute most of the day and I'll work there, so you'll have peace and quiet away from me most of the time.

The anteroom looks like this. It's a junk room for the time being, where I'm gradually putting things in order. I'm seriously worried about the bathroom problem. Since we might be moving to America soon or possibly to a larger apartment, I don't want to invest a lot of money in rebuilding. I still have to think about that.

Everything also depends on how various unresolved financial problems will be solved (Viennese securities, Brno house ownership, claims for advances).

The Berlin house issue should soon be solved to my satisfaction. This morning the agent through whom I bought the houses two years ago came to me; he knows a good Viennese buyer.

So I hope that this will be solved soon. Of course, there is still a bitter delay caused by the fact that we have to wait for the approval of the Berlin authorities.

But I don't want to give up hope that I will still be able to go to the country with you; perhaps somewhere where it is still warm in September and even October; it depends.

So don't pout anymore and show that you have learned to understand the rough reality of life.

A thousand kisses and heartfelt greetings from your Arthur

Arthur to Emma Vienna, August 20, 1924.
No. 81.

Bice darling,

Up to the hour, I have no news from Berlin but that's not why my report is shorter.

Today I was with two people who had advertised in the morning paper [*Neue Freie Presse*] about buying Berlin houses. One of them is the same person through whom I sold Landbergerstraße; he works for the realty company Dreher-Roske, which has just sold Landbergerstraße. This gentleman has now been sent to Vienna; he is to procure houses for interested buyers but too few houses are available. Today he again telegraphed to Dreher-Roske. I was also pleased to learn that the period between the conclusion of the contract and approval by the municipal authorities is usually only eight to fourteen days, so that the purchase price for Landsbergerstraße should arrive in Vienna any day.

This morning I also received an express letter from a Viennese gentleman who visited me at the end of July, wanting to buy Seesenerstr. As a precaution, I pretend that everyone is in a great hurry, that the house could be sold "tomorrow" elsewhere. Seesenerstraße will certainly be sold very soon (if not already) and so will Kriemhildstr.

Today I learned by letter from Prof. Lotmar in Bern that a feature article of mine has been printed in the *Berner Tagblatt* (I have already received the fee of 50 franc); another will be printed any day and so on, one every month. So it was not an empty promise and I am very happy to have this monthly additional income of 600,000 K.

A letter I sent to New York on July 26th was answered yesterday on August 19th with a reply dated August 7th. It has no content and is only an acknowledgement of receipt from the representative of the addressee who had left for Europe.[37] Also today, I have continued to make order in the apartment. It is a question of creating space everywhere, a matter of where what is in the cupboards to be reserved for you will now be accommodated.

Many fervent kisses and greetings from your Arthur

Obviously, Arthur Haas wants to row back to normality as soon as possible. The apartment description with drawings, according to his ideas and expectations, should lead women back from capricious rebellion to their sense of obligation and allow him to get back to his safe routine. His report from the house-sale front is meant to close the topic and move on. But Emma does not make it that easy for him.

Emma to Arthur Kapuzinerstraße 20, Munich, August 20, 1924.

Dear Arthur,

Your wish that I no longer sulk and instead show that I have learned to understand the harsh reality of life is well meant but superfluous. The misunderstanding that caused the present upset was far too serious to be dismissed with a childish pout. On the other hand, the "rough reality of life" today is probably misunderstood by very few people and, thank God, I do not belong among them.

For the injustice that you have done to me through the last letters, I have forgiven you today with all my heart. I am also trying to forget everything. I would prefer not to mention anything about your last three letters but after you have put together the budget with, I assume, great effort, and have asked for my opinion on it, I have the following opinion: I was already very convinced that I, or rather we, can get by with 8,000,000 crowns a month. No longer is this statement enough for you. To state it more clearly is probably not possible. But as far as the allocations of this budget are concerned, my dear Arthur, it is impossible for us to set it out, as you have tried, in such detail because we have not yet lived together. So I can neither approve nor disapprove of this plan now.

Once we have been married for three months, we will have experience and we will naturally base our expenses on it. I find it a little strange that you are already sending me my housewife duties in print and even suggesting what and how I should eat. In my opinion, the only way to discuss this is in person. I hope that our marriage will not consist of paragraphs. I hope

[37] Karl Georg Frank, representative for Siemens Halske in New York.

that I will not have to read my housewife duties from a typewritten piece of paper. If you ever have a bad experience with me after we are married, I will surrender to all your orders without the slightest objection but until then, I think it is better that every cobbler (you and I) stick to his task. This way, we can enjoy a much happier life together.

I do not mean to imply that I will be indifferent to your occupation in the slightest. On the contrary, I see in it my longed-for happiness - that I will be allowed to participate in the thoughts and ideas of a scholar and that I may be able to help him in some way. My ideal sense of living together is that each one seeks to complement the other spiritually. Isn't it sad that so many people, who are all created in the image of God, want to live only for their bodies and think so little of their subliminal minds?

But my thoughts lead me too far away from answering your letter. As far as the description of your home is concerned, I was naturally very interested. I am trying to imagine everything exactly as you described it. For the woman, her home is of utmost importance.

As often as I pass the kitchen, I make a small visit to the tabernacle and pray especially for you, that you will be able to sell the houses without further difficulties. I also pray that you will learn to see that a nature like mine becomes like soft wax in good dear hands but hard and inflexible like steel when you try to command it. Every man has his faults, you too dear Arthur, but I do not want to discourage you by pointing them out to you. Therefore, I ask you to turn a blind eye to mine; perhaps you will succeed in discovering a few virtues. I think that if we pay little attention to our bad sides and work with our good sides, life should be very comfortable for us in any case.

Do you think ... Dear Arthur, are you dealing with me as you did when you first sold the house? At first you thought all agents and buyers were phonies and you didn't trust anyone. Then you came to your senses and the whole thing unraveled to your happiness. Couldn't you do the same for me?

I hope you understand me a little better now. I don't want another serious misunderstanding like the last one to cause us unhappy hours.

I know that when I am quite happy, I behave like a bovine head and can seldom and only with the greatest effort write a decent letter, but when I feel that I am back on earth, I write very, very soberly - don't I! Best regards from your Bice

Arthur to Emma Vienna, August 21, 1924.
No. 82.

Dearest best Bice,

I haven't heard from Mr. Schirrmacher and I am very angry about it, but that doesn't diminish my optimism. It was once like that with the Landbergerstr.; I was already quite desperate because of the absence of news and then everything was fine. The majority of people are - at least in continental Europe - lazy pigs and in Berlin this seems to be especially true, despite its loud-mouthed inhabitants. Today I went to see a retired Prussian major who is buying houses for himself; he is currently in Vienna and has advertised. I am sure that everything will soon be taken care of very well.

[…]

-- Today an acquaintance of Mrs. Dessauer, a Dr. Herrlinger sent me a "Weekly Diary of Social Events" from May 26th - June 1st. On page 15, my lecture is listed as a "Social Event". Maybe you will be very, very happy about it; if so, then I will send it to you. Do you think that Miss Marjorie West, who is my neighbor in the brochure, is pretty? Hardly, otherwise she would have a more interesting topic.

--From my apartment, I want to tell you that I have some nice pictures, nice color reproductions by Romney, Botticelli, Tizian and the like. A beautiful Smyrna carpet. I also have very nice dining room, anteroom and kitchen furniture, and a beautiful Persian carpet that are not in the apartment. I continue to diligently tidy up the apartment. If it came about that before the end of August the other two houses are sold as well, how happy I would be.

I am also looking forward with great excitement to further news from Privy Councilor Wien in Munich (which, as you know, is worth 250 million crowns to me!).

Learn a lot of cute things in Munich! And see that you gain weight diligently! Be kissed fervently and greeted by your Arthur

Arthur to Emma Vienna, August 22, 1924.
No. 83.

My little Bice

Thank you very much for your so dear and so long letter from the day before yesterday, which the midday post brought me. I am very, very, very sorry if I hurt you with some of my letters; I really did not want to and did not even think of the possibility that it could be like this. I too do not want to harp about the misunderstanding. What I wrote, I only wrote out of conscientiousness, to save you from disappointment. I was afraid (based on a statement of yours) that you might have the wrong opinion that after the sale of the house all consequences of the May crash would be eliminated and everything would be back to normal. - And to save you from disappointment, I thought it necessary to tell you that unfortunately our way of life will have to be more modest than it was in April - and that a visit by "Lantos" or "Pelican," for example, could not be the rule and only a rare exception. - In any case, I believe that a certain division of the budget into categories is absolutely necessary because only one regularity of lifestyle can be achieved in each case.

You probably won't have the opportunity to act as a "housewife" either. We will probably eat out of the house, if only because I consider the smell of food in the living room or bedroom to be as terrible as you do - and therefore only odorless food (such as roast and ham and eggs and the like) would come into consideration. The apartment is too small for a maid - and we can't take a bigger one until there is clarity about America and until I have a better overview of assets and income. Also, according to the law, we are only allowed to have a three-room flat for two people - and you know how expensive it is to change the flat (over a hundred million that would be thrown out the window if I were to leave Vienna shortly afterward).

So all in all, you have hopefully forgiven me completely, as I was never angry with you at all; and hopefully you do not misunderstand my wish, expressed (and maintained) in earlier letters, that you might be so good as to think quite a lot about economic issues and talk about them with friends (especially married ones).

[…]

My brother arrives in Vienna tomorrow from Italy. He had to abruptly cut short his holiday because the managing director of a large stock corporation, which my father founded and for which Otto serves on the board of directors, suddenly shot himself.[38] Since my brother has invested a very large part of his assets in shares of this company, I fear for more than one reason that he may suffer serious financial damage. This all the more so as a whole group of companies, Otto's best client until now, will probably collapse. This all makes me quite sad.

If only I could send you a second happy telegram very soon. Best wishes and many warm kisses from your Arthur

Arthur to Emma Vienna August 22, 1924.
No. 84.

[38] This suggests new uncertainties because if Otto invested in this public company, founded by his father, then so did Arthur who discussed all of his financial transactions with his brother.

My Bice!

Dearest best Bice!

For twenty minutes I have been the happiest person in the world; for twenty minutes I have known that perhaps only a few days will pass before you will be mine forever. If there are no unexpected complications, the necessary conditions are met for us to marry with confidence. - You will have already received the telegram, when these lines reach you indicating the happy sale of the house in Kriemhildstraße. I came home at 8 [am] and 10 minutes later I received a telegram, "Kriemhild sells twenty-one thousand, Schirrmacher follows".

I was completely stunned with happiness; for you see, Kriemhildstraße had caused me sleepless nights. In the last few days, I have generally been told that it was worth far less than the other houses and that I could hardly sell it unless I dropped the price a great deal.

And now, although I had told Mr. Sch. that 20,000 marks was the lowest limit, it was sold for 21,000 marks complete net – the same price as Landbergerstraße.

I was far more interested in the sale of Kriemhild than in the sale of Seesenerstraße because I own Kriemhild completely, whereas Seesener and Tauroggen are only half mine. Seesener and Tauroggen are certainly more valuable, much more valuable than Kriemhild, and I have no doubt that I will receive much more than 21,000 Mk all together for my half of Seesener and my half of Tauroggen.

In the last few days, I have resolved to marry immediately when Landsberger and Kriemhild are sold. I only wait to see:

1) receipt of the money in Vienna from the sale of Landsbergerstr.
2) the details of the money from Kriemhild. If the buyer is a German Reich citizen, I am not afraid about permission.

The money for Landsbergerstraße is already due. I can telegraph you within the next few days, "Expect you". Then I will be ready for immediate marriage. The one to two days that you might have to wait, you can stay in Vienna in a hotel.

If you prefer the wedding to take place in the first week of September, then the wedding day would be September 5, 6, 7, 8, or 9, and afterward we could go traveling.

Please express your opinion about this immediately. Perhaps marriage on August 30th would still be possible but I am afraid to leave Vienna before I have invested the money properly.

After receiving the telegram, I went immediately to the main post office, taking the sheets of paper with me. I posted the telegram to you and then went into the illuminated Dominican church[39] next to the main post office with a grateful heart, then into the café to write you a letter, which I now carry to the main post office. Still quite stunned with happiness, your Arthur kisses you many times over and over again

The first letter of August 22nd, number 83 shows that despite Emma Huber's clarification regarding the role of women in marriage, Arthur Haas still insists on his budgetary plans. He apparently considers marriage a financial transaction that must be clarified before he can give his consent, a sentiment reminiscent of the arranged marriages still common in the nineteenth century. Nevertheless, he is full

[39] Although Haas discusses his Jewish heritage in his later memoire, his personal faith has only been mentioned when telling of his baptism in the faith of Christian-Protestant confession, as a student in Göttingen. So until 1924, not much information is available about where Arthur Haas stood in his faith and what caused his move to Christianity. With the economic difficulties, he seems to have found a certain inner support in God and his top priority in that moment was to visit the warmly illuminated, inviting Catholic Dominican Church in the Postgasse.

of happiness when the sequence of plans is finally complete for the financing of his wife and marriage, a mathematical arrangement that he so carefully orchestrated. Now he could move on to the next step of the transaction, the planning of the actual wedding.

There is no doubt about his inner happiness but his approach allows for insight into his thinking. He was more rigidly oriented toward preconceived, well-considered and well-formulated planning, which would not readily give way to the freedom of spontaneous development that Emma Beatrice Huber seems to have followed in her life so far. Nonetheless, Arthur did attempt to let Emma participate in the planning that he believed so necessary.

That same evening, he wrote letter number 85, which at first reflects a certain inner uncertainty about the state of affairs but is then again bubbling over with hope and plans.

Arthur to Emma Vienna, August 22, 1924.
No. 85.

My favorite little Bicella!

I have been so happy since last night and I know (whether you write it to me or not) that you share my joy.

You have received my telegram and my express letter. You know, the Kriemhildstraße was like that. Just imagine a mother has four daughters, three beautiful and one ugly. And she always thinks, "If I'd just get the one placed everything would be fine". Among my houses, "Kriemhild" was the one that was hard to place. That's why I was so happy when I succeeded.[40] Now I won't report on the houses in the future, since it's very bland - I was only interested because everything for us depended on the sales.

Unfortunately, I cannot tell you any more than I did in yesterday's express letter because I have to wait for Mr. Schirrmacher's report. In any case, I also have to wait for the money from the sale of Landsbergerstraße to arrive. It is already long overdue.

Placing the money in Vienna is also something that has to be done carefully, regardless of whether the money for Kriemhildstraße has already been transferred to me in Vienna or whether it will remain tied up in Berlin until it is approved. Depositing the money in Vienna has suddenly become more difficult because there are doubts about the stability of the Austrian economy. I don't want to leave the money un-invested while travelling with you for three weeks. This has to be arranged carefully beforehand, in a form that guarantees absolute safety against a possible fall of the crown.

That's why I'm in favor of waiting a little bit longer until it's sorted out. But I do not think you will have to wait any longer than September 6th.

I am also considering the possibility that you could already be in Vienna while I take the last steps and we could marry as soon as everything is sorted out. Much depends on whether and to what extent I could possibly entrust my brother with the arrangement of these details during my absence. But he himself does not yet know if and when and for how long he will leave Vienna again. I spoke to him only briefly this morning; at 8 [o'clock] in the evening I will meet him at the inn. He is sincerely very happy about the happy change in my life.

Today I have made great further progress in tidying up the apartment; I have again created new space by destroying a cubic meter of writings and manuscripts.

So Bicelein, just wait for my next letter; today I still know too little. But everything is now only a question of picking a date within the next few weeks, nothing more.

---

[40] An example that again reflects his attitude toward the role of women.

Oh, if I could just kiss you differently than by letter! Give my regards to your dear father, who I hope is now reconciled. Yours, Arthur.

Emma to Arthur Kapuzinerstraße 20, Munich, August 24, 1924.

My dear Arthur,

Today when I received your urgent letter of the 22nd, I did not think that it would contain such good news. The telegram that I received yesterday made me suspect something pleasant but I did not think that we are so close to our goals.

Dear Arthur, your last letters have made it a little difficult for me to give my joy in the natural expression that I am used to, which you have so often misunderstood. I couldn't bear it if even the slightest misunderstanding were to come at this very moment, which I am so infinitely happy about- that we will meet again soon.

Tomorrow I am going to the consulate to get my passport in order. The day after tomorrow (Tuesday) I will go to the Bavarian Forest for a few days; I have been invited there several times and since people are starting to be insulted because I have always cancelled, Dad advised me to go for at least a short visit. On Sunday at the latest I will be back in Munich. My address will be p.A. Mrs. Hedwig Höchstetter - Ausser-Rötzing, Bayr. Wald, Lower Bavaria.

As for our wedding day, my dear Arthur, I would like to ask you to choose September 8th. To marry before September 1st would be clumsy, and after September 1st you must give me at least six days' time. I would then arrive in Vienna on the evening of Saturday, September 6th. We would then have Sunday to make last minute preparations. You know, September 8th is a feast day of Our Lady. It is the birth of Mary. I would like it so much if we were to have our marriage on a Marian feast. Then the Mother of God will give us a special blessing.

The little booklet "Social Events" is quite interesting. I would like to know where you will be sent to next. (I almost wrote "Press") - well, where do you think that I will press you when I come to Vienna? Lovingly Bice

Arthur to Emma Vienna, August 24, 1924.
No. 86.

My dearest Bice!

The post office just brought me an express letter from Mr. Schirrmacher, from which I see the following highly gratifying facts:

1) The Landsbergerstraße building has been approved by the city authorities. On August 22nd, Mr. Schirrmacher wrote that the money should be transferred "immediately". Of course, I now urge the notary and Mr. Sch. to hurry up, in my outgoing express letters.
2) The house in the Kriemhildstraße has been sold to Germans so the permit is safe. (The buyer is a couple from Magdeburg.)
3) The money for Kriemhildstraße is at my disposal from September 1st, regardless of whether the permit has been issued or not by then (5,000 marks have been paid, which will default to me if the rest has not been paid by September 1st).

So everything is as good as one could wish. I hope there will be no difficulties with the transfer from Berlin to Vienna. I am also hoping that my Vienna investment will not cause any difficulties either. (I am thinking of a mortgage rate of about 5-20%, with an investment in farms in Lower Austria[41] that will be protected against the instabilities of the crown currency.)

[41] Arthur Haas has taken up his father's plan to invest in agriculture as a long-term secure investment, which Gustav Haas could not, however, carry out due to his untimely death.

In two to three days, I will be able to see more clearly and tell you more precisely when our wedding ceremony can take place within the period between August 30th and September 15th. (The time between the 1st and 5th of Sept. is probably out of the question.)

So, be glad that everything worked out so wonderfully well and don't lose your patience during the two weeks. I will certainly not make you wait any longer than necessary.

In any case, please fix the lock on the small suitcase that I lent you for your trip in May, but be prepared to leave within twenty-four hours if I ask you by telegram.

You can simply enter Salzburg at your border crossing and afterward you can forget about Bavarian authorities! An Austrian citizen (who will hopefully soon be a citizen of the U.S.A.) is not the business of the Bavarian authorities. Just don't forget the documents! I believe that when you arrive, the wedding ceremony can take place within forty-eight hours.

Many intimate kisses from your Arthur

Arthur to Emma Vienna, August 25, 1924.
No. 87.

My dearest Bice!

What a pity that you didn't write me a few lines the day before yesterday or yesterday after receiving my telegrams on Friday and the express letters that probably arrived soon afterward! But I don't want to reproach you unjustly for that.

The money from Berlin was not yet in the possession of the Verkehrsbank today; I deal with express letters every day and it must certainly be in Vienna in the course of this week. Unfortunately, I will have to wait until it arrives.

I have just received a letter from Schenectady N.Y., U.S.A. from Dr. R. Whitney,[42] the head of the Research Laboratory of the General Electric Company. The letter is dated August 13th and reads:

"Dear Sir!

I have your letter of July 29th, and have written to Dr. S.W. Stratton, President of the Mass. Institute of Technology, with the idea that he may wish to offer you a position. If he does, he will doubtless write you.

Yours very truly

W. R. Whitney m.p".

Boston University of Technology, which this is, has a good reputation and would certainly pay well. In connection with Prof. Dempter's statement that there is such a lack of theoretical physicists in America, perhaps a little (but not too much) hope may be drawn from Whitney's letter.

I am working like Hercules in the Augean stables, preparing the apartment. It has been "a very hard job" (as they say). I just managed to empty the writing table completely, after making room for the things that I had stored there. Before that and after careful examination and

[42] Willis Rodney Whitney (1868—1958) was an American chemist and founder of the research laboratory of the General Electric Company. In 1890 he received a Bachelor of Science degree from the Massachusetts Institute of Technology (MIT) and remained there for two more years, working as an Assistant Instructor of Chemistry. He then moved to the University of Leipzig where he received his doctorate in 1896 under Wilhelm Ostwald. Whitney returned to MIT and conducted research in application areas of electrochemistry but by 1900, he was employed as a consultant for the newly founded research laboratory of General Electric. In 1908, he left MIT and took a full-time position at the GE laboratories where he remained as director of the General Electric Research Laboratory until 1932. His responsibilities included electrochemical questions on corrosion, the development of vacuum and gas-filled lamps, the wireless telegraph and X-ray technology.

selection, several hundredweight of MS letters, drafts, brochures and the like were destroyed. And a box that stands in the hall, in which I store silver, real and fake, porcelain, glass and the like, is now nicely arranged and has a beautiful appearance.

While tidying up, I also found the photographs that I include: A part of my library, my parents' dining room (the furniture now belongs to me but is in storage), and finally the windows of my (ground-floor) bedroom from the garden.

Warmest kisses and greetings from your Arthur

When are the days when you can hardly travel because of the need for rest?

Arthur to Emma Vienna, August 25, 1924.
No. 88.

Dearest Bice.

I already wrote to you two hours ago but I would like to write to you again today (unlike you, who unfortunately writes to me too seldom).

After I wrote the letter to you and sent it off, I sat in the coffee house for dinner (Frankfurter) and came home again.

Suddenly I can't get the thing with Boston out of my head. I don't think I have sufficiently appreciated the meaning of the letter from Dr. Whitney. It could be a turning point in my life. But it would certainly be exaggerated optimism to already be sure that I have a "position" as professor at the Tech. University in Boston. However, Dr. Whitney's letter is of the utmost importance for another reason. I have written a long letter to Dr. Whitney to inform him in detail about myself, about my achievements and also about the reason that I am not getting a position in Germany and Austria. - He certainly knows the American situation very well - he is a very influential and respected man - and he would certainly not have written a letter to the president of Boston Tech if he did not have the general conviction that there is room for me at American universities. - And that is why I believe that his letter can be considered a significant event.

Best wishes and kisses from your Arthur

Emma to Arthur Kapuzinerstraße 20, Munich, August 25, 1924.

Arthur darling,

Just now I wanted to go to the post office to send you an express letter and, at the same time, I wanted to send a telegram to Lower Bavaria saying that I would not come because it is raining so hopelessly – but while there, I received your Saturday letter. Thank you very much for that and for everything that was in it.

Do you think it would be better if I came to Vienna a little earlier? I almost think so. But you have to be nice and well-behaved.

So, I am not going to the Bavarian Forest and am always at your disposal. I already have travel fever.

I want to ask you about my luggage. There are different things here that I would like to have in Vienna later on, like real crystal glasses etc. - and some of my books. I think it's better to have everything sent to me only after we've gotten everything in order in Vienna.

I am quite delighted with my calling card. But we could also print our wedding announcements if you agree with the 8th of September.

I recently had myself weighed: without everything, as I am = 104 pounds. I never weighed that much even in Vienna.

Many greetings from Papa and from me, too, and something in addition. Bice

P.S. I already wrote to the photographer a week ago and asked him about the price - but although I had enclosed a stamp, he has not yet given me an answer. If you happen to be close by, ask him what's going on.

Emma to Arthur Kapuzinerstraße 20, Munich, August 26, 1924.

My dear Arthur,

I haven't heard from you today. I am a little worried, as always, but I know it's only the mail's fault, which is again boring.

Yesterday I received a letter from you with contents for which I thank you very much.

I hope with all my heart that you will agree with the date of our wedding, which I like so much. I think it is better that I come to Vienna on Friday the 5th, then we will have Saturday free for errands that we might still have to do.

Dad already took my new business card - so please send me another one. Daddy has big business worries at the moment; he has a lot of money outstanding, but nobody pays him because nobody has money. But he is still in a good mood, that is, he always smiles when I talk to him and he sends you many warm greetings.

Where will we be on this day in fourteen days? As always Bice

Arthur to Emma Vienna, August 26, 1924.
No. 89.

My dearest Bice!

Thank you so much for your two dear letters from Sunday and Monday. Today it is a quarter of a year since a cruel fate separated us but, thank God, the days can be counted on our fingers until we are together again forever.

I am entirely of your opinion that Marian Day (September 8th) would be particularly suitable as a wedding day. However, the wedding ceremony can only take place in the afternoon and we could then leave Vienna in the evening. I am also in favor of you leaving Munich, if your state of health permits, on Thursday, September 4th (Friday, September 5th at the latest) at 8:30 am and arriving in Vienna at 7 pm. You might have some things to do and on Sunday everything is closed. I will take care of the lodging fee for you for the four nights.

I will also go to the pastor a few days before; maybe (but I'd rather not yet) send me the documents if I ask for them.

I strongly suggest that the wedding ceremony take place at your church in the fourth district. Otherwise there will certainly be complications since on Marian afternoon there will certainly be a large crowd.

As for your intention to bring other things for yourself, other than clothes, linen, etc., I remind you that only things that take up very little space will be necessary. I have almost everything but I could use just a few glasses. The only things I can provide are blankets, pillows and possibly (but I don't know yet) a gas stove. You have to keep in mind that we might be crossing the ocean in a few months!

The clearing out of the flat is really going well. Also, again today a huge number of letters and the like were discarded. It will be quite comfortable after I've finished. - But what will happen when we go to America? I think about it with shivers. There's not much that can be taken along.

You will have at your disposal: a big mirrored cabinet for folded things, half a hanging cupboard (I'm sure I don't have more than that), the writing chest, the "hat cabinet" (which can of course be used for other purposes as well) and I will also gradually clear all the drawers in the toilet table. Is that enough for you?

Tonight, I am with Mr. Dessauer after supper at his request. I immediately informed him and Mrs. D. of the happy turn of events.

Now just plan on gaining four pounds in one week! If you have a family doctor whom you've known for decades, maybe go and see him again before you leave! Healthwise, you are hopefully in good health after all!

I don't know where we're going after the wedding ceremony. I don't know for how long either. In the end, the American problem will already start in September!

Many, many kisses, my little Bicelein! And best regards to your dear father! Yours, Arthur.

Arthur to Emma Vienna, August 27, 1924.

No. 90.

My dearest Bice!

In spite of all my feelings of happiness, I am quite disgruntled today because, despite all of my most urgent demands, the money is still not in Vienna and no news has come. The only thing the Berliners are good at is mouthing off, otherwise they are a hundred times lazier and a thousand times slower than the Viennese.

Yesterday evening I was at Mr. Dessauer's and Professor Willner was also there. The three of us spent almost three hours chatting together. Mr. D. seems to be in a terrible state of constant misery; he has been away, looking for a job and cannot find one. And if he found one, it would pay only 1 ½ million crowns a month. You can see how well I am doing in relation to him.

If only the Boston thing would work out, too. You know, we can live reasonably well in Vienna without America. But, it's one thing to live on a pension income with casual work and it's quite something else to have a regular, regular, permanent, well-paid job with a nice sphere of activity.

Unfortunately, I can't take any steps regarding the summer retreat until the Berlin pigs have sent the money. There is a difference of 15 million crowns between getting it across through regular channels and having it smuggled across; that is exactly what I will need to arrange if the money hasn't arrived by September! The notary should be hanged. First, I have to spend a week needlessly in insane despair because instead of telegraphing me, the dog takes nine days to write a letter and now he is doing this new piece!

I enclose two pictures, one of which was taken in my parents' salon, the other in Strobl in 1916.[43] I discovered them during the "Großreinemachen," [cleaning up] which is successfully continuing.

As we are getting closer, write me diligently! Hearty kisses and greetings from your Arthur

Emma to Arthur Kapuzinerstraße 20, Munich, August 27, 1924.

My dear Arthur, you have already gotten my express letter on Monday. You can't complain that I'm not writing.

I've written to you every day since Sunday and I've never done that before. I am putting on steam - as the Americans would say. Today I received two letters from you. A thousand thanks for that and especially again for the enclosure. Of course, I showed the photos to Daddy, etc. and everybody is delighted with them.

I am of your opinion that the letter from Mr. Whitney is more important than it seems.

Well Arthur, you seem to be so clever at making things right that I almost fear you have missed your calling. Perhaps your talents can be exploited in some way later on.

[43] Strobl is a small village at the Wolfgangsee in the Salzkammergut where Haas spent his summer vacation in 1916.

I came home from the city today in a rage. You know what I had to pay for the visa and in taxes? 17 marks. That is simply stealing. But I had to get the visa because when I hand in my passport at the Viennese police office, they will find out exactly when I used it. With the money that you so kindly gave me, I bought myself some laundry [lingerie]. Prices are insane and I only bought the halfway cheapest. A nice combination alone costs between 30 - 40 marks. That is not the most beautiful by far. Of course, I bought much cheaper.

What should I wear to the wedding? That little costume with the blue hat? Gell [Ugh!], our wedding is on September 8th. Don't you think you should go to the Stephanskirche and reserve the date or at least ask for more details? Is your brother still in Vienna? Otto and Grete have sent me a very nice map of Italy; I would like to thank them for it. Hug and Kisses Bice

Arthur to Emma Vienna, August 28, 1924.
No. 91.

My dearest Bice!

Thank you so much for your letter from Tuesday. You are still such a child and worry about the mail on Tuesday, which inevitably will not be delivered.

Unfortunately I am angry because, due to the sloppiness of the Berliners, neither the money nor any news has arrived in Vienna today. Not to mention that, besides all of the aggravation this causes me, every day of delay means 300,000K in damages, for which I really do know a better use.

By the way, I have suspended my efforts to sell the other houses, as I expect housing prices to rise.

I am also very annoyed that I have not heard from Privy Councilor Wien. He is in the summer resort and wants to rest, but still!

Today I have changed the apartment with the help of my housekeeper. One room is now divided into four parts: dining room, library room, master bedroom and study; the other is divided into two parts: boudoir and bedroom. Yes, it is sad that there is so little space but on the other hand, it will feel cozy. In any case, I will try to make you as comfortable as I can in the moment.

Unfortunately, I can't make any concrete travel plans and the like, as long as the Berlin dogs continue to let me down. But since I expect money from two sides, about 20,500M from the notary (which he should have sent me fourteen days ago) and 20,000M from Mr. Schirrmacher (which he received yesterday), I will surely have at least one of the two shipments by the day after tomorrow at the latest, since I requested notification by telegram of the bank-to-bank transfer. Those dogs!

Yes, you want to know what the changes are: I put a baroque table in the living room, between the club furniture. Together with a tablecloth, this will be our breakfast table, so that with the armchairs it forms the dining room. - I moved the wardrobe out of the bedroom. I put the writing table between the windows and the mirror cabinet is where the wardrobe used to be. The table is also removed from the bedroom; in its place is a beautiful divan that was previously in the kitchen. A small copper table, currently between the club fauteuils, will be placed along the wall in the bedroom; on top of it I will arrange the hat cupboard so that a few things, needed for breakfast and snacks, can be placed in one of the compartments. It's hard to imagine that you never saw the apartment; maybe the sketch I sent you the other day will help.

My housekeeper thinks that there is too much empty space next to the bed (Is that malicious teasing?); I haven't yet notified her of my engagement but she knows for sure; of course, the preparation of the apartment is not finished.

That's right, Mr. Dessauer and also Hofrat Willner invite you to stay with them for the few days before the wedding. I'm not very much in favor, as both are without wives, and I want to look into a pension. My brother's place is unfortunately in a hopeless mess, as my sister-in-law has not made any provisions.

A thousand kisses and greetings from your Arthur

Emma to Arthur Kapuzinerstraße 20, Munich, August 28, 1924.

Dear Arthurchen,

Yesterday evening we had visitors again - all of them gentlemen (all friends of my brother - and former admirers of mine). In the course of the evening, we talked about the theory of relativity and later about the universe. My brother, who generally keeps quiet when nothing interesting is spoken, was very talkative. The other gentlemen mostly held the old views while my brother defended the newer ones. I was quite astonished at one point, after he had given us a short lecture on spectroscopic distance determination, and asked him where he had actually gotten all of this. He laughingly said: "von Haas". Thereupon your book was brought out and read from for the whole evening. At the end of the evening, I was like a balloon that needed to be punctured.

If you think that September 8th is a big public Marian day, then you are completely mistaken. Very few people know that September 8th is Mary's birthday. Even the Church does not celebrate it publicly but remembers it during Mass. So our wedding ceremony can take place in the morning.

If you are of the opinion that our wedding ceremony should take place in the church in the 4th district, because there could be complications in St. Stephen's Church, then please know that exactly the opposite will be the case - on our proclamation certificate it is expressly stated that we will be married in St. Stephen's Church. I am very certain that we can be married in absolute peace and quiet on Monday morning in St. Stephen's Church without any human being disturbing us. We will be married in a small chapel and I am convinced that it will not cost more than it would at my ugly church in the 4th district. But if you have an aversion to Stephanskirche, just choose the church where you want to get married and I agree. But, I would like to get married in Stephanskirche. Love Bice

Emma to Arthur Kapuzinerstraße 20, Munich, August 29, 1924.

Dearest Arthur,

These damn Berliners are trying to vex us right now! I could hang them all. Really true - right now I'm as angry at them as you are. But I'm equally morose with the post office. I didn't get a single line from you yesterday and this is the second day this week. I write you every day now. Have you noticed? But in yesterday's letter, I had already forgotten half of what I wanted to write.

First, I wanted to send you a nasturtium - it is one of the last ones still blooming on our porch.

Second, I wanted to tell you that I will be coming to Vienna on Thursday, September 4th. (Arthur darling, the thought makes me infinitely happy.)

Third, I wanted to ask you when I should address our wedding announcements.

Fourth, I wanted to tell you that I only want to bring the essentials with me. Everything else I will have sent later.

So - I forgot to say all that yesterday and now there is today's news.

For the letters (with contents) that I received from you the day before yesterday and today, I thank you many thousand times. Arthur, sweetheart, the photos that were in your letter today drove me crazy with joy. You are really in a wonderful mood. I could eat you up. Just the one where you read – this is exactly the way I want you to be. Dad and Georg are delighted too. Papa has already confiscated your dining room - but he can't have this last photograph. I think Daddy finds it quite difficult to let me leave again. But he loves me so much and only wants my happiness.

Arthur darling, if you have an aversion to St. Stephen's Church just tell me.

I'll be with you in eight days. But until then, write to me at least once a day. A thousand kisses, your Bice

Arthur to Emma Vienna, August 29, 1924.
No. 92.

My dearest Bice!

I received two letters from you today, for which I thank you most sincerely. You become so well-behaved in the end - even in writing letters - that I am more than touched.

Oh, my dearest Bice, my head is again full of worry and anger and grief. We've set the 8th of September but we didn't count on the Berlin rabble. I can't possibly keep to the 8th if I don't have the money from Berlin in Vienna by Tuesday at the latest, or maybe even Wednesday (2nd or 3rd of September) - if it isn't here, for God's sake, there's nothing else to do but postpone.

I have probably always known that it is the sputum of the scum of mankind that populates Berlin. But my worst fears are exceeded. The ragtag of a notary has collected the fees for the execution of the order but fails to carry it out. I write express letters to the notary and Schirrmacher every day and get no reply at all. Today I sent a telegram to the ragamuffin of a notary with a pre-paid reply and still no answer at all at this hour.

I don't know whether it is just plain laziness or sloppiness, or whether someone is deliberately withholding money in order to earn interest. Otto says that with all this sloppiness, there will be no other choice than for me to go to Berlin if the money is not here on Monday. I am as angry as I have never been.

Of course, I can't take any steps regarding the wedding ceremony, and none for the journey [honeymoon], because the date of September 8th is not yet certain under these circumstances. Also, you can't come to Vienna until I send you a telegram.

The world would be so beautiful if nine-tenths of the people were not rags.

But I think that even if you only find out on Friday (the 5th) that everything is all right and you travel through the night, marriage on the 8th would still be possible. I hope that I have clarity sooner.

It is so outrageous what I have to suffer and go through, for the second time, because of the notary's mischievous laziness.

I hope you are not too upset. The only thing at stake is a date, nothing more. As much as I am troubled by anger and rage, today's worry is as light as a feather against the worries I had three weeks ago.

Many fond kisses and greetings from your Arthur

Hopefully I will be able to finish the big tidying up in a few days. Hopefully I can tell you more pleasant things tomorrow.

Emma to Arthur Kapuzinerstraße 20, Munich, August 30, 1924.

Arthur dearest,

Thank you ever so much for your darling letter and the 50,000 crowns. Don't mind the Berlin nuts - they are all coo-coo. (Excuse the slang) And don't worry about the honeymoon trip.

If you and I can't get away - then we stay at home and look at Mars. Don't you think that would be fun?!

I have tried to imagine how you turned around your two rooms.

I really have so much to tell you - but goodbye sweetheart darling - in five days. Hugs and kisses Bice

Arthur to Emma Vienna, August 30, 1924.
No. 93.

My dearest Bice

Yesterday's sad letter is, thank God, overshadowed by a new message. My nasty telegram had the desired effect after all and the ragtag of a notary public has now finally sent the money.

While Mr. Dessauer is with me, I have just received the following delightful telegram:

"Both houses have just received a telegraphed transfer from Schirrmacher to the bank: Luedinghausen". (Luedinghausen is the notary!)

So, I expect you on Thursday (the 4th) in the afternoon (I think 7 o'clock) if that suits you. But you have to write to me, so that we have it exactly.

Today I was in the Stephanskirche. It is soon enough, if the wedding is scheduled on Saturday.

As for your outfit, your blue one is probably the most suitable. (Mr. Dessauer thinks the same.)

My brother will give us the complete furnishings for two beds as a wedding present, which I am telling you in advance so that you don't bring anything superfluous from Munich.

If everything goes well, I want to go with you to Portorose on the Adriatic Sea (near Trieste), after the wedding for about three weeks.

Since I want to accompany Mr. Dessauer, I will close with the most intimate kisses, happy to have you forever in Vienna in a few days. Many warm greetings to your father, your Arthur

Arthur to Emma Vienna, August 31, 1924.
No. 94.

My little Bice

Thank God this will be the third to last letter I write to you, another one on Monday afternoon and one express on Tuesday morning – I will get to No. 100 with my letters, but no more.

Do you know - I've calculated it - that our separation, if you don't count the Salzburg days of being together, will have been an entire hundred days in all.

I'm convinced that everything will be all right tomorrow, Monday at the bank because I cannot imagine that the notary would lie so boldly. He is only a "medium" crook and rascal who tried, by using all kinds of tricks, to extend the time in which the money, which I had deposited with him without interest, could earn him interest.

So, if you don't let me know anything else, I assume that you will leave Munich on Thursday, September 4th in the morning on D-train no. 204, which (at least these were the times in May, but they might have been changed a bit since then) leaves Munich at 7:35 [am] and arrives in Wien-Westbahnhof at 7:05 [pm]. I will of course meet you at the station, on the platform; if, contrary to expectations, the use of the platform is not allowed, I will wait in the hall at the exit. I assume that you will have lunch in the dining car and have dinner in Vienna. By the way, Mr. Dessauer said that Miss Weigl is also going from Munich to Vienna on Thursday.

I will not be able to solve the question of your accommodation until tomorrow, Monday and I will tell you about it then. Should my two later letters, contrary to all expectation and what is quite likely, not reach you and should you, contrary to all expectation, miss me at the station, go to my brother's apartment and wait there (but it is only with tenfold bad luck that we would miss each other).

I will get the Italian visa tomorrow, which costs for the two of us only about 3 ½ marks, because the Italians are not bandits like the authorities of other states. Your visa for 17M might prove to be superfluous.

Tomorrow I will also go to the church office to inquire about the dispensation.

Just don't forget to bring your documents with you and, apart from the ones you need for the wedding ceremony, also bring those that relate to your studies, so that you can be considered an extraordinary student at the University of Vienna. Vienna University has an extraordinary student program.

We will go to Semmering on September 8th at 4:30 pm, stay there for twenty-seven hours, then take the night D-train to Trieste, where we will arrive on September 10th at 9 am. From there we will go by boat (which will only take an hour) to Portorose. There we will find suitable accommodation, which of course we can easily find in the off-season.

I will get the announcements tomorrow. Please bring a completely finished list of names and addresses of the people you want to send cards to. I'll reserve about fifty announcements for you.

As far as errands for the journey are concerned, I assure you that you will have no use for a dirndl on the Adriatic Sea. If you have something wide (even a school hat) bring it with you. In Vienna we want to get a gauze-silk jersey dress, if you think so, just like your red-blue jacket. By the way, you can certainly get a lot of things that might later be necessary, more cheaply in Portorose or Trieste than in Vienna.

At the seaside, you will be able to swim daily (since you can do this until November), but you have to be reasonable!

Today only many intimate kisses from your Arthur.

Arthur to Emma Vienna, September 1, 1924.
No. 95.

My little Bice

The penultimate letter, then only one more and then there are no more letters to write.

First of all, many thanks for your two dear letters from August 29th and 30th. Today everything went according to your wishes. I went very early in the morning to the Verkehrsbank, which was informed at the same time, by the central office, about the credit of 40,545 marks to my account. I was able to access it immediately and I have paid off all the debit balances that were due on the various accounts for my securities holdings. If I don't do it before my departure from Vienna, when I return, I will invest the amount at 12% per year at the Postal Savings Bank (which is the absolutely safest).

Then I went again to the sacristan of St. Stephen's. Everything is in order and it is soon enough, if we present the documents to him on Friday. The wedding ceremony can also take place at 11 am.

Finally, the question of accommodation. If it suits you, the easiest thing would be to live in my brother's apartment for the four days, if you can excuse that, in my sister-in-law's absence, they cannot provide you with a cupboard for the few days; my sister-in-law has secured everything in storage while she is away. They would put your bed and washstand in the large basement room that would be at your disposal; you would also have to excuse Otto, who is terribly busy, if he doesn't take care of you as he would like to.

You would take meals with me in the inn.

I have an interesting bit of news to tell you: In Saturday's edition of the *New Free Press*, there was an advertisement from the Egyptian Ministry of Education. They are looking for three professors for the medical college in Cairo, one of them for physics, and they offer 900 – 1,140 Egyptian pounds as annual salary. Three-year contract, contract language English. Applications by September 30th. If I were to apply, I'm sure I'd have a good chance. An Egyptian pound is a little more than an English pound. That would be about $400 per month but life is certainly much cheaper than in the USA. What do you think about that? But in the USA, I would be much, much better suited. I can apply and if I find something better in the USA or Vienna, I can reject it later with any excuse.

Unfortunately, the apartment changeover is far from over. In the past, I had so many things that all the cupboards were not enough and everything was lying around everywhere. Now I not only want to get rid of everything lying around - moreover, I want to free the three wardrobes for you; it is not so easy! And a bathroom with a gas stove will be installed as well. It's all so small. You will be as comfortable as I can make it and you shouldn't have too much trouble with the organization.

And in 3x24 hours I'm going to West Station. May I hug and kiss you there in front of all the people? – "Of course you may!" Isn't it?

Many, many, many fondest greetings and kisses from your Arthur

Arthur to Emma Vienna, September 2, 1924.
No. 96.

My dearest little Bice
Thank God for the very last letter. Of course, nothing new has happened overnight since the last letter.

Do I need to tell you how overjoyed I am that the moment will come the day after tomorrow, which I have longed for with all my heart and which has been the focus of all my thoughts and striving in the last weeks and months?

But it remains the same: You leave Munich on Thursday morning with the D-train and arrive in the evening at 7 pm on Thursday in the Vienna Westbahnhof. I'll be waiting for you on the platform or if contrary to expectations, the public is not allowed to enter there, in the hall right where you come out after you have handed in your ticket - and you'll stay with my brother, without the comfort that one would like to offer you but cannot due to the absence of my sister-in-law.

Don't forget the documents and don't do anything stupid on the trip, so that you will not be ill on September 8th!

I won't send you any more kisses. I prefer to save them for the day after tomorrow. Many fond greetings, also to your dear father (to whom I wrote yesterday) from your Arthur

Emma to Arthur Kapuzinerstraße 20, Munich Sept. 2, 1924.

My dearest Arthur,

It's good that I received your urgent letter from Saturday before your sad one from Friday, otherwise I would have been quite disappointed. But it's all right now.

I didn't write anything yesterday or Sunday; I've been so terribly busy. I haven't forgotten about you for a single hour.

I was in town yesterday to get something for Lenchen [Otto's daughter]. But I can't get any doll that I would want to give her for under 30 marks. Daddy promised to get Austrian crowns for me, so with your help I will find something suitable for the little one in Vienna.

Fig. 11.15 Emma Beatrix Huber and Arthur Erich Haas at the time of their wedding in September 1924

> On Thursday I will leave Munich at 8:25 am and arrive in Vienna at 5:28 pm. I'm so tired from all the farewell cakes and celebrations that I'm looking forward to Thursday as though I will be in paradise. My only wish is to be able to sit quietly for a few hours and not have to talk or think and then to see you again.
>
> This letter today is the last I am sending you before we meet again. I wish I wasn't so terribly tired; I wouldn't have sent you such a lame letter. So goodbye, Huzerl, Thursday Bice

Contrary to Arthur Haas' fears, everything worked out as he had so carefully planned. The train was on time, the platform was accessible, and so was the bride for the welcome kiss. She had not forgotten any documents and they were accepted by the sacristan as promised. The dispensation for the wedding, since Arthur Haas was a Protestant Christian, was granted without hesitation and the wedding was scheduled for September 8th at 11 o'clock in the morning in a side chapel of St. Stephen's Cathedral. Even the four days' accommodation with Otto was easily survived with daily strolls through Vienna for last preparations, only interrupted for lunch in an inn. There was even time for taking pictures to commemorate the long-anticipated day, Fig. 11.15.

The wedding took place in the Stephans Cathedral (Fig. 11.16) as Emma Beatrice Huber, now Emma Beatrice Haas had wished. It was a small ceremony, just as Arthur Haas had wished. Karl Jellinek, the friend from Göttingen days who was staying with his family in Vienna at the time, and Arthur's brother Otto served as witnesses. The family was scattered, sister and sister-in-law were not present, domestic servants had not been notified and friends were informed by letter afterward.

The happy bridal couple left immediately after the ceremony, traveling to the first stop of their honeymoon, on the Semmering, in the afternoon. There is no record of where they spent the night on the Semmering, presumably at the luxurious

Fig. 11.16 St. Stephan's Church in Vienna in a picture postcard from the 1920s

Fig. 11.17 The Hotel Riviera in Portorose where the young married couple spent their two-week honeymoon. The room on the lake side is marked with an arrow

Südbahnhotel, with its convenient connection to the railway, where accommodation offered guests one of the first indoor alpine swimming pools. On the following day they continued their journey by the Südbahn to Trieste from where they reached their final destination in Portorose on the Adriatic Sea, as promised by Arthur. In Portorose, they were finally able to recover from the strains of their three-month separation.

They spent two relaxing weeks in the small Hotel Riviera, shown in Fig. 11.17, near the long beach promenade. Their room offered a wide view over the Adriatic Sea where they went every day, swimming in the still warm water and lying in the sun on the beach. Only a few excursions to nearby Trieste and by train to Venice interrupted the quiet days.

Although the restless life of the last few months, as well as worries about the future, seem to have been forgotten or more repressed, it was all waiting for their return. During this time, Otto took care of finances and future planning, made new bank deposits, settled claims and also had to send money (5 million) by bank transfer to Portorose, to finance the standard of living there. He also had to answer new inquiries about the sale of the remaining Berlin houses, mostly putting off those who were willing to buy until his brother's return.

Otto also answered his brother's urgent requests for information since, despite the apparently peaceful days in Portorose, he wanted to know whether mail had come from Prof. Donnan, from Privy Councilor Wien, from MIT in Boston or even from the Ministry of Education in Cairo. Arthur was already in the next planning phase. However, while reading the newspaper every day in the hotel or in the café on the

Fig. 11.18 The Neulingsgasse in Vienna's III district, in a contemporary postcard depiction. The arrow presumably indicates the location of the small apartment

promenade, he discovered a new announcement for a position in Vienna that seemed highly interesting to him and for which he wanted to apply immediately. But this had to wait until his return on September 22nd.

On September 21st, the married and happy young couple returned to Vienna to move into the newly renovated small apartment at Neulingsgasse 24, depicted in Fig. 11.18. There they found several letters and messages that would determine their future life.

Notes

i. Peter Altenberg, *Märchen des Lebens* (Berlin: S. Fischer, 1924).
ii. Walter Moore, *Schrödinger – Life and Thought* (Cambridge: Cambridge University Press, 1989).
iii. See C. F. Goodeve, "Prof. Frederick G. Donnan, C.B.E., F.R.S.," *Nature* 179 (1957): 235–236.
iv. Robert A. Millikan, "The Electron and the Light-Quant from the Experimental Point of View," accessed January 31, 2021, https://www.nobelprize.org/uploads/2018/06/millikan-lecture.pdf and Gerald Holton, "Millikan's Struggle with Theory," *Europhysics News* (2000): 12–14.
v. Kate Moore, *The Radium Girls: The Dark Story of America's Shining Women* (London: Simon & Schuster, 2016).
vi. See Edoardo Amaldi, *The Adventurous Life of Friedrich Georg Houtermans, Physicist (1903—1966),* (Heidelberg: Springer, 2012).

Part III
The Vienna Academy of Sciences from Science Administrator to Science Promoter 1925–1935

Chapter 12
Events at the Vienna Academy of Sciences

While Arthur Haas was busy in the summer of 1924 was busy with his marriage plans, negotiating his house sales, correcting his books, dreaming about and longing for America, changes were occurring in Vienna that would have long-term effects on the Haas family's fortunes. The actuary of the Vienna Academy of Sciences, physicist Professor Emil Kohl (1862–1924) had died unexpectedly in April of that year at the age of only sixty-two. Emil Kohl was a theoretical physicist who began his work in the realms of thermodynamics and the laws of radiation. He had also sought to study the theory of relativity but to little avail; his work was commented on by Max Planck in a letter to Wilhelm Wien, remarking that of all the equations set up by Kohl, not a single one showed a new relationship to measurable quantities. Emil Kohl had twice applied for a faculty position at the Charles University in Prague but in 1910 the offer went to Albert Einstein and, after the latter had moved to Zurich,[1] Philipp Frank was chosen as his successor in 1912. Emil Kohl remained in Vienna, as associate professor and actuary of the Vienna Academy of Sciences.

The Academy of Sciences, shown in a contemporary postcard in Fig. 12.1, was founded in 1847 as a scholarly society in the spirit of the British Royal Society, for scientific research and the dissemination of new knowledge in Imperial Austria. Its members covered a wide range of research disciplines in both the humanities and natural sciences. The Academy enjoyed international fame on the eve of the First World War particularly in physics through research at the Radium Institute. The position of an actuary, who as head of the office was also responsible for keeping the records and minutes, had been introduced in 1827. Two actuaries were responsible for the daily administrative tasks of the Academy's offices, one the humanities and the other for scientific matters. The actuaries were subordinate to the respective Secretary Generals of the Academy. In addition to the actuaries, there were a number of lower clerks, such as accountants, two academy servants and a porter, as well as a bookseller who was responsible for the distribution of academic writings.[i]

[1] Only two years later in 1914, Einstein moved to the University of Berlin at the instigation of Max Planck and Fritz Haber.

M. Wiescher, *Arthur E. Haas—The Hidden Pioneer of Quantum Mechanics*, Springer Biographies, https://doi.org/10.1007/978-3-030-80606-4_12

Fig. 12.1 The Vienna Academy of Sciences in a contemporary postcard

Emil Kohl had been a long-serving first actuary since 1900. The Germanist and music historian Viktor Junk was the second actuary, representing the humanities. (The numbering in the title did not imply any ranking of the two actuaries and was solely based on seniority at the Academy.) With the death of Emil Kohl and the retirement of the curator Dr. Karl Schwarz in July 1924, the office was reorganized and Viktor Junk, who had already served at the academy for more than twenty years, was appointed as first actuary. For this reason, a natural scientist was now being sought who would take over the office of second actuary.

The candidate was expected to fulfil the following conditions: "This position requires a civil servant with a full academic education in mathematics and science, knowledge of the main modern languages, library service and shorthand."[2] Dr. Schwarz's role as custodian was to be taken over by an administrative officer who would essentially be responsible for the bookkeeping and financial management of the Academy. This position was to be filled internally by a long-standing member of staff, Marie Calabek, while the position of second actuary was to be advertised.

[2] Archive of the Academy of Sciences, Vienna, Protocol to the Federal Ministry of Education, #405 Vienna 24 (June 1924).

The Chairman of the Presidium and the Secretary General of the Academy at that time (1911–1929) was Court Councilor Friedrich Becke,[3] who was also to head the selection process of the candidates. Friedrich Becke knew young Haas, his interests and activities and sent him a short note on September 16th with reference to the vacant position at the Academy. Haas submitted his application immediately after his return from his honeymoon in Portorose on September 22, 1924:

> His Excellency the Secretary General of the Academy of Sciences in Vienna
> My esteemed Court Councilor,
>
> Please accept my sincere thanks for your kind letter of the 16th of this month.
>
> I would like to submit my written application. I apologize for the delay of several days due to my being away from Vienna.
>
> Please allow me, Court Councilor, to express my excellent esteem and greatest devotion,
>
> Arthur Haas

Haas was not particularly eager to accept the position. He was still hoping for opportunities in the United States, from Caltech in Pasadena to Princeton and MIT, which had been promised to him by Prof. Donnan and Dr. Whitney. Haas had also just sent his application to the German University in Brno. In addition to these prospects, he had concerns about the position itself. He had been well acquainted with the former actuary Emil Kohl, especially since they had both worked in the field of heat and radiation theory. Haas' career had begun to take form with the successful publication of his books in various languages and in 1924 he considered himself to be academically far superior to Professor Kohl. Haas saw the position of actuary as a dead end, as it had been for Kohl. Furthermore, he considered it an end to any future career in science, which he was hoping for after all. However, new family conditions required permanent employment and this was a civil servant position that would provide a fixed and decent salary, thus offering a secure living situation for himself and his family.

So, he applied while still hoping for academic alternatives. He did not, however, hear from Prof. Donnan who must have returned from his extensive travels in the USA, nor from Prof. Dempster in Chicago, nor from Dr. Whitney about the opportunities at MIT, nor did he hear any news or offers from Cairo or South America. Haas also thought to secure the professorship in Brno, especially since he had grown up there and owned two houses in the area. To this end, he talked at length with his friend Philipp Frank about the possibility of succeeding Gustav Jaumann. But Frank explained to Haas outright that he had no chance whatsoever of being appointed to the position, especially since his former opponent Alfred Haussner, who had agitated

[3] Friedrich Becke (1855–1931) studied at the Technical University in Vienna. He turned to mineralogy and habilitated in the winter of 1880 on the subject of petrography. In 1882 he was appointed associate professor of mineralogy in Czernowitz and later to full professor. In 1890 he went to the German University in Prague and in 1898 to the University of Vienna. In 1911, Becke became Secretary General of the Vienna Academy of Sciences.

against Haas' habilitation many years earlier, now had far greater influence since being appointed rector of the German University in Brno.[4]

There were also other economic concerns to consider. The Currency Conversion Act of December 20, 1924 promised further stabilization of economic conditions in Austria. In effect from March 1, 1925, the Austrian crown, which had lost a great deal in value during the inflation years, was to be replaced by the schilling as the new currency unit. With this, 10,000 Austrian crowns were to be exchanged for a shilling, with the shilling being equivalent to 100 groschen as a subunit. This resulted in a new monthly income for the family of about 700 to 900 shillings. A fixed professional salary would greatly improve their financial situation and make it possible to escape the precarious housing situation on Neulingsgasse.

Time was pressing, especially as it soon became clear that his young wife was pregnant, so Haas postponed his dreams. On January 29, 1925 when the official offer for the position of II Actuary at the Academy of Sciences was made, Arthur Haas immediately accepted on January 30th. It was a position that, after six months' probation, was converted into a permanent position on July 1, 1925.

This decision was necessary because the loss of his fortune during the war and inflation years had left him with a considerable reduction in income. Haas now tried to compensate for these losses with the civil servant's salary and by publishing scientific and popular scientific works. This seemed to succeed over the following years, as his tax certificates demonstrate. While his gross income for 1923 had been set at 52 million crowns—compared to the new shilling currency of 5,220 ATS[5]—in 1924 his earnings were only 2,700 ATS, which mainly came from college fees for his lectures and from publishers' fees; in addition, there was an ATS 400 annual income from his investments. It is therefore not surprising that he was worried about the future financing of his marital status, with about half of his previous income now gone and two additional people depending on him.

Then in 1925 the economic situation improved considerably. In addition to his gross income of ATS 5,200 as an actuary, he received ATS 4,735 from author's fees for the books he had written over the past three years and ATS 4298 from capital income. The stock market had stabilized considerably and with this investment he had a handsome income of ATS 14,223, which increased to ATS 15,283 in the following year 1926. In 1927 he had some losses on the stock market, which he could balance with publishing income to reach ATS 15,298, and in 1928 he had a gross salary of ATS 16,253. This remained constant over the next few years, with his source of income being distributed over his steadily increasing civil servant salary, his author's fees, which ranged between ATS 4550 and ATS 5500 per year, and the income from his market shares. It was not until 1931 and 1932 that his gross income was noticeably reduced by the collapse of the stock markets following the Great Depression. So in

[4] Georg Jaumann's successor was the Brno private lecturer Erwin Lohr (1880—1951). See Pavel Ŝiŝma, *Zur Geschichte der Deutschen Technischen Hochschule Brünn* (Linz: Trauner Verlag, 2009), 37, 87–88.

[5] ATS is Austrian Shilling.

financial terms, Haas could have been satisfied with his regular income. This was not, however, the career that he had dreamed of and he missed science.

The position of II Actuary at the Academy of Sciences was not a scientific post per se, as he had always hoped for, but a more or less executive post in the administration of the Academy. In addition to planning and preparing meetings on various mathematical and scientific topics, he was responsible for keeping the minutes and files of these meetings. His duties also included editorial supervision of scientific topics for the various publications and statements of the Academy. In addition, he was responsible for the management of the library and the tasks associated with this role. He does not seem to have found any real pleasure in the job. As he wrote to his wife after only one month, while she was visiting her father in Munich:

> Today I am writing to you in the afternoon at the Academy because I have to be here for today's meeting. My life is very bland. From the office, where I could not do my own work because of too much typewriter rattling, I fled to Kärtnerstraße for half an hour in the morning. Otherwise, I wouldn't know what to tell you. After all, I don't experience anything outside of work and office.[6]

But only one week later he had better things to report:

> Since the stove was repaired and a humidifier installed, it's been bearable in the office, if only it wasn't for the horrible typewriter noise![7]

This was why he took measures to work together with I Actuary Junk, whom he seemed to get on well with at first, to improve the situation and escape the typewriter noise:

> Today I have agreed with my colleague that as soon as it gets warm, we will move into two really nice rooms, which are empty and very nicely furnished, so that everyone will then have his own room with a sofa! That will be fine.[8]

In addition to unloved administrative tasks, Haas also continued his lecturing activities as an associate professor of the university. The lecture in the winter semester of 1925 on atomic theory, based on his newly published book[9] on the same subject, seems to have been sparsely attended by only five gentlemen; ladies were now always absent, he reports. He, therefore, revised the lecture and gave it in the summer semester of 1925 as a series of ten general lectures on more or less well-defined topics: "Matter and Electricity," "The Building Blocks of Atoms," "The Quantum of Light," "Spectra and Energy Levels," "The Hydrogen Atom," "The Basic Materials," "The Atom as a Planetary System," "The Molecules," "Radioactivity," and "The Transmutation of the Elements." This series of lectures offered the whole spectrum of progress that had been achieved in atomic and quantum physics over the previous thirty years, in a generally understandable way.

---

[6] AEH to EBH (Vienna, February 26, 1925).

[7] AEH to EBH (Vienna, March 3, 1925).

[8] AEH to EBH (Vienna, March 4, 1925).

[9] Arthur Haas, *Atomtheorie* (Berlin: De Gruyter, 1924).

This same compilation of completed topics gave Haas the opportunity to develop various lecture series in the following years, through which he brought the complexity of modern science to a large public audience. He did not shy away from addressing and effectively presenting topics at the current forefront of science, topics whose physical explanations and solutions were still ten years in the future. For example, regarding the possible conversion of mass $\boldsymbol{m}$ to energy $\boldsymbol{E}$ according to Einstein's famous formula $\boldsymbol{E} = \boldsymbol{m} \cdot \boldsymbol{c}^2$ ($\boldsymbol{c}$ being the speed of light), Haas explained:

> The energy of a gram is 900 trillion ($3 \cdot 10^9$) erg. The largest ocean liners are operated with about 50,000 horsepower [...] The energy that such an ocean liner consumes when crossing the Atlantic Ocean would only be a mass equivalent of a quarter of a gram. [...] less than one gram would be sufficient to power an ocean liner, instead of perhaps 5,000 tons of coal.[10]

It was not until twelve years later, in 1938, that Otto Hahn and Fritz Strassmann discovered the phenomenon that led to the partial release of this intrinsic energy of matter—a process explained by Lise Meitner, Arthur Haas' former fellow student, as the splitting of heavy atomic nuclei.

Elsewhere, Haas described nuclear transformation in stars as a possible source of energy, a thought first formulated by the British astronomer Arthur Eddington (1882—1944).[11] Haas stated:

> A synthesis of one gram of helium from one gram of hydrogen would then generate the same amount of heat as the combustion of a twenty-ton wagonload of coal. [...] If, therefore, the enormous energy that all fixed stars constantly radiate only had its source in nuclear transformations [... this] could only be explained by the assumption that fixed stars, and not only atomic nuclei, are subject to constant transformation, and that also, in stars, an incessant destruction of electrons and protons takes place. These original components of matter in the fixed stars would disappear to bring light quanta of radiant energy into being, which in turn would perhaps be transformed back into matter in the cosmic nebulae.[12]

The last thought, albeit speculative, refers to the phenomenon of the pair production of electrons and positrons, which had just been discovered by Arthur Compton.[13] Haas himself took up this idea of Compton's and published two articles on the subject in 1926; the first was only a short note in which he suggested the possibility of the Compton effect for the conversion of light into matter.[14] In the second article, Haas presented a detailed calculation of the effect when photons gain energy by collision with relativistic particles.[15] He argues that at sufficiently high particle velocities, the

---

[10] Arthur Haas, *Die Welt der Atome– Zehn gemeinverständliche Vorträge* (Berlin-Leipzig: De Gruyter & Co., 1926), 121.

[11] Arthur S. Eddington, "The Internal Constitution of Stars," *Nature* 106, 2653 (1920): 14–20.

[12] Arthur Haas, *Die Welt der Atome*, 122–123.

[13] Arthur H. Compton, "A Quantum Theory of the Scattering of X-rays by Light Elements," *Physical Review* 21 (1923): 483–502.

[14] Arthur Haas, "Die Kosmische Bedeutung des Compton-Effekts," *Anzeiger der Akademie der Wissenschaft* 15 (Wien, 1926).

[15] Arthur Haas, "Über Frequenserhöhungen von Lichtquanten durch Zusammenstöße mit rasch bewegten Materieteilchen," *Sitzungsberichte der Akademie der Wissenschaft in Wien* 135, 10 (1926): 647–653.

upward scattered energy could be sufficient for pair production, i.e., the conversion of photon energy into matter. This effect is now known as inverse Compton scattering and is indeed a well-known phenomenon in astrophysics. The effect causes the conversion of light into high-energy X-rays, i.e., an increase in frequency, which is achieved by the scattering of relativistic electrons in the corona of black holes or neutron stars. In today's nuclear physics laboratories, inverse Compton scattering is used as an important tool for generating high-intensity photon beams for a variety of applications. However, the cosmological relevance for the production of matter in the universe, as Haas imagined it, is not given according to current understanding.

Three years after Haas' publication of these theses on energy production in stars, the first qualitative theory for understanding energy production in our sun was formulated in 1929 by Bob Atkinson (1898–1982) and Fritz Houtermans (1903–1966).[ii] The quantitative physical–mathematical derivation of this idea was not delivered until ten years later, in 1936, by the theoretical physicist Hans Bethe (1906–2005)[16] who had emigrated from Germany to the United States. With the formulation of the so-called pp-chains as a possible reaction sequence for the energy production in our sun, he had worked out a coherent model for the nuclear fusion processes.[17] The understanding of the formation of heavy elements, as Haas imagined in fixed stars, would have to wait decades before it could become a reliable quantitative form.[iii]

Arthur Haas took the latest ideas, formulated new physical concepts and then presented their possible practical consequences, for both development and technology, through his lectures and popular-science books. It was an approach that he had been developing for years and in which he saw his future and career. He did not want to just administer science; he wanted to make science and popularize the world of science.

The time for this was favorable. There was broad public interest and in the first post-war years, the social democratic government in Austria supported the expansion of popular education. This was especially true in "Red Vienna"[18] in the 1920s, when numerous government-supported education centers for adults were opened.[iv]

During these interwar years, interest in new technical achievements and opportunities remained in the fore and was widespread among all sectors of the population; popular education was intended to meet this interest. Interest also extended to popular literature where authors with a scientific or technical background, such as the German engineer Hans Dominik (1872–1945), made an enormous contribution with their futuristic novels based on the latest scientific findings. Dominik, with novels including *The Fire of Khufu's Pyramid* (1927) and *Atomic Weight 500* (1935), literarily portrayed the natural scientists' vision of an unlimited energy source (and its dangers)—often highlighting this work as the German-national mission—while

---

[16] Regarding the life of Hans Bethe, see Silvan S. Schweber, *Nuclear Forces—The making of the physicist Hans Bethe* (Cambridge: Harvard University Press, 2012).

[17] Hans A. Bethe, Charles L. Critchfield, "The Formation of Deuterons by Proton Combination," *Physical Review* 54 (1938): 248–254.

[18] Nickname for Vienna when under a social democratic government.

Haas remained on the scientific side. His writings and lectures did, however, increasingly offer such futuristic perspectives. This was especially true on the question of possible uses for the largely unknown nuclear energy, the release of which was still a secret.[19]

As a popular science forum for their thoughts and lectures, university teachers including Haas traditionally preferred the Urania, as it probably offered the widest range of popular science courses. Haas also lectured for the Philosophical Society, which provided an audience for his intellectually more sophisticated lectures. The Urania annual report from 1926 to 1927 gives the following figures:

> Urania counted about 57,000 members. The adult education courses included 562 courses with 14,662 participants, together with 9,501 lectures with 239,316 visitors. In the public lecture courses, 108 shorter courses with 598 lectures were attended by 26,000 people, 20 lecture series with 157 lectures by 16,000, 556 individual lectures by 57,800 people. Forty-seven different films were shown to 229,000 visitors on 536 afternoons and evenings, while 31 Urania events with music, etc. were attended by 101,800 people on 235 evenings.

The Philosophical Society at the university was, on the other hand, a more privileged circle of university members as well as members of the educational community, intended to facilitate intellectually stimulating exchange between the new scientific disciplines.[v]

As noted earlier, upon Haas' return to Vienna and before his marriage, he gave a lecture on "The Harmony of the Spheres of the Atom" in May 1920 and a lecture on "The Problem of Primordial Matter in Modern Physics" in December 1921. However, no lectures are recorded for 1922 and 1923.[vi] These were the years in which Haas devoted himself primarily to the completion and publication of his books, but this was also the time when inflation had put him in great economic difficulty. He tried to compensate for this by engaging in time-consuming speculations about the stock market and housing market. Although he remained in contact with the Urania and probably also took part in various film projects, he had reduced his lecturing work. Nor did he give lectures in the Philosophical Society, except for a talk much later on "The Objectives and Human Physics" in 1926.

The year 1924 was devoted to securing his new life with Emma, as discussed earlier, while the years 1925 and 1926 were devoted to his new family, to writing and revising his increasing number of textbooks, to working on new scientific ideas and projects, and probably also to the obligations of his new position at the Academy.

The new edition of his textbook on *Theoretical Physics* was generally well received. In May 1925 the pedagogical journal *Der Mittelschullehrer* recommended to teachers: "The whole work will be of the best service to all those who want to keep up to date in theoretical physics," while the *Zeitschrift für Flugtechnik* noted at the same time:

> For the practically minded engineer, however, only a few chapters are of direct interest; but if we want to learn more about quantum and relativity than from the popular writings, we will be pleasantly surprised. For Haas is an excellent educator who works without excessive use of mathematics and has made a careful selection from the great wealth of material.

---

[19] Arthur Haas, "The Atom as Source of Energy," *Scientific Monthly* (February 1928): 140–146.

The *Physikalische Zeitschrift* praises:

> The presentation, which consistently makes use of the descriptive vector notation, is concise, clear and elegant. The overall impression of the book [...] demonstrates the love and skill of the teacher at every turn.

This is the tenor of the reviews in all of the journals that focused on engineering and the natural sciences, a general praise that the author gladly accepted. The acclaim also promised him a necessary long-term source of income.

His wife Emma Haas spent considerable time in Munich during February and March of 1925, staying with her father whose declining health worried her. A further reason for being away from Vienna was that she too felt the stress not only of her pregnancy but also of the small apartment, which was not a pleasant stay. During her absence Haas tried to exchange apartments with families he knew but was unsuccessful.

Since it had become clear over the winter that the apartment on Neulingsgasse was simply too small, especially since a child was to be added, Arthur often met with his brother in order to clarify the question of the remaining houses. Housing prices had risen and the economy had stabilized. The housing market had also improved and there were offers to buy. From Haas' side, interest in selling was primarily driven by the fact that he needed something bigger than his converted bachelor home. However, despite the Social Democrats' aggressive program in Red Vienna for building and providing affordable housing, rental space was still limited. Haas was thinking of a house that he could afford from the profits made by the sale of the Berlin apartment buildings.

> Yesterday I had to work all afternoon with Otto on the compilation for the Berlin riffraff until my head was swollen to head size 80.[20]

He made the first preparations in March during the absence of his wife, since he did not want to burden her with his housing concerns. During this time of being a bachelor, as his friends scoffed, he again resumed the tradition of the daily letter to Emma in which he reported the successes and failures of his daily life. However, the tenor of these letters was different, no longer about despair and worries but about hopes and successes. He reported:

> And when I think of all the wretched letters that I sent to you in Munich at that time and think of today, the unspeakable happiness that this half year has meant to me becomes so clear. Now I may hope that it will always remain like this or perhaps even get better.[21]

Besides this, he worked.

The reviews of his book *The Atomic Theory* were brilliant; *Pharmaceutical Monthly* wrote in April 1925:

> The author has brilliantly succeeded in making the whole complex of modern atomic physics understandable to the non-physicist without the help of higher mathematics.

---

[20] AEH to EBH (Vienna, March 21, 1925).

[21] AEH to EBH (Vienna, February 25, 1925).

Physics Journal judges: "Considered as a whole, the book can be called a complete success." Even Arnold Sommerfeld writes in *Deutsche Literaturzeitung*: "Speaker is convinced that an elementary description of atomic physics cannot be written better and more carefully than it has been done here." Haas seems to have found his destiny as a popularizer of the complex interrelationships of modern physics for the scientific public.

By the end of 1924 he had already completed his corrections to the English edition of *Atomic Theory*, which was published in February 1925 by Constable & Co in London. He was then free to fully concentrate on the *Handbook of Experimental Physics*. In November 1924 he had signed a contract with the Akademische Verlagsgesellschaft, as initiated by Privy Councilor Wilhelm Wien, to write this work. However, it was not the hoped-for multi-part contribution on mechanics, optics, and topics of modern physics that he had anticipated but rather, as Wilhelm Wien had already announced, the book was to only include mechanics and traditional optics, such as general optics, interference, and diffraction.[22] This was much less than Haas had originally expected. According to the contract, the work should be completed by Easter 1926. While Haas did not have much time,[23] he was also disappointed. He had hoped for at least one topic on modern physics, such as radiation or atomic physics, which he had planned for and had spent so much time on. The proposed content resembled a topic with which he had inwardly long since finished, but he had to complete the contract for political reasons—he did not want to spoil his relationship with Wilhelm Wien—and also for financial reasons. He himself wrote about the situation a year later when he was working on his new book *The World of Atoms*:

> They will be very good, unlike my Mechanics book, whose paternity, alas, fills me with shame. The fact that it was written for Atzi's sake is the only consolation.[24]

So it was the need for income that drove him back to the relatively unloved subject of mechanics, and this need included a larger home.

In May 1925 after Emma's return, they managed to find a house at Baumgartenstraße 58, as seen in Fig. 12.2. It was just outside the city center, in the XIII district of Penzing, still to the west of Schönbrunn Palace, but there was a tram connection and the garden seemed important for the children's upbringing. Finally, the furniture belonging to Arthur's parents, which had been stored with his siblings Otto and Margarethe, could be brought to fill the new house. This was not without arguments but finally the siblings agreed on the question of respective furniture ownership.

Shortly after this successful move, on July 21, 1925, Atzi—the son Arthur Gustav Haas, named after father and grandfather—was born, whereupon the Ministry of Education granted the II Actuary Haas an annual child allowance of 60 shillings.

---

[22] Contract between Prof. Dr. Arthur Haas, Vienna and the Akademische Verlagsgesellschaft, Leipzig, dated October 29, 1924.

[23] W. Wien, F. Harms, *Handbuch für Experimentalphysik* Band 2: "Mechanik der Massenpunkte und der Starren Körper. Von Prof. Dr. Arthur Haas" (Leipzig: Akademische Verlagsgesellschaft, 1926.).

[24] AEH to EBH (Vienna, March 21, 1926). Atzi is the pet name for their first son.

Fig. 12.2 The house at Baumgartenstraße 58, where today a faceless block of apartments stands

Emma Haas was weak after the birth; she lost weight and Haas was constantly worried about her. She spent a week in the sanatorium in Bad Gastein, a popular and successful radium bath[25] where Emma took baths and inhalations to regain her strength. This popular sanatorium has large, magnificent hotels that offer accommodation for spa guests. In fact, when Emma returned to Vienna, she had gained weight and looked visibly recovered.

However, a cold winter followed and Emma's father invited her to come back to Munich in March to be pepped up by Munich beer. Atzi stayed at home with his father who pampered him in the evenings. During the day, Atzi was taken care of by the housekeeper or his aunt. Haas spent a lot of time with him and the majority of the letters deal with the growth and progress of the baby, as is common among young parents.

> So Atzi gets cuter every day. How cute he was in the bath yesterday cannot be described, and he continues to say Daddy and when I come to him, he grabs my hand with both hands.

[25] Measurements from the time indicate an activity of 80,000 Bq per liter of water in the baths. The respective bathing and inhalation sessions were limited to two hours and there was no fear of possible negative consequences from the radiation. Hermann Sieveking, "Die Radioaktivität der Mineralquellen," *Berl. Klinische Wochenschrift* 4 (June 23, 1906): 779–780 and no. 24 (June 11, 1906): 809–811, and K. Becker, "One Century of Radon Therapy," *Int. J. Low Radiation* 1/3 (2004): 334–357.

> In bed, he is already crawling around so energetically that he can get from one end to the other in a minute.[26]

Even so, Haas had not given up his dreams of America. He wrote a letter to Augustus Trowbridge, Professor for Natural Sciences at Princeton, asking about possibilities for an invitation to America. Trowbridge was chairman of the International Education Board in Paris, which supported development projects in Europe such as the founding of the Bohr Institute in Copenhagen, the new International Mathematical Institute for Hilbert, and Courant in Göttingen.[vii] But the only response was a rejection:

> Today's mail brought me a letter from Prof. Trowbridge in Paris that almost devastated me. I see that at present an invitation by the International Education Board is out of the question. If I went to America, even temporarily, I believe it would only be possible if I were invited by a single university.[27]

But Haas also used the time to work. Work on the *Experimental Physics Handbook* was finished and he feverishly wrote the popular science booklet *The World of Atoms*, which summarized the nine lecture topics given during the 1925 summer semester. In the letter to his wife of March 20th, he wrote:

> I am now working diligently on the book *The World of Atoms* and after two years, I cannot tell you how happy it makes me to be able to work on something that comes from the heart once again.

The work in the "office" of the academy does not seem to have interested him very much:

> In the morning there was a lot to do in the office today [...] from 2:00 to 5:00, I was in the café and danced around the city like a bachelor [...] Now I am sitting again in the Academy in the meeting [...]

His attention was completely focused on his writing:

> I'm usually at home in the afternoon, and the work on the book *The World of Atoms* is going very well. I have completed two of the eight chapters, a third one partially.

He had new book plans and hoped for a better income from his writing:

> Yesterday I wrote to the publisher about the new books I plan to write. I believe that now, when the price of books has risen so insanely, instead of going through the pain of writing for the *Handbook*, I could earn money more easily and pleasantly, and perhaps more abundantly and without annoyance, with other books that are very dear to my heart. I hope to see whether this is true before Easter.[28]

For the time being, he concentrated mainly on his popular-scientific presentation of the *World of Atoms*. Before the end of March, he had already completed several

---

[26] AEH to EBH (Vienna, March 20, 1926).

[27] AEH to EBH (Vienna, March 15, 1926).

[28] AEH to EBH (March 24, 1926).

chapters and in April 1926, he again signed a contract with De Gruyter & Co. for publication.

Regarding his dreams of America, Haas was aware that his lack of success resulted from the fact that he did not have the name recognition necessary for invitations from institutions and individuals in the United States. Apart from his initial work on the atomic model, he had made few new or groundbreaking contributions to physics and this first contribution had remained almost unknown in the shadow of Niels Bohr.[viii] Haas therefore planned to have his books published in English and tried to secure invitations to some well-known universities. He did not feel comfortable with his professional activity at the Academy; he had countless new ideas that he could not put into practice. And so, despite the futile efforts of his 1924 inquiries, he had new hopes and plans:

> Tonight I received a very gratifying letter from Professor Donnan. He writes, "I saw Professor Sommerfeld and spoke to him about getting a suitable university post for yourself. He seemed quite sympathetic and said he would do what he could […] I had written to my friend Prof. L. M. Dennis,[29] Prof. of Chemistry at Cornell University [Ithaca, N.Y.] suggesting that you should be invited to Cornell University to give a course of lectures. He has just written to say that this matter is now receiving the favorable consideration of his physical and mathematical colleagues. Let us hope that they will take some action!" If only something would come about! The office is at its most disgusting at this time of the year, with the holidays coming up.[30]

But Haas was also worried about his wife in Munich. She often lay ill and seemed to continue losing weight despite the cure. She was pregnant again and her loss of weight, when she merely weighed 47 kg, was a great worry for him, for her, and also her Munich doctors. Her main concern was for the child who might suffer because of the situation. In each of his daily letters, Haas asked about her health and weight. She hardly ever answered and when she did, she avoided his urgent questions. As one letter detailed, he was planning to visit her for Easter:

> My dearest heart!
>
> Thank you very much for your dear letter, which arrived this morning. It is quite long, but unfortunately you do not write a word about how you are doing, what you are doing, whether you are putting on weight, and that worries me very much.
>
> Due to my private conspiracy agreement with Dr. Junk, I will secretly skip the academy on Good Friday and Holy Saturday and therefore arrive in Munich early on Good Friday. But on Easter Monday evening, as I am only unofficially absent, I must leave Munich again, as I absolutely must be back in office on Tuesday morning. Otherwise the whole business will come out, which I could only arrange because the Secretary-General is leaving tomorrow and I will just call in sick on Good Friday.[31]

[29] Louis Monroe Dennis (1863—1936) was professor of inorganic chemistry at Cornell University. He originally worked in the field of gas chemistry but, after World War I, he extended his studies to the field of metal chemistry, working in particular with his students on the chemistry of germanium and gallium. Dennis was one of the founders of the Baker Laboratory in the Department of Chemistry, which he successfully directed for thirty years.

[30] AEH to EBH (Vienna, March 19, 1926).

[31] AEH to EBH (Vienna, March 26, 1926).

Fig. 12.3 Sattendorf at the Ossiacher See in Carinthia on a contemporary postcard

As the letter explained, the trip was only possible because his colleague Junk was privy to the secretive absence, indicating that at that time the two actuaries were still on good personal terms.

Even before Haas' departure, however, Emma wrote that the doctors had diagnosed gallstone inflammation and recommended a stay in hospital. Yet, Emma Haas refused to spend 25 marks on daily-care costs despite severe pain. She stayed in Munich, under Marie's care, to fight pain and stones with hot compresses and medication before returning to Vienna.

In July 1926, the family was able to go on a summer vacation for the first time, escaping the oppressive sultriness of Vienna. Since Emma Haas was pregnant and still had to recover from her biliary colic, they chose Sattendorf in Carinthia as a quiet retreat and Arthur's sister Grete found them a small place to stay.

Emma traveled from Vienna to Sattendorf (Fig. 12.3) as early as July 5th, but Arthur had to spend another week in hot Vienna because of his official business. So, he spent the days working, mainly finishing *World of Atoms* but also turning to new topics.

Haas went to his office at the Academy in the morning and to the Physics Institute in the afternoon. In addition, he often went to the Radium Institute (Fig. 12.4) of Stefan Meyer and Karl Przibram to discuss projects and to talk about ideas regarding Compton scattering and radioactive decay, both topics discussed in detail in the book. Sometimes these discussions forced him to rethink his thesis, as he writes:

> The discussions I had at the Radium Institute led me to delay the publication that should have been presented to the Academy the day after tomorrow.[32]

[32] AEH to EBH (Vienna, July 6, 1926).

Fig. 12.4 The Radium Institute of the University of Vienna, established in 1910

But finally on July 10th, Arthur Haas also left Vienna, traveling to Sattendorf. The family spent peaceful days there but soon Arthur had to return because, unfortunately, there was a lot to do at the Academy. At the institute he found a letter dated June 29, 1926 from Max Born, to whom he had previously written for help.

> Dear colleague!
>
> Thank you very much for your kind letter. Perhaps it will indeed be possible for me to fulfill your wish and do something for you in America. I myself have received several invitations for permanent relocation during my trip to America, but I have now decided to decline them all. Perhaps I can take this opportunity to recommend you.
>
> With best regards, your very devoted M. Born

This letter from Max Born put Haas back into a heavenly, highly jubilant mood, full of hope about entering the land of dreams:

> Now I believe that my chances of going to the USA permanently are already greater than 50%. [...] I now confidently believe that in a year at the most I will be permanently liberated from all local existence.[33]

Notes

i. Stefan Sienell, *Das Verwaltungs- und Dienstpersonal der Akademie der Wissenschaften 1847 bis 1960. Eine Sozialgeschichte* (Wien: Austrian Academy of Sciences, 2019).

ii. Robert E. Atkinson, Fritz G. Houtermans, "Zur Frage der Aufbaumöglichkeit der Elemente in Sternen," *Zeitschrift für Physik* 54 (1929): 656–665.

iii. E. Margaret Burbidge, Geoffrey R. Burbidge, William A. Fowler, Fred Hoyle, "Synthesis of the Elements in Stars," *Review of Modern Physics* 29, 4 (1957): 547–650.

iv. See Klaus Taschwer, "Wissenschaft für Viele," (dissertation, University of Vienna, 2002), 225–242.

v. Robert Reiniger, ed., *50 Jahre Philosophische Gesellschaft an der Universität Wien 1888—1938* (Wien: Philos. Ges., 1938).

vi. Ibid., 35–38 and J. Blackmore, R. Itagaki, S. Tanaka, eds., *Ernst Mach's Vienna 1895—1930—Or Phenomenalism as Philosophy of Science* (Dordrecht: Kluwer Academic Publishers, 2001), 293–296.

vii. Constance Reid, *Hilbert Courant* (New York: Springer Verlag, 1986), 326–327.

viii. See Armin Hermann, *Arthur Erich Haas.*

[33] AEH to EBH (Vienna, July 7, 1926).

Chapter 13
The Institute of International Education (IIE)

Haas had, however, been full of anticipation and hope before. As a result of the many disappointments, he had experienced over the past two years in his efforts to secure a permanent position at an American university through recommendations, he also developed alternative avenues for reaching his goal.

The Institute of International Education (IIE) was known to Haas through Professor Paul Leo Dengler, who had founded the Austro-American Institute (AAIE)[1] in Vienna in 1926. The IIE was an American foundation established in 1919 after the war to create a better understanding and relationship between former enemy countries through the exchange of students and scholars. The Institute financed and organized lecture tours in the US for well-known Central European scientists and also awarded scholarships to talented students for temporary research positions at American institutions. However, it was also a program with the aim of attracting promising young scientists from Europe to America—a selection mechanism, as it were, for American universities. In any case, it was a program from which both sides benefited, as many talented young scientists from impoverished Europe were given the chance to build a new career and it also provided a first step for the United States, a means for bringing the mostly backward science systems of smaller universities and colleges up to the standards of the young twentieth century.

In a letter dated August 17, 1926, Dengler recommended Haas for two lecture topics, "Objective and Human Physics" and "The Atom as Source of Energy" and negotiated a scholarship for an extended lecture tour in the spring of 1927. Dengler also agreed to help Haas organize and coordinate the trip. The letter from Max Born had, however, put Haas back into such a hopeful mood that he now thought he could leave sooner. Yet, the fact remained that he was largely unknown in the United States and the invitations from the various US universities that went to Max Born were not forwarded to Arthur Haas. So, Haas returned to Sattendorf in August and spent the rest of his summer vacation there with his family.

---

[1] The AAIE was founded in 1926 as an Austrian cultural institute by the Austrian pedagogue Prof. Paul Leo Dengler. The task of the non-profit association was and is cultural exchange between Austria and the United States, especially in the educational and cultural fields.

M. Wiescher, *Arthur E. Haas—The Hidden Pioneer of Quantum Mechanics*, Springer Biographies, https://doi.org/10.1007/978-3-030-80606-4_13

Back in Vienna on September 7, 1926, a second son Georg Arthur Haas was born, whereupon the Ministry of Education increased the child allowance for II Actuary Haas by a further 60 shillings annually.

New additional income also came in autumn with publication of *The World of Atoms,* which Haas had completed over the summer. The book's first edition of 3,000 copies was published for the contractually set retail price of 4 to 6 ½ reichsmark and Haas made a profit of 1,400 to 1,500 reichsmark. In addition to thirty free copies, this corresponded to an author's share of approximately 12%.

And again there were brilliant reviews, not only in the professional world but also in journals that otherwise had little concern for the developments in modern atomic physics; there were reviews from the *Zeitschrift für Feinmechanik* to the *Zeitschrift für den mathemathischen Unterricht*, and included the *Blätter für Bücherfreunde*, of the *Ars Medici* and the *Journal of the Verein deutscher Eisenbahnverwaltungen*, the *Pharmazeutischen Monatshefte* and the *Annalen der Philosophie*, as well as the geography journal *Die Erde* and the geology and mining journal *Kali, Erz und Kohle*. When general newspapers in the German-speaking world took up the subject, from the daily newspaper *Tageblatt* in Brno to the *Kölnische Volkszeitung*, the actual specialist journals did not want to be left behind: the *Physikalische Berichte* as well as the *Physikalische Zeitschrift*, and the *Zeitschrift für angewandte Chemie* up to the *Zeitschrift für angewandte Physik* were all united in their praise of the work. *The World of Atoms* was considered an excellent example of an efficient popular-science approach. Further editions followed and, at Haas' insistence, contacts with American publishing houses were initiated, which led to a contract in 1927 with Van Nostrand Company, a small publishing house for scientific journals and books in New York.

The writing of this small book led to new thoughts and plans, which Haas wanted to translate into new books and also into reports for the Academy. He wanted to be a part of the scientific breakthroughs and discoveries that were occurring. He was bubbling over with ideas but family and the Academy demanded their time.

Autumn was a quieter time of year. The Haas family on Baumgartenrstraße was mainly busy with their two small children and with rearranging the house after the restless summer months; a housekeeper and maid were employed, as the income and economic situation had significantly improved.

In November 1926, Haas received an invitation from the Saxon Academy of Sciences to participate in the preparation of the next edition of the *Poggendorff Handbook* in Leipzig. His experience as a former editor had been highlighted as a special reason for the invitation. Haas asked the Secretary General of the Academy, Privy Councilor Becke for permission to take an eight-day business trip to Leipzig and Berlin in the beginning of December, which was granted.

From 1914 to 1919 in Leipzig, Haas had participated in work on the fifth volume of the *Handbook*, which had finally been completed in 1926. Now, Haas was to take part in this meeting regarding procedures for the sixth volume, which was to cover the period from 1923 to 1932. In Berlin, he was invited to give lectures for the local Urania Society. Haas also planned to take this opportunity to dissolve his de forma unpaid employment with the University of Leipzig. He now had plans other than to return to the editor's task of the *Poggendorff Handbook*—at the end of November, at

Paul Dengler's renewed request, the IIE had officially invited Haas to give a series of lectures in the United States.

So at the beginning of December, Haas travelled to Berlin and Leipzig; his wife wanted to accompany him but he declined for reasons of cost. He had invitations to give two lectures at the Berlin Urania on the topics "The Atom as Energy Source" and "Objective and Human Physics," so this was a dress rehearsal in German for the upcoming American tour. Besides these lectures, Haas also conducted negotiations with representatives of the UFA[2] for one of his long-planned and hoped-for projects at the Urania, the development of documentary films on scientific topics. With De Gruyter, he negotiated new editions of his very successful books in order to create more favorable conditions for himself. Everything seemed to be working out and it looked as though orders for new and reissued books would secure him an income for the decades ahead.

Overwhelmed by his own success, he wrote enthusiastic letters home. Even though two years earlier Haas had so hated Berlin, his current experiences seemed to give him new energy and the courage to face life.

> I am perhaps the biggest idiot not to have shaken the Austrian dust off my feet long ago. Here, I see that Vienna is a paradise for spending money but a hell for a capable person who wants to achieve something.[3]

In Leipzig he visited his former colleagues and was warmly received. He had invitations for lunch, coffee and cake, and dinner. Including his two patrons Professor Wiener and Privy Councilor Sudhoff, all came together to meet with Haas. No mention was made of his aversion to German nationalists and warmongers; for the sake of peace, friendships seemed to have finally been restored after ten years. Haas gave a colloquium and conducted further negotiations with the Akademieverlag and his old friend Jolowicz for the publication of a mathematics book.

The response from Vienna to the detailed descriptions of his experiences, successes, and plans was cool. Emma wrote:

> I sincerely congratulate you on your success and I am glad that you have been relieved of worries that your journey might bear a deficit. [...] You are successful, you see other people, other areas, you are invited, you can exchange your thoughts with others, in a word, you can live as every human being is entitled to live.
>
> On the other hand, I feel as if I am completely outcast from life. [...] But you will never understand the sacrifices I have made during the last two years - because you have neither the time nor the interest. The Berlin trip would have done me so much good, even if you wouldn't have had much time for me. It would have been a break from this terrible loneliness and monotony. I would have returned home a new and satisfied person. But for some reason you didn't want to, because I can't believe that a possible deficit in travel expenses was so problematic that you would rather see your wife ruined than have a deficit in your travel expenses. [...] I have decided to go to Munich on Monday, December 13th and stay there for eight days. I must have a little change. But I would have preferred to have it with you than alone.[4]

---

[2] The UFA (Universum-Film AG) in Berlin was the leading German film production company at the time. Many UFA stars of the 1920s and early 1930s later emigrated to the United States and had a second career in Hollywood.

[3] AEH to EBH (Berlin, December 6, 1926).

[4] EBH to AEH (Vienna, December 6, 1926).

The answer again shows how strongly Arthur Haas was rooted in patriarchal thinking; he did not understand the arguments that Bice put forward in her much more detailed letter. When he replied three days later by express letter from the Park Hotel in Leipzig, his reaction was that of an offended and wounded man:

> This afternoon I received your letter of Monday evening, which of course I was extremely upset about. Must you really write such absurd letters? [...] As if the anguish of this journey, which I undertook only for the material and idealistic benefit of my family, had anything to do with pleasure. But I did not expect that you would stab me in the back with such a letter during my terribly difficult struggles. I confidently hope that I will find a sensible woman on Saturday after my return. If you were really to go to Munich the day after my return, I would not be able to forgive you. So be reasonable![5]

The rest of this letter was again dedicated to his success stories about the lecture in Leipzig, his successful negotiations for his planed book on higher mathematics, and especially the fact that he had been proposed as a candidate for the chair of theoretical physics at the University of Königsberg.[6] From Leipzig he returned to Berlin, to continue negotiations with the UFA. Her follow-up letter was even cooler:

> If you can, arrive Sunday at noon, because I'm leaving on the 11 o'clock train on Monday.[7]

We don't really know if Emma Beatrice went to Munich. No personal letters have been found from the following weeks, only official letters and requests that were made by Arthur Haas on December 13th, as explained below. It is reasonable to assume that she went to Munich for a week because after that the relationship changed. The change was not so much in their warm and deep relationship with each other, which remained, but in the freedom with which Emma shaped her life through travel and study over the following years. She formed herself into an independent woman and personality whose expectations had already been made clear in her earlier letters.

Arthur Haas remained a physicist whose ambition was to make his contribution to science, to be recognized, and to achieve his highest goal of living in his dream country of the United States. The first step had been made with the IIE's promise to finance his lecture tour.

[5] AEH to EBH (Leipzig, December 9, 1926).

[6] He must have been suggested as the successor of Paul Volkmann (1856—1938) who retired in 1924. It did not work out; Königsberg decided to make the offer to Richard Gans (1880–1954) from the National University of La Plata in Argentina, where he had been the successor of the late Emil Bose (formerly of Göttingen).

[7] EBH to AEH (Vienna, December 9, 1926).

Chapter 14
Lecture Tour in the USA

The program for the American trip was in the hands of Mrs. Waite who, as Executive Secretary of the IIE in New York, was the decisive force in the organization. She had put together a two-month travel and lecture program for Haas, during which he was to visit various universities in the USA.[1] His first stop in Baltimore was planned for the second week in February 1927, with arrival on February 13th. On February 15th, two lectures at John Hopkins University on the two proposed topics, "Objective and Human Physics" and "The Atom as Source of Energy," were scheduled. On February 17th, a lecture on the atomic energy source was planned at the National Academy in Washington D.C. On Friday, February 18th a lecture on "Objective and Human Physics" at Columbia University was to be followed by a stay of several days in New York. On February 23rd, there was a lecture at the Brooklyn Polytechnic Institute and on the following day the longed-for lecture at Princeton on the topic of nuclear energy sources was also scheduled.

A week later he was to speak on this same topic at Yale in New Haven, at Wesleyan University in Middletown, Connecticut, at Brown University in Rhode Island, and at Wellesley College in Massachusetts where he would finally be able to stay, from February 4th to 8th, as a guest of the Massachusetts Institute of Technology. By way of well-known Smith College for female students, Haas was then to travel to Syracuse University in New York, from there to the research center of the General Electric Company in Schenectady, and on to Cornell University where he was to stay for two days, from March 15th to 16th.

From the East Coast, Haas was to travel to the University of Toronto in Canada. He was then scheduled on a train from Toronto to Ann Arbor and the University of Michigan, and on Sunday, March 20th, he was going to Chicago. But already on March 22nd, a lecture for doctoral students was planned at Iowa State University and two days later at the University of Wisconsin, followed by his return to Iowa State University. From there it was planned that Haas go back, via Chicago and Pittsburg, to give lectures at Holyoke and Swarthmore Colleges. The return trip to Vienna was

[1] Mary Waite of the IEE, "Itinerary of Dr. Haas," (New York, December 29, 1926).

M. Wiescher, *Arthur E. Haas—The Hidden Pioneer of Quantum Mechanics*, Springer Biographies, https://doi.org/10.1007/978-3-030-80606-4_14

scheduled for April 20th. It was an exhausting travel plan, on long railway lines in Pullman sleeping-car trains, and Haas was enthusiastically looking forward to the trip. At last his hopes were fulfilled and he was traveling to America, the land of his dreams.

The request for a three-month leave of absence (from January 26th to April 26th), which Haas submitted on December 13, 1926, had already been approved on December 17th. His obligations were to be taken over by I Actuary Junk as well as two academy employees (Fig. 14.1), Ms. Calabek and Ms. Neumann who in return would receive a temporary increase in monthly salary. Official approval was then given on January 8th, in which the Secretary General also stated that the Academy would take over the costs for the burden of the additional employees. This underlines the fact that the Academy was positively disposed toward the scientific work that Haas was doing and perhaps even more so toward the international scientific recognition that he had received, as this could also be seen as appreciation for the Academy itself.

Haas thus left behind orderly working conditions and also orderly plans for the family; his wife and two children were to remain in Vienna supported by a maid and housekeeper. With this trip he did, however, make provisions to limit unhappiness! Since Emma Beatrice had complained that she was all alone during his stay in Berlin and had not received any visits or invitations from friends and family, it was now more the opposite. The visits did not stop and there was hardly any peace in which to reflect on being left behind in Vienna. The relationship between the couple actually seems to have become closer, as can be seen from the intimacy of their letters.

Arthur Haas left Vienna in the last days of January, 1927, travelling by train to Hamburg where he embarked for New York on February 1st on the HAPAG passenger ship *Hamburg*, newly built in 1925 and pictured in Fig. 14.2.

The crossing was extremely stormy. The German-language *New York State Newspaper* wrote on February 10, 1927:

> As Captain Schwamberger reported, the *Hamburg* had its worst voyage since it was put into service, on the just completed crossing. Hurricanes of wind force 12 were constantly on the agenda, and on February 2 and 3 [the ship] had to be turned for seven hours each time, because the violence of the storm could not be countered. The overturning mountains of water as high as a house smashed several cabin windows and a first-class passenger, Mr. Udo Bintz from Berlin, was caught by a wave and thrown onto the canopy.[1]

As the New York newspapers dramatically reported, Bintz had to be treated by the ship's doctor because of his numerous injuries; the ship itself arrived two days late because of these storms. These were probably terrible days for Arthur Haas but as he reported in a telegram from New York to Vienna after his arrival, he had survived the turbulence without seasickness. Obviously, nothing could dampen his anticipation of America.

The journey was indeed an enormous success. Unfortunately, none of his almost fifty letters written during this trip to America are available so we are not able to know his initial, personal impressions of the country of his dreams. Fortunately, Emma's numerous replies to the different stages of his trip, in which she reflects on

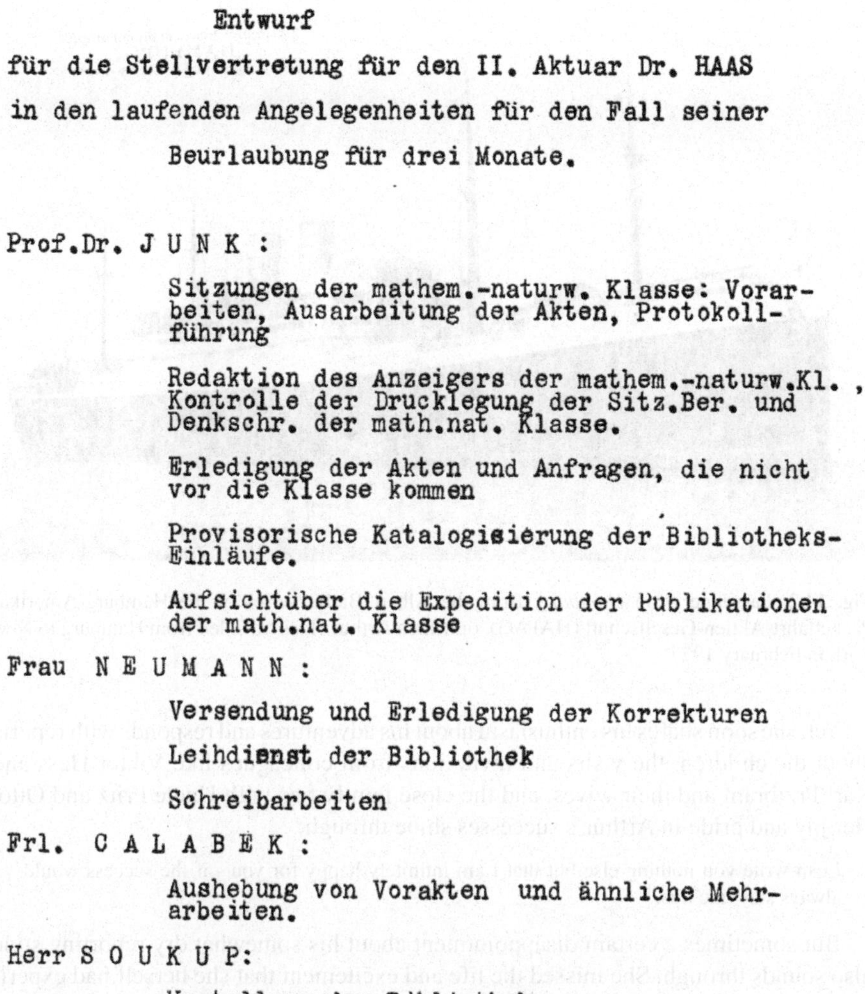

Entwurf

für die Stellvertretung für den II. Aktuar Dr. HAAS in den laufenden Angelegenheiten für den Fall seiner Beurlaubung für drei Monate.

Prof.Dr. J U N K :

Sitzungen der mathem.-naturw. Klasse: Vorarbeiten, Ausarbeitung der Akten, Protokollführung

Redaktion des Anzeigers der mathem.-naturw.Kl. , Kontrolle der Drucklegung der Sitz.Ber. und Denkschr. der math.nat. Klasse.

Erledigung der Akten und Anfragen, die nicht vor die Klasse kommen

Provisorische Katalogisierung der Bibliotheks-Einläufe.

Aufsichtüber die Expedition der Publikationen der math.nat. Klasse

Frau N E U M A N N :

Versendung und Erledigung der Korrekturen

Leihdienst der Bibliothek

Schreibarbeiten

Frl. C A L A B E K :

Aushebung von Vorakten und ähnliche Mehrarbeiten.

Herr S O U K U P:

Umstellung der Bibliothek.

Fig. 14.1 Distribution of tasks at the Academy for the period of Arthur Haas' three-month absence, detailing the extent of his usual work load as II Actuary (Personnel files, Arthur Haas of the Vienna Academy of Sciences)

his experiences, clearly demonstrate that he had reported enthusiastically about his successes, but her letters contain less about his impressions. She does write about her own impressions at the beginning of his absence:

> I am desperately trying not to think about you staying away for three months, or that you have left me at all. […] As often as I go into the study, I miss you then the most because now everything remains so nicely in order. I just hope that time will pass quickly.[2]

[2] EBH to AEH (Vienna, January 27, 1927).

Fig. 14.2 Passenger ship *Hamburg,* from the Albert Ballin series of the Hamburg-Amerika-Packetfahrt-Aktien-Gesellschaft (HAPAG), on which Arthur Haas travelled from Hamburg to New York in February 1927

Yet, she soon shares his enthusiasm about his adventures and responds with reports about the children, the visits and invitations from colleagues like Viktor Hess and Karl Przibram and their wives, and the close family ties with Uncle Fritz and Otto. Her joy and pride in Arthur's successes shine through:

> I can write you nothing else but that I am infinitely happy for you - if the success would always stay like this.[3]

But sometimes a certain disappointment about his somewhat dry reporting style also sounds through. She missed the life and excitement that she herself had experienced so strongly a decade earlier and she also reflected on, perhaps in connection with her own memories, how much she had influenced him with her reports about America:

> If your letters were a little more exciting in some ways, I wouldn't mind. […] That's why it hurt me a little today when I read your letters and they were so objective.[4]

At the beginning of March, Bice went to Berchtesgarden for a week with her father on a mountain hike and then stayed in Munich for a few days. Uncle Fritz and the housekeeper took care of the children. Her sister-in-law Maria Huber returned with Emma to Vienna, to personally get to know and love the children. She is then replaced by a visit from sister-in-law Grete Gerstenberger of Sattendorf. So it was a time of

[3] EBH to AEH (Vienna, February 22, 1927).

[4] EBH to AEH (Vienna, March 2, 1927).

comings and goings in the Haas house at Baumgartenstraße 58, an atmosphere that Emma sometimes felt was too much for her but through which much closer family ties were built than had been previously possible:

> Otto and I have become very good friends. I must tell you a lot about it. [...] Uncle Fritz and I are also the very best of friends. Last time he called me his treasure. I'll be back for lunch with him on Sunday.[5]

For Emma Haas, the three months in Vienna passed more quickly than initially feared.

For Arthur Haas, time resembled a whirlwind with lectures and railway journeys all over the East Coast and deep into the Midwest. He gave great, well attended public and academic lectures at the various, most-often already famous American universities. His lectures were also well documented in often exuberant press reports (Fig. 14.3). Hearing that a Viennese professor was to lecture about "the atom as an energy source" was a local sensation that attracted many listeners from the public. His lectures on "Subjective and Objective Physics" were less sought after by the audiences and so as his journey progressed, Haas limited himself more and more to discussion of the atom as a unique energy source.

Haas' statement, "Earth [is] to last for untold Ages," although incorrect from today's point of view, reassured people. As the *Post Standard* in Syracuse writes: "Nervous need have no fear for quintillions of years, Dr. Haas assures."[ii] *The Chicago Herald Examiner*, with a bold headline on March 30, 1927, points out the solution to all energy problems:

> World-Running Power sought in Atom's Frozen Energy—Harnessing of tides and sun is aim. [...] One gram enough to heat every building in Chicago would supplant coal and oil, but Haas also points to future alternatives: What other sources of power will man use in the future? We may expect him to harness the energy of the sun, of the tides, of the winds and of the heat of the earth's interior. The winds will be an especially favorable source of power.[iii]

This was an interesting statement for 1927, at a time when the last sailing ships disappeared from the oceans and the last windmills were replaced by coal-fired power stations.

Due to his successful lecture activities, the American journal *Science* approached Haas with a request to publish his two lecture topics as articles. He was happy to do so and in November 1927 the first article on "Objectives and Human Physics" was published.[6] This was probably the last work in which he summarized his scientific-historical and philosophical views, from then on devoting himself entirely to modern physics.

The *Science* article starts with Democritus and discusses the differences in the way that nature is understood, when considering it either from the point of view of the observer or from the knowledge of the thinker. If Democritus saw the concept of an atomic world as a necessity that resulted from his conclusions, then he needed means other than by the observations made in daily life. Therefore, such a conclusion

[5] EBH to AEH (Vienna, April 1, 1927).

[6] Arthur Haas, "Objectives and Human Physics," *Science* LXVI (1927): 463–469.

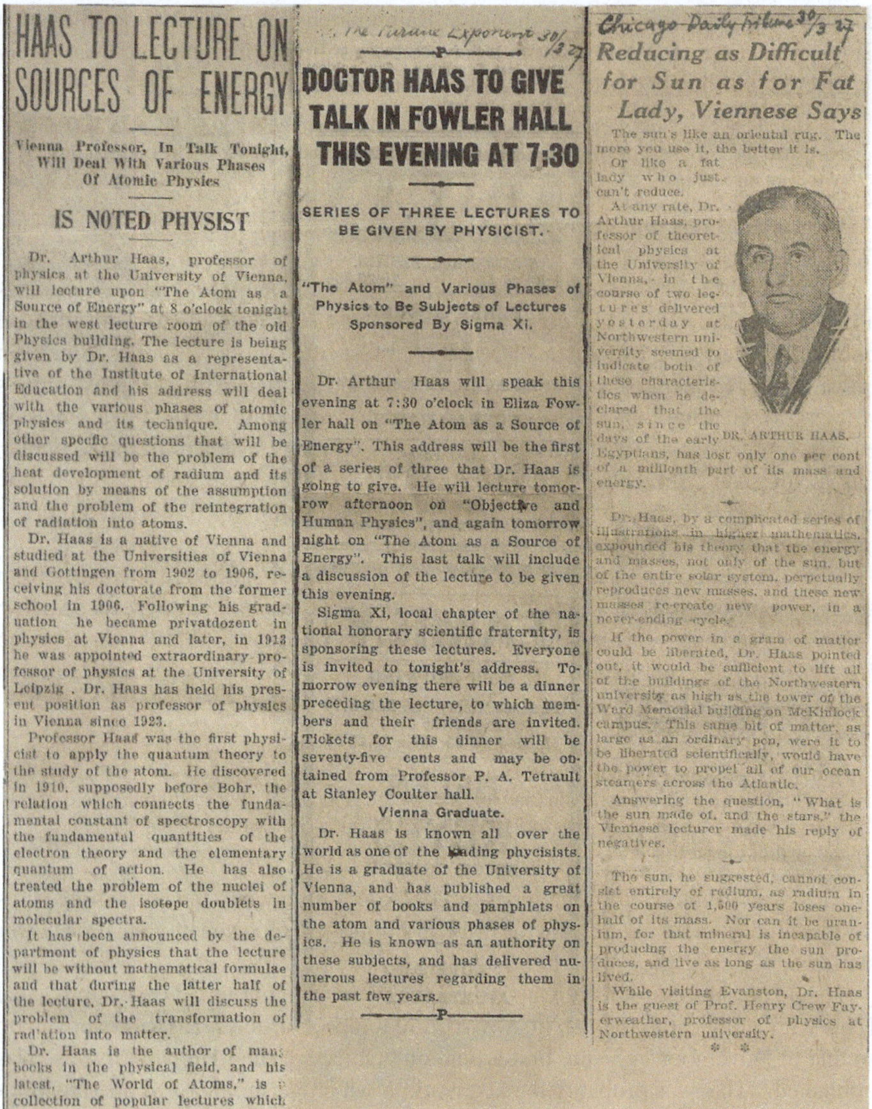

HAAS TO LECTURE ON SOURCES OF ENERGY

Vienna Professor, In Talk Tonight, Will Deal With Various Phases Of Atomic Physics

IS NOTED PHYSIST

Dr. Arthur Haas, professor of physics at the University of Vienna, will lecture upon "The Atom as a Source of Energy" at 8 o'clock tonight in the west lecture room of the old Physics building. The lecture is being given by Dr. Haas as a representative of the Institute of International Education and his address will deal with the various phases of atomic physics and its technique. Among other specfic questions that will be discussed will be the problem of the heat development of radium and its solution by means of the assumption and the problem of the reintegration of radiation into atoms.

Dr. Haas is a native of Vienna and studied at the Universities of Vienna and Gottingen from 1902 to 1906, receiving his doctorate from the former school in 1906. Following his graduation he became privatdozent in physics at Vienna and later, in 1913 he was appointed extraordinary professor of physics at the University of Leipzig . Dr. Haas has held his present position as professor of physics in Vienna since 1923.

Professor Haas was the first physicist to apply the quantum theory to the study of the atom. He discovered in 1910, supposedly before Bohr, the relation which connects the fundamental constant of spectroscopy with the fundamental quantities of the electron theory and the elementary quantum of action. He has also treated the problem of the nuclei of atoms and the isotope doublets in molecular spectra.

It has been announced by the department of physics that the lecture will be without mathematical formulae and that during the latter half of the lecture, Dr. Haas will discuss the problem of the transformation of rad'ation into matter.

Dr. Haas is the author of many books in the physical field, and his latest, "The World of Atoms," is a collection of popular lectures which was published last summer.

The Purdue Exponent 30/3 27

DOCTOR HAAS TO GIVE TALK IN FOWLER HALL THIS EVENING AT 7:30

SERIES OF THREE LECTURES TO BE GIVEN BY PHYSICIST.

"The Atom" and Various Phases of Physics to Be Subjects of Lectures Sponsored By Sigma Xi.

Dr. Arthur Haas will speak this evening at 7:30 o'clock in Eliza Fowler hall on "The Atom as a Source of Energy". This address will be the first of a series of three that Dr. Haas is going to give. He will lecture tomorrow afternoon on "Objective and Human Physics", and again tomorrow night on "The Atom as a Source of Energy". This last talk will include a discussion of the lecture to be given this evening.

Sigma Xi, local chapter of the national honorary scientific fraternity, is sponsoring these lectures. Everyone is invited to tonight's address. Tomorrow evening there will be a dinner preceding the lecture, to which members and their friends are invited. Tickets for this dinner will be seventy-five cents and may be obtained from Professor P. A. Tetrault at Stanley Coulter hall.

Vienna Graduate.

Dr. Haas is known all over the world as one of the leading phycisists. He is a graduate of the University of Vienna, and has published a great number of books and pamphlets on the atom and various phases of physics. He is known as an authority on these subjects, and has delivered numerous lectures regarding them in the past few years.

Chicago Daily Tribune 30/3 27

Reducing as Difficult for Sun as for Fat Lady, Viennese Says

The sun's like an oriental rug. The more you use it, the better it is.

Or like a fat lady who just can't reduce.

At any rate, Dr. Arthur Haas, professor of theoretical physics at the University of Vienna, in the course of two lectures delivered yesterday at Northwestern university seemed to indicate both of these characteristics when he declared that the sun, since the days of the early Egyptians, has lost only one per cent of a millionth part of its mass and energy.

DR. ARTHUR HAAS.

Dr. Haas, by a complicated series of illustrations in higher mathematics, expounded his theory that the energy and masses, not only of the sun, but of the entire solar system, perpetually reproduces new masses, and these new masses re-create new power, in a never-ending cycle.

If the power in a gram of matter could be liberated, Dr. Haas pointed out, it would be sufficient to lift all of the buildings of the Northwestern university as high as the tower of the Ward Memorial building on McKinlock campus. This same bit of matter, as large as an ordinary pen, were it to be liberated scientifically, would have the power to propel all of our ocean steamers across the Atlantic.

Answering the question, "What is the sun made of, and the stars," the Viennese lecturer made his reply of negatives.

The sun, he suggested, cannot consist entirely of radium, as radium in the course of 1,500 years loses one-half of its mass. Nor can it be uranium, for that mineral is incapable of producing the energy the sun produces, and live as long as the sun has lived.

While visiting Evanston, Dr. Haas is the guest of Prof. Henry Crew Fayerweather, professor of physics at Northwestern university.

Fig. 14.3 Press reports on Arthur Haas' lectures (from left to right): University of Michigan in Ann Arbor; Purdue University in Lafayette, Indiana; and Northwestern University in Chicago, Illinois

required a considerable expansion beyond the means of observation, in order to attain the rather abstract concept of Democritus—for such an achievement the perception of the thinker was required. The problem poses a classical argument in the German world of thought, an argument that Friedrich Schiller had already formulated in his

Wallenstein: "It is the mind that builds the body."[7] And yet, this argument had been rejected by Ernst Mach who opposed the idea of the atom with the question, "Did you ever see one?".

Today, this question has been answered—or perhaps not, if we want to consider modern field theory as a continuation of Mach's claim that energy, rather than the atom is the basis of all things. However, Haas' argument maintained that it is not nature that is complicated, it is only our way of making it understandable that is complicated. He presented these points of view with many examples from the microscopic world, a world that, although it has always existed, has only today been made accessible and measurable by new means of observation.

As already mentioned, this lecture topic from the German intellectual world was less attractive to the American audience, whereas his second topic on the atom as an energy source caused a great stir and was reproduced in an article in *Science Monthly*.[8]

But what was the physics behind these spectacular promises of a powerful energy source? The physics itself came from a first interpretation of nuclear energy, which was based on available knowledge about radioactivity and its relation to atomic nuclei. This was not the nuclear energy of today's understanding, which results from two possible sources. Today we have the fusion of light nuclei to heavier nuclei, an idea that Haas actually supported in his discussion of nuclear fusion as a solar energy source as suggested by Arthur Eddington.[9] Today's second source for nuclear energy is the idea of fission, the splitting of heavy atomic nuclei under the release of fission energy, which results from the mass difference of the heavy nuclei and the two lighter fission products. The latter only became known as a possibility for energy production through the observation of nuclear fission by Otto Hahn and Fritz Strassmann in Berlin in 1938 and its interpretation as such by Haas' former fellow student in Vienna and, at that time, refugee in Stockholm, Lise Meitner and her nephew, the theoretical physicist Otto Frisch (1904–1979).[iv]

Arthur Haas based his proposals on observations of radioactivity that were being made in the first two decades of the twentieth century, which provided an important basis for the idea of the atom at that time. It was known that the atom consisted of a heavy, positively charged nucleus whose mass was almost equal to the mass of the whole atom and that, according to the Bohr model, electrons on well-defined quantum orbits circled the nucleus. This was also the theory that Haas advocated in his books on atomic theory. However, this theory did not yet include neutrons, which had been proposed by Ernest Rutherford as a thought model but were not experimentally proven until 1932 by his student James Chatwin. Without the knowledge of neutrons and their importance to the atomic nuclei, Haas' models remained speculation. But they were, none the less, correctly based on the observation of radioactivity as a heat source—the decay of long-lived radioactive nuclei emitting high-energy alpha particles that transfer their energy by scattering processes in the surrounding material.

---

[7] Friedrich Schiller's *Wallenstein'sDeath*, act 3, performance 13 of *Wallenstein* (1799).

[8] Arthur Haas, "The Atom as a Source of Energy," *Scientific Monthly* 26, 2 (1928): 140–146.

[9] Arthur S. Eddington, "The Internal Constitution of Stars," *Nature* 106, 2653 (1920): 14–20.

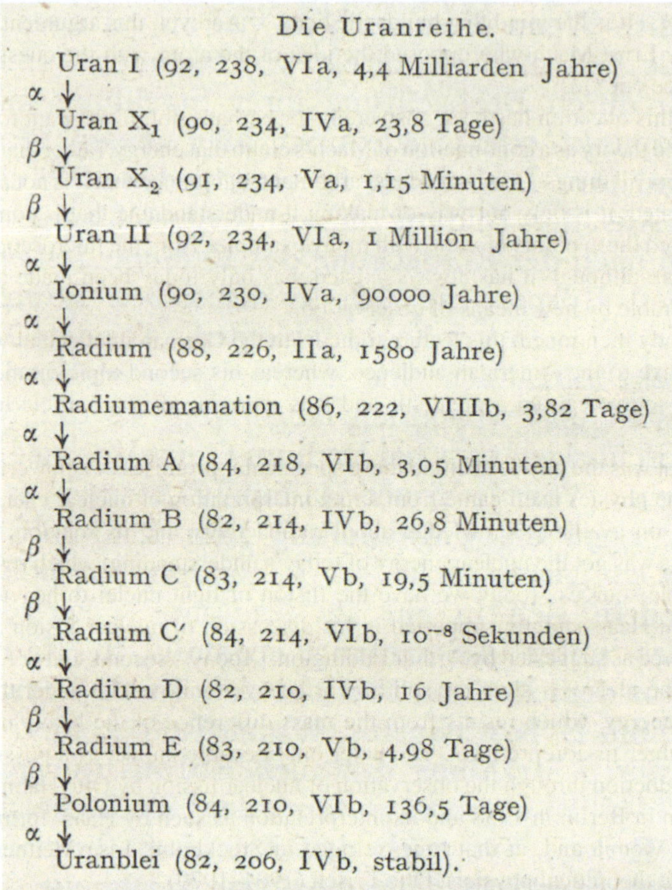

Die Uranreihe.

Uran I (92, 238, VIa, 4,4 Milliarden Jahre)
α ↓
Uran X_1 (90, 234, IVa, 23,8 Tage)
β ↓
Uran X_2 (91, 234, Va, 1,15 Minuten)
β ↓
Uran II (92, 234, VIa, 1 Million Jahre)
α ↓
Ionium (90, 230, IVa, 90000 Jahre)
α ↓
Radium (88, 226, IIa, 1580 Jahre)
α ↓
Radiumemanation (86, 222, VIIIb, 3,82 Tage)
α ↓
Radium A (84, 218, VIb, 3,05 Minuten)
α ↓
Radium B (82, 214, IVb, 26,8 Minuten)
β ↓
Radium C (83, 214, Vb, 19,5 Minuten)
β ↓
Radium C′ (84, 214, VIb, 10^{-8} Sekunden)
α ↓
Radium D (82, 210, IVb, 16 Jahre)
β ↓
Radium E (83, 210, Vb, 4,98 Tage)
β ↓
Polonium (84, 210, VIb, 136,5 Tage)
α ↓
Uranblei (82, 206, IVb, stabil).

Fig. 14.4 The decay series of uranium as presented by Arthur Haas in 1926. Most of the decay components were chemically unknown at the time and were therefore collectively referred to as daughter elements of uranium and radium. From Arthur Haas, *The World of Atoms* (Berlin-Leipzig: De Gruyter & Co., 1926)

Thus, a heat source stronger than coal and oil that originated from inside atomic nuclei was what Haas and many others had in mind.

Haas demonstrated this idea with the classic example of heat release resulting from the decay of radium, known from many experiments by Marie Curie and others. Radium is a decay product of the so-called uranium series pictured in its interpretation at that time in Fig. 14.4, a sequence of alpha and beta decays following the decay of the long-lived uranium-238 atom with a half-life of 4.4 trillion years.[10] This radioactive decay chain acts, among other things, as the inner energy source of the earth, which,

[10] The half-life is the time over which a radioactive element decays to 50% of its original quantity. Today, 4.5 million years after the formation of our planetary system, the original amount of natural

as we know today, is responsible for a large part of the earth's internal heat and thus for a series of dynamic geological processes such as the convection of the earth's mantle and the slow drifting of the continents.[11]

Enormous energies were rightly suspected in these decay chains, but how they could be released, apart from natural decay, was still largely a matter of speculation. Arthur Haas had already questioned the origin of the decay chains in earlier works: "Is it really the heavy actinide nuclei, such as U-238 and Th-232, or are these themselves only products of heavier elements?" This was a thought that Haas pursued in 1912 in his philosophical work on the infinity of the world.[12] He had proposed that even though these ultra-heavy elements or trans-uranium[13] elements may have long since decayed, their artificial production would promise a perhaps greater source of energy than the faint reflection of radium generated as one of the last decay products on the way to stable lead.

Indeed, the idea of superheavy elements for energy production was widely popular, as demonstrated by the large number of attendees at Haas' lectures on the subject. This train of thought has been pursued in popular scientific-utopian literature—now known as science fiction. In particular, the genre frequently took up this idea of an enormous, germinal energy source. It was clear that in order to produce energy, energy must be invested and so in the corresponding books, secret machines were constructed by reclusive scholars, which could be used for the good but often also for the ruin of mankind. Dr. Frankenstein, in the romantic novel by Mary Shelley, was one of the forefathers of these often fanatical, fictional scholars, many of whom wanted good and created evil; they became the symbol of utopian novels by Jules Verne (1828–1905) in France and Herbert George Wells (1866–1946) in England. Thus, these characters also became a symbol for and representative of modern science in popular German futuristic literature of the early twentieth century.[v]

uranium-238 has thus been reduced to half, making it a weaker source of the geothermal energy that is being generated inside the planet.

[11] The continental drift had been postulated in 1912 by the German geologist and meteorologist Alfred Wegener (1880–1930) but was not generally accepted by the scientific community until fifty years later. While Wegener himself still suspected external forces—such as winds caused by the Earth's rotation or the gravitational pull of the Moon—to be the reason for the slow shift of the continents against each other, today we know that it is the internal slow convection of the Earth's mantle, driven by the decay heat of radioactivity, that moves the continents apart and against each other along their fault lines.

[12] Arthur Erich Haas, "Ist die Welt in Raum und Zeit unendlich?" *Archiv für Systematische Philosophie* 18 (1912): 167–184.

[13] Modern theoretical calculations regarding the stability of heavy atomic nuclei do indeed confirm the possibility of such superheavy elements. If found, they would not only provide important insights into the structure of nuclear matter but also contribute greatly to the understanding of element synthesis in stars and stellar explosions. The search for such elements is therefore still today an important research area in nuclear physics, in which especially the work at the German Helmholtz Gesellschaft für Schwerionenforschung GSI in Darmstadt plays a leading role worldwide, next to Dubna in Russia and Berkeley in the United States.

On the one hand, these works warned against the excesses of a technically dominated future, as in the well-known novel *Metropolis,*[14] which was later successfully made into a film. On the other hand, in the dynamically rising German society of the twentieth century that was being propelled by pragmatic engineers, ideas of technical possibility and growth were the driving forces of the future. This explains the success of the utopian novels by the young engineer Hans Dominik in these years before the Second World War, novels that were largely motivated by the idea of new inventions—powerful machines for energy production and weapons technology. Dominik also took up the idea of superheavy elements as a future energy source in his novel *Atomic Weight 500,*[15] which popularized the mysterious energy inside the nucleus. Figure 14.5 shows the cover of this popular novel.

Haas met the mood of the time with his popular scientific writings and his examples of enormous energy sources from atomic nuclei that could propel ocean liners across the seas and lift whole cities into the sky. His writings and numerous lectures on this subject reached audiences not only in Germany but also in the great young industrial power of the United States, where precisely such examples caused a sensation and found echoes in press reports. He was one of the most important spokespersons for the popularization of Einstein's formula $\boldsymbol{E = m \cdot c^2}$—mass is equivalent to energy and converting a small mass of matter into energy would change the future of mankind for the better. As Haas wrote in *Scientific Monthly*[16]:

> This is, indeed, an inconceivably great amount of energy. The quantity of energy, which is hidden in the mass of a single gram, would be more than enough to lift a weight as heavy as all the buildings in New York City put together to the height of the highest skyscraper. Or, to give another example, this amount of energy would also be sufficient to raise the temperature of all the rooms of all the houses of New York City by several degrees. If we possessed some means by which we could extract from matter its total energy, then, instead of thousands of tons of coal or oil, one gram of matter would be enough to propel the biggest of all ocean-going liners across the Atlantic.[17]

Mathematically this was correct but even with today's nuclear methods, only a small portion of the energy source promoted by Haas can be used. Haas proposed two methods for retrieving this energy. One method was the conversion of the decay heat of long-lived radioactive isotopes into energy. This is a method that is used today in the form of atomic batteries. However, since the energy produced is proportional to the radioactivity stored in the battery, potential radiation dangers set limits for the power generation of such batteries. In 1924, the idea of radiation protection and safety was not yet widespread, although the dangers of a large dose from the use of X-rays were already well known. Today, this energy source is mostly used for low-power nuclear batteries in the operation of microelectronics. However, large plutonium batteries are

[14] Thea von Harbou, *Metropolis* (Berlin: August Scherl Verlag, 1926).

[15] Hans Dominik, *Atomgewicht 500* (Berlin: August Scherl Verlag, 1935).

[16] *Scientific Monthly* was a scientific magazine published by the American Association for the Advancement of Science (AAAS) from 1915 to 1957. In 1958, *Scientific Monthly* was incorporated into the AAAS' journal *Science.*

[17] Arthur Haas, "The Atom as Source of Energy," *Scientific Monthly* (February 1928): 140–146.

Fig. 14.5 Cover picture of Hans Dominik's novel *Atomic Weight 500*, demonstrating the forces that were assumed to be necessary when forging heavy elements together. It was correctly thought that fusion could only be achieved by using high-voltage electrical fields

used in space travel, especially to maintain energy supply during long interplanetary journeys. A well-known example is the use of decay energy in the operation of the Mars rover *Curiosity*.[vi]

However, Haas also discussed another energy source, namely the fusion of hydrogen into helium. This is an idea that he had already formulated in his book

The World of Atoms and he then brought it to America in his lecture series. Here he had the sun as a daily shining example of efficient energy production:

> For example, the atomic weight of helium at 4.00 is about 0.03 less than the multiple of the atomic weight of hydrogen (1.008: based on oxygen=16). It follows that the synthesis of the helium atom on four hydrogen atoms would release an amount of energy whose mass equivalent is three percent of the mass of the hydrogen atom (i.e., 45 millionths of a yield) [...] A synthesis of one gram of helium from one gram of hydrogen would then generate as much heat as the combustion of a twenty-ton wagonload of coal.[18]

This was the birth of the now almost 100-year-old dream of hydrogen as an inexhaustible energy source. Arthur Eddington was the first to identify hydrogen as a possible source of energy for the life of the sun and other fixed stars and Arthur Haas promoted its application for human energy needs. But it was his Austrian compatriot Edwin Salpeter,[19] following in the footsteps of his mentor Hans Bethe at Cornell University in the 1950s, who succeeded in deciphering the complex reaction mechanism and enabling a reliable calculation of the reaction.[20]

It is the reaction sequence of the so-called pp-chains that makes this fusion possible. This sequence cannot take place in fusion reactors since the first step, the fusion of two hydrogen atoms, is infinitely slow—thus guaranteeing the long life of the sun—but other reactions of these chains could be used for energy production in the fusion reactor. However, due to the enormous technical challenges, this has not yet been achieved.

Even so, optimistic theoretical physicist Haas considered these technical difficulties to be manageable in principle and continued to describe his vision of the energy production of the future.

On April 20th, Arthur Haas returned home from the land of hope. His admiration for America and his dream to live there had only increased. In the preface of the 1928 American edition of *The World of Atoms,*[21] he described his impressions of the "imposing laboratories" and the "great and ardent" spirit of American scientists.

But now he had returned from the land of unlimited possibilities to his small country of Austria, which was being shaken by political and social turmoil fueled by conservative, if not right-wing politicians and influential clerics. This was especially true at the university where normal teaching had become more difficult. Already in February, Emma had told him about university brawls:

> At the University of Vienna, they are now only fighting - even professors were beaten up yesterday, they say. Today the university is closed and a strong police force is on guard. But

[18] Arthur Haas, *Die Welt der Atome*, 122.

[19] Edwin E. Salpeter (1924–2008), who died November 26, 2008 at his home in Ithaca, NY, belonged to the "second wave" of Jewish scientific refugees from Nazi-dominated Europe, those who left as children just before the onset of WWII and so completed their educations elsewhere. Salpeter was born in Vienna on December 3, 1924 and arrived with his family in Australia in 1939; his father was a physicist and a close friend of Erwin Schrodinger. Virginia Trimble, obituary.

[20] Edwin Salpeter, "The Reaction Rate of the Proton-Proton Chain," *Astrophysical Journal* 116 (1952): 649–650.

[21] Arthur Haas, *The World of Atoms—Twelve Non-Mathematical Lectures* (New York: D. Van Nostrand Company, Inc., 1928).

I don't think it was of any use because the *Abendblatt* is full of reports again. Be glad you're not here.[22]

The reports told of German-national anti-Semitic student groups and their attacks, which were especially directed against the Jewish anatomy professor Julius Tandler and his institute.[vii]

The life of young Jewish lecturers at the university had become more difficult. Anti-Semitism was more and more apparent and indirectly promoted by university authorities. Young promising natural scientists such as Otto Halpern, Hans Schiller, Franz Urbach, Paul Weiss and many others were refused the habilitation under various pretexts. Franz Urbach reported on this change:

When I began my studies (1921), an academic career in Austria was somewhat difficult for a Jew, but certainly accessible; when I finished it (1926, long before the outbreak of German National Socialism), it was already almost impossible, and two years after the doctorate, when I was ready to do my habilitation, it was out of the question. At that time, a state position had become almost unattainable for a Jew, and the university and technical college in particular were centers of the sharpest anti-Semitism. I was so at the mercy of silent Austrian university anti-Semitism.[23]

Haas was aware of this from his own experience; he did not directly see his position at the Academy as endangered, but his success in the United States had strengthened his hopes of being offered a position there and of then being able to turn his full attention to science. In the months following his return, he submitted applications to Browne University in Rhode Island and Northwestern University in Chicago, Illinois, where he had been acknowledged for giving successful lectures and where he was well received.

But before tending to his career, Haas had to adjust to his daily service work at the academy and to his personal life with his wife and children at Baumgartenstraße 58. He played tenderly with the children again, the now almost two-year-old Arthur junior and Georg who, according to his mother's reports, had developed into a strapping, enterprising boy, as pictured in Fig. 14.6. But alongside this, when back at his father's familiar oak desk, Haas worked on the new ideas and impressions that had come to him on the trip and in discussions with colleagues.

Haas had already received a salary increase at the beginning of 1927, but his official obligations at the Academy and his private life with two small children reduced his personal, innovative publishing and other scientific activities in the following years. He primarily concentrated on writing and publishing that involved implementing the latest scientific results for textbooks and popular-science publications; only when his interest was aroused by some spectacular new results did he try to pursue other, as yet unexplained or uncorrelated aspects of the field of modern physics and draw further conclusions. He prepared his new books mostly from lectures at the University of Vienna, which remained well attended. These lectures also provided the topics for

22 EBH to AEH (Vienna, February 5, 1927).

23 Franz Urbach in a letter at the end of 1934, quoted after Johannes Feichtinger, *Wissenschaft zwischen den Kulturen. Austrian university lecturers on emigration 1933–1945* (Frankfurt: Campus Verlag, 2001), 140.

Fig. 14.6 Emma Haas with her two sons Arthur and Georg around 1927 and Arthur Haas with young Arthur in around 1926 before his first journey to the United States

his public lectures at the Urania and other forums. He earned well during these years and could offer his family a solid and prosperous life. His son George[24] remembers:

> My father spent many, many extra hours writing textbooks on physics, chemistry, mathematics, etc. in addition to teaching. But it was the textbooks that provided the money for us to live quite comfortably in Vienna. (We enjoyed the luxury of a full-time maid and live-in governess).[viii]

After his popular-science work on the world of atoms, Haas turned his attention to the hypothesis of young Louis de Broglie (1892–1987). De Broglie postulated that particles could behave like waves,[ix] and Haas tried to correlate these predictions with other elements of modern physics. In 1927 he published a short paper[25] in which, starting from the relationship between the front velocity and the group velocity of a wave packet of energy $\boldsymbol{h{\cdot}\nu}$, he derived the time dilation and mass increase for relativistic particles as predicted by special relativity. While it is not necessarily a great discovery, Haas did demonstrate the coherence of modern approaches to physics.

During the hot summer months, the family traveled back to Sattendorf to relax in the Villa Seefried, which was near the new house of Arthur's sister and brother-in-law with their two children. For a long time, there were hardly any limits on the length of vacation for actuaries and other employees at the Academy; a certain restriction was

---

[24] Georg Haas adopted the name George after the family had moved to the United States.

[25] Arthur Haas, "Über die Ableitung der fundamentalrelativitäts-theoretischen Sätze aus der Brogliesсhen Hypothese der Phasenwellen," *Physikalische Zeitschrift* 28 (1927): 632–634.

only necessary when considering and balancing the vacation wishes of colleagues. In return, the respective employees had to declare their willingness to return to Vienna immediately if necessary. "Four to six-week vacations were the rule rather than the exception for actuaries; four-week vacations were regularly taken by the rest of the firm's staff."[x]

But Haas only put off work duties for a short time, travelling with his family to the summer resort and returning to hot Vienna after only a few days; in his letters, he then reported on his work overload and lamented the scorching heat, but in summer it was mostly quiet at the Academy[xi] and one can assume that Haas took advantage of the silence in the large house to write his scientific publications and his books. Neither official obligations nor family stood in the way of his creative urge. He usually returned to his holiday resort at the end of July but traveled back to Vienna only a few weeks later, long before his family who stayed until September. After his return during the summer months of 1927, Haas probably worked on his articles for *Science* and for the *Physikalische Zeitschrift*, the latter providing the motivation for his new book on wave mechanics.

However, in the late summer months, the actual work at the Academy began again. Actuaries and staff were back; the office and the meetings were up and running again. Haas kept the minutes of the meetings, processed the writings and files of the natural science department, and prepared the Academy's publications for printing, while always including one of his own contributions; and, he had again taken over the management of the library. He, his colleagues and co-workers had once again integrated themselves into the daily work routine, which he personally did not like at all and about which he may have thought of the words of the German poet, Baron Joseph von Eichendorff:

> Spending nights at dossier reading,
> prattling, as the custom is,
> and the giant treadmill speeding
> like an ox: I could do this.
> Thinking, though, that this baloney
> were not just a bunch of crap,
> but high heaven's testimony:
> I could never manage that.[26]

After commission and academy meetings, Haas usually went to the coffee house to read his newspapers or to work in peace, far away from the clatter of the typewriters. He could not leave science and writing behind.

He was particularly fascinated by the idea of matter waves that was emerging in the new results and in the work of his friend Erwin Schrödinger on wave mechanics. This world of thought was the focus of his new book, providing an introduction to the wave mechanics of Schrödinger and the matrix mechanics of Werner Heisenberg. For this publication, Haas made a contract with the publishing house of his friend Leo

---

26 Joseph Freiherr von Eichendorff, "The Curmudgeon," trans. Walter A. Aue (1841).

Jolowicz in Leipzig, with whom Haas had already had good experience in publishing his contribution to the *Handbuch der Experimentalphysik*. Haas prepared this new book as usual, by first giving lectures during the winter semester 1927/28 at the university. In the preface, he thanked a colleague in Vienna, Eugen Guth (1905–1990). Guth was a young physicist from Budapest with whom he would stay closely connected for a long time. Although Haas wrote in the preface that he, the author believed,

> that there might be a desire for a shorter presentation, for a presentation that does not presuppose a high level of mathematical knowledge, that only reproduces the essence of the thought processes of de Broglie, Schrödinger and Heisenberg, that considers the inner coherence of these ideas,

the book is, nevertheless, a writing that does not skimp on mathematical formalism and that seeks to present, to the layman, the mathematical formalism of matrix theory as well as that of Schrödinger's differential equations.

Haas does not shy away from the mathematical demonstration of the equivalence between the Heisenberg-Born-Jordan matrix theory and Schrödinger's wave formalism. Having grown up with the theory of Boltzmann's heat statistics, Haas was particularly keen to introduce the reader to the new quantum statistics for photons. He discussed the work of Satyendra Nath Bose (1894–1974) from which the fundamental Planck's formula can be derived[27] and also made a first attempt at introducing the young Enrico Fermi's new statistics for quantum gases[28]—with regard to particles obeying the Pauli principle.[29] This was undoubtedly a great concern for Haas, who set himself the task of following these rapid developments and summarizing the developments to January 1927. The book was a considerable challenge even for the educated lay reader and yet, these were the things that fascinated Haas and made the bureaucratic work at the Academy bearable. The book was published by the Akademische Verlagsgesellschaft[30] in the summer of 1928 and *Materiewellen und Quantenmechanik* became a great success.

While Haas wrote and worked over the winter and into the spring, he was increasingly concerned about the lack of response from America. He had placed great hope on his application to Brown University. He was also thinking again about the possibility of a position in his home town of Brno. Personal letter writing resumed at the beginning of May when Emma was in Munich to look after her diabetic father who was getting worse and worse, as her sister-in-law Maria faithfully reported. Arthur Haas stayed in Vienna with the children and tried to entertain them with activities,

---

[27] Satyendra Nath N. Bose, "Planck's Law and Light Quantum Hypothesis," *Zeitschrift für Physik* 26 (1924): 178–181 and "Heat Equilibrium in a Radiation Field in the Presence of Matter," *Zeitschrift für Physik* 27 (1925): 384–394.

[28] Enrico Fermi, "Zur Quantelung des Idealen Einatomigen Gases," *Zeitschrift für Physik* 36 (1926): 902.

[29] The Fermi statistics take into account that no two particles may have the same quantum configuration, which is usually called the Pauli principle.

[30] Arthur Haas, *Materiewellen und Quantenmechanik* (Leipzig: Akademische Verlagsgesellschaft, 1928).

at least on weekends. After Emma's return in mid-May, the couple was planning a trip to Brno together. She had never been there and Arthur Haas wanted to take the opportunity to show her the place of his origins. In each of his daily letters to Munich in which he set out these plans, he also expressed his everyday concerns:

> The Academy is desperately busy. The letter that I expect from Providence, I look forward to with feverish anxiety. If only I could already see clearly into the future.

And:

> Today I have been terribly busy for the Academy. From –four to six commission meetings, where I had to do everything, then the Academy meeting [...] I am a bit nervous because I haven't heard anything from Providence yet. Whether I didn't just mess something up with my application.[31]

So spring passed and the trip to Brno became just a weekend trip together on May 19th and 20th, as Emma had only just returned from Munich on May 17th. In the midst of their concerns, the trip wasn't a much-needed break but instead an exhausting tourist's weekend. The rest of the month as well as June flew by and the warm summer of 1928 began. Already at the end of June, Emma decided to go to Lake Ossiach again with the two boys, not to Sattendorf this time but to the nearby village of Annenheim where they found a cheap guesthouse. Arthur accompanied the family with their summer luggage but had already returned to Vienna on July 2nd, from where he wrote a first letter of complaint to Emma in Annenheim:

> Now I'm all alone again in the "big apartment," which I can only revive with thoughts of you and the boys. It is so lonely, i.e., it would be lonely if I were ever in the apartment for more than a minute at noon.

But he was working; upon his return, following a train ride of several hours from Carinthia to Vienna, he was in the office. He later reported:

> I quickly went to the academy, where there was a lot to do; at noon I had little time for lunch; then, for the last time, lecture and seminar and then to bed.
>
> Today I worked very diligently from 5 am to 8:15 am and now I am sitting in the academy again with a lot of work.[32]

He also reported on personal letters from new acquaintances in America, although there was no news about a possible offer; he dismissed this with a light remark:

> I have heard nothing new from the U.S.A.; undoubtedly, it is that the matter has been postponed until October. So there is still hope, but it is getting easier and easier. If it is not Brown, it will be Northwestern and if it is not, it will be something else. The thing with Yale proves that the good Lord means well with us.[33]

The last remark refers to a new possibility from his old friend Professor Donnan of London. Donnan recommended Haas for a project at Yale—to become a member

---

[31] AEH to EBH (Vienna, May 9/10, 1928).

[32] AEH to EBH (Vienna, July 3, 1928).

[33] AEH to EBH (Vienna, July 9, 1928).

of the so-called Gibbs committee and write a summary commentary on the work of former Yale professor J. Willard Gibbs (1839–1903), on thermodynamics. This project lasted several years and was not completed until 1936 with a two-volume edition,[34] but Haas was enthusiastic even in this early initial phase and was justifiably honored by his election. Haas was a perfect choice, not only because was he scientifically qualified in the field of the thermodynamics of the late nineteenth century but also because he had gained, through his work on the *Poggendorff Handbuch,* considerable experience in glossary publication and the analysis of scientific works.

While Emma was at the summer retreat and wrote in her letters about hiking and swimming with the children, Arthur reported on American visitors. These included Professor Heindler from Iowa and Professor Colby from Ann Arbor in Michigan, whose acquaintances Haas had made on his tour of America; both now paid Haas a return visit while on a summer trip to the old continent. The American visitors attracted attention and rumors developed at the academy that the visits would be followed by an offer from America. Such an offer was, of course, Haas' hope but it was not fulfilled that summer either. Instead, Haas worked on new editions of his books, especially the theory of atomic energy:

> My work is progressing at a rapid pace. When I leave in a week [to Annenheim], I will have four-fifths of the work on the new "atomic theory" ready. And I have terrible work to do; it will not get any better in the foreseeable future. I have to add a lot of new chapters to my *Atomic Theory*, terribly difficult.[35]

The book was a complete revision of the first edition released in 1924. Rapid development characterized the field of atomic physics in those few years and Haas not only had to expand and update the individual chapters but also had to add new chapters. He included a chapter on the basics of quantum mechanics, which by then had been accepted as the theory of the atomic model. In addition, he added a chapter on the new discoveries of the Compton effect, a phenomenon that he suspected would play an important cosmological role. He also added a chapter on the quantum description of molecules, a new and important field for the theoretical explanation of experimental discoveries in the field of molecular spectroscopy in the infrared range. He would continue to pursue this latter topic after completing of this second edition of *Atomic Theory.*

Nonetheless, Haas went to Annenheim on July 20, 1928 to spend some of the summer holidays with his children and wife. He stayed for three weeks to swim in the cool waters of Lake Ossiach but eventually returned from the cooler mountainous region to summer-hot Vienna on August 10th, to continue his work on the books. It was not only the new edition of *Atomic Theory* that was to be completed but also a new edition of the recently published first work on matter waves that the publisher

---

[34] *A Commentary on the Scientific Writings of J. Willard Gibbs, Ph.D., LL.D., formerly Professor of Mathematical Physics in Yale University* vol. 1: "Thermodynamics," eds. Prof. F. G. Donnan and Prof. Arthur Haas, xxiii, 742; vol. 2: "Theoretical Physics," ed. Prof. Arthur Haas, xx, 605 (New Haven: Yale University Press; London: Oxford University Press, 1936), 45*s*. net.

[35] AEH to EBH (Vienna, July 13/18, 1928).

had requested. It was a difficult summer for Haas who spent his time between work and the beach, in order to be able to bear the summer heat. In between, American visitors again arrived, including President Kolbe[36] from Brooklyn College in New York with whom Haas again discussed his possibilities in the United States. On August 24th, however, he traveled back to Annenheim where he spent another three weeks before finally returning to Vienna on September 16th. The family planned to remain in Annenheim until the end of September.

However, bad news arrived from Munich. Michael Huber was not well, as cook Marie and brother Georg Huber reported; Emma's father had a severe lung disease with shortness of breath:

> Father is severely diabetic and lung sick. Whether the lung disease is due to tuberculosis or cancer, Lampé [the family doctor] could not immediately tell me with certainty. Probably "tube" is to be assumed [...] Help is no longer possible. The left lung is completely destroyed and the right one is a bit damaged. [...] Lampé believes he can at least give him another fourteen days.[37]

According to initial reports, the disease was believed to be tuberculosis and therefore very contagious. This caused Arthur Haas extreme concern when Emma decided to travel directly to Munich, leaving the children with Grete in Sattendorf. His letters expressed special concern for his wife. He was particularly upset about the possible risk of infection from tuberculosis:

> He [the Haas' family doctor, Dr. Löwenstein] warns you in the most decisive way, since your father's illness is absolutely contagious. I am to advise you with the utmost determination against going to Munich. Even if there are duties to your father, you must not commit a crime against the children. But if all warnings are of no use, then you should only go to be with your father for as short a time as possible; you should not live in the house under any circumstances and you should not eat anything in the house under any circumstances![38]

Strong words but Arthur knew his wife. As expected, Emma went anyway—but lived in a hotel—and Arthur brought their two sons back to Vienna at the beginning of October.

Arthur Haas' concern for the family's health was quickly replaced by financial concern when he heard from Emma that her father, under the erroneous assumption that Arthur had become rich by selling the houses in Berlin, had left his fortune to his son and business successor Georg Huber and thus only a small compulsory portion was to fall to Emma. In actuality, Arthur was not rich nor had his anticipated appointment in the United States materialized during his year of waiting. Now, added to these deep disappointments, his family seemed to be excluded from the rich inheritance, a circumstance that worried Haas very much:

> I'm beginning to be quite pessimistic about the future and I believe it is your duty to the children (for whom, God forbid, this might be their only support in the event of my premature

[36] Parke Rexford Kolbe will play an important role in Haas' later life and will be introduced in more detail on that occasion.

[37] Georg Huber to EBH (Munich, September 20, 1928).

[38] AEH to EBH (Vienna, September 24, 1928).

death) to rectify Papa's awful error (which was only caused by our generosity). If Papa's condition still allows it, you are to inform him that his opinion, as you heard it from Georg that I had "suddenly become very rich," is based on a misunderstanding that is incomprehensible to you. On the contrary, the fraud perpetrated on me from Berlin [April 1928], with the retroactive revaluation, has considerably worsened our financial situation. And, the absence of the hoped-for offer from America also fills us with serious economic worries.[39]

It was probably not quite that bad, as seen from analyses of his tax return. His income had risen steadily, with the next salary increase already scheduled for January 1929. In addition, his books were considered bestsellers—at least for scientific publications. This, together with the assets from the sale of his houses, gave him a regular and good income. But the trauma of inflation and the loss of his father's and mother's fortune, which had once been so great, did not pass Haas by without leaving a mark; the thought of another loss of fortune filled him with panic. He announced a trip to Munich for the following weekend, to see for himself what was going on. But Emma answered immediately:

I have enough worries now and would be happy if another one didn't come along.[40]

But in the case of each concern, infection and disinheritance, the all-clear was soon given. It was lung cancer that increasingly prevented Michael Huber from being able to breathe and "disinheritance" was only a fiscal paraphrase by the company, which allotted Emma her full share as a partner. Nevertheless, Arthur Haas travelled to Munich on October 13th to see his father-in-law for the last time, since his death seemed to be foreseeable. Haas was also planning a personal visit to Arnold Sommerfeld's Institute of Physics and a visit to the Deutsches Museum. As it turned out, Sommerfeld was on world tour in India, from where he planned to travel via Japan to the United States as an honored guest.[xii] Therefore, the visit with Sommerfeld did not happen. Only one day later, on October 14th, Arthur and Emma traveled back to Vienna with the children but Emma received daily letters telling her about the state of Father's health from her sister-in-law Marie Huber.

Warned by these letters, Emma traveled back to Munich on November 14th; the end was imminent and only days later Michael Huber died on November 23, 1928. He was widely loved and a respected member of the Alpine Club, the Schiller Lodge, the Stoanernes Krügl skittle club, and the Drahrer table society. He was also a member of several military and veteran associations, all of which were represented on November 25th by brass and music bands for the funeral at Munich's Ostfriedhof cemetery, giving their deceased member the appropriate last escort. Arthur Haas went back to Munich for the funeral but only stayed for two days. Emma stayed a few days longer to finally settle inheritance matters. She did not return to Vienna until December.

The death of her father, painful as it was for Emma Haas who had been very attached to him, brought her financial independence, an independence that she increasingly exercised in the years that followed. She traveled on her own, something

[39] AEH to EBH (Vienna, October 5, 1928).

[40] EBH to AEH (Munich, October 6, 1928).

that she had always dreamed of and that gave her a sense of balance relative to her husband's business and lecture tours, since Arthur never seemed inclined to take her with him on his trips.

Arthur returned to his work, to his business meetings, which he never discussed in detail, and to his scientific writing, which was increasingly demanding of his time. But he also devoted a great deal of time to their fast-growing children and did a great deal with them, especially during periods when Emma was absent.

In addition, new theoretical and experimental findings forced him to frequently modify his writings:

> Yesterday afternoon I worked at the Institute of Physics. Again, I have to rework all kinds of things that I thought I had already finished. But I hope to be finished with the new edition of *Atomic Theory* in mid-December.[41]

He finally completed his revision of *Atomic Theory* before the end of the year and the book appeared in print in the spring of 1929.[42]

The new results had also generated new ideas and thoughts, and Haas was somewhat overwhelmed. His book on matter waves was so successful that Akademische Verlagsanstalt requested a new edition just a few months after the first. It was to be another exhausting spring in which he would concentrate on these new projects. During these months of winter and spring in 1929, he devoted himself to the task of creating a comprehensible description of quantum physics, also for chemists. This project probably went back to his original discussions with George Donnan, who had pointed out to Haas the necessity of introducing chemists to this field. As preparation, Haas gave a series of lectures, once again in front of the Physical–Chemical Society in Vienna. Based on these lectures he published a booklet on quantum chemistry with the Akademische Verlagsgesellschaft in Leipzig in July 1929.[43] This was the first work to use the term "quantum chemistry," now a widely used term although the subject and definition have changed over the decades.[xiii] The book covered topics on chemical periodicity, chemical bonding, electron configurations, and molecular structure—topics that had already been expanded in the book *Atomic Theory* but this time without much mathematical formalism.

Over the course of this work, it had become summer again in Vienna and with the warmer days, plans for the summer retreat were awakened. The family decided, mainly for the sake of the children, to again spend the summer in Annenheim, in Carinthia at the Ossiacher See. The opportunity for time with brother-in-law and sister Gerstenberger heightened the anticipation. The great heat wave in June of 1929, lamented in the letters, accelerated departure to the cooler mountain climate of Carinthia. Emma Haas and the two boys traveled by train, via St. Veit and Feldkirchen, to their lodgings on Lake Ossiach. Arthur stayed behind to work but also, as he wrote, to keep an eye on the new subtenant Peter who, as a young student from a friendly Munich family, was living in the attic and, to the annoyance of the landlord, always kept the stair light on.

---

[41] AEH to EBH (Vienna, November 15, 1928).

[42] Arthur Haas, *Atomtheorie* (Berlin and Leipzig: De Gruyter & Co, 1929).

[43] Arthur Haas, *Quantenchemie* (Leipzig: Akademische Verlagsgesellschaft, 1929).

Probably the more important concern keeping Haas in Vienna was the completion of work that he planned to present to the Academy of Sciences on June 20th. In this paper he followed the fundamental effects of the particle wave concept, applying it to the statistical concept of thermodynamics and entropy.[44] He was hoping to maintain his scientific reputation but the originality of his contributions was no longer the same. He worked out the mathematical formalism, comparing different relationships, but did not draw any significant conclusions. This gives the impression that while he followed the rapid developments in the field, he had stepped back from the front line; Haas' focus was only on reporting and interpreting results and developments for his more popular science books.

While summer heat continued, Haas wrote a letter to Prof. Duggan of the IIE in New York to apply for another lecture tour of various universities in the United States. In this letter, he also pointed out the growing number of books that he had translated into English, including the 1928 editions of *Introduction to Theoretical Physics*, *Atomic Theory*, *The World of Atoms*, and *Wave Mechanics and the New Quantum Theory*, all published by Van Nostrand & Co in New York. The short time span between the German and American first publications demonstrates how successful Haas was as an author and how well he was considered by publishers. His application for a second tour was actively supported by the Nostrand publishing house in a letter dated July 11, 1929, which included a list of universities that had adopted *Introduction to Theoretical Physics* as an official textbook for their students: University of Arizona, Brown University, University of California, California Institute of Technology (Caltech), University of Indiana, Rensselaer Polytechnic Institute, Syracuse University, Utah Agricultural College, and the University of Utah. Undoubtedly, this was an impressive list.

Notes

i. *New Yorker Staats-Zeitung*, "Wiener Physiker zu Vortragsreise hier." February 10, 1927, 12.

ii. *The Post Standard*, "Concerned people need have no fear for quintillions of years, assures Dr. Haas." Syracuse, March 12, 1927.

iii. *Chicago Herald Examiner*, "World-Running Power sought in Atom's Frozen Energy." March 30, 1927.

iv. Otto Robert Frisch, *What Little I Remember* (Cambridge, Cambridge University Press, 1979).

v. See Darko Suvin, *Poetik der Science Fiction* (Frankfurt: Suhrkamp, 1979), 155–280.

vi. Martin LaMonica, "Nuclear generator powers Curiosity Mars mission," *MIT Technology Review* (2012), accessed January 31, 2021, https://www.technologyreview.com/2012/08/07/184595/nuclear-generator-powers-curiosity-mars-mission.

---

[44] Arthur Haas, "Die Ableitung des Boltzmann'schen Entropiegesetzes mittels der Vorstellung der Materiewellen," *Sitzungsberichte der Akademie der Wissenschaft in Wien* 138, 5/6 (1929): 459–468.

vii. Klaus Taschwer, *Hochburg des Antisemitismus—Der Niedergang der Universität Wien im 20. Jahrhundert* (Wien: Czernin Verlag, 2015), 91–93.
viii. George and Mary Jane Haas, "Letter to our Grandchildren, Remembrances of our Family," (unpublished manuscript, Haas family collection, 1965), 43.
ix. Louis de Broglie, "Ondes et quanta," *Comptes Rendus* 177 (1923): 517.
x. Stefan Sienell, *Das Verwaltungs- und Dienstpersonal,* 163.
xi. Ibid., 162.
xii. Michael Eckert, *Arnold Sommerfeld—Science, Life and Turbulent Times 1868–1951* (Heidelberg: Springer Verlag, 2013), 309–315.
xiii. Ana Simões, "Dangerous Liaisons or Unavoidable Associations: Quantum Chemistry at the Crossroads of Chemistry, Physics and Mathematics," paper presented at the 6th International Conference on the History of Chemistry, (Leuven, Belgium, 2007), 35–52; Helge Kragh, "Haas, Arthur Erich" in *Eight Physicists and Astronomers: Biographical Portraits*, accessed January 31, 2021, arXiv, http://arxiv.org/abs/1208.3114.

Chapter 15
Holiday Stress and Other Sorrows

Hot weather and his longing to again swim in the clear and crisp waters of the lake drove Arthur to Annenheim on June 23rd. There he spent a quiet week without work, although the weather had turned a bit cooler:

> I don't want to work during the holiday week. I will read *Dodsworth*[1] and maybe something else.

Haas returned to Vienna at the end of the month to continue his work. He had also promised to replace I Actuary Junk at the Academy during the first two weeks of July. But no sooner was he back in Vienna than it was hot again.

Therefore, he interspersed his work in the stifling hot academy offices with "frequent breaks in the nearby café house". There he was busy calculating the salary necessary for his life in America, the details of which can be seen in Fig. 15.1. This exercise was based on a cost-of-living study that he had found, prepared by the University of Berkeley. Despite his good income and pleasant life, the longing for America was never far away as the last sentence of his letter to Emma testifies: "If only I had some good news. I haven't longed for America in a long time as I do now."[2]

Despite his constant complaints about the rapidly changing weather—heat, cold, rain, sultriness—and the Shuffler Peter, as Haas called the tenant with his constant disorder, there were also reports of walks and excursions to the various beaches and pools in the area. Haas was also occupied by his two summer projects: the new edition of his *Introduction to Theoretical Physics*—it was the fifth edition, one of his most successful books—and the expanded third edition of the book on matter waves. These were progressing steadily.

As he wrote in a letter to Annenheim on July 7th:

---

[1] *Dodsworth* is a satirical novel by the American writer Sinclair Lewis, first published by Harcourt Brace & Company in March 1929. Its theme is the difference between American and European intellect, manners, and morals. Undoubtedly interesting and instructive reading for Arthur Haas who wanted to deepen both his knowledge of the language and his understanding of life in America.

[2] AEH to EBH (Vienna, July 3, 1929).

M. Wiescher, *Arthur E. Haas—The Hidden Pioneer of Quantum Mechanics*, Springer Biographies, https://doi.org/10.1007/978-3-030-80606-4_15

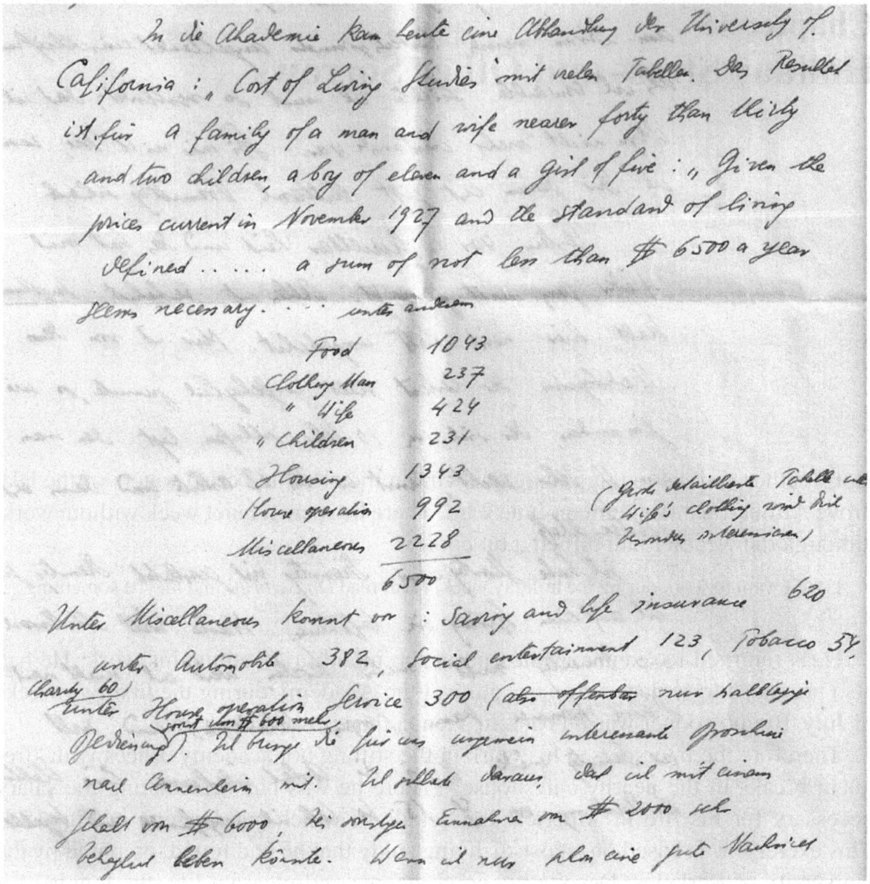
In die Akademie kam heute eine Abhandlung der University of California: „Cost of Living Studies" mit vielen Tabellen. Das Resultat ist für „a family of a man and wife nearer forty than thirty and two children, a boy of eleven and a girl of five: „Given the prices current in November 1927 and the standard of living defined ..... a sum of not less than $ 6500 a year seems necessary. . . : unter anderem

Food	1043
Clothing Man	237
" Wife	424
" Children	231
Housing	1343
House operation	992
Miscellaneous	2228
	6500

(Ganz detaillierte Tabelle über Wife's clothing und [illegible])

Unter Miscellaneous kommt vor: Saving and life insurance 620, unter Automobile 382, Social entertainment 123, Tobacco 54, Charity 60) unter House operation Service 300 (also offenbar nur halbtägige Bedienung). [illegible] Ich bringe die für uns ungemein interessante Broschüre nach Annenheim mit. Ich schließe daraus, dass ich mit einem Gehalt von $ 6000, bei mäßigen Einnahmen von $ 2000 sehr behaglich leben könnte. Wenn ich nur schon eine gute [illegible]

Fig. 15.1 Longing for America, Arthur Haas calculated the cost of living and the salary required

> The manuscript for the first 240 pages of the new edition of my *Introduction to Theoretical Physics* is now also finished and will leave tomorrow.[3]

And only three days later he notes in a subordinate clause:

> I must continue to work overtime for the academy because there is so much to do. Today I sent the manuscript for the new edition of *Matter Waves*; and I have so much of all kinds to do![4]

His work at the Academy was probably focused on his books, since *Matter Waves* is a largely revised and expanded new edition that must have taken a lot of effort.[5]

---

3 AEH to EBH (Vienna, July 7, 1929).

4 AEH to EBH (Vienna, July 10, 1929).

5 Arthur Haas, *Materiewellen und Quantenmechanik* (Leipzig: Akademische Verlagsgesellschaft, 1929), 90.

But only two days later he fell into despair. Like many others during these years, Haas had speculated with shares on the market in order to earn the anticipated fast money. Things had gone well in the years of the upswing that followed the currency reform but the first signs of the world economic crisis were already evident in the early summer of 1929; suddenly falling share prices led to losses in his speculative investments. Black Thursday on October 14, 1929, which heralded the world economic crisis with a complete collapse of the stock market, was still months away but the first signs of unrest in Germany and also in Austria were already visible, with the traded prices of many shares being inflated far above their real value. This was largely ignored, however, since the mindset was one of eternal growth. Haas, too, held fast to the ideal and blamed his stock advisors for his first losses:

> I'm completely worn out by grief about the financial catastrophe that is afflicting us thanks to the stupidity of Steppan and Dr. Hönig. Half of the huge sum I had invested in weapons is lost; I am making desperate attempts to save the other half. But I can't attach my files, values are falling too fast. If I do manage to sell the shares today, I will have a loss greater than two years' salary at the Academy. I can only hope that this terrible loss will perhaps be compensated by Munich's favorable development. It is dreadful, year after year a new catastrophe. And then there are the terrible rye stock losses[6]; I hope that this can be mitigated.[7]

Haas, like many, believed that temporary fluctuations and short-term, albeit painful losses in some of his stocks and shares would in all likelihood, soon be offset by gains in others. He had no idea that this development would turn into a long-term crisis that would change Europe forever and have a drastic effect on his life. Initially, his hopes seemed to be confirmed. He had acquired arms shares in 1927 and they had brought him 900 shillings in the first year, with 1800 shillings in the second year, double the amount. However, in 1929 the value fell and he had only 900 shillings left. The loss in value of the rye stocks also remained moderate; in 1928 they brought him an additional income of 122 shillings but in the crisis year this was reduced to 77 shillings.

In order to absorb his losses, he sold his League of Nations papers, which had earned him 988 shillings each year, and invested in Portland Cement shares. This reduced his losses but then a gradual decline in his stock assets set in. Whereas in 1927 he still had annual capital income of 3,032 shillings, in 1928 it was only 2,770 shillings, which fell to 2,272 shillings in 1929. His loss was therefore relatively small and more than compensated for by his salary increase in January 1929.

His complaints were also unjustified at the time inasmuch as his total income had increased steadily since 1925 from 14,223 schilling annual income to 16,540 schillings in the crisis year 1929, of which an average of 5,500 schillings was income from fees for his publications. He was therefore financially well, but he was tormented by constant worries about his assets.

---

[6] Agricultural stocks were issued in Germany in the 1920s. These were for a defined number of hundredweight of rye. The interest and redemption payments were linked to the official rye-price quotes on the Berlin commodity exchange. The collapse of these stocks was due to the agricultural crisis caused by the overproduction and sales problems of the grain suppliers, which followed the development of new areas of land with an associated increase in cultivation and yield.

[7] AEH to EBH (Vienna, July 12, 1929).

Fig. 15.2 Left is Arthur Haas with Atzi at the end of the 1920s; right, Arthur and Emma Haas with tennis rackets, with Atzi on the shore of Lake Ossiach, together with members of the Gerstenberger family in the early 1930s

In spite of the financial misery of the stock market, which he lamented quite a lot about, Haas calmed down only one day later. He travelled back to Annenheim on Saturday to forget his financial worries by swimming in Lake Ossiach and hiking with the family in the Carinthian mountains (Fig. 15.2).

Arthur Haas did not return to Vienna until August 15th. But already in his letter the next day, he complained about his workload and the heat in the city. He made up for this by visiting the various open-air swimming pools in the area—from the Hietzinger beach pool to the pool in Klosterneuburg, and with trips to the Danube spring and the Gänsehäufel.

Emma replied somewhat coolly to the lament, which extended over several letters:

> I'm glad that you can go to the Danube beach despite your crazy work load, but I wouldn't mind if you didn't complain so much in your letters. I certainly believe that it is hot in Vienna and that you have a lot of work, but still...!

This warning was of no use. The weather and an excess of work at the academy dominated his life and his letters. As far as work was concerned, he was primarily occupied with his new assignment as editor for the Gibbs committee at Yale University. It was actually a task that he took pride in and was diligent about, bringing modern concepts such as matter waves into harmony with Gibbs' ideas and models on thermodynamics and then publishing these concepts at the Academy.[8]

---

[8] Arthur Haas, "Die Ableitung des Boltzmann'schen Entropiegesetzes mittels der Vorstellung der Materiewellen," *Sitzungsberichte der Akademie der Wissenschaften* 138 (1929): 459–468.

His letters also describe meetings with former students such as Karl Horowitz, a professor at Purdue University in Lafayette, Indiana, who had returned to Vienna for the summer:

> Early next week I will meet with former student Karl Horowitz and his wife. They knew that I was in New York and invited me. He has been a professor at Purdue University in Lafayette, Indiana, for a year.[9]

Horowitz left Austria after his habilitation in physics was blocked in 1923 by the Bear Cave cabal at the university.[10] He had received a fellowship from the Rockefeller Institute, with which he travelled to America to do research at the University of Toronto, the University of Chicago, the Rockefeller Institute and Stanford University, until he was appointed professor at Purdue University in 1928.[11]

Shortly after this visit, Dr. Paul Epstein,[12] Professor of Theoretical Physics at Caltech in Pasadena, announced his intention to combine his stay in Vienna with a working meeting with Haas, who commented on this:

> I will now have about three very restless days. But I am very glad that I will be able to meet with Prof. Epstein. He will be of great help to me as he has already been of use to me before.[13]

Haas had high hopes for Epstein, whose appointment at Caltech through Millikan and subsequent successful career provided an enviable example. It was a pleasant visit, however, Haas reports less about their discussions of physics but more about the evening visits to the Volkstheater and the Johann Strauss Theater. After Epstein left, Haas decided to return to Lake Ossiach until September 10th, spending the last days of the summer there. In his last letters, he complained about further losses of shares and reduced book income; he even threatened the family with restrictive budgetary measures for the coming months.

Autumn of 1929 brought catastrophe with the collapse of the American stock exchange in New York. The crises spread quickly to Europe because American banks, until then so willing to lend to European banks and countries, stopped lending. This

---

[9] AEH to EBH (Vienna, August 21, 1929).

[10] Peter Goller, "Ein starkes Stück. Versuchte Habilitation eines kommunistischen Juden—Universitäten im Lichte politischer und rechtlicher Willkür am Beispiel des Habilitationsverfahrens von Karl Horowitz (1892–1958) an der Wiener Universität 1923–1925," *Dokumentationsarchiv des österreichischen Widerstandes, Jahrbuch* (1998), 111–134.

[11] Horowitz, who later called himself Lark Horowitz, built the Physics Department at Purdue University into a successful institution, particularly in the new fields of nuclear physics and solid-state physics, accessed January 31, 2021, http://www.physics.purdue.edu/about/prizes_awards/horovitz.html.

[12] Paul Sophus Epstein (1883–1966) was born in Warsaw and studied physics in Minsk and Moscow before moving to Munich in 1910 to study theoretical physics under Arnold Sommerfeld. During the war, Sommerfeld protected him so that Epstein could continue his work without being disturbed. After the war, Epstein went to Leiden as an assistant to Lorentz and later Ehrenfest. In 1921, Epstein was appointed by Millikan to Caltech, working mainly on problems of quantum mechanics in atomic physics.

[13] AEH to EBH (Vienna, August 22, 1929).

affected Germany in particular, with its lack of money and gold reserves. But with only a one-year delay, neighboring Austria was devastated as well and some banks had to file for bankruptcy in 1931. Haas watched this development with concern, although his income as a civil servant at the Academy was secure and the sale of his books was only marginally affected. He was worried about his capital income, which in fact fell by 40% between 1929 and 1931.

He again tried to obtain a position in the United States through the IIE but with the rapid loss of university assets, which had been invested on the stock exchange, the doors of American universities were closed to new hires. Princeton refused with regret:

> I regret to say that there is no possibility of a vacancy in our department for a man of Dr. Haas' caliber. [...] I know Dr. Haas' book on theoretical physics very well and have found it extremely useful in my first-year graduate course. A couple of years ago Dr. Haas lectured here and I met him at that time. I am very glad to hear that he wishes to come to America and hope that he will be able to find a suitable position. Regretting my inability to be of more assistance.[14]

December 1929 did, however, bring better news in the form of a renewed invitation from the IIE, inviting Haas to the United States in the spring of 1931 for another lecture tour. He was offered $75 per lecture, a little less than his previous tour due to the economic crisis. In the following months travel plans were arranged and this took most of Arthur Haas' time. He himself contacted many of the universities and physicists he knew, wanting to receive as many invitations as possible. He hoped to be successful and from this series of visits to receive a firm offer. He decided to travel alone again because the costs of a family trip to America, which would not be covered by the IIE or other lecture fees, seemed to be too high.

Perhaps in reaction to this decision, in March his beloved Bice decided to fulfill her life's wish and travel to Italy for a month. With her inheritance from her father, she was financially independent and she prevailed over distraught Arthur's wishes while he had to stay at home in Vienna with his two sons. Her reasons for this decision can only be a topic for speculation: Was the trip in reaction to Arthur's planned-for, long trip to America, throughout which she would be left alone again for months with the children—a reason that she had already referred to four years earlier when she spontaneously fled from Vienna to Munich? Or, was it just the longing to escape from wintery cold Vienna to sunny Italy, a dream of many Germans since time immemorial? Or was Emma's need to escape her husband and his constant financial worries and existential fears? One does not know and many of these aspects may have contributed to her need for a holiday. The decision came suddenly and spontaneously, leaving Arthur very surprised as can be seen from the correspondence that he resumed in a slightly patronizing tone:

> My dearest Bicerl!
>
> Well...

[14] Letter by Prof. Henry D. Smyth (Palmer Physical Laboratory, Princeton University) to Prof. Stephen P. Duggan (IIE, New York).

> Well...
>
> Well, I'm writing to you in Milan today so that you can see that I still think about you a lot.[15]

This introduction, from his first letter, sounds like a father who has forgiven his petulant daughter her stubbornness, and the second letter also starts with double-negative reproaches:

> Dearest Bicerl
>
> So the accusation that you write too much can really not be made. You really could have let me know if you arrived safely in Villach.[16]

Finally, a postcard from Milan reaches him and Arthur seems a little relieved in the fourth letter. He also seems to have come to terms with his fate:

> I hope that my "unhappiness" about your absence is at least compensated for by the fact that you are really enjoying the trip. When I saw the splendor of the Milan Cathedral on your postcard, I became really envious.[17]

Emma Beatrice Huber-Haas had never been a daily writer; she wrote when her heart was full and she had something to say. A daily exchange of trivialities was rather repugnant to her, as she had told him on many occasions. She had taken the opportunity to travel with her brother-in-law Edi Gerstenberger in an open coupe from Vienna to Villach in Carinthia and from there she took the train to Milan. After a detour to the Côte d'Azur with a visit to Monte Carlo, she was planning a round trip through the whole of Italy down to Sicily.

Arthur was rather grumpy when only now and then a postcard arrived, in which she regularly talked about her enjoyment of or even overindulgence in Chianti but provided none of the precise travel information that he so appreciated. He did, however, continue to write his daily report about life with the children in Vienna. She wrote of Monte Carlo and Nice (Fig. 15.3), the first stops on her tour, whereupon Arthur immediately advised against Monte Carlo and the lower class of people who traveled there. He pointed out the better public in Nice and praised the choice of a cheap hotel but not without adding mocking instruction:

> Otto had warned you about Monte Carlo with regard to the audience, because there is only the "sputum," while the better people go to Nice or Cannes. I am not surprised that you find the audience so horrible. It is gratifying that your hotel room is so cheap. The only thing I can't believe is that you can see the Adriatic Sea from your window, because it is somewhere else. And since Monte Carlo has never been on the Adriatic Sea, but has always been on the Ligurian Sea, I don't think that this has suddenly changed.

He also criticized her about other things, mostly that he hadn't received enough postcards and that some most likely wouldn't arrive:

> That I get your cards at all, if you write an illegible "Vienna" and "Allemande" (Germany is "Allemagne" and Vienna is not in Germany) is a miracle in itself.[18]

---

15 AEH to EBH (Vienna, March 22, 1930).

16 AEH to EBH (Vienna, March 24, 1930).

17 AEH to EBH (Vienna, March 26, 1930).

18 AEH to EBH (Vienna, April 1, 1930).

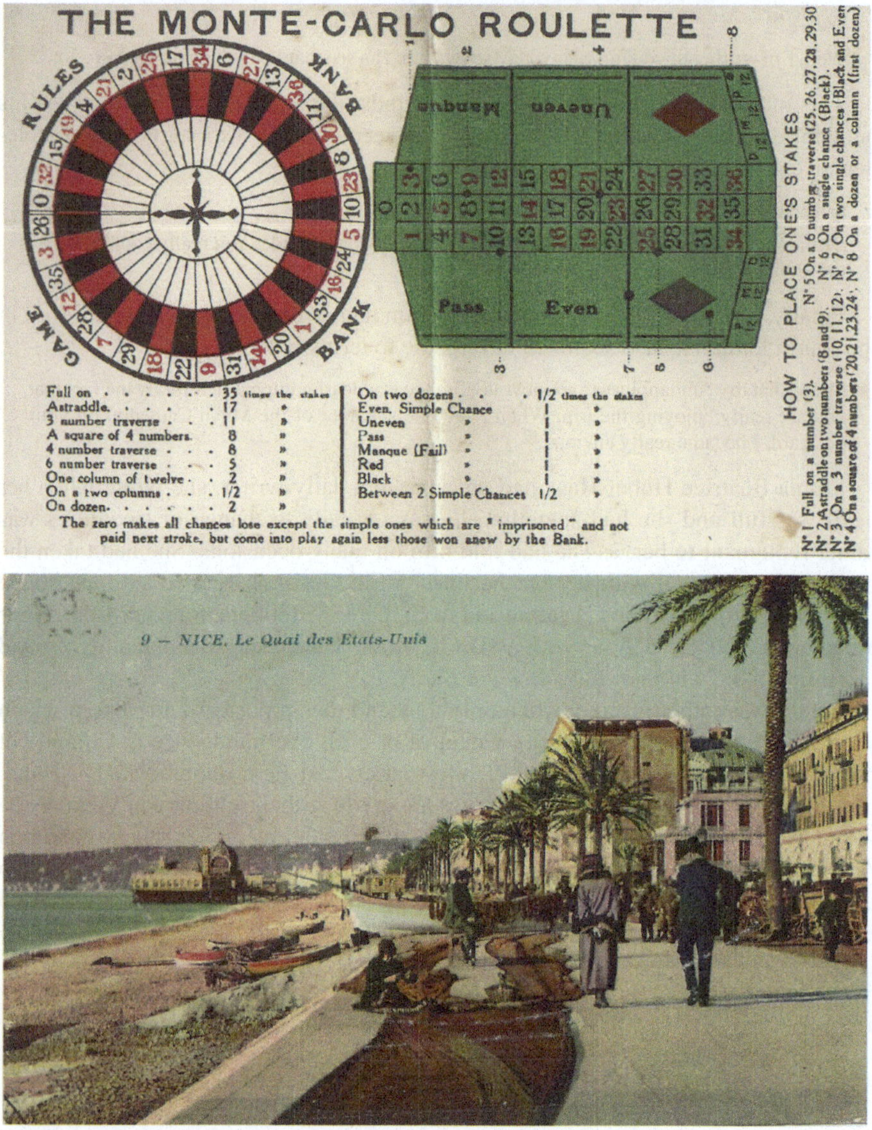

Fig. 15.3 Emma Beatrice Haas' impressions of Côte d'Azur for Arthur

The fact that Emma had moved Vienna to Germany in the address' of her letters to Arthur seemed especially annoying to him, as this faux-pas was mentioned in several subsequent letters. However, Emma did not seem to be bothered by this and she travelled on, via Genoa to Florence where she stayed for several days and enjoyed the city and its surroundings. It was early April and her first letter from Florence was posted on April 6, 1930.

Fig. 15.4 Florence from the postcard collection of Emma Haas

In each of the cities she visited—Florence, Rome, Naples, Syracuse, Taormina—she bought postcards. They depicted the Italian cities and even more so their churches, mostly in colorful collage, just as Emma Haas met them, traveling in the early season of a time when tourism was still an exception. She was enthusiastic about Florence, its museums and art treasures (Fig. 15.4). She spent several days in the city before she traveled on to Rome, not without first sending Arthur her itinerary and giving him instructions to send his letters poste restante to the respective destinations.

He, on the other hand, told of excursions with the boys and of invitations from his colleagues—the physicist Felix Ehrenhaft, the theologian Karl Beth (1872–1959)[19] with his wife, the lawyer Mrs. Marianne Beth (1889–1984),[20] and the meteorologist Viktor Conrad[21] with his wife—who all seem to have taken care of the "widower."

[19] Karl Beth was a theology scholar who was particularly interested in the religious views of Eastern Christianity. Originally from Saxony, he was Dean of the Faculty of Theology in 1938 when he was forced into early retirement. He emigrated with his wife and took over a professorship at the University of Chicago until his retirement in 1945.

[20] Born Marianne Weisl, she changed from the Jewish faith to Protestantism when she married Karl Beth in 1906. Marianne decided to study law in 1908 but was not admitted; she then studied Orientalism, a course she completed with a doctorate. In 1919 after the war, Marianne began to study law and received her doctorate in 1921. From 1928 she worked as a lawyer in Vienna, representing mainly women. In 1938 she emigrated with her husband to the United States.

[21] Victor Conrad (1875–1972) was born in Vienna. From 1896 he began studying biology at the University of Vienna but then moved on to physics. He studied with Franz Exner, Viktor Lang and Ludwig Boltzmann before turning to earthquake research and geodynamics in 1901. In 1901 he was employed as a university assistant at the Central Institute for Meteorology and Geodynamics where

Haas described these evenings in his letters as terribly boring but seems to have enjoyed himself nonetheless. Emma Haas' journey must have been an interesting topic of conversation, especially for the ladies of society who wanted to know more. Arthur asked Emma to send postcards to the respective ladies, something she faithfully did. Prof. Felix Petritsch[22] from the Vienna University of Technology also became a new friend, as they both worked on a committee to identify a suitable candidate for a job in communication engineering at Brooklyn College.

During these years, Arthur Haas seems to have been firmly established in the Viennese scientific community and appears to have done well in its social environment. Yet, during Emma's absence, his mind was on his wife's trip to Italy and he sent her daily instructions and suggestions about where to visit and what to see. He asked her several times to skip Taormina in Sicily and stay for a longer period in Rome. Whether she followed these many suggestions is uncertain; she did at least go to Taormina.

> I would like to suggest to you once again, to shorten your stay in Palermo and Taormina by twenty-four hours each, or at best to cancel Taormina completely. I would also urge you to avoid the gruesome visit to the Capuchin crypt in Palermo (where the dead are buried standing freely) at all costs, although travelers love to visit these sights. One day in Palermo should be enough, as well as one evening and one morning in Taormina.[23]

She visited the catacombs in Rome, Syracuse and Palermo; postcards arrived from Rome and Naples and the views increased Arthur's longing to get to know these cities. He had never been further south than Florence and that was before the war. He began to long for distance again and felt dissatisfaction with his own position:

> Today in the office was the most disgusting day ever. From America there are direct friendly invitations for lectures but mainly I don't hear anything and everything here is so hateful, so hateful that I can't describe it.[24]

He did not write about what it was that he hated, not as much as he had in the past. He did not express any direct criticism in his letters and only complained about his

---

he was engaged in research tasks in the field of physical meteorology. In 1938 with the annexation of Austria, Conrad emigrated to the US where he first worked at Pennsylvania State University. He moved several times, working in the California Institute of Technology at New York University, at the University of Chicago, and finally at Harvard University in Cambridge, Mass.

[22] Felix Petritsch (1878–1951) studied physics at the Vienna University of Technology. In 1911 he began working in the Austrian state telegraph administration. After the collapse of the monarchy, he left Austria and in 1918 became head of the experimental department for cables and amplifiers with the Dutch telegraph administration. Then in 1928, he was appointed professor for telecommunications engineering at the Vienna University of Technology where he built up an exemplary research and training facility for telecommunications engineering in the following years. Petritsch was a technician and, in addition, had strong literary and philosophical interests. He believed that connecting distant cities and countries through communications engineering was an important step toward world peace. In 1938, Petritsch was considered a proponent of the corporative state by the National Socialists and was retired against his will. Only after the Second World War could he resume his work.

[23] AEH to EBH (Vienna, April 13, 1930).

[24] AEH to EBH (Vienna, April 14, 1930).

general uneasiness with his situation. Was it the work or the working atmosphere that troubled him? Haas did not yet realize that his opportunities for going to America had disappeared with the beginning of the Great Depression and universities no longer had the means to attract European researchers to the States.

The friendship with Professor Petritsch, who lived nearby in Hietzing, intensified and visits were made. The two men understood each other in their technically applied interests and also maintained similar political opinions. They were deliberately conservative, rejected the Social Democrats and leaned more toward the Christian Social Party. They got along with each other and in a letter Haas even equated Petritsch with Professor Donnan from London.

Finally, on April 29th, Emma Haas returned from her long journey. It had grown warmer in Vienna and Arthur handed over the parenting role, which he had so successfully organized and executed with excursions and tours for the three Haas men. In addition, Emma was immediately commissioned to organize the summer retreat, which this time would not be to Annenheim but in Reifnitz on Wörther Lake. For this, an apartment had to be found for the summer season.

Arthur worked and also planned his lecture series for the IIE. He had no major book projects but he was working on new editions; he was also following new ideas, which lead him back to cosmology. The lectures that he was preparing for his American tour reflected some of these new interests and also incorporated his abiding interest in the wave theory of particles and its impact on relativity. In many of his thoughts, he followed Arthur Eddington, whom Haas considered to be the greatest living scientist and whom he himself placed above Einstein. Membership in the Gibbs committee also required attention, which was happily given since it encouraged hopes for the future.

At the end of June, Emma traveled with the two children to Reifnitz on Wörther Lake. The lake offered wonderful bathing opportunities and the thermal baths were particularly tempting as Emma suffered from constant discomfort and aching limbs. Warm baths promised hope for relief.

Arthur planned to follow only a few days later. His friendship with Professor Petritsch had deepened and Haas was often invited to dine there and also with Professor Conrad. Petritsch tried to find a teaching position at the Technical University so that Haas could give lectures and talks there on nuclear physics. Petritsch also suggested that Haas apply to the Technical University in Karlsruhe, with which Petritsch himself had excellent connections. A special Institute for Theoretical Physics was to be established in Karlsruhe with the additional task of instructing physics students in experimental work in the field of radio technology.[i] This opened new possibilities and Haas postponed his vacation trip. In the meantime, work at the academy became more and more unsatisfying:

> It's disgusting in the office. My mood is very bad. If you had to experience the deception that I have had to experience, it's no wonder. I start to think only in English. It's the only thing left after such disappointments.[25]

[25] AEH to EBH (Vienna, July 14, 1930).

He did not provide any details or explanations about the deceptions that were being committed. It is possible that the relationship with Actuary Junk had deteriorated. While the two actuaries worked very well together in the beginning, Junk was a secret member of the forbidden National Socialist Party in Austria at that time, a political attitude that he shared with other colleagues[ii] and could have easily led to anti-Semitic statements and taunts. There is no direct evidence for this assumption; all written statements and correspondence involving Junk point to collegial cooperation despite conflicting political opinions.

Also, the hope of a new position in Germany was quickly dimmed. Karlsruhe developed into a professorial cabal between different interest groups and factions, each of whom wanted to claim the newly created chairs for their own interests. Arthur Haas described the ups and downs of feelings and hopes in his letters to Emma in Reifnitz:

> Bad news just came in from K. A thirty-five-year-old snot boy, Viennese by birth and a friend of Smekal's, who has loose connections to the Technical University, was - of all people! - appointed to serve as evaluator of my scientific accomplishments and in his report, he spoke out against my appointment. Naturally there is hardly anything left to hope for, except that something quite unexpected happens.[26]

But then:

> You must have received my telegram, which informed you of the much more favorable assessment of the situation in K.
>
> Prof. Abel[27] today received a letter from Prof. Mark[28] (who was wrongly suspected of taking a stand against me, it seems) in which Prof. Mark resolutely speaks out for me and agrees to argue on my behalf. Mark now has a tremendous influence on the question of my appointment. The actual decision, he writes, would have to be made by the electrical engineers but they seem to be favorably disposed toward me.
>
> Abel, who has always been absolutely pessimistic, told me that he now sees things very favorably (unlike before).[29]

Arthur had to wait as things developed:

> In K., the decision will probably come soon. I am not sure whether Prof. Abel's optimism is justified.[30]

---

[26] AEH to EBH (Vienna, July 11, 1930).

[27] This is most likely Emil Abel (1875–1958), professor and director of the Institute for Physical Chemistry at the Vienna University of Technology. He was a specialist in electrochemistry. Abel was dismissed after the annexation of Austria by the German Reich and emigrated to England where he was taken in by Prof. Donnan at the University College in London, as a research assistant until 1940. Afterwards he worked as head of the research department of the dry-battery manufacturer Ever Ready Company Ltd.

[28] Hermann Franz Mark (1895–1992) was a professor of chemistry in Vienna and one of the founders of polymer chemistry. As a conservative Jew, he held the wake at the coffin of murdered Chancellor Dollfuß. Mark fled to the United States after the Anschluss in 1938, where he became professor at the Polytechnic Institute in Brooklyn.

[29] AEH to EBH (Vienna, July 18, 1930).

[30] AEH to EBH (Vienna, July 21, 1930).

And then:

> Yesterday I was with Petritsch for supper. He had driven from Steinach to Kitzbühel and met with Thirring (apparently on business) and learned by chance through Thirring that Mark is now also coming to Kitzbühel. He then informed Thirring about the K. affair and Th. will talk to Mark in detail and report to Petritsch at the end of this week when he comes to Vienna.[31]

Finally,

> I just learned from Herzfeld[32] that Sommerfeld recommended me in K. on his own initiative, even before Herzfeld had written to him.[33]

But despite this support from electrochemists and theoretical physicists, it again didn't work out. Instead of Haas, Walter Weizel[34] was appointed, who set up a group for atomic spectroscopy in Karlsruhe before he was forced into early retirement in 1933 because of his critical attitude toward National Socialism.

There is no comment from Haas on this decision; he may have been disappointed that he was denied another professorship but on the other hand, he had never had much hope and also no great desire to go to Germany. As always, he had his dreams of America in mind as the final destination.

Haas spent August with his family in Reifnitz where he was hiking and swimming in Wörther Lake, picturesquely depicted in Fig. 15.5.

Notes

i. Franz Wolf, "Aus der Geschichte der Physik in Karlsruhe," *Physikalische Blätter* 24 (1968): 388–400.
ii. See Stefan Sienell, *Das Verwaltungs-und Dienstpersonal,* 211–325.

---

[31] AEH to EBH (Vienna, July 29, 1930).

[32] Karl Ferdinand Herzfeld (1892–1978) worked in areas of theoretical quantum physics, statistical physics and physical chemistry. He received his doctorate from Hasenöhrl in 1914, so Haas was well known from his early private lecturing period in Vienna. Herzfeld was an assistant to Sommerfeld for several years and was appointed to John Hopkins University in Baltimore in 1926. He was probably, like so many of the former Viennese in America, on an annual visit to Vienna, accessed January 31, 2021, https://www.aip.org/history-programs/niels-bohr-library/oral-histories/4668.

[33] AEH to EBH (Vienna, July 29, 1930).

[34] Walter Weizel (1901–1982) was a theoretical physicist who received his doctorate in Heidelberg and completed his habilitation in Rostock in 1929. After first being with the BASF doing industrial chemistry work, he received a Rockefeller stipend to Chicago, from where he returned to accept the position in Karlsruhe.

Fig. 15.5 Reifnitz at Wörther Lake on a contemporary postcard

Chapter 16
The Second Journey to America

The vacation may have provided Haas with some relaxation, even though his thoughts were probably on his next big US trip. Before his departure to Reifnitz, he had compiled a list of all his invitations to American universities. It was to be a full round trip.

He had a total of eleven confirmed invitations and fifteen possible or expected invitations throughout the United States, listed in Fig. 16.1. This time he hoped to travel not only to the East Coast and the Midwest but also to give lectures in Western states such as Colorado and Utah, and to even fulfill several invitations on the West Coast. Particularly Caltech in Pasadena had remained an unmet dream since his application to Millikan for a position there had been denied.

In addition, there were other desirable candidates for lecturing opportunities (Fig. 16.2) and Haas was relying on the persuasiveness of Mrs. Waite, the Executive Secretary of the IIE, to open the way to these various physics departments.

Haas had prepared a series of lectures for the trip based on his latest books and writings but he was uncertain about which lectures to give at the respective institutions. He well remembered the problems of his last tour, when a topic he had chosen was less appealing to the audience. So, he sent a list of proposed lectures to the IIE, offering two general, more popular-science lectures that were also related to his current interests: "Modern Physics and our Views of the Universe" and "Light-corpuscles, Material Waves, and the Laws of Physics (from the standpoint of De Broglie's theory)." The first topic was related to the new direction of his reflections, which considered symmetries between the atomic world and cosmology. Like Eddington, he saw these symmetries hidden in the fundamental constants of physics and in the relations they shared. He hoped to make this mystery fathomable by comparing in detail the constants of their relations.

The second topic was the classical quantum-physical problem of the duality between particle and wave, which had received its first mathematical interpretation in 1927 by Niels Bohr and Werner Heisenberg in the so-called Copenhagen

M. Wiescher, *Arthur E. Haas—The Hidden Pioneer of Quantum Mechanics*, Springer Biographies, https://doi.org/10.1007/978-3-030-80606-4_16

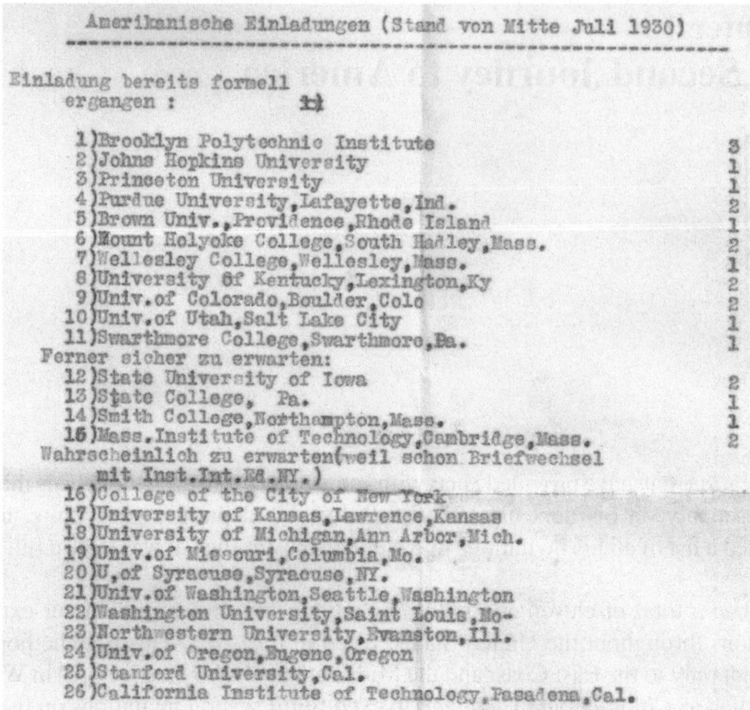

Amerikanische Einladungen (Stand von Mitte Juli 1930)

Einladung bereits formell
ergangen : 11

1)Brooklyn Polytechnic Institute 3
2)Johns Hopkins University 1
3)Princeton University 1
4)Purdue University,Lafayette,Ind. 2
5)Brown Univ.,Providence,Rhode Island 1
6)Mount Holyoke College,South Hadley,Mass. 2
7)Wellesley College,Wellesley,Mass. 1
8)University of Kentucky,Lexington,Ky 2
9)Univ.of Colorado,Boulder,Colo 2
10)Univ.of Utah,Salt Lake City 1
11)Swarthmore College,Swarthmore,Pa. 1
Ferner sicher zu erwarten:
12)State University of Iowa 2
13)State College, Pa. 1
14)Smith College,Northampton,Mass. 1
15)Mass.Institute of Technology,Cambridge,Mass. 2
Wahrscheinlich zu erwarten(weil schon Briefwechsel
mit Inst.Int.Ed.NY.)
16)College of the City of New York
17)University of Kansas,Lawrence,Kansas
18)University of Michigan,Ann Arbor,Mich.
19)Univ.of Missouri,Columbia,Mo.
20)U,of Syracuse,Syracuse,NY.
21)Univ.of Washington,Seattle,Washington
22)Washington University,Saint Louis,Mo.
23)Northwestern University,Evanston,Ill.
24)Univ.of Oregon,Eugene,Oregon
25)Stanford University,Cal.
26)California Institute of Technology,Pasadena,Cal.

Fig. 16.1 List of invitations, issued and expected, from various American universities and colleges. Arthur Haas compiled the list in the spring of 1931, planning to give lectures on his present studies and thoughts on physics

interpretation.[1] Haas had sought to reformulate this duality in his last book *Matter Waves and Quantum Mechanics* and he posed the question in his lectures:

> At the center of Schrödinger's theory is the concept of wave, at the center of quantum mechanics the concept of corpuscles. What does the electron mean from Schrödinger's point of view and what does a matter wave mean from Heisenberg's point of view?[2]

This is a question that Haas tried to explain in his lectures and books in a simplified way, with regard to the Copenhagen interpretation. He also sought to interpret its effect on the classical principle of causality in physics:

---

[1] The Copenhagen interpretation was formulated around 1927 by Niels Bohr and Werner Heisenberg in Copenhagen and is based on the interpretation of the wave function as a statement of probability, as proposed by Max Born. The interpretation posits that the particle and the wave are not realities but, rather, are probability states of the same phenomenon, which causes an uncertainty in the absolute certainty of classical complementary parameters, such as location and momentum of a particle or even energy and time of a wave.

[2] Arthur Haas, *Materiewellen und Quantenmechanik* (Leipzig: Akademische Verlagsgesellschaft, 1928), 98.

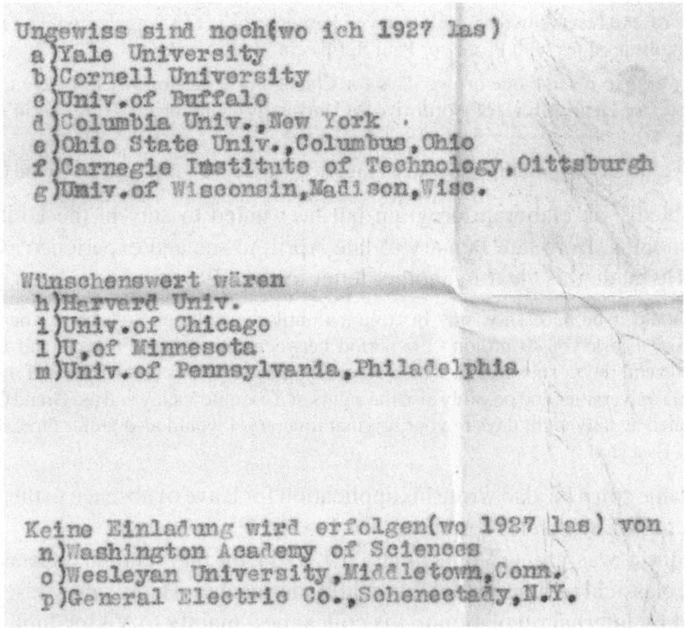
Ungewiss sind noch(wo ich 1927 las)
a)Yale University
b)Cornell University
c)Univ.of Buffalo
d)Columbia Univ.,New York
e)Ohio State Univ.,Columbus,Ohio
f)Carnegie Institute of Technology,Oittsburgh
g)Univ.of Wisconsin,Madison,Wisc.

Wünschenswert wären
h)Harvard Univ.
i)Univ.of Chicago
l)U.of Minnesota
m)Univ.of Pennsylvania,Philadelphia

Keine Einladung wird erfolgen(wo 1927 las) von
n)Washington Academy of Sciences
o)Wesleyan University,Middletown,Conn.
p)General Electric Co.,Schenectady,N.Y.

Fig. 16.2 List of further possibilities for invitations, in preparation for the 1931 trip

> If now [according to Heisenberg's uncertainty principle] a precise description of the atomic event in the classical sense is impossible in itself, the causal principle loses its importance for physics. For this principle, according to which precise knowledge of the present enables an exact calculation of the future, becomes irrelevant if precise knowledge of the present is unattainable. According to the quantum mechanical view, causality is to be negated for the elementary processes of physics and only affirmed for the probabilities that can be assigned to these individual processes for statistical reasons.[3]

In addition to these two popular-science lectures, he also offered two lectures for advanced students and doctoral candidates. One was on his new area of work with the Gibbs committee at Yale University, "J. Willard Gibbs and Modern Atomic Theory." The second was an extension of his work in which he had derived the natural consequence of relativistic relations for mass and time from the laws of wave mechanics,[4] "The Relations between Wave Mechanics and the Theory of Relativity."

In a letter to the IIE on October 18, 1930, he asked for special stimulus targets and dates:

> With regard to the itinerary, permit me to put the following requests to you:
>
> 1. In any case to reserve two days for a visit to New Haven where I have to confer with the Gibbs committee of Yale University.

---

[3] Ibid., 103.

[4] Arthur Haas, "Über die Ableitung der fundamentalrelativitäts-theoretischen Sätze aus der Brogliesсhen Hypothese der Phasenwellen," *Physikalische Zeitschrift* 28 (1927): 632–634.

2. In any case to reserve two days for a stay in Pasadena or in Los Angeles, which I need for important discussions with Professor Paul S. Epstein.

3. In any case to reserve one or two days for Chicago or Evanston where I have to confer with Professor Henry Crew (of Northwestern University) with regard to the World's Fair in Chicago.

4. To Give Me an Opportunity to See San Francisco, Yosemite Valley and the Grand Canyon.

Undoubtedly an elaborate program but he wanted to stay in the United States for three months, from late January to late April, to see and experience as much as possible. He made this clear in another letter to the IIE:

> I can, should it be necessary, stay in America until about the end of April. Therefore, I would like to ask you to prolong the period between my visits to Kansas and Utah by at least several days. This period, which would include my visits to Arizona and to several California universities and possibly also the sights of Yosemite Valley and the Grand Canyon, is estimated as only eight days in your original itinerary. I would also prefer three days for Chicago-Evanston.[5]

In the same spirit he also wrote his application for leave of absence to the Academy on October 23rd, which is reproduced for Fig. 16.3.

The request was granted on January 19, 1931 by the Secretary General of the Academy, classical philologist Dr. Ludwig Radermacher. Haas' salary reserves were distributed by internal regulation to his colleagues, mainly to Viktor Junk who took over most duties.

The IIE had announcements printed (Fig. 16.4) and made travel arrangements for visits to thirty-five universities and research institutions in the US and Canada, with Haas committed to giving two lectures at many of these locations. The tight three-month program included the requested vacation days at the Grand Canyon but Yosemite Park was cancelled; in the end, Caltech had not submitted an invitation. More than a third of this time would be spent on long railway trips, covering the eastern states, the Canadian provinces of Quebec and Ontario, the Midwest, across the western and southwestern states of Colorado, Utah, and Arizona, to the states along the Pacific coast from California to Oregon and Washington. Haas had planned an intense program for this trip, which included several face-to-face meetings with colleagues to exchange ideas and work together.

January 1931 was filled with preparations, final projects, and the handover of work at the Academy. This time he began his trip by traveling first to Berlin on January 28th, to finish some meetings with his publisher De Gruyter, and from there to Bremen where he embarked on the SS *Bremen,* of Norddeutscher (North German) Lloyd (NDL), to New York on January 31st.

Pictured in Fig. 16.5, the *Bremen* was put into service in 1929 and was considered the fastest ship on the transatlantic route. The voyage duration was five days, instead of the eight-plus days of the last voyage, and Haas enjoyed the luxury that the ship offered.

> I am wonderfully accommodated on the ship. The cabin II Class is as beautiful as on the *Hamburg* four years ago. The social areas II Class are not inferior to those of the *Hamburg* I

[5] AEH to IIE (Vienna, October 22, 1930).

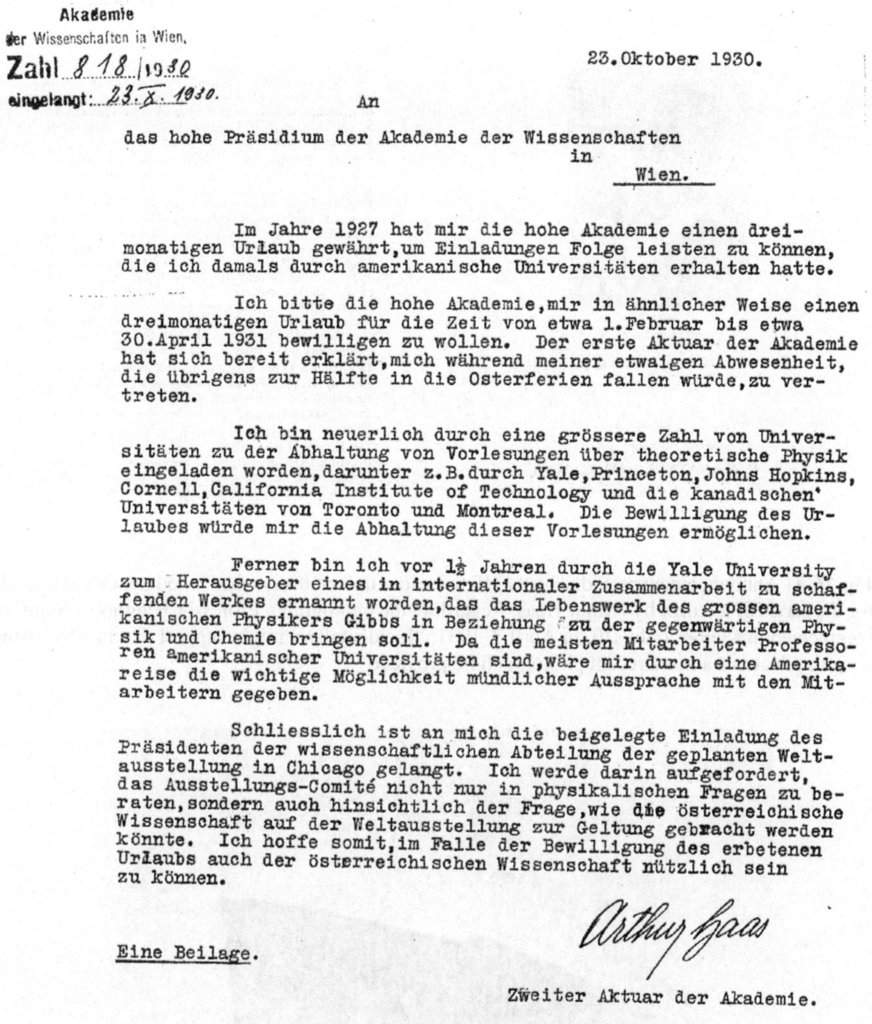
Akademie
der Wissenschaften in Wien.
Zahl 818/1930
eingelangt: 23.X.1930.

23.Oktober 1930.

An

das hohe Präsidium der Akademie der Wissenschaften
in
Wien.

Im Jahre 1927 hat mir die hohe Akademie einen dreimonatigen Urlaub gewährt, um Einladungen Folge leisten zu können, die ich damals durch amerikanische Universitäten erhalten hatte.

Ich bitte die hohe Akademie, mir in ähnlicher Weise einen dreimonatigen Urlaub für die Zeit von etwa 1. Februar bis etwa 30. April 1931 bewilligen zu wollen. Der erste Aktuar der Akademie hat sich bereit erklärt, mich während meiner etwaigen Abwesenheit, die übrigens zur Hälfte in die Osterferien fallen würde, zu vertreten.

Ich bin neuerlich durch eine grössere Zahl von Universitäten zu der Abhaltung von Vorlesungen über theoretische Physik eingeladen worden, darunter z.B. durch Yale, Princeton, Johns Hopkins, Cornell, California Institute of Technology und die kanadischen Universitäten von Toronto und Montreal. Die Bewilligung des Urlaubes würde mir die Abhaltung dieser Vorlesungen ermöglichen.

Ferner bin ich vor 1½ Jahren durch die Yale University zum Herausgeber eines in internationaler Zusammenarbeit zu schaffenden Werkes ernannt worden, das das Lebenswerk des grossen amerikanischen Physikers Gibbs in Beziehung zu der gegenwärtigen Physik und Chemie bringen soll. Da die meisten Mitarbeiter Professoren amerikanischer Universitäten sind, wäre mir durch eine Amerikareise die wichtige Möglichkeit mündlicher Aussprache mit den Mitarbeitern gegeben.

Schliesslich ist an mich die beigelegte Einladung des Präsidenten der wissenschaftlichen Abteilung der geplanten Weltausstellung in Chicago gelangt. Ich werde darin aufgefordert, das Ausstellungs-Comité nicht nur in physikalischen Fragen zu beraten, sondern auch hinsichtlich der Frage, wie die österreichische Wissenschaft auf der Weltausstellung zur Geltung gebracht werden könnte. Ich hoffe somit, im Falle der Bewilligung des erbetenen Urlaubs auch der österreichischen Wissenschaft nützlich sein zu können.

Arthur Haas

Eine Beilage.

Zweiter Aktuar der Akademie.

Fig. 16.3 Leave application by Arthur Haas to the Vienna Academy of Sciences on October 23, 1930; Yosemite Park and the Grand Canyon are not mentioned. The chief actuary Viktor Junk had again declared his willingness to take over Haas' official business for the time of his absence

> Class in terms of splendor and comfort. The ship sails very quietly and so far, thank goodness, I have had no trace of seasickness.[6]

Haas was not the only scientist on board. Giving him company for the journey, he met the chemist Prof. Karl Freudenberg from Heidelberg and the theoretical physicist Eugen Wigner from Berlin, both of whom were on their way to the United States for

[6] AEH to EBH (The *Bremen,* February 1, 1931).

ARTHUR HAAS
Professor of Physics, University of Vienna

Lecturing under the auspices of
THE INSTITUTE
OF INTERNATIONAL EDUCATION
2 West 45th Street
New York City

PROFESSOR ARTHUR HAAS

Upon receiving the degree of Doctor of Philosophy from the University of Vienna, Professor Haas became a lecturer there. He was then called to the University of Leipzig and since 1923 has been Professor of Physics at the University of Vienna. In 1924 he lectured at University College, London.

Dr. Haas is the author of "Introduction to Theoretical Physics," in two volumes, which has been published up to the present time in six German and two American-English editions, and which is being used as a text in many of our colleges and universities. Other works translated into English are: "Atomic Theory," "Wave Mechanics and the New Quantum Theory," "Quantum Chemistry," "The New Physics (third English edition, 1930"), "The World of Atoms." Dr. Haas is one of the Collaborators of the "Handbuch der Experimental-physik" and many papers by him have dealt with atomic and quantum theory, wave mechanics and the theory of relativity. He was the first physicist, even before Bohr, to apply the quantum idea of atoms.

In 1928 the Gibbs Committee of Yale University appointed Dr. Haas co-editor, with prominent American, English and Dutch scientists, of the Commentary on the Works of the great American physicist, J. Willard Gibbs.

Dr. Haas will be available for lectures from February to April, 1931, in any part of the country. This coming tour will be his second, as he lectured in the chief educational centers when he was here in 1927.

Fig. 16.4 Announcement provided by the IIE to various universities in the United States and Canada for the series of lectures by Arthur Haas. Regarding the photograph, Professor Higgins of Stanford commented in a letter to the IIE on April 7, 1931: "Incidentally, I might suggest that in advertising a visiting lecturer a more recent photograph be provided"

Fig. 16.5 The SS *Bremen* passing a lightship at the mouth of the Weser estuary

several months of research. On February 6th, the ship arrived in New York harbor on schedule. Haas wrote home with enthusiasm:

> So, for a few hours I have been back in the wonder city of New York. The tallest skyscrapers I saw four years ago appear like dwarfs next to the new ones, which are about 100 stories high;

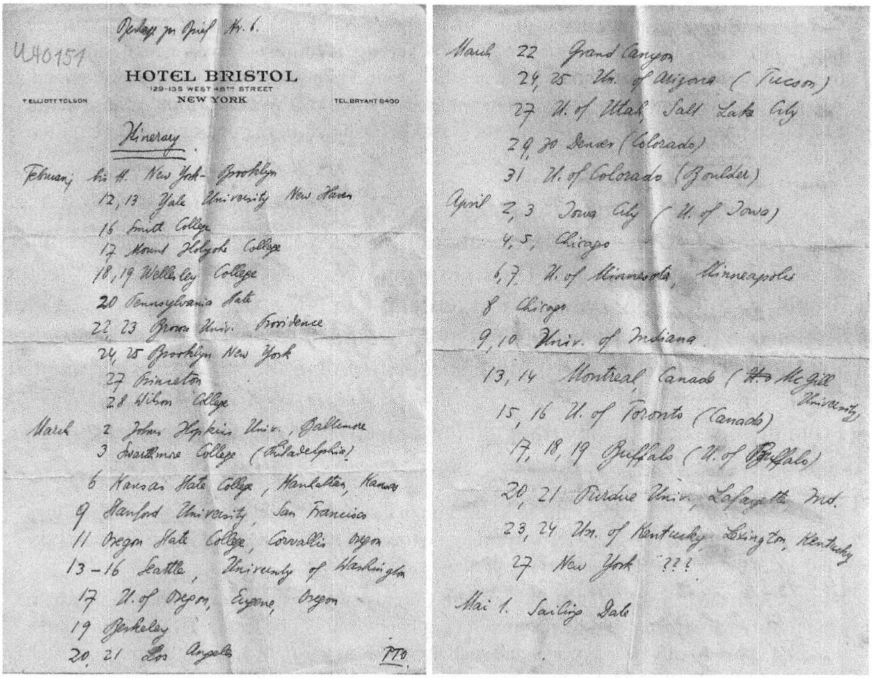

Beilage zu Brief Nr. 6.

U40151

HOTEL BRISTOL
129-135 WEST 48TH STREET
NEW YORK
T. ELLIOTT TOLSON
TEL. BRYANT 0400

Itinerary

February bis 11. New York - Brooklyn
12, 13 Yale University New Haven
16 Smith College
17 Mount Holyoke College
18, 19 Wellesley College
20 Pennsylvania State
22, 23 Brown Univ. Providence
24, 25 Brooklyn New York
27 Princeton
28 Wilson College
March 2 Johns Hopkins Univ., Baltimore
3 Swarthmore College (Philadelphia)
6 Kansas State College, Manhattan, Kansas
9 Stanford University, San Francisco
11 Oregon State College, Corvallis Oregon
13–16 Seattle, University of Washington
17 U. of Oregon, Eugene, Oregon
19 Berkeley
20, 21 Los Angeles
PTO.

March 22 Grand Canyon
24, 25 Un. of Arizona (Tucson)
27 U. of Utah, Salt Lake City
29, 30 Denver (Colorado)
31 U. of Colorado (Boulder)
April 2, 3 Iowa City (U. of Iowa)
4, 5 Chicago
6, 7 U. of Minnesota, Minneapolis
8 Chicago
9, 10 Univ. of Indiana
13, 14 Montreal Canada (McGill University)
15, 16 U. of Toronto (Canada)
17, 18, 19 Buffalo (U. of Buffalo)
20, 21 Purdue Univ. Lafayette Ind.
23, 24 Un. of Kentucky Lexington, Kentucky
27 New York ???
Mai 1. Sailing Date

Fig. 16.6 Arthur Haas' travel and visitation plans after his arrival in New York

> the magnificence of the skyscrapers, which are illuminated in the evening, is unbelievable. [...] It was a tremendous experience when I rode up the tower of the Chrysler Building (71st floor) and had an indescribable view over the sea of skyscrapers.[7]

In the meantime, further invitations had arrived and Haas sat down to make a careful list, which he immediately sent home to Vienna as seen in Fig. 16.6. He also calculated that the proceeds would be about $3,000, almost double the amount of his first visit.

He spent the first few days sightseeing and plotting, with great expectation of further offers. However, the weather was terrible; it was raining avalanches of snow in the form of sleet, which made it difficult for him to go out. But he immediately had a familiar visitor at the hotel:

> On Monday (Feb. 9) Dr. Doermann[8] visited me at the hotel and brought Dr. Halpern from Vienna (who is now also working at NY University) and (despite protests from my side)

[7] AEH to EBH (New York, February 6 and February 11, 1931).

[8] Fritz W. Doermann was a mathematician who received his doctorate in Vienna. He was offered a professorship in physics at New York State University and worked closely with Otto Halpern on the interpretation of optical phenomena and spectroscopic studies of atoms. Doermann joined the Sperry Gyroscope Company in 1942, combining his interests in electronics with those in nuclear physics. The company developed at the time to one of the largest suppliers for weapons production in the United States during World War II. Doermann's trace is lost there. See Benjamin Bederson,

they bought dinner for the three of us in the hotel [...] Halpern makes a good carrierè here.[i] On Tuesday morning I visited [...] the Rockefeller Institute of Experimental Medicine; there I met Nobel Prize winner Landsteiner[9] from Vienna and was warmly welcomed. Afterwards, President Kolbe[10] invited me - together with Prof. Weber[11] from Vienna, whom I had recommended there, and Professor Beach - for lunch in the distinguished Hamilton Club.[12]

After these initial conversations, his expectations were, however, dampened and he realized that it was not the bright America of four years ago but, rather, a country shaken by the economic crisis. The following Wednesday, he gave his first lecture at Brooklyn College on "Modern Physics and Our View of the Universe," a clear indication that his interests had shifted more toward questions of cosmology, a field that at the time was considered more speculative than the serious, experimentally verifiable physics.

From there he travelled by train to New Haven where he was received as a guest of honor:

Friday, I had many meetings in the morning and in the afternoon, I attended an official Gibbs committee meeting. I had looked forward to the meeting with some trepidation and was very glad when everything was over to my greatest satisfaction.

At 6 o'clock the Departments of Physics and Chemistry gave me a very festive dinner with about thirty participants and then came the big event: my Gibbs lecture, which was a huge success. I have hardly ever experienced such applause, at least not at any other place of such high scientific standing.[13]

Haas admired Yale and the scientific atmosphere that exists and grows there:

Yale has added some magnificent new buildings since my last visit, including a wonderful library. The warmth with which I was received was immense. It is a pity that there seems to be no possibility to accommodate me there.[14]

H. Henry Stroke, "History of the New York University Physics Department," *Phys. Perspect.* 13 (2011): 260–328.

[9] Karl Landsteiner (1868–1943) discovered the blood group systems. See Friedrich Stadtler, ed., *Vertriebene Vernunft I—Emigration und Exil österreichischer Wissenschaft 1930–1940* (Münster: Lit Verlag, 2004), 366–367.

[10] Parke Rexford Kolbe (1881–1942) was an American university lecturer and president of three American universities: University of Akron, Polytechnic Institute of New York, and Drexel University. Haas knew him well, since Kolbe had been in Vienna the previous year and they had worked closely with Petritsch on Weber's appointment.

[11] Ernst Weber (1901–1996) graduated from the Vienna University of Technology in 1924 and received his doctorate in engineering from the University of Vienna in 1926 and from the Vienna University of Technology in 1927. From 1924, he carried out development work at Siemens-Schuckert in Vienna. In autumn 1930 he was appointed professor at the Polytechnic Institute of Brooklyn. Weber was later, for many years, head of the Polytechnic Institute of New York University, where he was president from 1957 to 1969. In 1945 he founded the Microwave Research Institute there, which was later renamed Weber Research Institute in his honor.

[12] AEH to EBH (New York, February 11, 1931).

[13] AEH to EBH (New Haven, February 14, 1931).

[14] AEH to EBH (New Haven, February 15, 1931).

In the following days he visited three Massachusetts-based girls' colleges, Smith College, Mount Holyoke College and Wellesley College, where he spoke about the "Atom and Universe" to about 300 students each time. In addition, Haas enjoyed lunch and dinner with the professors, who made a great impression on him: "You don't see as many gorgeous old ladies anywhere else as in America."[15]

After a three-day detour by train—via New Haven and New York to Tyrone and Pennsylvania State College, where he had given three lectures—Haas traveled the same route back to Providence in Rhode Island and Brown University where he hoped to meet friends and awaited job offers. Professor Palmer, the then Director of the Physics Department whom Haas knew and appreciated from his previous visit, hosted him in his home. In addition to the lectures Haas gave, several receptions were held in his honor; but as Mrs. Palmer informed him on one of these occasions:

> [...] they had tried so hard (the last few years) to get me into Brown University, but the funds were not available. At Yale University everyone also talked about how nice it would be if I could come to America. But these are all just theoretical phrases. The first concrete offer has not yet been made and is most likely to be expected in the Midwest, which has the least pleasant climate.[16]

Although his lectures were also a success at Brown University, Haas felt his expectations had been betrayed and he returned from Providence to New York on the evening train. He spent two days enjoying the big city, the crowds and the night lights before leaving for Princeton and his next lecture. It was only a short, albeit very friendly visit since the IIE had booked him for further lectures at Wilson College in Pennsylvania.

> Last night I had the greatest success I have ever had here. Wilson College is a charming women's college, with exquisite "material," only 600 female students, but I think much smarter and more likeable than at Smith, etc. The female students surround me all the time and of course I enjoy that very much. The surroundings are very beautiful; they remind me of Semmering and yesterday the physics professor, Miss Weeks[17] (PhD; otherwise still young) drove me around in her car on a wonderful spring afternoon.[18]

But after this peaceful weekend, Haas went straight on to Baltimore where he was welcomed at John Hopkins University by his former Viennese colleague Karl Ferdinand Herzfeld:

> I [...] attended a lunch that Herzfeld gave in my honor (with the physics professors), gave my Light-Corpuscle Lecture at 4 o'clock with great success, then had dinner with Herzfeld

[15] AEH to EBH (Providence, February 18, 1931).

[16] AEH to EBH (Providence, February 24, 1931).

[17] Dorothy Weeks (1893–1990) received her doctorate from MIT and was Professor of Physics at Wilson College from 1930 to 1956. See Joanna Behrman, *The Other Physicists* (PhD thesis, John Hopkins University, 2020) and, accessed January 31, 2021, https://www.aip.org/history-programs/niels-bohr-library/oral-histories/4943.

[18] AEH to EBH (Baltimore, March 1, 1931).

> and was later with him at Prof. Hubbard's,[19] who had tried so hard to get me into Brown University in 1928.[20]

From there he went west, first to Chicago and from there for seven days in a sleeping car to San Francisco, but he had not taken a direct train. He had to travel on local trains to lectures at smaller colleges in the Midwest, including Iowa State in Iowa City and Kansas State in Manhattan, Kansas.

> I am terrified by the seven consecutive sleeper nights with incessant changing to local trains. But it is the only way to not renounce $100 from Iowa State and $50 from Kansas State College.[21]

But he loved Iowa City where he met a Viennese acquaintance Arthur Steindler,[22] professor and head of orthopedics at the university, who was the brother-in-law of Haas' friend and colleague Felix Ehrenhaft.

> At 7:30 pm, I gave my second lecture (it was a premiere) on "Relations between Theory of Relativity and Wave Mechanics." Both lectures were a huge success. I spent the end of the evening with Prof. Steindler (Ehrenhaft's brother-in-law) and wife. Prof. Steindler took me to the train at midnight. I spent the third of seven consecutive sleeper nights in good sleep, changed trains in Omaha at 8:00 and arrived in Lincoln (Nebraska) at 9 a.m. where I stayed for half an hour. Then it's off to Manhattan, Kansas by motor coach with several changes.[23]

The visit to Kansas State University also went successfully, following the usual pattern of lunch, lecture, and dinner that included everyone in the professorial circle. As always, the welcome was cordial as one does not always have a professor from Vienna visiting. At midnight, Haas continued for three more days on the *San Francisco Limited* from where he wrote his letter on the way through Utah and Nevada:

> The ride toward the West is wonderful. The abundance of wonderful and unusual landscapes! You often have the feeling of being on the moon. Bizarre colors, high snow-covered mountains, then hours of riding through the Salt Lake, a long ride through the completely desolate Salt Lake desert. They are wonderful, hardly describable panoramas. [...] Today, as the train split up (Los Angeles and San Francisco), I am again lonely and very absorbed in scientific thoughts.
>
> My main problem is of course the question of a permanent professorship. My main hope is the University of Minnesota in Minneapolis. This university is without a doubt one of the ten most respected in America. The climate is generally described as extremely healthy,

---

[19] John C. Hubbard (1879–1954) had a successful career in physics, with professorships at New York University and Clark University in Worchester, Massachusetts, before becoming director of the Radiation Laboratory at John Hopkins University from 1927 to 1947.

[20] AEH to EBH (Chicago, March 3, 1931).

[21] AEH to EBH (Chicago, March 4, 1931).

[22] Arthur Steindler (1878–1959) studied medicine and orthopedics at the University of Vienna. He emigrated to the United States in 1907. His sister Olga Steindler remained in Vienna and married Felix Ehrenhaft. After initially difficult years in Chicago, Steindler founded an orthopedic surgery clinic at the University of Iowa in Iowa City and over the course of thirty-four years managed its growth into a national and international orthopedic center.

[23] AEH to EBH (Manhattan, March 6, 1931).

> although dry and cold in winter. The salaries are good, the city is extremely pleasant, life is not expensive.
>
> Then one might expect an offer from the University of Washington in Seattle. Climate there very mild, scenic splendor, popular destination for pleasure travel.
>
> Then there is the possibility of Un. of Indiana in Bloomington, Indiana.
>
> Then the Institute of Internal Medicine ... Ed. first steps (most confidential) [...] toward a position at the newly founded Duke University in South Carolina,[24] also paradise with ridiculously cheap living, about eighteen-hour train ride south of New York, with high salaries (up to $12,000 for faculty).
>
> Give God that at least one thing will come to something![25]

Finally, the train arrived in California. Arthur Haas believed himself to be at the destination of his dreams and longings:

> I woke up Monday morning in the sleeping car and thought I was dreaming. A paradisiacal landscape - in front of me the waters of the Pacific Ocean, many palm trees and, after all the icy weather of the last days, summery warmth. I was in California. It became more and more beautiful. From the train I took the "ferry" (a floating station and restaurant) to San Francisco (indescribably beautiful sight) and there, Professor Higgins from Stanford University was waiting for me. [...] We first drove through San Francisco [...] Then we took an indescribably beautiful highway through the most beautiful landscape (wonderful view of the ocean) for the one-hour drive to Palo Alto, the seat of Stanford University. And suddenly, after the worst of Siberia, I was walking around without a coat as if in a summer resort.[26]

But to his greatest regret, the stay in the paradise of Palo Alto was short. After giving only two lectures, he was already rushing back to San Francisco and taking the ferry to the station on the opposite side of the bay. From there he took the train north to Oregon and the next lectures that were waiting for him. It was another strenuous twenty-four-hour train journey before the train reached the small town of Albany in Oregon. He was picked up at the station and taken by car to nearby Corvallis and Oregon State University. His lecture "Atoms and Universe," in front of 400 students, was a success. Mrs. Ullmann from Vienna was his dinner partner but then he continued by car to Eugene, the seat of the University of Oregon.

> I have never had such triumphs as in the West. [...] yesterday at the University of Oregon, I spoke to more than 600 listeners. My visit and my lecture were a great event for the city of Eugene, a charming little villa town of about 20,000 inhabitants and two giant hotels. The surroundings here are wonderful, Semmering-like. Not far from primeval forests with giant bears. If only I had one more day.[27]

But there was no time. From Eugene he was taken by car on a drive of several hours along the wide valley of the Willamette River toward the north and Portland, the largest city in Oregon. There he would give his next lecture at Reed College, a small but very selective institution. Another triumph, he reports, but he was in a hurry

---

[24] Haas is wrong here; Duke is in Durham, North Carolina not South Carolina.

[25] AEH to EBH (train to San Francisco, March 8, 1931).

[26] AEH to EBH (Palo Alto, March 10, 1931).

[27] AEH to EBH (Portland, March 13, 1931).

to continue his journey by night train further north, crossing the Columbus River and then along the coast to Seattle in Washington State.

In his letters Haas described his impressions in enthusiastic tones:

> The car trip through the Portland area (a city of about 300,000 inhabitants) was overwhelming. Nothing as beautiful as in the surroundings of the city, which is at sea level with snow-covered mountains rising to the height of 4,000 m!
>
> And in Seattle, just from the beautiful university you can see the indescribably magnificent Mount Rainier, which also rises almost 5,000m from the plain. And this splendor of the lakes and the sea's bosom around Seattle and everywhere the most beautiful forests.[28]

He enjoyed Seattle; he strolled through the city in the morning before he gave his Gibbs lecture in the afternoon and evening, followed by his successful lecture on the atom and the universe. Another success and:

> In Seattle everyone was thrilled with me and I think I would have been offered a professorship immediately - if the state hadn't cut the university's budget by a third just a month earlier. It's enough to make you cry because I also liked Seattle a lot.[29]

It wasn't meant to be. And so, he had to get back on the train and make the long return journey to California because his next three lecture dates were waiting there in Berkeley. And again, it was a paradise, not only scenically but also in terms of his lectures on March 19th; one lecture was for almost 300 listeners at the university, followed by a specialized lecture for the physics department and then a third lecture in front of the Phi Beta Kappa student union. Haas spent three days there admiring the university and he also managed to get himself to San Francisco where he strolled through the city and especially through Chinatown, which impressed him in its strangeness.

However, there were more appointments waiting. With Professor Epstein from Caltech in Pasadena, Haas had organized a working meeting and he went by train along the coast, suffering from the constantly rising heat. Further south in Glendale, at that time a small town outside of Pasadena, he was picked up by Epstein. They drove by car to Pasadena where Haas was comfortably accommodated in the Caltech guest house, the Athenaeum located at the end of the relatively small campus. However, in his letters Haas was constantly complaining about the heat and lack of shade. Daytime in Southern California didn't appeal to him but then as evening fell and it grew cooler, he reported more sympathetically about the blossoming flowers on the campus and in Pasadena.

Yet the next day, the ordeal began again when he had to be driven an hour and a half by car, while formally dressed in a tuxedo, to Claremont where he gave a lecture at Pomona College on his most coveted topic, the "Atom and the Universe." He proudly reported 700 students ("… the largest auditorium I've ever had").

The next day he dared to travel by tram and bus from Pasadena to the coast, to swim and cool down in the cold waters of the Pacific at Ocean Park beach in Santa Monica. It was indeed quite a trip because Pasadena is located at the foot of the San

---

28 AEH to EBH (Seattle, March 15, 1931).

29 AEH to EBH (Seattle, March 17, 1931).

Gabriel mountain range and far from the coast. The trip itself took several hours and although he enjoyed the swim, the cool breakers with the strong surf, he then had difficulty finding his way back to Pasadena.

Epstein had arranged a seminar for Haas at Caltech after all, which he gave the next day, but this is only mentioned in passing as part of his further travel plans:

> Today I'm reading at 2:00 pm at the California Institute for Technology, then I will be leaving Los Angeles for Tucson, Arizona at 6:00 am.[30]

The following report on his impressions of the city and the university, which he sent from Tucson to Vienna, was distinctively negative:

> Yesterday I left Pasadena and Los Angeles, respectively, and was happy to escape from this hell. In my memory, both cities will live on as murderously heated by sunlight, completely shade free, endless seas of houses without any public transport, ringing with the incessant sound of painfully shrill bells, for regulating the car traffic, and filled with thousands of faces, all chewing gum, which, when they speak, cannot be understood by anyone.
>
> The three days in Pasadena offered me two beautiful things: my lecture at Pomona College, which was a huge success, and the wonderful bath in the Pacific Ocean - but otherwise everything was hell.
>
> Also, apart from Prof. Epstein, nobody cared about me, my lecture was completely useless - at 2 o'clock in murderous heat. Almost nobody came and those who came fell asleep in the insane heat. I was completely exhausted. I doubt it was worth the pain (and no fee, by the way).[31]

He was also not very impressed by the University of Arizona in Tucson, as he wrote in the same letter:

> The university seems to be very insignificant but has 3,000 students; the city has 20,000, the whole huge state of Arizona only 400,000 - the majority of them Mexicans. It is not far to the Mexican border.

Even though Haas was not impressed, suffered from the heat and complained about the high prices, which in his opinion were twice as high as in New York, his lectures were successful, which as always filled him with satisfaction:

> In Tucson everyone was enthusiastic about me and they asked discreetly if I would like to take over a professorship with a salary of $5,000-6,000. I didn't say no and I didn't say yes, but I can't accept because life in Arizona is twice as expensive as life in New York or Boston or Berkeley. But if an offer comes up, it would certainly help me.[32]

His wife Emma was less happy about this, accusing him of making his decision all too quickly. She answers him from Paris:

> I don't understand why you turned down the professorship in Tucson - just because it's hot there? It's hot elsewhere. Just because it's expensive there? It is also expensive elsewhere. No, you were just too bad-tempered and as I know you, in that kind of mood you just don't appreciate anything. Maybe that was our only chance to get away from Vienna. A

[30] AEH to EBH (Pasadena, March 23, 1931).

[31] AEH to EBH (Tucson, March 24, 1931).

[32] AEH to EBH (Denver, March 26, 1931).

> second chance doesn't seem to come anymore. I'd like to go to Tucson right now, in spite of everything.[33]

Two Swiss professors from Tucson's physics department, Prof. Walter Soller and Prof. Buerler, took Haas to the night train on which he started the long journey back to Los Angeles. There he had to stay a few more hours, which did little to improve his opinion about the city but rather solidified it, and then he left that unloved place by train for Salt Lake City, where the next lecture date awaited him at the University of Utah. He did give two lectures with the usual success but in his letters he sounded disgruntled. Haas had neither a firm offer nor the great financial benefit hoped for from the trip; the hotel costs in the West were much higher than he had estimated and so were the costs of the tickets for his train journeys.

In Utah, Haas finally received the long-awaited letter from his wife who told him about the situation at home in Vienna, but she also told him that she had used the opportunity of his absence to visit Paris and London. As he made clear to her in a strongly worded letter, he was not prepared to pay a share of the travel expenses as these were, in his opinion, unnecessary excursions.

He was obviously still stuck in his bad mood. Even cities like Denver, where he had to spend two days doing nothing after the train trip from Salt Lake City, did nothing to improve his bad mood. In Boulder at the University of Colorado, he also complained about a lack of organization but appreciatively observed that numerous listeners had come from other universities and colleges in Colorado to meet with him (Fig. 16.7).

Haas' thoughts were, however, already in Minneapolis where he hoped for a permanent job offer. Yet in his letters, he already showed a certain purposeful pessimism:

> I'm very nervous about Minneapolis but rather pessimistic. There are no signs that I am eagerly expected there.[34]

But first something else was waiting him, a letter from Emma. She informed Arthur that she planned to extend her stay in Paris to six weeks and to then be back in Vienna in time for his return. Arthur wrote a long, horrified letter, reciting for her the financial losses that he had experienced on his current trip to America, a concern that robbed him of every night's sleep in the hotel and on the train.

He had hoped for additional income and a firm offer, and neither had come. Costs were growing, revenues were shrinking, and the many offers that Ms. Waite had temptingly promised him during the initial discussions with the IIE had not materialized yet, due to the financial plight of the universities. Now Bice, who had not yet repaid her allowance for last year's trip to Italy, had new and, as he expected, undoubtedly expensive plans for an extended stay in Paris. All of this while he, helplessly at a distance and without influence, had to watch these, in his opinion, unnecessary escapades.

---

[33] EBH to AEH (Paris, April 12, 1931).

[34] AEH to EBH (Denver, April 1, 1931).

Fig. 16.7 Arthur Haas in March 1931, standing in the center with the faculty of physics and chemistry at the University of Colorado in Boulder. From left to right: geologist J. B. Ekeley, chemist F. C. Walz, physicist W. B. Pietenpol, A.E. Haas and two physicists, O. C. Lester and M. C. Hylan

> If I do not exceed the normal budget in any way in 1931, I have a deficit of 3,500 schillings, which robs me of my sleep all the time and fills me with despair. [...] I am speechless and horrified that you think you are entitled to spend 2,100 shillings on a pleasure trip without having paid off your debt from last year. I am so horrified that I will certainly not be able to close my eyes today.[35]

This nervous breakdown was followed by a professional one. Haas liked Minneapolis and the university extraordinarily well. The professors were friendly, his lectures were successful, and he even heard that a professorship was up for grabs. All of this put him in a hopeful happy mood until he learned to his horror and despair that the department was thinking of giving the position to a physicist in the field of acoustics, in order to establish connections with the rapidly growing field of the radio industry. During recent weeks, Haas had placed all of his hopes on Minneapolis and they had collapsed within hours.

Universities' expectations had changed in recent years. In the early 1920s, theoretical physicists with profound knowledge and spectacular insights on relativity and quantum theory were desperately sought after. This market had, however, been saturated with young people, many of them from Austria and many of them Arthur Haas' former students. Now in times of shrinking resources, universities were looking for academics with applicable knowledge that might open new connections to industry and novel financial opportunities. Haas had already realized this during his previous

[35] AEH to EBH (Minneapolis, April 2, 1931).

Fig. 16.8 Impressions from the Chicago World's Fair 1933: a picture of the family in the *Chicago Herold* and a poster from the exhibition

visit to the US, when he had lectured to half-empty rooms on the more philosophically evaluated aspects of subjective and objective physics; conversely, he had filled the classrooms to bursting with the promise of usable atomic energy. His lectures on matter waves and corpuscles in quantum physics also met with approval but the direct benefit was not recognized. Perhaps, after returning home, it was this realization that prompted him to change his writing, switching from his previously preferred topics in fundamental physics to topics of physics application.

He left Minneapolis disappointed but full of expectation since he had been invited to be a technical advisor in preparation for the 1933 Chicago World's Fair. His expertise would be used to build and exhibit atomic models. On his very first day in Chicago (Fig. 16.8), he found himself pleasantly surprised:

> Well, I've already experienced a lot but now this. I was sitting on the Elevated [EL] in Chicago and a strange man across from me was reading a local newspaper [*The Chicago Herald*]. I, by chance, realized that a photo of you and the children was facing me from the newspaper sheet. - That was something new.[36]

The *Chicago Herold* ran an article about the "famous" atomic physicist from Austria but to emphasize the personal element, the family was photographed instead of Arthur. The Chicago Treasurer and Chairman of the Chicago Exhibitions Committee, Charles S. Peterson,[37] was very much impressed and excited about the idea:

---

[36] AEH to EBH (Chicago, April 9, 1931).

[37] The exhibition was an idea of Peterson's, who wanted to use this venture between Art Deco and nuclear physics to improve Chicago's diminished reputation as a killer and gangster capital. *New York Times*, April 5, 1930, 7.

> At noon, Peterson gave a big lunch in my honor in the exhibition building. During the welcome, he hugged me three times in front of everyone and gave a big speech in which he mentioned, among other things, that I am married to one of the most beautiful girls of Austria. Then he said that, "his wife is much more charming" than myself [...] I have to put up with that.

On the second day of the visit, Haas worked with Gordon S. Fulcher,[38] head of physics for the exhibition, on the planning and development of the various demonstrations that the exhibit's brochure details:

> For instance, the exhibits include a working model with steel balls instead of molecules showing how pressure is due to bombardment of the walls by molecules, which have the speed of rifle bullets. [...] The electron and the proton, tiniest of particles, can be seen individually, but when given speeds of 100 to 100,000 miles a second they are called cathode, canal, alpha, or beta rays, and produce effects which can be seen. Exhibits show luminous effects, due to cathode and canal rays, in vacuum tubes, also tracks of single alpha rays from radium, and the properties of X-rays which are produced when cathode rays strike a target. Finally, a "hodoscope" will show the paths of individual cosmic rays by means of flashing neon lamps.[ii]

Haas' remaining time was, however, devoted to science. He went to a meeting of scientists and had the opportunity to visit both the University of Chicago and Northwestern University:

> I spent the afternoon and evening at the University of Chicago. Prof. Arthur Compton, Nobel Prize winner but not yet forty, showed me around his wonderful laboratories (wonderful research). [...] Afterwards Prof. Harkins showed me his laboratories.

With this visit Haas made his first contacts with Compton and Harkins, which would lead to further cooperation in the future.

Professor Crew from Northwestern then took Haas to Union Station, from where he "was painfully roasted" on his trip to Bloomington, Indiana "in a relentlessly sun-irradiated steel wagon". Although his spirits were high when he left Chicago, his subsequent visit to Indiana University in Bloomington was again less promising. Resignation set in about his chances for getting a permanent professorship in the United States. In addition, from the onset he had had little hope regarding the two Canadian universities, McGill University in Montreal, in Quebec and the University of Toronto in Ontario. He liked the French atmosphere in Montreal but had little sympathy for the condescension with which Canadians looked on neighboring America.

During these last days of his journey, Haas wrote less and less about his impressions and reflected more and more on his long absence, with worries about his two boys and whether his wife would be back from Paris in time to meet him at the station. With his thoughts directed toward his return home, he had obviously written

[38] Gordon S. Fulcher (1884–1971) was an expert in glass chemistry and physics. He became rich through his inventions in the glass industry. In addition to his technical developments to improve the quality of glass, he also worked as an editor of physics journals and published on economic theory and psychology. (https://www.corning.com/worldwide/en/innovation/the-glass-age/inspiration/glass-heroes/gordon-scott-fulcher.html).

off the possibility of receiving an offer from any of the places he still had to visit. The enthusiasm was gone and the letters became litanies of complaint about his lack of success, the cost of the trip to America, and the cost of her trip to Paris; these recitations were repeated often word for word in each of his letters. It was probably all too much for Emma as well because her letter from Paris, on his birthday shortly before his return, ended with:

> Please, when you come back, don't start whining again. I can't take it anymore. I now know exactly how bad your financial situation is and I will be very modest. But leave me alone.[39]

After Toronto, he returned to the US; his next destination was Buffalo, the city on Lake Erie. He had already visited there in 1927 but he still went to Niagara Falls again—with a Dr. Olmsted, in his car—and then took the night train to Indianapolis.

Haas counted the days and only thought of the journey home:

> The steamer (S.S. *Europa* of Norddeutscher Lloyd) arrives in Bremen on May 8th and if I don't telegraph otherwise, I will arrive via Prague on Saturday at noon, May 9th at the Franz Joseph Station. There you can expect me [...] The nicest thing about a trip to America is that it comes to an end.[40]

This was almost literally what his former mentor Ludwig Boltzmann had said nearly twenty years earlier upon his return to Vienna from Berkeley.

But Haas' return was still a long way off. He described Indianapolis, with its rich German-immigrant colony, as interesting but he spent only the weekend there before going to the university town of Lafayette, a little further north, to give his lectures at Purdue University as he had four years ago. However, the language of the Hoosier, as the inhabitants of the state are generally called, sounded exceedingly unpleasant to him:

> The Indiana dialect is getting on my nerves. These are the Saxonians of America. When I learn that schweveral Trainsch go on the Pennschylvania Railroad, I get a nervous shock.

But he was looking forward to Purdue because there had been, since his last visit, an important change:

> The director of the Physical Department is now (since two years) a Viennese, Karl Lark-Horovitz, a thirty-seven-year-old man and "student" of mine. You may remember that he and his wife (an artist) hosted me in New York four years ago right after my arrival.
>
> He has created great things here in a short time. Of course, I was also received here with great cordiality. I gave all four lectures with great success.[41]

And on April 22nd after a last lunch with Lark-Horowitz, Haas took the train to Cincinnati in southern Ohio where he only stayed overnight. He then continued on, crossing the Ohio River to Lexington in Kentucky. There he was again warmly welcomed and even invited to attend one of the famous horse races, the Lexington Derby. But what impressed him even more was that the university offered him a temporary professorship.

---

[39] EBH to AEH (Paris, April 19, 1931).

[40] AEH to EBH (Buffalo, April 19, 1931).

[41] AEH to EBH (Lafayette, April 22, 1931).

Fig. 16.9 Arthur and Emma Haas back in Vienna

> The University of Kentucky already invited me to spend a few weeks in Lexington on a third trip to America, with a fee of $300 per week. What do you say to that? Isn't that great? It's a pity that Un. of Kentucky is too poor to offer me a sufficiently well-paid professorship at the moment.[42]

But then it was back to New York for a lecture at Bell Telephone Laboratories and the annual, traditional April Conference of the American Physical Society, which Haas didn't want to miss because he planned to deepen his new friendships. But he was already thinking about his journey home on the fast steamer *Europa* from Bremen.

It had been a long journey and, on the other hand, Emma had spent several weeks in Paris where she had visited museums, primarily the Louvre but also numerous churches. It had been an exhausting time for both parties and they had to make the adjustment back into the daily life of Vienna, as seen in Fig. 16.9. The office, with its daily administrative work, and his two sons were already waiting for Arthur.

Notes

i. Klais Taschwer, "Der verlorene Schlüssel des Otto Halpern – Zur Karriere eines theoretischen Physikers, den man in den 1920er-Jahren auf infame Weise aus Österreich vertrieb und nie mehr rehabilitierte," *Der Standard*, 2012.
ii. *Official Guide, Book of the Fair 1933*, 33–35.

[42] AEH to EBH (Lexington, April 25, 1931).

Chapter 17
From Inner Emigration to Thoughts About the Universe

Haas returned to an Austria that had, like its big neighbor Germany, turned away from the social-democratic traditions that had dominated the 1920s. Due to the economic crisis, Austria now followed more and more the nationalistic as well as radical, right-wing ideas and parties. While the universities had already proven themselves to be a bastion of right-wing extremist and anti-Semitic ideas, this mindset increasingly spread to the wider population. The Academy was no exception.

Despite his numerous letters, Haas almost never took a stand on political events and developments. In particular, the countless anti-Semitic demonstrations by right-wing student organizations in 1931 and 1932[i] remained unmentioned in his notes and letters.[1] His statements and remarks were mainly about his family, his finances, his science and the career hopes associated with it. Nonetheless, it is clear that he not only wanted to leave Austria and Germany but further, that there were no pleasant prospects for him; his wishful thinking was directed toward America, which increasingly seemed to him to be the ideal country for his family and his science.

Haas was a conservative man from the upper-middle-class of Austria and while he maintained liberal elements in his thinking—particularly regarding the co-determination of individual ethnic groups in the Habsburg Empire—he was critical of post-war social-democratic politics and he detested radical elements in politics. Even about the murder of the left-wing Jewish journalist and writer Hugo Bettauer[2] by the National Socialist Otto Rothstock, Haas commented only briefly but with a certain benevolence:

> ... Dessauer's situation seems to be slowly improving.
>
> The happy event of Vienna is that an assassination attempt was made on Bettauer, whose condition seems to be hopeless.
>
> I'm working very hard now...

---

[1] One should be aware that only letters from the time of separation from his wife are available.

[2] Hugo Dessauer (1872–1925) published a satirical magazine and became particularly well known for his novel, *Die Stadt ohne Juden* (A City without Jews) published in 1922, in which he described the devastating economic consequences for Austria if all Jews were expelled. See Hans Frey, *Aufbruch in den Abgrund* (Berlin, Memoranda Verlag, 2020), 303–305.

M. Wiescher, *Arthur E. Haas—The Hidden Pioneer of Quantum Mechanics*, Springer Biographies, https://doi.org/10.1007/978-3-030-80606-4_17

He doesn't seem to have had much sympathy, immediately turning to other topics. At least in his letters, he avoided taking a stand on the political issues of the day or even mentioning the constant unrest at the university. As some side remarks show, he had come to terms with the situation and saw the only possibility and hope in emigration—if not external, then at least internal emigration regarding political questions and disputes. He was well established, was earning well, and had set his hopes on time. But the economic and thus the political difficulties that the world economic crisis brought for Austria led to radicalization of the political parties. The political situation in Austria soon caught up with him.

On May 11th, only three days after Haas had returned to Vienna, one of the largest Austrian banks threatened to collapse under its debt burden and could only be kept liquid with international support from banks in England and France. This led to further political crises that gradually strengthened the conservative and right-wing parties, which as a coalition set the government against the social-democratic opposition. The financial crisis increased in Austria with disastrous consequences for the economy and for unemployment rates. Haas' letters and notes contain no indication of this. He had retreated into his science.

It was not easy to find his way back into the daily grind after three months of railway travel through the North American continent, but mountains of work had accumulated despite the help of his colleague Viktor Junk. Moreover, Haas had not been able to do any kind of scientific work during the entire trip; new ideas for projects and books had accumulated during his time in America, projects that he now wanted to take up. These were based on thoughts that he had presented and expanded upon in his lectures on the atom and the universe, and which he wanted to develop further in theory. He also wrote several commentaries and papers for the Gibbs committee.

In addition, he had to work on the new editions of various books, something that became his first concern. In particular, a new edition of his book *Naturbild der Neuen Physik* was overdue. The first edition had been published in 1920, followed by a 1924 edition, and in the years since then physics had been revolutionized. De Gruyter, who watched with unease as Haas published more and more with the Leipzig competition Akademische Verlagsgesellschaft, pressed for a new edition of the book, which had been successful for so long but was now outdated and had to be completely rewritten. Haas undertook this task in the summer and autumn months of 1931, before turning to new work in the winter months.

Emma was also busy; she had decided to take her final exams at the Gymnasium for the Matura[3] and was working on the preparations.

At the end of June, the time of summer retreat began in Salzkammergut at Mondsee (Moon Lake). Here the family had rented a small holiday home near Unterach, directly on the lake shore so that everyone could indulge their passion for swimming. Arthur went with their two sons while Emma stayed in Vienna for the first two weeks to continue her studies. This was followed by three weeks together before Arthur returned to work at the Academy during August.

[3] US degrees were not accepted in Europe at the time.

Several visitors came from Vienna to the Mondsee and joint car tours were made to Ischl, Salzburg and the nearby Wolfgangsee. Otherwise, the rest of the family enjoyed themselves with mountain climbing and swimming. However, toward the end of August the weather became cold and rainy and the last weeks of the summer holiday threatened to become unpleasant. In Vienna, Arthur Haas had finished his manuscript for the new edition of the *Naturbild der Neuen Physik.*

> I work with incredible efficiency. On Wednesday morning I sent an entire MS [manuscript] to America. Since then I have completed another sheet [sixteen pages] in the raw stenographic manuscript, sent guidelines to Prof. Epstein for his new contribution, and today the MS for the new edition of my "Nature Picture of New Physics." For today and tomorrow I have nothing urgent to do.[4]

Haas went to Salzkammergut for another two weeks. On September 16, 1931, he returned to Vienna; the family remained at Mondsee for the rest of the month. Once back, he was again gripped by the restless need to turn to his new ideas and projects.

Since his early doctoral studies, Haas had been fascinated by the combining of various concepts and ideas. This was already the case in his first work where he had succeeded in applying the formalism of Thompson's early atomic model to spectroscopy and had derived his first ideas about the dimensions of the atom. He was also fascinated by the duality of particle and wave, two concepts that had to be reconciled in order to describe in a meaningful way many of the ideas and observations in the microscopic world. But now he was reaching for something higher; in the similarity of the structure of the atom and the structure of the universe, he sought to recognize a fundamental common principle underlying the workings of the microcosm and the macrocosm.

Haas saw certain symmetries in the laws and fundamental constants of the various forces and models with which the physical world was described. He searched for a meaningful correlation of, as well as the secret relationships between, these constants. This was true not only for Haas but also for other theorists, including Paul Dirac and Albert Einstein, who were looking for a formalism—the world formula that includes all forces as part of a uniform basic principle and whose coupling is likely to be found in the respective constants. The publication of Einstein's alleged world formula by the Prussian Academy of Sciences on January 30, 1929 had been a bestseller and, in the following days, was translated and printed in its entirety by both the *New York Times* and the *London Times*. This was the formula of the unification of forces that Einstein himself presented in an interview in the *Daily Chronicle*:

> But only now do we know that the force that makes the electrons move on their ellipses around the atomic nuclei is the same force that makes our Earth move its wretched orbit around the sun, and brings us the light and heat that make life on Earth possible.[5]

The British astrophysicist Arthur Eddington, a close friend of Einstein's and an idol of Arthur Haas', was also looking for the world formula or "The Theory of

[4] AEH to EBH (Vienna, August 31, 1931).

[5] "Einstein's New Theory," interview in *Daily Chronicle*, January 26, 1929, quoted after Albrecht Fölsing, *Albert Einstein* (London: Penguin Books, 1998), 687.

Everything."[ii] Eddington was one of the leading astrophysicists of his time and also the first to propose nuclear reactions as the energy source for the burning of stars. He was also considered an outstanding expert on the theory of relativity and provided the first experimental proof by demonstrating the diffraction of light from distant stars by the gravitational field of the Sun. His theory was based on the relationship between elementary quantities, such as the dimension or radius of the electron—the only truly proven elementary particle at the time—and its mass.[6]

In the ratios of natural constants, Eddington saw the ratios of the micro-cosm and macrocosm reflected and he sought to extract them through complex mathematics. He understood the fundamental constants as the building blocks of the universe and compared them to the notes that make up a musical scale:

> We may ... look on the universe as a symphony played on seven primitive constants, as music played on the seven notes of a scale.[7]

Another feature of Eddington's interpretation of physical quantities and constants was the fundamental importance that he attributed to mind and consciousness. Physics, he argued, could never reveal the objective nature of things—that is, a truth—but was concerned only with the relationships between objects of observation, which are subjectively chosen by the human mind. These views are reflected in Haas' lectures and writings on objective and subjective physics.

In the years after 1924, Haas had mainly focused on questions of atomic physics and then on the problems of wave mechanics and corpuscle theory, and he had published several papers and books on these subjects. Now his attention again turned more to questions of cosmology, which he had considered from a philosophical point of view twenty years earlier. In those early years of his academic career, he had used physical arguments—such as the concept of entropy and the considerations of the then newly discovered radioactive decay chains—to demonstrate the finiteness of the universe.

The event that drew Haas' attention back to cosmology was a publication by the American astronomer Edwin Hubble in 1926. In his work on the extragalactic nebulae[8] that he had observed, Hubble used the Einstein equations straight from Haas' textbook on theoretical physics to calculate the dimensions of a finite universe. With the discovery of extragalactic systems, the question of relative velocity to the Milky Way became relevant and in 1929, Edwin Hubble published the linear correlation between these velocities and the distance of extragalactic systems from the Milky Way, the so-called Hubble relationship, which points to an expanding universe.[9]

---

[6] Arthur Eddington, *The Mathematical Theory of Relativity* (Cambridge: Cambridge University Press, 1923), 167.

[7] Arthur Eddington, *New Pathways in Science* (Cambridge: Cambridge University Press, 1935), 227.

[8] Edwin Hubble, "Extra-Galactic Nebulae," *The Astrophysical Journal* 64 (1926): 321–369.

[9] Edwin Hubble, "A relation between distance and radial velocity among extra-galactic nebulae," *Proceedings of the National Academy of Sciences* 15 (3), (1929): 168–173.

Arthur Haas argued in the summer of 1930 that the size and mass density of the universe could be determined by comparing them with the gravitational energy of the universe:

$$E_{grav} = -G \cdot \frac{m^2}{2r}$$

According to Einstein's relationship, the total energy corresponds directly to the mass ***m*** of the universe:

$$E_{tot} = m \cdot c^2$$

where ***G*** is the gravitational constant, ***c*** the speed of light, ***m*** the mass of the universe and ***r*** its radius. Since the gravitational energy in the universe must be smaller than or comparable to the total energy in the universe, the following applies

$$|E_{grav}| \leq E_{tot}$$

From this the average density of the universe can be calculated if the radius is known. Haas took values that were derived from Hubble's and other astronomers' observations at the time and calculated an upper limit for the average density of the universe: $\rho \leq 10^{-25}\,\text{g/cm}^3$ with an upper limit for the radius $r \leq 5 \times 10^{26}$ cm.

In this way Haas derived the dimensions of the universe and presented these results in 1930, in a paper for *Nature*.[10] Haas had derived the expansion velocity from the mass of the universe $M \approx 1.8 \times 10^{57}$ g a, using the Einstein equations. For the mass of the universe, he used the number given by Hubble in 1926. The value obtained for the speed of expansion of the universe from $v_{exp} \approx 2000\,\text{km/s}$ is comparable to the expansion speed of the most distant nebulae at that time.

In all cases calculated by Haas, the proposed values differ considerably from what is generally quoted today. He calculated for the mean density $\rho \approx 10^{-30}\,\text{g/cm}^3$, for the radius of the universe: $r \approx 4.5 \times 10^{27}$ cm, for the masses of ordinary visible matter $M \approx 1.5 \times 10^{53}$ g and for the expansion velocities of the galaxies closest to us $v_{exp} \approx 70\,\text{km/s}$. The differences are due to the dimensional assumptions Haas made on the basis of the observational data available at the time. Changing the assumed parameters to the values known today brings his predictions much closer to what is known from the observational data in current use: $\rho \leq 10^{-29}\,\text{g/cm}^3$, for the radius of the universe $r \approx 5 \times 10^{28}$ cm. In addition, the expansion speed is different because Haas started from a much higher value for the mass and did not include what today is known as dark matter or dark energy in his considerations.

Like Einstein and Eddington, Haas fundamentally believed in the existence of a higher being, and this thought provided the basis for the conviction shared with

[10] Arthur Haas, "A Relation between the Radial Velocities of Spiral Nebulae and the Velocity of Dissolution of Matter," *Nature* 126, 3184 (1930): 722.

Eddington[11] that all fundamental constants must be correlated and therefore can be explained by each other.[12] Eddington summarized this idea and interpretation again in detail in 1946, in a theory of natural constants. Unfortunately, he was not able to finish this work but in it, he explained once again all of his thoughts and convictions about the fundamental correlations between the constants and from which their values should result—within the correlation of the constants hides the rules that determine the dimensions and fate of the universe and all of its components. However, these views were interpreted at the time by many of Eddington's colleagues as a slip into numerology and mysticism. For his part, Arthur Haas also sought connections and correlations but he used an approach that seems rather arbitrary from today's perspective, driven by a conviction that borders more on religion than on a systematic development of thought. He continued working on these emerging concepts after his return from America.

In a 1932 paper, Arthur Haas considered the possibility of explaining the Planck constant by cosmological parameters.[13] In his 1910 work, he had tried to give the Planck constant a physical meaning by associating it with the dimensional parameters of the atom, which he considered to be the fundamental unit—in the tradition of the Greek "intellectual world" or philosophical thought. With this approach he had provided a first connection between quantum mechanics and spectroscopy. Now, twenty-two years later, he was looking for an extended explanation of the Planck constant—the fundamental dimensions and parameters of the universe.

Following the arguments of Arthur Eddington,[14] Haas took the ratio $\boldsymbol{A}$ of the radii of the universe $\boldsymbol{R}$ and that of the electron and derived the mass of the proton $\boldsymbol{m_p}$ from the square of the ratio and the mass of the universe $\boldsymbol{M}$:

$$A = \frac{R}{a} \quad A^2 = \frac{M}{m_p} \quad m_p = \frac{M \cdot a^2}{R^2} = 1.66 \times 10^{-24}\,\text{g}$$

He derived his numerical parameters for the dimensions of the universe $R = 9.6 \times 10^{26}$ cm and $M = 2 \times 10^{55}$ from the formalism of the theory of Belgian priest and physicist Canon Georges Lemaître (1894–1966), which describes the expanding universe with an empirical Hubble constant of 560 km/s Mpc.[15] Haas' resulting value of the proton mass, regardless of how well it agreed with the measured value, does naturally depend very much on the dimensions of the universe that he had adopted for his calculations. Adopting the parameters that are being used today—$R = 8.8 \times$

[11] As a practicing Quaker, Eddington was convinced that faith and the practice of science were not contrary ideals and schools of thought but, rather, indivisible components of existence. See Mathew Stanley, "Eddington's Limits of Knowledge: The Role of Religion," in *Information and Interaction,* eds. Ian T. Durham, Dean Rickles, (Cham: Springer, 1917) 35–44.

[12] Arthur Eddington, *Fundamental Theory* (Cambridge: Cambridge University Press, 1946).

[13] Arthur Haas, "Wirkungsquantum und kosmische Konstanten,“ *Die Naturwissenschaften* 50 (1932): 906.

[14] Arthur S. Eddington, "The Expanding Universe," *Proceedings of the Physical Society London* 44 (1932): 1–16.

[15] Today the Hubble constant is about 70 km/s per megaparsec (1 Mpc $= 3.09 \times 10^{19}$ km).

10^{28} cm and $M = 1.5 \times 10^{56}$ g for visible matter or $M \approx 3 \times 10^{57}$ g as total mass (including dark matter and dark energy)—the postulated relation of the mass of the universe to the mass of the proton cannot be confirmed by Haas' calculations since the predicted mass of the proton would be substantially smaller than measured today.

Haas continued the argument of the close relationship between the macroscopic dimensions of the universe and the microscopic dimensions of the particle components of the atom by predicting that the cubic value of the radius ratio $\boldsymbol{A^3}$ should describe the relationship of a cosmic universal constant ***M-R-c*** and the Planck constant ***h***:

$$A^3 = \frac{M \cdot R \cdot c}{2.14 \cdot h}$$

The relationship, using the available numerical parameters, required a correction factor of 2.14, which he did not explain further but attributed to the inaccuracies of his input parameters.[16]

With these arguments Arthur Haas sought to find his way back to science, a path which he would continue to follow regardless of political and personal interruptions. In 1932, cosmology was still largely speculation, as the number games of Haas and Eddington underline.[iii] Yet, they had entered a new world of thought, according to which the cosmos had become subject to physical conditions. Only six years later, these thoughts would lead to the first international conference on cosmology at the University of Notre Dame with more than 100 participants, thus attributing scientific relevance to this field.[iv]

On the other hand, while Haas was engrossed in his world of physics, Emma was also trying to find her way by studying for her Matura. This year it was she who sacrificed her summer retreat with the family because, after passing the external entrance examination, she was admitted to the Grillparzer Realgymnasium and had to take part in the necessary preparatory classes. Arthur traveled with his two sons to familiar Sattendorf to find a holiday guesthouse on the lakeside near his sister. However, an express letter from his wife, with news of a court summons, soon called Haas back to Vienna. The summons concerned inconsistencies at the academy that are not further described, but which he dismissed in his reply with the short remark:

> The court summons, of which you enclose a note, unfortunately again concerns the lunatic who persecutes me at the Academy. [...] Indeed, I left Vienna in a very distressed state, and I am only happy that I did not have to join in the final hullabaloo of the Academy in the heat of madness that presently reigns in Vienna.[17]

The "lunatic" was most likely the war-invalid Julius Matula (1885–1972), who served as a minor official at the Academy. In 1931/32 he was accused not only of drunkenness at the Academy but also of embezzling correspondence and letter postage.[v] Based on the January 22, 1932 testimony of Arthur Haas, who had lost

[16] Arthur Haas, "Über die Möglichkeit einer kosmologischen Deutung des Wirkungsquantums," *Sitzungsberichte der Akademie der Wissenschaften Wien* 69 (1932): 256–257.

[17] AEH to EBH (Sattendorf, July 9, 1932).

several mailings due to Matula's embezzlement, Matula was punished with early retirement [...] along with a 2% deduction from his pension.[vi] An extraordinarily lenient sentence but, with his retirement, Matula also had to vacate his official residence in the academy building in April 1933, a circumstance for which he most likely blamed Arthur Haas. We have no further written information about the outcome of the matter from Haas.

Only on July 20, 1932 did the correspondence with his wife resume. Emma had retreated to the mountains in order to find relaxation and opportunity for high-altitude air treatment at the Kanzelhöhe in Carinthia above Lake Ossiach. Arthur stayed in Vienna with the boys, making day trips with them and mainly devoting himself to their swimming lessons. The joint summer retreat in Sattendorf was then planned for August.

Although Arthur resumed work at the Academy, he devoted most of his efforts to preparing a new series of lectures since he had applied to the IIE for a third trip in the following year, 1933. With this new attempt, he hoped to finally get a permanent position across the ocean and far from the Vienna riots.

> I now have a lot of worries about my American lectures. The task I set myself and have to set myself is incredibly difficult and I often despair whether I will be able to do it. Hopefully I will get the necessary ideas in Sattendorf. I have to take a lot of books with me.[18]

However, there would only be a week of summer vacation together because, as Emma explained, she had to return to Vienna to study. This was a decision that Arthur was initially very bitter about; instead of being bent over his books in preparation for his lectures, he now had to swim in the lake with the boys and hike in the mountains. Yet as the following letters demonstrate, this did him a great deal of good and he reported in enthusiastic tones about their joint undertakings.

He also wrote enthusiastically about his correspondence with Ms. Waite from the IIE; numerous invitations had been received, many from institutions he had not yet visited, and some invitations such as the one from Seattle extended over several weeks. His popularity was growing and the many positive reviews that his lectures had received in the previous year were now part of the IIE's documentation. Planning for 1933 was progressing well and Arthur Haas was already expecting good additional income. He assured Emma in each daily letter that a position in America would almost certainly become available next year.

But then despite the positive feedback and the many initial engagements that had been secured, the total number of invitations remained low relative to a year ago. As Ms. Waite wrote in October 1932:

> I must warn you that because of the severe depression in this country, that you cannot expect to receive as many invitations as you did, when you were here in 1931.[19]

Even a month later, things did not look any better:

---

[18] AEH to EBH (Vienna, July 30, 1932).

[19] IIE to AEH (New York, October 11, 1932).

> In response to our letters a number of institutions which you visited, when you were here previously, have written that due to lack of funds they will not be able to invite you. We have not yet heard from Duke University and the University of Washington. As we have written several letters with no response, I fear this means that these engagements are not coming through this year.[20]

In his reply letter, Arthur Haas canceled his travel plans; the invitations he received would not have covered even half of his travel expenses:

> To my deep regret, I do not see any other possibility than to ask you to cancel my engagements for this season. I feel unhappy that I shall appear impolite to the colleges that honored me by their invitations. I am especially unhappy about all the trouble that I caused you, which now appears fruitless, at least for this season - but I hope you will understand my very difficult situation and forgive me.[21]

Only two years later, under President Roosevelt, would the economic situation in the United States improve to such an extent that American universities could provide new funds for foreign visitors. During these two years there were, however, dramatic changes in the political situation in Germany and Austria, changes that would have a great influence on Professor Haas' future.

The political developments during that time are largely known and will therefore only be summarized here. In the German Reichstag (Parliament) elections of January 1933, the National Socialists under Adolf Hitler achieved a favorable election result with 33% of the vote. President Hindenburg then appointed Hitler as chancellor of a minority government on January 30, 1933. Declaring the arson at the Reichstag (German parliamentary building) on February 27, 1933 an act of communist terror, Hitler successfully pushed the Enabling Act through German Parliament. The new law came into force on March 24, 1933, giving Hitler more or less dictatorial power. This was the end of the Weimar Republic and democracy in Germany and the beginning of National Socialist dictatorship.

In neighboring Austria, the Christian-Social Chancellor Engelbert Dollfuß also ruled with the help of Enabling Acts, some of which had been established during the wartime period around 1917 and were initially intended to stabilize the Habsburg imperial government. The Acts were, however, being implemented to stabilize the conservative government *against* the Social Democrats. Dollfuß, who came from a small-farming background, had been appointed chancellor by Austrian President Miklas in 1932, after a rapid political rise at the age of only thirty-nine. Dollfuß represented the Catholic conservative bloc of parties but he also supported the Heimwehr (Homefront), which later became the Austrian National Socialist Party; the government had only a one vote majority in parliament. Then on March 4, 1933, under the pretext of a crisis in the Rules of Procedure, Dollfuß declared the National Council incapable of action. He dissolved the Parliament and its members were prevented by police from entering chambers.

Dollfuß then took advantage of the rapidly developing unrest and fighting that threatened to turn into a civil war between the various parties' troops. He banned

[20] IIE to AEH (New York, November 11, 1932).

[21] AEH to IIE (Vienna, November 28, 1932).

the Social Democratic Party in February 1934 and established a Fascist-Catholic regime closely aligned with Mussolini's Italy. Dollfuß not only fought the Social Democrats with his political maneuvers but also tried to prevent a takeover by the Austrian National Socialists, from whom he sought to clearly distinguish himself. The latter, encouraged by their party's seizure of power in neighboring Germany, sought to force a takeover in Austria.

None of the Haas family's letters from 1933 have survived to shed light on the political position of Arthur Haas during these events. However, given his conservative attitude, he most likely supported Dollfuß politics. There is no doubt that Haas only saw hope in a strong Catholic conservative bloc against the threat of Nazi dominance in Austria, as later comments demonstrate. While there are no letters from 1933, his comments three years later in a letter to a priest of German origin in the United States, underscored Haas' political stance. In this letter he cited his political views as the main reason for his emigration:

> The concern was that Austria would soon become the prey of the Nazis, which for me would have been the destruction of my existence if only because I had always been a zealous member of the Vaterländische Front [Patriotic Front] founded by Dr. Dollfuß.[22]

In a later chapter we will return to this letter's recipient, Reverend Sigmar who played a somewhat shady role for Haas.

The Vaterländische Front had been proclaimed a rallying movement by Dollfuß, following the example of the fascist movement of Mussolini in Italy, to mobilize all patriotic and pro-state Austrians. This "Austrian Front" was intended to arm the state against threats from Marxists (both Social Democrats and Communists) and Nazis. Haas' friend Professor Petritsch of the Technical University was a leading representative of this movement, now considered fascist, which formed the backbone of the Austrian corporative state that gave dictatorial power to Chancellor Dollfuß. Arthur Haas apparently also became a member. This civilian movement was founded with a military component, which included the Heimwehr and other associations capable of military action. The civilian organization was based on the participation of professional organizations from agriculture, commerce, trade and industry. These organizations, or corporations, were to become the bearers of political will and form the basis of the corporative state.[23]

Haas recognized the growing political threats; this is not necessarily demonstrated by political activity or politically motivated statements but, rather, by his financial transactions. Since 1931 he had been investing the income from his lectures and from foreign, mainly British and American publishers into an account with the Swiss Bank Corporation. This explains his increased complaints about a lack of money, since he had not only invested this part of his income permanently in Switzerland but he had also, in the economic and political crisis years after 1932, liquidated his various share packages from Polish and Czech stocks. He also invested these funds in Switzerland,

---

[22] AEH to Rev. Julian Sigmar, C.S.C. (Brunswick, April 15, 1936).

[23] With the collapse of the corporative state and the annexation of Austria by the German Reich in 1938, the functionaries of the Vaterländische Front, if unable to flee, were arrested and taken to concentration camps; the assets of the movement were confiscated by the new rulers.

mainly in dollars, so that he could have access from the United States. In view of the constantly tightening foreign exchange regulations in both Germany and Austria, these actions provided him with secure backing, which he continuously sought to expand.

In the early 1930s, the Academy had maintained political neutrality, although political views naturally diverged internally. But Haas kept himself out of these discussions and was able to get along well with his colleague Junk, who was a supporter of the National Socialists. And so in 1933, Haas continued to concentrate on his scientific and journalistic work. Already in the autumn of 1932, he had signed a contract with the Akademische Verlagsgesellschaft in Leipzig with the aim of writing a book on cosmological questions and developments in physics. In spring 1933, he offered a course at the University of Vienna in which he tried to summarize his thoughts and considerations on the physics of the universe. He translated his course notes into the new book, during the summer and autumn of that year. In the autumn of 1933, after a successful summer retreat in Sattendorf, Arthur Haas completed the manuscript on cosmology, which was then published in the spring of 1934.[24] Haas' *Cosmological Problems of Physics* is one of the first popular-science books published on cosmology, alongside the writings of Arthur Eddington.[25]

The thoughts and conclusions resulting in this book were to shape the scientific work of Arthur Haas until the end of his life. Much of the work is, however, misconception. The conclusions are, like Eddington's, explained by a thorough understanding of the theory of relativity but are ultimately lost in number speculation due to philosophical and aesthetic considerations. It is interesting that both Eddington and Haas were among those who unconditionally represented and defended the reality of the cosmological constant resulting from Einstein's calculations. This is a prediction that Einstein himself described as the greatest folly of his life, but today this is identified as dark energy, a central point in the modern understanding of cosmology.[vii]

The first five chapters of Haas' book are primarily aimed at the generally interested reader and are therefore free of mathematics. They provide a more vivid, easy-to-read summary of the recent observations regarding extragalactic nebulae, the expansion of the universe, the age of the universe, and the experimental and observational methods used to determine it. He dedicated the second chapter to the question of the expansion of the universe as observed by Hubble, which is the central point of Eddington's book, *The Expanding Universe.* The third chapter is dedicated to the question regarding the age of the Universe; a large amount of observational data is quoted, which he combined with arguments about speed and duration of the observed expansion. Based on these arguments he concluded that the age of the universe must be more than two billion and up to several billion years.[26]

In the fourth chapter Haas presented the problem of the energy balance of the universe. He identified the fusion of four hydrogen atoms to form helium as the

[24] Arthur Haas, *Kosmologische Probleme der Physik* (Leipzig: Akademische Verlagsanstalt, 1934).

[25] See Arthur Eddington, *The Expanding Universe* (Cambridge: Cambridge University Press, 1931).

[26] Based on modern observation, the universe is considered to be 13.787 ± 0.020 billion years old.

most probable source, following earlier proposals by Eddington[27] and more quantitative predictions by Atkinson and Houtermans.[28] Haas' main interest was, however, the phenomenon of cosmic rays, which were first detected by his friend Viktor Hess in balloon experiments and identified as incredibly high-energy particles. Haas suspected an extra-galactic origin, perhaps a signal from the very first moment the universe was created.[29]

In the fifth chapter Haas presented discussion and arguments for a finite universe. He kept his arguments general, in easy-to-understand language but then followed, in the next two sections, with the mathematical derivation of the Einstein model and the generalized Friedmann-Lemaitre universe model. From these Haas derived the dimensional parameters of the universe such as size, volume and mass, but also the cosmological constant, which he, like Eddington, considered to be essential, although Einstein had begun to doubt the necessity of parameters in his equations.

The following eighth, last chapter was devoted to Haas' own thoughts on the correlation between the fundamental constants. With this he tried to formulate a simple scaling symmetry between the macroscopic and the microscopic worlds, in order to use these correlations to derive the dimensions of the universe from the dimensions of the quantum world. In the concluding words of the book, Haas justified his approach with a quotation from Goethe:

> If in the *infinite* you want to stride,
> Just *walk* in the *finite* to every side.[viii]

Haas was guided by this motto and came in the final chapter to his conclusion:

> How strange it would be if it were really possible to recognize the size of the universe by passing through an unimaginably tiny atom in one's mind.

Already in July 1934, shortly after appearing in print and although written in German, the book was reviewed in *Nature*.[30] In the review, Haas' *Kosmologische Probleme der Physik* was described as an "excellent and objective summary" of the topic "for both experts and non-experts." The last part of the book, in which Haas presented his own ideas on the correlation of the constants of the micro- as well as the macrocosm, is mentioned with a slightly dismissive sounding side note, without any evaluation: "A concluding chapter is devoted to Eddington's and Prof. Haas's own speculations on the connection between the cosmological constant and the constants of atomic physics." Unfortunately, the name of the reviewer is not known.

The reaction of the wider scientific community was similar; Arthur Haas' ideas and suggestions did not really catch on, although he published updated versions in the

---

[27] Arthur S. Eddington, *"The Internal Constitution of Stars,"* 14–20.

[28] Robert E. Atkinson, Fritz G. Houtermans, "Zur Frage der Aufbaumöglichkeit der Elemente in Sternen."

[29] Georges Lemaitre, Manuel Vallarta, "On Compton's Latitude Effect of Cosmic Radiation," *Physical Review* 43 (1933): 87–92.

[30] *Nature* 134, 3378 (1934): 125.

Anglo-Saxon literature.[31] Similar to Arthur Eddington's arguments about the correlation between the Sommerfeld fine-structure constant and the number of protons in the universe, more precise measurements of these constants have shown that these arguments were based on random similarities rather than on basic connections between macro- and microcosm.

Arthur Haas' book on cosmology was, apart from various new editions of his older works on theoretical physics and atomic theory, his last publication with a purely scientific purpose. The book signals a change in Haas' life plan—he had given up hope of a secure academic career in theoretical basic research and put his hopes on a career in close contact with industry, in popular scientific descriptions of the technical concepts of the modern age. This is underlined by the fact that during these years he actually signed two contracts for book projects in this application-oriented direction; in 1932 he promised the Akademische Verlagsgesellschaft a book entitled *Metal Theory*, which he never wrote and, in 1933, he contacted with Julius Springer Verlag with whom he published *Physik für Jedermann* (Physics for Everyone) in December.[32]

With this new book Haas tried a completely new approach by presenting the whole of modern physics to the public in a generally understandable way. Haas saw hope for a rich technical future in the development of modern physics, which was increasingly portrayed and denigrated as "Jewish physics" by the radical right-wing parties in Germany and Austria. This phenomenon had increased greatly with the National Socialists' rise to power in Germany in January 1933. Haas therefore attempted to provide a broader understanding of the new approach to physics in order to counter pseudo-scientific ideas, which had been strongly promoted by the National Socialists in particular.

Especially the theory of relativity, formulated by Einstein, was a thorn in the side of the new leadership in Germany, despite experimental confirmation. Numerous pseudo-scientific ideas were formulated and presented as scientifically proven alternatives to the theory of relativity and the resulting statements regarding astronomy and cosmology. The authors of these alternative ideas were often engineers and other technically educated writers who based their work primarily on the mathematical and physical tools of classical physics. This reliance on classical models can mainly be attributed to a general lack of understanding of the complex new theories, such as the theory of relativity and quantum physics, but was later also politically motivated. Since it was predominantly Jewish scientists who were involved in the development of these theories, together with all of the numerous Jewish scientists who were proponents of them, attempts were made to discredit these developments as Jewish physics and to support alternative-world thought. The best-known examples of the alternative-world models are the Hollow Earth theory[33] and the World

[31] Arthur Haas, "A Relation between the Fundamental Constants of Physics," *Physical Review* 49, (1936): 635 and "The Size of the Universe and the Fundamental Constants of Physics," *Science* 84, 2191 (1936): 578–579.

[32] Arthur Haas, *Physik für Jedermann* (Heidelberg: Julius Springer Verlag, 1933).

[33] The Hollow Earth or Inner Earth theory claims that we do not live on the surface of a planet orbiting the sun but on the inside of a hollow sphere that surrounds the sun as its center. While this

Fig. 17.1 Contemporary publications for the propagation of the Hollow Earth theory by the engineer and astrologer Johannes Lang[35] and on the World Ice theory by SA officer Rudolf Elmayer-Vestenbrugg,[36] who published numerous writings for dissemination of the Hörbiger theses in national and Nazi circles

Ice theory,[34] both of which received great support from the party apparatus of the National Socialists (Fig. 17.1). This led to bizarre excesses and consequences for science in Germany during the Third Reich, which Haas wanted to counteract with his popular scientific description of modern physics.

Arthur Haas' book, *Physik für Jedermann*[35] can thus be interpreted as an answer to these pseudo-scientific theses. This book developed into a great commercial success

largely religiously motivated theory of American physician Cyrus Reed Teed (1839–1908) could only be made possible by relatively complex coordinate transformation within the framework of classical mathematics, it contradicted the predictions of modern physical theories such as the theory of relativity, which has since been confirmed by a large number of experiments. In the 1930s, the hollow-world theory was presented to the public as a "new world view" by the writer and astrologer Johannes Lang (1899–1967), among others. Even in 1963 the Austrian engineer Johann Dolanski (1889–1966) published a paper in which he wanted to present the hollow-world theory on a scientific basis. The Apollo 11 landing on the moon, which according to the hollow-world theory should be a sphere of 3000 km within a sphere of 12,000 km diameter, can probably be interpreted today as an absolute refutation of the theory.

[34] The Austrian engineer Hanns Hörbiger (1860–1931) developed the World Ice theory in 1913. It states that the moon, stars and the entire Milky Way band consist of glittering lumps of ice, which were formed during the explosion of a primordial planet after collision with or penetration by a gigantic primal sun. In doing so, he contradicted the idea of an infinitely expanding universe, as represented by modern science, and with it, Arthur Haas' cosmology.

[35] The title of the book may be an allusion to the well-known play, *Jedermann* by the contemporary Austrian writer Hugo von Hofmannsthal, which Haas regularly saw at the opening of the Salzburg Festival.

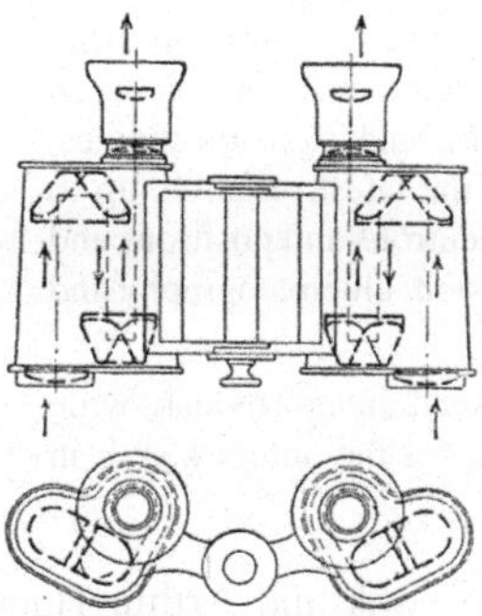

Abb. 20. Prismenfernrohr.

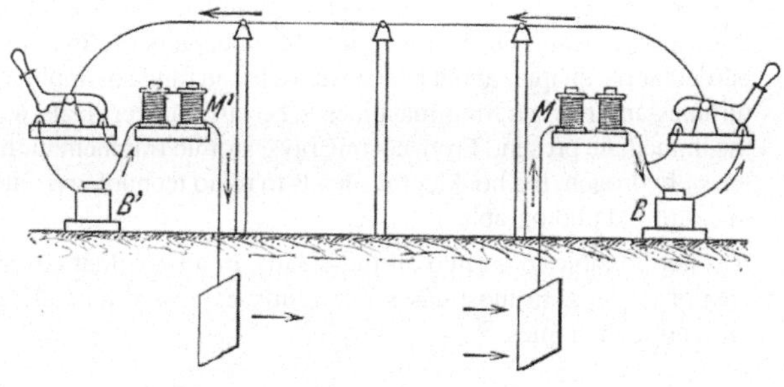

Abb. 46. Schema der Telegraphie.

Fig. 17.2 Two schematic representations from *Physik für Jedermann*. On the left is a depiction for the function of a prism telescope, in which incident light is directed to the eye. On the right is a display of radio telegraphy, the preferred international communication medium at the time

for the publishing house Julius Springer, which at the time was aggressively leading the scientific-literature market.[ix] The book covered the entire field of physics; for example, in contrast to the *Naturbild der neuen Physik*, this latest work included the classical physics of optics and electricity. What was different about the text was that it was not based on the philosophy of science but rather on practical applications; Haas included many examples, technical drawings, and photographs with which he especially intended to attract readers interested in technology. Figure 17.2 shows some of these drawings.

Physik für Jedermann was published in several editions and also several translations during the following years. The specialized press as well as the general press praised it as a generally understandable presentation for educated non-scientists, although some of the reviewers turned their noses up at this popular scientific approach.

In terms of content, the book covered, apart from classical mechanics, all of modern physics in vivid language. The Berlin mathematician Paul Kirchberger wrote in a review:

> It takes a man who can draw from the full, whose knowledge is on the one hand broad enough to oversee this immense field and, on the other hand, deep enough to select these few words with all the more precision because a few clues, for very difficult questions, must suffice. That is what has happened here in full.
>
> The book is not intended to be a textbook, and still less a first introduction to the scientific study of physics, for it deliberately pushes back the methodological aspects and instead puts the results of research in the foreground.

> The fact that the discussion of mechanics and acoustics has been omitted is probably what most readers want. Of the remaining four chapters on light, electricity, heat and matter, in particular the chapter about heat covers a broad range of applications, such as steam engines, turbines and refrigerating machines. The last chapter deals with the latest research, up to the immediate present. From cosmic rays, atomic fragmentation, neutrons and positrons and heavy hydrogen, the booklet thus leads to radio technology, television, cinematography and the ultra-red photograph.
>
> It is not possible to go into all the details, in a book that covers such an enormously wide area of physics. In most cases I can only express my admiration for the author's skill in presenting the topics.[38]

Other reviewers also expressed exuberant praise for the book and Arthur Haas proudly put together the first eight reviews for his publisher. This list is provided here, in Fig. 17.3.

Notes

i. Klaus Tachwer, *Hochburg des Antisemitismus* (Vienna: Czernin Verlags GmbH, 2015), 142–158.
ii. Helge Kragh, "Eddington's Dream: A Failed Theory of Everything," in Dean Rickles and Ian Durham, eds., *Information and Interaction: Eddington, Wheeler, and the Limits of Knowledge* (Heidelberg: Springer Verlag, 2018), 45–58.
iii. Helge Kragh, *Higher Speculations—Grand Theories and Failed Revolutions in Physics and Cosmology* (Oxford: Oxford University Press, 2011), 110.
iv. Helge Kragh, *Cosmology and Controversy—The Historical Development of Two Theories of the Universe* (Princeton: Princeton University Press, 1999), 96.
v. Cf. Stefan Sienell, *Das Verwaltungs-und Dienstpersonal*, 270–271.

---

[35] Johannes Lang popularized the mathematically and geometrically complex theory in the 1930s. Leading representatives of the National Socialist Party supported this thesis, which was partly ideological and also partly involved military strategical thinking—because life in the high sphere would make it easier to locate enemy fleets and troop movements—and in 1942, with the support of Göring and Himmler, radar was installed on the island of Rügen to underpin the hollow-earth thesis. The experiment to measure the opposite hemisphere by the reflection of radar waves failed. Cf. Hans Frey, *Aufbruch in den Abgrund: Deutsche Science Fiktion zwischen Demokratie und Diktatur* (Berlin: Memoranda Verlag, 2020) and Louis Pauwels and Jaques Bergier, *Aufbruch ins dritte Jahrtausend—Von der Zukunft der phantastischen Vernunft* (Berlin-Darmstadt-Wien: Deutsche Buchgemeinschaft, 1965), 356–364.

[36] Rudolf Elmayer-Vestenbrugg (1881–1970) was an Austrian engineer from Graz and a distinguished author of National Socialist texts. As SA Obersturmführer, he published several popular depictions of the theses of Hanns Hörbiger. As a convinced representative of the theory of world ice, he was in the camp opposite the hollow-world theory proponents, so that he fought bitterly against not only the theses of the theory of relativity but also the competing theory of the hollow-worlders. Cf. Rudolf Elmayer-Vestenbrugg, *SA Männer im feldgrauen Rock* (Leipzig: Hase & Koehler, 1941).

[38] Paul Kirchberger, "Rezension von Arthur Haas: Physik für Jedermann, it besonderer Berücksichtigung der modernen technischen Anwendungen," *Deutsche Literaturzeitung* 5 Jahrgang, /7 (1934): 328.

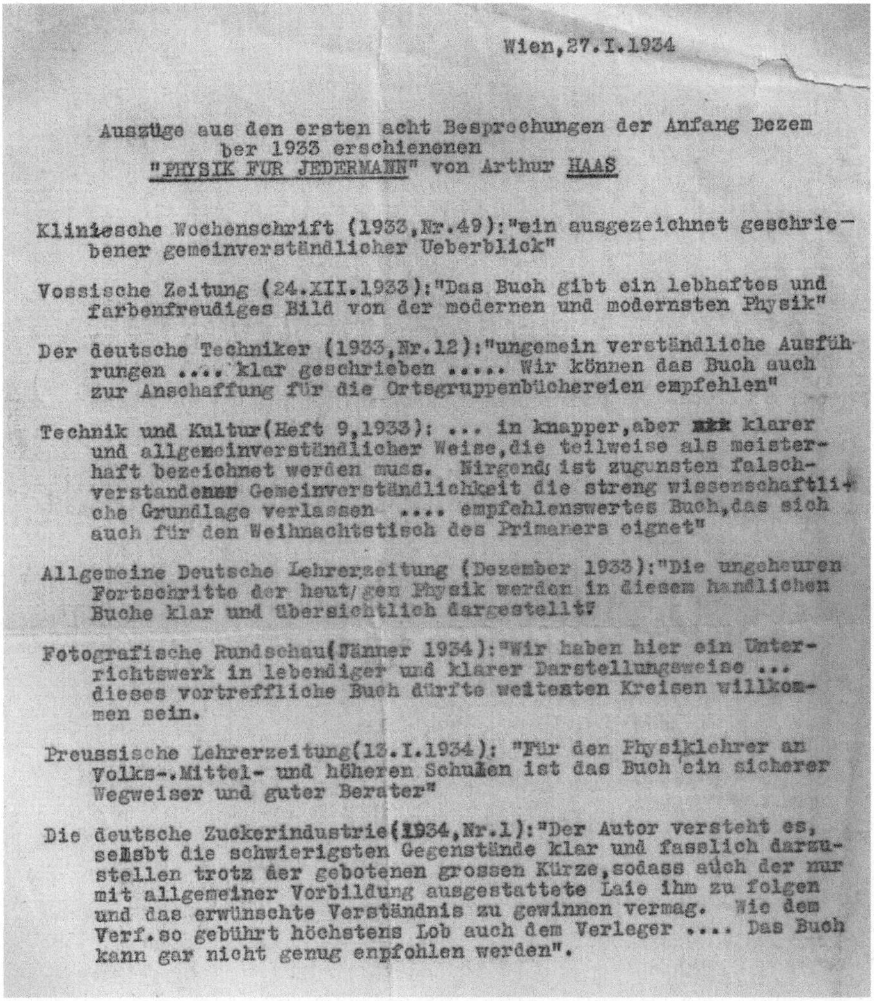

Wien,27.I.1934

Auszüge aus den ersten acht Besprechungen der Anfang Dezember 1933 erschienenen
"PHYSIK FÜR JEDERMANN" von Arthur HAAS

Kliniesche Wochenschrift (1933,Nr.49):"ein ausgezeichnet geschriebener gemeinverständlicher Ueberblick"

Vossische Zeitung (24.XII.1933):"Das Buch gibt ein lebhaftes und farbenfreudiges Bild von der modernen und modernsten Physik"

Der deutsche Techniker (1933,Nr.12):"ungemein verständliche Ausführungen .... klar geschrieben ..... Wir können das Buch auch zur Anschaffung für die Ortsgruppenbüchereien empfehlen"

Technik und Kultur(Heft 9,1933): ... in knapper,aber klarer und allgemeinverständlicher Weise,die teilweise als meisterhaft bezeichnet werden muss. Nirgends ist zugunsten falschverstandener Gemeinverständlichkeit die streng wissenschaftliche Grundlage verlassen .... empfehlenswertes Buch,das sich auch für den Weihnachtstisch des Primaners eignet"

Allgemeine Deutsche Lehrerzeitung (Dezember 1933):"Die ungeheuren Fortschritte der heutigen Physik werden in diesem handlichen Buche klar und übersichtlich dargestellt"

Fotografische Rundschau(Jänner 1934):"Wir haben hier ein Unterrichtswerk in lebendiger und klarer Darstellungsweise ... dieses vortreffliche Buch dürfte weitesten Kreisen willkommen sein.

Preussische Lehrerzeitung(13.I.1934): "Für den Physiklehrer an Volks-,Mittel- und höheren Schulen ist das Buch ein sicherer Wegweiser und guter Berater"

Die deutsche Zuckerindustrie(1934,Nr.1):"Der Autor versteht es, selbst die schwierigsten Gegenstände klar und fasslich darzustellen trotz der gebotenen grossen Kürze,sodass auch der nur mit allgemeiner Vorbildung ausgestattete Laie ihm zu folgen und das erwünschte Verständnis zu gewinnen vermag. Wie dem Verf.so gebührt höchstens Lob auch dem Verleger .... Das Buch kann gar nicht genug empfohlen werden".

Fig. 17.3 A list of reviews compiled by Arthur Haas for his book *Physik für Jedermann* (Physics for Everyone)

vi. Protokoll der Sitzung der Disziplinarkommission am 10. Februar 1932. AÖAW, PA Julius Matula, No. 767/1931.

vii. Mario Livio, *Brilliant Blunders* (New York: Simon & Schuster, 2013), 221–245.

viii. Johann Wolfgang Goethe, "Gott, Gemüt und Welt," in *Gedichte. Ausgabe letzter Hand* (Stuttgart & Tübingen, Cotta, 1827).

ix. David Rowe, "Einstein's Allies and Enemies: Debating Relativity in Germany, 1916–1920," in *Interactions: Mathematics, Physics and Philosophy, 1860–1930* (Berlin Heidelberg: Springer Verlag, 2006), 235–237.

Chapter 18
The Dark Years

After completing *Physik für Jedermann* in the fall of 1933, Arthur Haas began a new project that he had carried around with him for a long time, namely to write a small textbook on the physics of sound film. Both acoustics and the medium of film had interested him for many years and he had participated in various documentary film projects at the Urania. With the premiere of the movie *The Jazz Singer* in 1927, sound movies began to gain commercial acceptance. Haas signed a contract with the publishing house Teubner in Leipzig and he developed a lecture on acoustics and sound film for the winter semester 1933/34, which he then translated into five popular-science lectures at the Urania in Vienna and Berlin. These lectures formed the basis for his little booklet. It is notable that this was not a book on basic scientific topics from the atom to cosmology but, again, a book on the practical application and implementation of physical principles. Here his connections to the industry seem to have had an effect because cooperation with the film industry (Klangfilm G.m.b. H.) and the electrical industry (Siemens & Halske A.-G.) is explicitly recognized in the acknowledgements.

After an introduction to the physics of acoustics, the following chapters are mainly devoted to technical issues such as the conversion of acoustic signals into electrical signals, storage on data carriers, and the retransformation to acoustic signals by loud-speaker systems, whereby the effects of sound and space on the quality of sound are also discussed. The booklet was largely popular and even without the mathematical framework of these physical questions, it covers a surprisingly wide range of techno-logical details. Haas presented this information to the reader by means of technical illustrations and drawings as well as graphic representations of the interconnections between disciplines and applications.

This book again has a general technical approach, which Haas now seems to have adopted. His interest in the fundamental questions of modern physics had given way to a new vocation—creating understandable presentations of modern physical principles and their possible technical applications. These books are specifically

M. Wiescher, *Arthur E. Haas—The Hidden Pioneer of Quantum Mechanics*, Springer Biographies, https://doi.org/10.1007/978-3-030-80606-4_18

Fig. 18.1 Taormina in Sicily as displayed on one of the postcards that Emma wrote home to Vienna

addressed to technically interested and educated citizens like engineers and physicians, professionals who seemed to appreciate the pseudoscientific arguments and theses promoted by the new political Right.

While Haas was continuously working on this book, Emma Haas fled again, at the end of March during the Holy Week of 1934. She migrated from cold Vienna to the south, taking a four-week tour to her dream country Italy. She had passed her Matura and had already scheduled first lectures at the University of Vienna in art history and architecture; she was ready for a break and sought warm refuge in southern Sicily where she had found much enjoyment two years earlier. She did not make this trip alone but with a Viennese travel company.

Arthur faithfully wrote his daily letters, for which he received only occasional but detailed and passionate reports from Rome, Taormina and Naples. She wrote enthusiastically about the long train journey via Rome to Taormina (Fig. 18.1), during which she became friends with another lady from the travelling party, was admired by polite Italians, and was guarded in her sleeping car by young fascist officers. She often wrote about the philosopher and member of the Viennese Circle,[i] Moritz Schlick[1] who also travelled through Italy[ii] and visited her at her hotel in Sicily.

[1] Moritz Schlick (1882—1936) was born in Berlin on April 14, 1882. He studied physics and completed his studies in 1904 with a doctorate under Max Planck on a topic related to light reflection. He spent the following year 1905 in Göttingen as an assistant to Woldemar Voigt, where Schlick might have met Arthur Haas. Schlick stayed only one year in Göttingen and then began to study philosophy at the Swiss Federal Institute of Technology in Zurich. From 1911 to 1921, he had a lectureship in Rostock and Kiel but in 1922, he moved to Vienna where he took over the chair of philosophy of inductive sciences. He succeeded the philosophers and physicists Ernst Mach

> Schlick also wrote me that he will probably arrive in Syracuse on the 10th (tomorrow) [...] Anyway, we will form again a funny and nice company. If the weather is good, Moritz will also stay in Taormina for a few days and take a room in the Hotel Naumachia.[2]

The Haas and Schlick families seem to have been good friends with common interests in natural philosophy and America. Schlick's wife, Blanche Guy Schlick née Hardy (1879—1964), was an American from a Methodist pastor's family in Massachusetts and with her it was easy to practice English.

Actually, Emma Haas and her new lady friend had planned to travel together from Taormina to Tripoli on the tourist ship *Ozeania*, since Emma very much wanted to go to the new continent of Africa. But Moritz Schlick successfully advised her against this plan.[3] The travel group did, however, engage in numerous excursions and undertakings in Sicily itself, about which Emma reported to Vienna in detail (Fig. 18.2). Schlick took part in some of these tours but then traveled back to the Amalfi Coast where he had his vacation spot.

Haas' letters, on the other hand, reported of loneliness and work, as well as academic progress and his excursions with their two sons. He also told of his meetings and dinners with Professor Petritsch and a younger colleague Guido Beck[4] who had

---

and Ludwig Boltzmann. During this time, Moritz Schlick dealt with questions of logic, aesthetics and ethics. In addition, he published numerous works on philosophical questions of mathematics and natural science, especially on Albert Einstein's theory of relativity and on philosophical epistemology. In 1924, Schlick founded a scientific discussion circle from which the Vienna Circle later developed. He undertook numerous journeys abroad, especially to the United States. Together with Philipp Frank in Prague, he published writings on the scientific world view in which the neo-positivist theses of the Vienna Circle were represented and disseminated. In 1930, Moritz Schlick's work was published with the title *Fragen der Ethik*. In this work, he unequivocally denied the existence of absolute ethical values. Nor did he consider the conventional notions of norms to be desirable. In contrast, he set individual goals as the "natural inclination" of man. In his ethics, Schlick more often referred to the concept of "happiness" as one of man's goals. In the early 1930s, Schlick worked with Ludwig Wittgenstein, with whom he maintained an intensive correspondence. Moritz Schlick was murdered in Vienna on June 22, 1936 by one of his former students Hans Nelböck on the way to university.

[2] EBH to AEH (Taormina, April 9, 1934).

[3] Tripoli had served since 1911 as the capital of the Italian colony of Tripolitania, which had just been expanded by the occupation of the Cyrenaica by Mussolini in 1933–1934 to form the larger colony of Libya.

[4] Guido Beck (1903–1988) studied physics at the University of Vienna and received his doctorate on the theory of relativity in 1925 under Hans Thirring. Shortly afterwards Beck wrote a paper on the Compton effect, which also influenced Haas' thinking. Beck worked first at the University of Bern, then again in Vienna under Ehrenhaft, and from February 1928 for four years as first assistant to Werner Heisenberg in Leipzig. In 1930/31, Beck worked at Cambridge with Ernest Rutherford, then in Copenhagen, and also in Prague with Philipp Frank before he moved to Kansas in 1934. However, he did not stay there for long but taught at the University of Odessa from 1935 to 1937, at the invitation of Jakow Iljitsch Frenkel. When the political situation in the Soviet Union darkened, Beck left via Copenhagen to France. The conquest of France by Nazi Germany endangered him as a Jew and he went to neutral Portugal in 1941 and from there to Argentina in 1943, where he taught at the University of Buenos Aires. He was an important advisor to Argentinean President Juan Perón and played a major role in the development of nuclear physics research in South America.

Fig. 18.2 Left: The Austrian ladies in their traditional garb in Sicily. Right: The *Ozeania* departing for Tripoli. (pictures taken by Emma Haas)

just been appointed to a professorship in Kansas. In this context Haas again mentioned the possibility of a one-year American guest professorship.

However, in the course of her tour, Emma's social excursion came to a scandal and Moritz Schlick's wife asked Haas for a personal conversation, in private. Arthur reported about it in a long letter:

> My joy about your letter would be much greater if I did not have to get terribly upset about things that your and M. S.' indescribable thoughtlessness has unfortunately given rise to. Although it is not easy for me to write you a letter that could cloud your good travel mood, which I grant you from the bottom of my heart, I must do it nevertheless in order to prevent greater misfortune over time.
>
> When Mrs. Harris recently held a large reception at her house, to which you were of course invited, I am sure I heard general gossip about the fact that you were seen with M. in Italy - without your husband, without his wife - and that you went on a trip together. If it is really necessary to talk so wantonly, is it then justified to take the stand that it is only a clear conscience that matters - when you have a family, children and a husband who is struggling so hard for its existence? Was it not frivolous of M. to compromise you by visiting Taormina for several days, even if not the slightest injustice has occurred - as I firmly believe?
>
> I met Blanche today at her urgent request; you are still very much in her good graces and she has no trace of resentment against you. But she is completely horrified by the disaster that M. has caused through his unworldly childishness; he has acted unspeakably ruthlessly against you and has also done tremendous harm to his own reputation. At the university they only say that he doesn't care about his duties at all anymore, he leaves everything, doesn't care about the beginning of lectures, doesn't review dissertations anymore, so Blanche is very worried about his position. And if the silly rumors concerning you are being circulated

> (apparently involving especially Mrs. Annelise F.-J. and Ann Tizia[5]), he will be finished there.
>
> Do you really think you know the world so well that you think you can ignore all my urgent warnings? If we were so independent that we didn't have to worry about anyone, maybe your point would be right - but even then, only if you didn't have children!
>
> Blanche very clearly asked me not to write to you about our conversation; I am only doing so to prevent - for God's sake - you showing yourself somewhere with M., wherever in Italy. The question is whether the disaster, already caused by the evil gossip, can still be reversed. In any case, Blanche wants to demonstratively show herself to be a good friend with you, to prove that there is nothing to the rumors. But she is just as horrified as I about the possibility that M. could compromise you in his childishness by accompanying you from Naples (where he is going again unnecessarily) to Florence or Venice or from there even to Austria. If he is in Venice in the end, please say goodbye to him immediately. Since I know that you have done nothing wrong, I will not hold any grudge against you but I hope that you will listen to my advice and possibly that of Blanche.
>
> You must not tell M. about this letter. I just can't understand how a man of his age and intelligence can be so thoughtless. And if you think a little bit about it, you'll understand why I had to write you this letter - and then we both want to forget all this and just make an effort to refrain from anything that could harm your reputation, for the sake of your children, even if it is unjustified. But in this respect, you must proceed with tenacious consistency, regardless of whether you offend M. or not.[6]

Haas sent more letters in quick succession, suspicion accumulated, and there was no reply from Bice who was travelling through Italy. The last report from her had been about the beauty of the Amalfi Coast and amusing encounters in Naples, which seem to have turned suspicion into certainty; fear that Moritz Schlick would accompany Emma home turned into terrible certainty, on Arthur's part, that Moritz and Bice were travelling the canals of Venice in a gondola. In Arthur's letters, which increased to panic, the friend Moritz became the "irresponsible M.," "the old donkey M." and finally, "the donkey" became a miserable, raggedy rogue (Haderlump). Reproaches followed ultimatums and the marriage was questioned:

> But there is no compromise. Only a yes or a no. And if you say yes, I will completely forget all your stupidities - because I hope they were only such - and you will be my Trixerl[7] again as always. If you can't give me the complete promise by telegram, a birthday party is out of the question![8]

It was a storm in a teacup; the suspicion that Bice and Moritz were travelling alone through Italy turned out to be wrong and a consequence of false assumptions and conclusions. Arthur had to apologize in a telegram for his suspicions, a telegram that

---

[5] Ann Tizia Leitich (1896—1976) was an Austrian reporter who lived in the US and Germany from 1921 to 1928, where she became known for her witty social reports. In 1928 she returned to Vienna where she wrote more than fifteen books on various socio-political, historical and geographical topics.

[6] AEH to EBH (Vienna, April 21, 1934).

[7] Pet form of Beatrice.

[8] AEH to EBH (Vienna, April 22, 1934). The birthday party was the delayed celebration of Emma Beatrice Haas' birthday, which Arthur wanted to prepare with the children for her return.

was not without reproaches regarding the naivety and intolerable unconventionality of his wife, qualities to which he attributed this social debacle.

It was finally one of those scenes that Emma had remarked upon in an earlier letter:

> If you could just give up your periodic panic attacks, life would be really good. Maybe you'll try a little bit more, because you often spoil the best days for you and me with these hideous attacks. And that is actually so unnecessary.[9]

Emma returned from Italy to Vienna. She was safe and sound and she had her birthday party. The family was reunited. Arthur continued to write his booklet on sound film and Emma began her first semester at the University of Vienna. The crisis was over.

At the beginning of June 1934, Arthur travelled with the two children to Sattendorf where he hoped to find peace in the summer resort at Lake Ossiach. Emma stayed behind in Vienna until the end of the semester, so that she could concentrate on her studies and exams for another week without any distraction from the boys. This was a new development that Arthur did not really like but had to agree to because Emma was striving for an independent career.

The daily letters from Sattendorf to Vienna included, in addition to reports on his and the children's swimming successes and admonitions to come soon, Haas' first comments on political events. These comments presented more clearly than before his political thinking:

> Today the newspapers report that near Passau one hears heavy cannon thunder from the direction of Munich. If only Rupprecht would take the helm![10] And if there were blue and white flags again![11]

Crown Prince Rupprecht of Bavaria was in those days one of the hopes of the conservative opposition in Germany. The conservatives were not, however, striving to preserve what they saw as the failed democracy of the Weimar Republic. Rather, they supported the establishment of an authoritarian corporative state, similar to the state formed by Dollfuß in Austria, based on the authoritarian leadership of the nobility, the church, and the conservative middle classes. The interests of the people were to be regulated in this corporative state not by parliament and parties but directly, through negotiations between these estates and professional groups while excluding trade unions. Similar ideas had already provided the basis for fascist Italy and would later form the government in fascist Spain. However, the principle of the corporative state was not in the interests of the all-German parties in Austria. Nor was it in

[9] EBH to AEH (Rome, April 5, 1934).

[10] These might have been the last fights of the so-called Röhmputsch, albeit that only light weapons were used. Crown Prince Rupprecht of Bavaria (1869—1955) was a last hope for opponents of the National Socialists. His followers fought to prevent the National Socialist state by re-establishing the Kingdom of Bavaria as a corporative state. Rupprecht did indeed have contacts with SA leader Röhm but could not decide to resist. Rupprecht narrowly escaped the purges following the Röhm coup in 1934, maintained contact with opposition groups and went into exile in Italy.

[11] AEH to EBH (Sattendorf, July 4, 1934).

the interests of the National Socialists on either side of the border who strove for a more nationalistically oriented Greater Germany encompassing all German-speaking areas. Eventually, after Röhmputsch and the associated elimination of socialist and also bourgeois conservative elements in the party, the nationalistic goal of annexation of Austria could be pursued.

On July 25, 1934, a coup took place in Vienna. The National Socialists, who had been prepared in Germany for a long time, killed Dollfuß in an exchange of gunfire at the Chancellery. The dying man's requests for a doctor and priest were denied. The National Socialists had hoped for support from the Austrian Armed Forces who, however, remained loyal. The murderers of Dollfuß were arrested and later executed. The Dollfuß successor in the Austrian Federal Chancellery was Kurt Schuschnigg, a representative of the Patriotic Front who initially continued the policy of his predecessor.

The years of authoritarian government led by federal chancellors Dollfuß and Schuschnigg (1933/34—1938) were, however, either denounced as Austrofascism by the Left wing of the party or defended by the Right wing as a Christian corporative state. Today this is interpreted as the struggle of the conservative parties of the Center-Right who were trying to maintain national independence from Hitler's Germany. This struggle culminated in the attempted coup of 1934 but finally failed four years later with the annexation of Austria by the German Reich.[iii]

At the time of the first coup attempt and the murder of Dollfuß, Haas was back in Vienna at the Academy and Emma was in Sattendorf with the children; in his letter of July 26th, he was not forthcoming but on July 27th a more detailed report came from Vienna in which Arthur Haas made clear his shock and horror, but also his relief at the failure of the coup:

> As you can imagine, I was terribly upset by the political events, but I hope that things will now improve. My grief for Dollfuß is sincere. The refusal of priestly assistance by the murderous beasts will hopefully destroy the last sympathies of the people. [...] The danger of an Italian invasion of Carinthia, which was great on the day of the murder, no longer seems to exist.

In the same letter he also reported about his own difficulties and problems at the Academy. For the first time, he identified his opponents and critics in the office.

> I am terribly upset about the conditions in the office. Mrs. Rep.[12] cannot come into the office because her husband was imprisoned as a Nazi. Mr. L.[13] has to represent her and is angry

---

[12] This is Maria Calabek who had been married to Ludwig Repototschnig (1891—1945) since 1931. While her husband was a member of the banned NSDAP, she was considered a sympathizer of the movement. Cf. Stefan Sienell, *Das Verwaltungs- und Dienstpersonal,* 221–223.

[13] According to the description of personal infirmities, this is the accountant Leopold Liegler (1882—1949) who had worked in the accounting department at the Academy since 1914. He had applied unsuccessfully in 1924 to succeed Kohl as first actuary. The fact that the position was taken, in his place, by Arthur Haas was probably one of the reasons for the internal dispute. The obituary in the *Österreichische Zeitung* of October 13, 1949, probably did not take into account the resulting anti-Semitism of the deceased but rather portrayed Liegler more as an intellectual resistance fighter, due to his past as a social democrat: "Anyone who had the opportunity at the turn of the years 1944 and 1945 to meet with the now deceased Leopold Liegler could not have believed that this terminally ill man with his weakened, crippled body would live for several more months.

about it. He takes all ordinariness out on me and when he turned to anti-Semitic vulgarities today (an experience I have never had in my long life so far), I lost my patience. I told him that he should consider himself to be slapped in the face by me, and that it was only thanks to his being a cripple that he did not get the slaps in reality.

With that, the conversation ended without any witnesses and without him being able to reply. You can imagine what a hell I have to consider my "service" to be from now on, when I have to stay in the same office with such a poison-breathing reptile crippled by nature. If only a kind fate would finally deliver me from this circle of hell - or let him die of his stomach ulcers.[14]

Six days later, however, Haas wrote in a subsequent letter:

I'll have peace and quiet again at the academy. To my great surprise I found a letter from Mr. L. on my desk yesterday in which he humbly asks for forgiveness, which I then granted him. There is a plaque on Mrs. Repototschnig-Calabek's shop[iv]: "The shop belongs to a member of the Dollfuß Murderer Gang." We won't hear from her for a while.[15]

In subsequent letters he described neighbors and acquaintances, assigning them various party affiliations, but Haas also pointed out that he was still being greeted politely by members of the now-banned National Socialists. The situation calmed down and Italian troop movements on the Austrian border stopped worrying Haas.

For the following week, he continued his work on various book projects until he returned to Sattendorf on August 11th for a family vacation. They were only all together for a week because Emma left for Millstatt on August 28th to join the family of von Blanche and Moritz Schlick, who were spending their summer vacation there. She planned to return to Vienna together with the Schlicks.

The promised postcard, which should indicate your good arrival in the circle of the Schlick family, unfortunately did not arrive today. I hope you feel comfortable as a "straw widow." [...] Many greetings to Mrs. Schlick and Moritz![16]

This probably embarrassing visit was most likely a tactical step to demonstrate to Viennese society that the rumors, as described earlier, were false and that a true friendship existed between the families. The "raggedy rogue" was called "Moritz" again, indicating that the personal relations between Moritz Schlick and Arthur Haas had improved.

How close the relationship really was, personally and intellectually, is not clear. There is no direct evidence that Haas was involved in the Vienna Circle. He belonged to the Philosophical Club in the 1920s, now called the First Viennese Circle,[v] and was friends with the physicist Philipp Frank as well as with the mathematician Hans Hahn, who both belonged to the close Viennese Circle founded in the early 1930s.

But in this organism dwelt a tenacious, strong spirit, sustained by the passionate desire to live to see the end of the hated Nazi rule, and this dogged will not only saved Liegler, but also caused the death candidate to recover, blossom and appear younger after the liberation of Vienna. With fiery zeal he threw himself into the work of building a new state system, which was pressing from all sides." Cf. Stefan Sienell, *Das Verwaltungs- und Dienstpersonal,* 266–269.

[14] AEH to EBH (Vienna, July 27, 1934).

[15] AEH to EBH (Vienna, August 2, 1934).

[16] AEH to EBH (Sattendorf, August 29, 1934).

Haas was also friends with the mathematician Karl Menger (1902—1985), as will be discussed in more detail later. It was only natural that Schlick and Haas knew each other; both had a keen interest in the physical and philosophical questions of relativity and modern physics. Their membership in the American Club had probably also led to closer social, if not friendly relations that may have been somewhat clouded by this episode.

It is possible that Haas was more politically than philosophically active at the time, especially since his friend Ernst Petritsch played an influential role in the cultural organization of the Vaterländische Front.[vi] Like Haas, close members of the Viennese Circle such as Kurt Gödel (1906—1978) and Hans Thirring were also members of the Vaterländische Front[17] and even Moritz Schlick expressed certain sympathies.[vii]

After his return from Sattendorf on September 8th, Haas took up work as usual. The political unrest of the summer seems to have been forgotten or at least was no longer noticed, for he had a new book project and he was organizing the associated preparatory lectures. He had already signed a contract with De Gruyter in May 1934 to complete a manuscript for a textbook, *Zerlegung und Umwandlung der Grundstoffe* (Decomposition and Transmutation of the Elements), by March 1935.

Although many of Haas' thoughts on modern physics were incorporated into this book, it was again much more practical than his earlier works. The structure of the book was defined by a sequence of lectures, which he had presented during the winter term 1934/35 with each lecture resembling a chapter. The first chapter is on the materialization of light and mainly explains, along with the basic dualistic idea of light as waves and photon particles, that it is possible to visualize photons and other particles with newly developed detector techniques.

In the second chapter he speaks about the definition and characteristics of the elements, as well as their analysis and identification by X-ray and mass spectroscopy. The third chapter is dedicated to the means of atomic disintegration, which in today's nomenclature is the study of nuclear reactions. After introducing Ernest Rutherford's classical concept of using radioactive elements as radiation sources for high-energy particles, Haas describes the development of new concepts for particle accelerators. His discussion includes the Cockcroft Walton Linear Accelerator in Cambridge and the cyclotron in Berkeley developed by Ernest Lawrence (1901—1958). Haas also talks about newly discovered neutrons and the technical possibilities of carrying out experiments with them. The actual experiments and their results are presented in the following chapter, where he provides an overview of the first generation of nuclear physics experiments using particle accelerators.

In the fifth and final chapter, he returns to radioactivity, about which revolutionary insights had been gained over the previous decade. Haas summarizes these

[17] Membership in the Vaterländische Front was requested by the government for academic and government related positions, consequently every third Austrian joined the movement. Cf. Karl Siegmund, John Dawson, Kurt Mühlberger, *Kurt Gödel – das Album* (Wiesbaden: Vieweg Verlag, 2006), 48.

in a concise way, especially focusing on the production of artificial radioactivity—the manufacture of new elements—which he hopes will have a wide range of applications. With this in mind, he ends the chapter and book with these words:

> And yet, in view of the rapid scientific progress of our century, who could deny any possibility that perhaps only a few decades could separate us from the industrial exploitation of such knowledge, which our age collects in the field of raw material conversions in a beginner's way?[18]

These were prophetic words indeed because today, almost 100 years later, artificial radioactivity plays an important role in numerous industrial and technical applications, such as the analysis of matter for production processes, in diagnostic and therapeutic medicine, in forensics and archaeometry, and finally in energy and space applications.[viii] Added to this are applications in the realms of security and defense policy, which had ominous beginning with the discovery and identification of nuclear fission by Otto Hahn, Fritz Strassmann, and Haas' former fellow student Lise Meitner.

A sixth chapter was originally planned but fell victim to a lack of time due to illness, as Haas writes in a letter to his wife during her visit to Munich in November 1934:

> I am feeling a little better, and I hope to be able to work fully again in a week's time. After all, my indisposition has the unpleasant consequence that I have to skip the planned 6th lecture for my book on the transformation of chemical elements, which unfortunately reduces my fees by 200 - 300 shillings.[19]

Unfortunately, Haas did not comment on what topic, of the then modern field of nuclear physics, he would have covered in the planned sixth chapter.

Emma had travelled to Germany for several weeks in early December. Her first goal was to visit her brother and friends in Munich and Eugenie Grau at the Ammersee. She was also to settle the status of her and her brother's partnership in the company, including securing her share of the business income that was to be paid out to finance a trip to England. From there she planned a trip via Rothenburg, Nuremberg, and Bayreuth to Berlin where she wanted to stay for three days. Then she was to travel to Dresden and Leipzig, where she wanted to spend a week on business. Afterwards she had scheduled a skiing holiday to Garmisch-Partenkirchen on the Zugspitze, from where she planned to return to Vienna via Kitzbühel on December 22nd to be home for Christmas.

This trip was not only for pleasure. Emma Beatrice travelled from Munich to Berlin and Leipzig on behalf of her husband, in order to obtain royalties from the various publishers. Arthur Haas as a Jew did not want to expose himself to such a trip in the "New Germany" and his wife had agreed to act as his business representative. However, he commented somewhat critically on the extended travel plans when he learned of them through her letter from Munich:

> Your letter yesterday [...] surprised me very much with your far-reaching plans. Although I do not agree with everything, I do not want to talk you into anything. I only hope that you

---

[18] Arthur Haas, *Die Umwandlungen der chemischen Elemente* (Berlin: De Gruyter, 1935), 104.

[19] AEH to EBH (Vienna, November 29, 1934).

> will be successful and that you will feel comfortable at Ammersee. [...] I find myself in the most serious worries and I only hope that you will receive orders.[20]

His worries were mainly due to the fact that the new German government had imposed strict restrictions on the transfer of funds abroad and, consequently, the publishers did not want to transfer the promised author's fees to Haas in Austria.

In the further text of the letter and in accordance with his old habit, Arthur listed a full program of recommendations and visitations in which he named important people and institutions in Munich, Berlin, and Leipzig. He also asked Emma to draw attention to the imminent publication of his new book (*Umwandlungen der Chemischen Elemente*). For this, he recommended a very short visit to Prof. Gerlach in Munich:

> Please visit Prof. Dr. Walther Gerlach at the Institute of Physics in Munich, very briefly, and tell him that you only wanted to visit him very briefly, without disturbing him, in order to convey my greetings to him personally at my request.

The visit lasted longer than planned, however, because Walther Gerlach[21] personally gave Emma a guided tour through the institute:

> So my new friend Gerlach is a character in his own right. What I did not learn yesterday about physics is not worth the effort to learn. Have you ever created the Fata Morgana artificially, or the Halo's ring on the moon? I did, yesterday. Have you ever been allowed to sign the famous guestbook of the Phys. Inst. directly beneath the signature of the Archduke Joseph Franz (or Franz Joseph)? - I did, yesterday. Have you ever been guided through the institute with all honors etc., etc., etc.?[22]

She spent her whole day there, which delayed her departure to Berlin. However, in Berlin she also reported on interesting encounters, although not at the Berlin museum sites so highly recommended by her husband.

> I spoke to the Prussian Prime Minister personally last night - yes, to Göring. Even Dr. Jolowicz[23] found him charming. He was delightful. Then I gave him 50 pfennigs (which I quickly borrowed from Dr. Jolowicz) for the "Winterhilfe" [winter aid]. All this took place

[20] AEH to EBH (Vienna, December 2, 1934).

[21] In 1929 at the instigation of Arnold Sommerfeld, Walter Gerlach (1889–1979) was appointed from his position as chair of experimental physics in Tübingen to succeed Wilhelm Wien in Munich, where Gerlach established a research field in applications of spectroscopy methods in materials analysis and also in chemistry, biology and medicine. Later he became head of the secret German uranium project, which aimed at building a nuclear reactor. After the war and his imprisonment and internment at Farm Hall by the British, Gerlach served as rector of the University of Munich from 1948 to 1951.

[22] EBH to AEH (Berlin, December 5, 1934).

[23] This was presumably Walter Jolowicz (1908—1996), son of the aforementioned Leo Jolowicz (1868—1940) who owned the Gorch Fock antiquarian bookshop and was a good friend of Arthur Haas from his Leipzig days. Leo Jolowicz had founded the Akademische Verlagsgesellschaft (Academic Publishing Company) where Haas mainly published, along with De Gruyter in Berlin. Until the 1930s, Akademische Verlagsgesellschaft was the publishing house with the most Nobel Prize winners in the field of natural sciences in Germany. Walter Jolowicz together with his brother-in-law, the chief editor Kurt Jakoby, ran the Akademische Verlagsgesellschaft at the time of Emma's visit. But already in 1938, the publishing company was "aryanized" and Walter Jolowicz emigrated

in the elegant hall of the Hotel Esplanade. Then we (by we, I mean from now on always Dr. Jolowicz and I) went to the theatre (lodge of course) to *Lady Windermere's Fächer* [fan]. Then we decided to go to a very exclusive and unbelievably elegant ball given by the tennis club "Rot Weiß." At 2 o'clock I staggered home.[24]

Emma turned out to be a skillful negotiator in representing the interests of Haas, as she testified in the same letter:

I also casually met another publishing bookseller, Kayn, who is an intimate friend of De Gruyter's boss and will introduce me there tomorrow morning. I am going there on the advice of Dr. Jolowicz. Maybe I can help a little bit. Should this Mr. Kayn also have connections with *Research and Progress*,[25] then I will also kindly say "hello" there.

From Berlin she planned to travel on to Leipzig where the publishers were already making preparations for the reception, which she painted out for him in the same letter:

Anyway, I will drive to Weimar by car (Mercedes) on Tuesday or Wednesday, because Dr. J. wants me to see this wonderful little town. Tomorrow evening, guests are already invited in my honor, including Heisenberg, I believe. Debye[26] is not in Leipzig at all; he is currently somewhere else. I feel so well here that I can hardly believe that one can still feel so well in the world. We've already become real moochers in Vienna. Here you see elegance and glamorous life that makes you feel like you're from the provinces.

You can use powder and lipstick as much as you want, nobody says anything and the people are very friendly and obliging. Especially the guards or S.A. soldiers answer so patiently and nobody said "Heil Hitler" to me. Nobody greeted me with the Hitler salute, not even Göring. I remain completely calm, do and say nothing that has to do with politics and am happy that the people here are also happy and content. Everyone makes that impression.

No doubt an interesting passage in the detailed letter in which she unbiasedly described "positive" impressions of her visit to Berlin. Was it naïve good faith and hope for positive developments after the turbulent last years of the Weimar Republic, an optimism still present among many citizens, that are reflected in her letter? Was this confidence due to the National Socialists' quite successful economic development, which impressed many of the German citizens?[27] Or was Emma merely teasing

---

to the United States where he called himself Walter Johnson. There he founded the Academic Press New York together with his brother-in-law Kurt Jakoby. However, in 1934 the academic publishing company did everything to make Emma Haas' stay, as the wife of their successful author, as pleasant as possible; they even provided the car for her tours and trip from Berlin to Leipzig. Cf. Klaus G. Saur, *Verlage im "Dritten Reich "* (Frankfurt am Main: Vittorio Klostermann, 2013), 10.

[24] EBH to AEH (Berlin, December 9, 1934).

[25] *Forschungen und Fortschritte* (Research and Progress) were the annual reports generated by the Prussian Academy of Sciences.

[26] Peter Debye (1884—1966) was a Dutchman who studied under Arnold Sommerfeld at the TH Aachen and followed him to Munich. After the death of Otto Wiener in Leipzig in 1927, Debye became his successor and then became director of the Kaiser Wilhelm Institute in Berlin from 1934 to 1939. He emigrated to the United States in 1939 and became director of the Chemistry Institute at Cornell University, where he remained until the end of his life.

[27] These measures did, however, frequently follow concepts already developed in the days of the Weimar Republic, concepts that had only become enforceable through dictatorial measures. Such

her husband in reaction to his accusations and suspicions during her trip to Italy? Probably the latter because, while she told him about Hotel Esplanade where she met Dr. Jolowicz, her accommodation was actually the hospice at the train station Friedrichstrasse where her single room cost between 3 and 7 reichsmark and breakfast was RM 1.50. Nevertheless, she did not remain unimpressed by the glittering life in Berlin:

> Today I met the most charming and kindest person, the boss of De Gruyter, Mr. Cram.[28] You'll be paid your fee of 17 hundred and several shillings by next week at the latest. [...] I can't tell you how charming he was to me. I could fall in love right away this time. He also regretted that I was leaving Berlin so soon. I've thought about whether I should stay in Germany for the time being, until you're well again. Aunt Anna has built herself a wonderful villa and I could live there splendidly.[29]

But Arthur Haas replies calmly to this teasing:

> As for you, I'm very happy that you're having such pleasurable weeks, which unfortunately can't be offered in Vienna. At least you take part in the glamorous life far away from your home.[30]

However, Arthur had other concerns as well. During Emma's trip to Germany, their sons were enrolled in the nearby Trazerberg daycare center[31]; extra childcare was needed in order to give their father the opportunity and time to finish his book by deadline. Arthur (Atzi) and Georg (Schorschi) were to continue attending their usual school, a catholic private school that was attached to a monastery. The daycare, on the other hand, was a largely liberal educational institution with a progressive approach to education and society, which particularly promoted free contact and exchange between students of different social classes and different talents. Emma had also chosen the daycare for medical reasons as she was always concerned about

---

appreciation could only be possible because many closed their eyes to anti-Jewish measures and decrees.

[28] Herbert Cram was the son-in-law of Walter De Gruyter, the founder of the publishing house, and took over management after his death in 1923. Cram successfully managed the publishing house through the years of the Third Reich, building success through adaptation. In 1962, Cram tried to re-establish business relations with Emma Haas, which she found less charming after these developments. For more on Herbert Cram, cf. Angelika Königseder, *Walter De Gruyter – Ein Wissenschaftsverlag im Nationalsozialismus* (Tübingen: Mohr Siebeck, 2016).

[29] EBH to AEH (Leipzig, December 10, 1934).

[30] AEH to EBH (Vienna, December 12, 1934).

[31] This was the Villa Auersperg at Trazerberggasse 6, a single-story villa that was most probably built around 1855 after the demolition of the previous building; Vinzenz Carl Fürst v. Auersperg was the owner of this property. He was followed as owner by the industrialist family Schrantz (later Hutter & Schrantz), followed by the Jewish pediatrician and writer Dr. Otto Almoslino who established a home for Jewish children there. So it was Jewish children, often Eastern Jewish children, who found a home there and Arthur Haas thus expressed his prejudices against the increasing immigration of Polish and Galician Jews to Vienna. One year after the Anschluss, in 1939, the villa fell to the city architect Siegfried Schlosser. Dr. Otto Almoslino emigrated via Istanbul to Palestine. For further details see Ober St, Veit An Der Wein, accessed January 31, 2021, https://www.1133.at/document/view/id/1218.

the children's low weight. She liked the head physician of the institution, Dr. Otto Almoslino, and his medical ideas.

It was in this new setting that the boys first came into freely innovative contact with children other than their cousins, children from households that were not shielded by a maid and housekeeper. This led to problems, especially in the strictly hierarchically structured Catholic school. There was also a conflict between the respective educational views of the two institutions, in which the parents became involved. Thus, Arthur worriedly described the inflammatory situation between the two educational directions:

> I went to the children's school this morning. The headmistress as well as the teachers expressed their horror at the state of complete neglect in the children's manners since their move to Trazerberg. The headmistress was most outraged by the fact that children of a Catholic school were prevented from attending church – because of playing football etc., since this was considered more important. [...]

The headmistress had reported that,

> Atzi had been completely and suddenly changed. The well-behaved child had become a sassy, extremely cheeky and naughty boy who, especially in religious instruction, made the most impudent remarks, worse than she had heard from other children. She wanted to warn you about the fact that there are all these imbecilic children at the Trazerberg. In particular, I was told terrible things about the son of the opera singer Nemeth.[32]

Finally, Arthur has himself observed:

> I recently saw a horrible cretin there, learned that he was young Nemeth, and was terrified when I heard from Schorschi that he was his and Atzis' best friend, [...] I think it was a terrible idea to put two healthy children among sick people and cretins. [...] Whatever bad habits they heard from this Eastern European child's behavior, they picked up. Atzi's beautiful fountain pen was also stolen by a kleptomaniac classmate and has disappeared.[33]

These were not necessarily enlightened views that Arthur expressed but rather were the views of the conservative bourgeoisie of his own upbringing; but Bice took a more relaxed view:

> I think the teacher is exaggerating and Atzi would have been a big rascal even without the daycare - he's at that age now. And as far as friendship with the "cretin" is concerned, I wish you had a more serious description.[34]

But Arthur was extremely worried about leaving his boys exposed to such dangerous social ideas and influences. The headmistress of the school had highlighted all the possible dangers for the future development of his sons, so he cancelled the daycare without further discussion. The director of the Trazerberg Home, Dr. Almoslino, wrote directly to the boys' mother when he saw his nutrition and education program interrupted. While this letter is presented in Fig. 18.3, it was when each

[32] Maria Nemeth (1897—1967) was one of the leading opera voices (soprano) in Vienna during the interwar years. She was a member of the Vienna State Opera from 1924 to 1942/46.

[33] AEH to EBH (Vienna, December 4, 1934).

[34] EBH to AEH (Berlin, December 5, 1934).

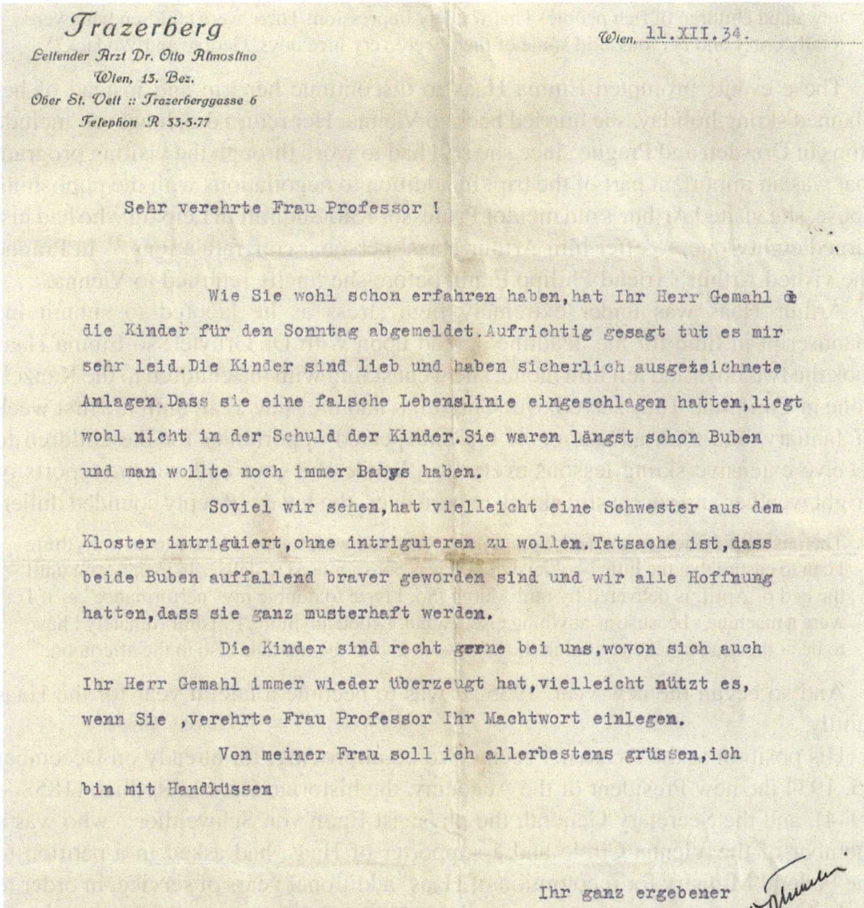

Trazerberg
Leitender Arzt Dr. Otto Almoslino
Wien, 13. Bez.
Ober St. Veit :: Trazerberggasse 6
Telephon R 53-5-77

Wien, 11.XII.34.

Sehr verehrte Frau Professor !

Wie Sie wohl schon erfahren haben, hat Ihr Herr Gemahl die Kinder für den Sonntag abgemeldet. Aufrichtig gesagt tut es mir sehr leid. Die Kinder sind lieb und haben sicherlich ausgezeichnete Anlagen. Dass sie eine falsche Lebenslinie eingeschlagen hatten, liegt wohl nicht in der Schuld der Kinder. Sie waren längst schon Buben und man wollte noch immer Babys haben.

Soviel wir sehen, hat vielleicht eine Schwester aus dem Kloster intriguiert, ohne intriguieren zu wollen. Tatsache ist, dass beide Buben auffallend braver geworden sind und wir alle Hoffnung hatten, dass sie ganz musterhaft werden.

Die Kinder sind recht gerne bei uns, wovon sich auch Ihr Herr Gemahl immer wieder überzeugt hat, vielleicht nützt es, wenn Sie, verehrte Frau Professor Ihr Machtwort einlegen.

Von meiner Frau soll ich allerbestens grüssen, ich bin mit Handküssen

Ihr ganz ergebener

Fig. 18.3 Letter from Dr. Almoslino to Emma Haas with his request to keep the two boys in his school. The wife of Dr. Almoslino was Ilse Almoslino, who ran a limited partnership with Mrs. Lola Reitz and was the actual owner of the properties

of his sons appeared with a bad grade in their report card that Haas put a final stop to the attempt in modern-life education.

Reports from other students do, however, seem to confirm Haas' concerns:

> They put my sister, who was two or three years old at the time, and me, I was seven or eight, in a children's home in Ober St. Veit, to the Doctor Almoslino [...] It was a huge complex and very beautiful. I think it borders on the Lainzer Tiergarten. It was like that on a hill. And Doctor Almoslino was majorly into health, and the older children had to sleep in barracks at night with the front open for fresh air ventilation. That was in 1928; that was the coldest winter ever, it was 25 degrees below zero. And I slept there, with six quilts, because every night I was sick from the cold. But nobody cared. Once or twice a grandfather or grandmother came. We were pretty poor then. And there were, I can remember, all these

> unwanted children of rich people. That was my impression. There were children who were totally crazy and nervous, and some of them were very nice boys, George and Wodack.[ix]

These events prompted Emma Haas to discontinue her trip and instead of her planned skiing holiday, she hurried back to Vienna. Her return did, however, include stops in Dresden and Prague since she still had to work through the visiting program that was an important part of the trip. In addition to negotiations with the publishing house, she visited Arthur's old mentor Professor Karl Sudhoff in Leipzig who had just turned eighty-one, to offer him Arthur Haas' personal congratulations.[35] In Prague she visited Arthur's friend Philipp Frank before she finally returned to Vienna.

Arthur Haas was under extremely high stress as he labored to submit his manuscript in time for the deadline agreed upon with De Gruyter, so Emma Haas took the two boys and left him alone; she went skiing with the children to the Kanzelhöhe in Carinthia. They stayed over Christmas and the New Year, until the first week of January, and took advantage of the unexpected opportunity for the children to receive extensive skiing lessons every day. While she gave enthusiastic reports of bright weather and successful skiing experiences, the letters of reply sounded duller:

> There is nothing new from me. I work terribly hard and with success. I have to do everything I can to ensure that the little booklet, which I was not supposed to deliver in manuscript until the end of April, is delivered by mid-March. So, I have to double my "performance" as if I were a machine - because as anything else, I am not good anymore. [...] unfortunately I have to be at the academy quite a lot in the new week and in the institute also in the afternoon.[36]

And so began the new year, 1935. It was to become a fateful year for the Haas family.

His position at the Academy seemed to be improving, for already on December 15, 1934 the new President of the Academy, the historian Oswald Redlich (1858—1944), and the Secretary General, the physicist Egon von Schweidler[37] who was a member of the Vienna Circle and a supporter of Haas, had asked in a petition to the Federal Ministry for recognition of Haas' additional years of service, in order to make better retirement conditions possible. Although Haas was only fifty-one years old and felt in the best of health, it was an active step on the part of the Academy's management to express their appreciation for his ten years of service to the Academy.

[35] Karl Sudhoff had become a member of the NSDAP just one year earlier, in 1933. This was an opportunistic step to save his institute and life's work, a step taken by many German scientists in those days. With his eighty-first birthday, he retired for good and died four years later in 1938 in Salzwedel. Cf. Thomas Rütten, "Karl Sudhoff and 'the Fall' of German Medical History," in *Locating Medical History: The Stories and their Meanings* (Baltimore: Johns Hopkins University Press, 2004), 95–114.

[36] AEH to EBH (Vienna, January 5, 1935).

[37] Egon von Schweidler (1873—1948) studied physics and mathematics at the University of Vienna after graduating from the Schottengymnasium in 1890. He received his doctorate in 1895 and became an assistant to Franz Exner. In the following years he worked with Stefan Mayer investigating the properties of various radioactive substances, in particular the effects of external magnetic fields. In 1911 he was appointed Associate Professor in Vienna but in the same year he moved to Innsbruck where he remained until 1926, before returning to Vienna as full professor at the Institute of Physics. From 1929—1933 he was secretary of the Austrian Academy of Sciences, from 1933—1938 he was secretary general and from 1939—1945 he held the office of vice president.

It may be that this submission arose from attempts by anti-Semitic groups who sought to remove Haas from his position at the Academy. These attempts against Haas are evidenced in a letter written by one of the members of the Bear Cave group at the university. Musicologist Robert Lach[38] wrote, in a letter dated January 6, 1934 to the German musicologist and influential party member Erich Müller von Asow,[39] suggesting that Haas be deprived of his influence and be replaced by a colleague who was politically more sympathetic to the party. In this context, Lach formulated a truly spiteful letter:

> Speaking of Jews! Allow me to turn to you in one matter, with the request that you, by your position and alliances in the party, might be able to bring about a beneficial change! The journal *Forschungen und Fortschritte* (organ of the Prussian Academy of Sciences in Berlin [...]) employs for some years (apparently arranged by the former ratty Jewish regime of Kestenberg [Leo Kestenberg (1862—1962)]) some Dr. Arthur Haas as trustee and consultant for Austria. He is a private lecturer for physics at the University of Vienna and actuary at the Academy of Sciences [...].
>
> Now Haas is one of the most disgusting and horrible Jews imaginable, a genuine Jewish busybody, and a striver of the most obnoxious sort, and it is a real irony of fate that at these

---

38 Robert Lach (1874—1958) is considered a lesser-known musicologist and composer who studied under the founder of musicology in Vienna, Guido Adler (1855—1941) and became his successor in 1927, as professor at the University of Vienna. Lach's most famous contributions to music history were recordings and descriptions of the songs of Russian prisoners of war from the First World War, which he published in three volumes through the publishing house of the Austrian Academy of Sciences between 1926 and 1931. In doing so, he worked closely with anthropologists Rudolf Pöch and Otto Reche mentioned above, who sought to confirm the superiority of the Germanic race in their anthropological studies. [Cf. Monique Scheer, "Captive Voices: Phonographic Recordings in the German and Austrian Prisoner-of-War Camps of World War I," in *Doing Anthropology in Wartime and War Zones*, eds. Reinhard Johler, Christian Marchetti, Monique Scheer (Bielefeld: transcript Verlag, 2010), 295–302.] Lach was a close friend of Othenio Abel, whose poems he set to music for voice and piano accompaniment between 1930 and 1940. Through this collaboration, an anti-Semitic and pro-national-socialist attitude probably developed. Lach had been a member of the NSDAP since 1933 and, along with Otto Reche, belonged to a secret academic association called Bärenhöhle (Bear Cave). Organized by Othenio Abel, Bärenhöhle members pursued anti-Jewish university policies and used racist arguments to denigrate colleagues; Abel also agitated against other people of Jewish origin in his personal environment and in public life.

39 The following is recorded in the list of estates held by the Austrian National Library concerning Erich Müller von Asow: Müller von Asow (1892—1964) studied musicology, ethnology and education at the University of Leipzig from 1912 and received his doctorate in 1915 with his dissertation "Die Mingottischen Opernunternehmungen." He then worked as an assistant director at Leipzig's New Theatre and in 1917 became artistic director of the First Modern Music Festival in Dresden. From 1926, he taught musicology as a lecturer and from 1931, he headed the music department at the Central Europe Institute in Dresden. Between 1933 and 1945 he worked as a freelance musicologist and after 1945 founded and directed the International Musicians' Letter Archive. Müller von Asow was a specialist in musician epistolography; he initiated the Heinrich Schütz Society (1922) and the German Chopin Society (1959). He published, among other things, a *Deutsches Musiker-Lexikon* (1929) as well as *Die Musiksammlung der Bibliothek zu Kronstadt* (1930) and edited letters from Handel, Bach, Brahms and Reger. Less well known is the fact that Erich Müller von Asow often used material from the texts of Jewish colleagues for his numerous music-historical writings without seeing the necessity to quote or mention the authors. In 1934 he was co-author of *Handbuch der Judenfrage*, where he wrote the chapter on Jews in music. In this, he identified numerous composers and musicians as Jews and spoilers of German music. He also did not shy away from so identifying

Fig. 18.4 Left is composer and music historian Robert Lach and, on the right, Germanist and music historian Viktor Junk. It is interesting to note that in Vienna there has been a street named for each of them, for their services to Austrian science: in 1968, the Robert-Lach-Gasse in Vienna-Floridsdorf in the 21st district and in 1955, the Junkgasse in Vienna-Essling in the 22nd district

> new times the regime in Germany allows such a creepy Jew to handle eminently important matters of culture for the German Empire.
>
> The matter is all the more grievous, since Haas is a younger colleague of our common friend Junk at the Academy [Viktor Junk (1875—1948) was the actuary for the humanities. Haas for the natural and mathematical sciences] and Junk even wrote a book about the bearers of the Nobel Prize. That is to say, in all respects, if it were a matter of appraisal and factual appeal, he should have been the first to be appointed to that office. But the Jew Haas who is a deadly enemy of all Aryan feeling and ambitions, and of all ethnic aspirations, was entrusted with this function by the former Berlin Jewish clique of Kestenberg.
>
> Couldn't you, my esteemed and dear friend, make arrangements through your connections so that the disgusting Jew would be dismissed, and Junk would be entrusted with this function instead of him?[40]

The wording of this declaration shows the nature of the hateful spirit that dictated the opposition to Jews in general and Jewish scientists in particular, and it clearly justifies Haas' concerns and underscores his longing to leave the country.

The relationship between Robert Lach and Viktor Junk went back a long way (Fig. 18.4). Junk, as a permanent contributor to the music journal *Der Merker*,[41]

German musicians who were personally not suited to him. Cf. Toby Thacker, *Music after Hitler 1845—1955* (New York: Routledge, 2006), 173–174.

[40] Letter of January 6, 1934 from Robert Lach to Erich Müller von Asow, reprinted in Gerhard Oberkofler, Manfred Mugrauer, *Georg Knepler: Musikwissenschaftler und marxistischer Denker aus Wien* (Insbruck: StudienVerlag, 2014), 69–70.

[41] *Österreichische Zeitschrift für Musik und Theater*, published by Richard Batka and Ludwig Hevesi from 1909 to 1922.

benevolently reviewed Lach's *Oriental Songs for Voice and Piano*[42] written in 1917.[43] This probably resulted in a joint work, the publication *Songs of Russian Prisoners of War.*[44] This project was compiled between 1927 and 1931 at the Academy and the I Actuary served as the editor. Junk was also familiar with the history of music, as he received his doctorate in 1899 on the literary and historical significance of Mozart's *Magic Flute*. In a letter to Müller von Asow, dated August 24, 1933, Lach writes:

> A few days ago, I received almost simultaneously from my dear old friend Junk and from the editorial staff of the *Deutsche Musik* a booklet in which, in an extremely kind manner, an infinitely benevolent and amiable review is dedicated to yours truly. Since I know that I owe this to your friendly attitude toward me, it is naturally a matter of the heart for me to thank you from the depths of my soul for this new proof of your friendship. [...] etc., etc., etc.[45]

This somewhat embarrassingly ingratiating letter at least proves that there was something like a friendship or a relationship through common interest between the three musicologists.

As a student, Junk belonged to the All-German (Schönerer) Party and even before the Anschluss he was considered an "old fighter" and member of the (illegal) NSDAP (membership number 1.616.309). As former secretary and later president of the academy Heinrich von Srbik[46] —who had already helped Robert Lach to his chair as successor of Guido Adler—wrote in October 1942 in an assessment:

> Professor Dr. Viktor Junk is a convinced National Socialist. He allows himself to be guided exclusively by this spirit in all tasks of his ministry and in his dealings with his followers [...]. During the Austrian System Age, he helped two party comrades who were persecuted by the Dollfuß government to escape across the German border.

It is not quite clear what the reason was for Lach's demand that Arthur Haas be replaced and his duties dealing with the Prussian Academy be handed over to Junk. It is clear enough that Robert Lach wanted to use his German connections to get his friend Junk a better position at the academy, which was under the relatively liberal and Haas-benevolent leadership of Egon von Schweidler. It is quite conceivable

[42] Robert Lach, *Orientalische Gesänge* vol. 1 (Wien: Adolf Holzhausen, 1917).

[43] Victor Junk, "Orientalische Gesänge für eine Singstimme, mit Klavierbegleitung, von Robert Lach," *Der Merker* IX/8 (1918): 303; "Moderne Bewertungen sind allerdings vernichtend," in Peter Revers, *Das Fremde und das Vertraute – Studien zur Musiktheoretischen und Musikdramatischen Ostasienrezeption* (Stuttgart: Steiner Verlag, 1997), 95–97.

[44] Robert Lach, *Gesänge russischer Kriegsgefangener* vol. 1, 1929—1940, vol. 2, 1930—1939, vol. 3, 1928—1931 (Wien: Österreichischen Akademie der Wissenschaften).

[45] Letter by Robert Lach to Erich H. Müller (Berlin, August 24, 1933). (Quelle: ONB, Musiksammlung, Lach Robert).

[46] Heinrich Ritter von Srbik (1878—1951) was an historian at the University of Vienna. He held the office of Austrian Minister of Education from 1929—1930, an office that he used to appoint Robert Lach to the chair of musicology. Srbik was also a member of the anti-Semitic professors' association Bärenhöhle, already known to us. Srbik joined the NSDAP at the beginning of National Socialist rule in Austria from 1938 to 1945. Also, from 1938 to 1945, Srbik was president of the Austrian Academy of Sciences.

that Junk had complained to Lach about the work overload that had resulted from his colleague Haas' trips to America, although there is nothing in writing about a personal or even political dislike between the two actuaries. It is not known whether Haas knew about this letter from Lach regarding Junk, at least there is no reference to it in Haas' letters. Whatever its purpose, this letter was not successful; Schweibler continued to support Haas as a request for an increase on behalf of the latter testifies.

A further positive event for Haas was an invitation from the president of the Drexel Institute (today Drexel University), Dr. Parke R. Kolbe, to teach in Philadelphia for three months as Visiting Professor. Haas had become good friends with Kolbe, who had already visited twice in Vienna and had invited Haas several times to the Brooklyn Polytechnic Institute in New York, where Kolbe had been president until 1932. He was an important patron and a great hope for Arthur Haas.

This sabbatical visit was planned for January through March 1936 but Haas had already formulated his application for a leave of absence from the Academy in February 1935, and he tried to again convince the Academy of the benefits of such international connections:

> This visiting professorship is to be combined with lectures at the University of Pennsylvania, in Philadelphia, and at Yale University. [...] On the basis of an exchange of letters, it is also certain that I will be invited to lecture at Princeton University, John Hopkins University and the Royal Astronomical Society of Canada, as in previous years.
>
> Provided that I would be granted a three-month leave of absence, I would consider it a pleasant duty to accept the invitation - especially since I am convinced that every Austrian who fulfills such an assignment, even in the most modest sphere of activity, can thus benefit his fatherland a little. I am also convinced that, through my relations with Austrian groups in New York and Philadelphia as well as with the Rockefeller Foundation in New York and the Carl Schurz Foundation[47] in Philadelphia, I might be able to obtain funding for one of Austria's scientific institutes in one case or another.[48]

A handwritten note under the application remarks:

> Basic agreement to accept the invitation! Decision in the presidential meeting of February 22, 1935 Schweidler

Arthur Haas was delighted with the presidium's quick approval and immediately used connections with Ms. Waite from the IIE to organize further invitations. This time he hoped to take the family with him to the United States; in his opinion, the two boys were now old enough and Haas nourished his secret hope of being able to find a permanent professorship in the US within this three-month period. For this

[47] The Carl Schurz Foundation was established in 1930, named in honor of the late German revolutionary who was later an American senator and Secretary of the Interior. The founders were German immigrant industrialists who had built a fortune in the United States, including Ferdinand Thun, Gustav Oberlaender, Henry Janssen, and Hanns Gram, who each wished to promote and improve the teaching of German language and culture, and to foster friendship between the United States and German-speaking countries. CF. Horst Heidermann, Klaus Vollmer, *Millionäre und Mäzene* (Wuppertal: Köndgen, 2014).

[48] Request by Arthur Haas to the High Presidium of the Academy of Sciences in Vienna (February 19, 1935), (Akt 135/1935).

purpose, he also contacted the Carl Schurz Foundation, as he had committed with the Academy to do, in order to make new connections and offer his services.

Due to his intensive planning and preparation, with countless letters to America, Haas was probably unable to write a new book. He did, however, work on the third edition of his *Atomic Theory*, as well as on a substantially extended English edition. *Physik für Jedermann* was a great success at Springer and Haas negotiated with several foreign publishers for Polish and Finnish editions of the book. He also negotiated with a Yugoslavian publisher but, since he could not get permission to leave the country, the negotiations came to nothing.

And so, summer was quickly approaching. Because of Emma Beatrice's studies, summer plans changed. The retreat in June was cancelled and, in view of plans to go to the United States together in January, the summer held different priorities. Emma wanted to see her family again in Munich and her friends at Ammersee and in the Bavarian Forest. She traveled to Munich in early July with her youngest son Georg, who was now almost nine years old, while the older Arthur Gustav stayed with his father in Vienna. Georg settled in Germany very well, went sailing with his mother on the beloved Ammersee and was admired by everyone. Meanwhile young Arthur enjoyed swimming and making long excursions into the surrounding area with his father. The success and growth of the children was exchanged in regular letters.

As in past years, Professor Kolbe announced a European visit for the summer of 1935. Haas wanted to use this opportunity to meet with him and explore more long-term possibilities at Drexel. It therefore became Haas' most important task to help in planning a vacation for the Kolbe couple, at least for their stay in Munich and Vienna. He repeatedly told Emma to be especially friendly and hospitable, and to ensure a high-level visit for Kolbe in Munich. Kolbe and his wife were due to arrive in Munich on July 23rd from Göttingen, where they planned to stay for three days before traveling via Salzburg to Vienna. There they wanted to visit both Arthur Haas and Professor Petritsch before traveling by ship on the Danube to Budapest.

Emma did her best in Munich—from the Hofbräuhaus to the Deutsches Museum, all the sights were covered. She even drove with the Kolbes to the Ammersee, on the "magnificent new Reichsautobahn"[49] in a car borrowed from her friend Eugenie Grau. A trip to Augsburg was also organized to visit the Fuggerei, the former seat of the late medieval banking family of the Fugger. Emma showed her most hospitable side and Arthur appreciated that:

> I am very touched that you took so much trouble with Kolbe. I am looking forward to their coming visit with great excitement and, in the spirit of your admonition, I will make every effort not to embarrass myself. Even if Dr. Kolbe wants to have me in a permanent position, it is in keeping with his nature that he will not open his mouth until he has seen how I prove myself as a visiting professor.[50]

The Kolbe couple's visit to Vienna was also a great success. Arthur Haas showed them around the city and invited them to dinner in his home. After dinner Haas

[49] EBH to AEH (München, July 29, 1935).

[50] AEH to EBH (Vienna, July 30, 1935).

went alone with Professor Kolbe to the Heurigen[51] in Grinzing, where he took the opportunity to raise the subject of his move to the United States.

> He was quite surprised when I told him in the course of conversation that, as nice as Vienna is, I would like to leave it forever to find a new sphere of activity in the USA. He said he was sure that this would not be difficult to find for me, after all. Then he became thoughtful and asked whether I was also aware that I would then have to teach physics at a much lower scientific level than I was used to in Vienna. I said that I would not mind and that I would take great pleasure in teaching. He became thoughtful again - then we talked about something else.[52]

Further, the question of the three-month stay in Philadelphia, for which Haas was to arrive at the beginning of January 1936, was discussed. During the days of Kolbe's visit, Haas was already busy looking for favorable ship connections. German ships were no longer an option for him and he looked for opportunities to travel on the passenger ships of the British Cunard Line. Finally, he had worked everything out. His plan was,

> to leave Vienna on December 23rd or 24th, spend 1 to 1½ days in Paris and leave Le Havre early on December 27th with the *Franconia* to land early on January 4th with this ship of the Cunard-White Star Line.

However for the remainder of the summer, he was planning to spend August with Atzi at Mondsee in the Salzkammergut, with Emma eventually coming to join them from Germany. But it all turned out differently as he suddenly wrote in excitement to Emma:

> I still don't know what I'm going to do with all this joy. Tears are still running from my eyes in happiness. Today, as I believe, the fate of not only ourselves but also that of my children and, God willing, of my grandchildren, has decidedly led to happiness and success. I believe we are saved. I am still too excited to write you beautifully.
>
> The cable is as follows: Are you available visiting professorship Bowdoin College October June Salary four thousand Dollars opportunity fulfill lecture engagements Drexel. signed Sills President Bowdoin.
>
> I contacted Kolbe immediately. He said I had a good chance of making it all the way to the top if I let everything else take a back seat to this opportunity. He firmly believes that the matter will become permanent if I want it to. Thinks $4,000 is very little. Advises immediate acceptance, even if it means ditching Drexel.
> I'll wire acceptance tomorrow.[53]

This letter certainly reflects the enormous sense of relief that Arthur Haas must have felt and indicates the extent of his worries about his family. That same day Haas wrote an extended request for leave to the Academy Presidium, claiming no salary for the time he would be absent. He had also already been in contact with I Actuary Junk and reported:

---

[51] Heuriger is the name given to taverns in Central Austria at which local winemakers serve their new wine–also called Heuriger–under a special license, in alternate months during the growing season.

[52] AEH to EBH (Vienna, August 6, 1935).

[53] AEH to EBH (Vienna, August 8, 1935).

> Professor Junk, with whom I had a detailed conversation today, agreed to replace me in any case during my absence, if the high presidency allows it. [...] I took the liberty of writing to the Secretary General in the same terms.[54]

Approving answers from Academy President Redlich and Secretary General Schweidler came from their vacation spots, both assuring Haas that they would vote in favor at the next Academy meeting. The member of the Presidium, Professor von Sbrik, also congratulated Haas warmly and declared agreement for the extended leave. And so it happened, the request was granted and one-third of Arthur Haas' salary was allocated to the employees who took over his duties during his absence.

The rest of the year passed with hectic preparations, although he did still spend a few weeks with Atzi in the summer resort at the Hotel Krone am Mondsee. There Haas indulged his passion for swimming again but otherwise mainly devoted himself to planning for the future. He prepared a budget to see if the $4,000, plus his book and lecture royalties, would be enough to allow him and his family to live in the US. Another problem was school. Should he take the children with him and send them to school in the US? Because what if, contrary to expectations, he did not find a permanent job? These are problems that would concern anyone who was considering going to America with family for an extended stay as Haas planned to do.

As happened every year, some of the Viennese academics who had emigrated to the US were back in Austria. They provided Haas with a good opportunity to obtain first-hand information. Otto Halpern from New York assured him "that life in Brunswick would be incredibly cheap," and others agreed. Haas calculated the cost of travel and daily expenses, took into account the exact exchange rate of the dollar, fought with every penny to make the budget cheaper, and sent all these calculations in long letters to Bice who was staying with Gocki (as the second son Georg was now called in the letters) in the Bavarian Forest. She was also supposed to find out if she would be able to use German money to pay her travel expenses on German ships, for herself and the children. Could she use money from her German account, which held the income from the partnership with her brother? All of these thoughts flew back and forth in Arthur's head and in his letters.

Further, negotiations with Cunard representatives were dragging on; contrary to earlier promises, they did not want to give a discount. In frustration, Arthur even considered travel on German ships:

> The Cunard line is a bunch of crooks. Orally, they promised a 5% discount and half-fare for Atzi, and now they deny everything. Of course, I prefer the German line if Atzi gets half price and Cunard doesn't come through. Also important is the free ride from New York to Boston on the daily night ships.[55]

Finally, the planning was finished and Arthur was moving forward. For him, a cabin on the *Berengaria* of the Cunard line (Fig. 18.5) was booked for September 28th. He wrote again from Paris, where he realized that it would be a final farewell:

---

[54] AEH to President, Court Councilor Dr. Oswald Redlich (Vienna, August 8, 1935).

[55] AEH to EBH (Vienna, August 17, 1935).

Fig. 18.5 The R.M.S. *Berengaria* in the port of New York, after an advertising poster of the Cunard line

> The trip from Salzburg to the Austrian-Swiss border was beautiful. I haven't seen anything so beautiful for a long time. [...] When I left Austria, my heart was aching despite everything and I was particularly depressed and sad as I rode through the Vienna woods, past all the places that I used to show proudly to the children.[56]

A few weeks later the family followed, also on a ship of the Cunard line. Arthur's tenacious negotiating tactics with the Cunard representatives had been successful after all.

The apartment on Baumgartenstraße was sublet to an English family, Dunlop and was to be kept in impeccable order, as sister-in-law Grete Haas was supposed to watch for from Vienna:

> Here, everyday life goes on unchanged and with unchanged difficulties.[57]

For the family of Arthur Haas, new life began in a new country where they soon felt at home.

Notes

i. Friedrich Stadler, *Der Wiener Kreis—Ursprung, Entwicklung und Wirkung des Logischen Empirismus im Kontext* (Frankfurt: Springer Verlag, 1997).
ii. Friedrich Stadler, Hans Jürgen Wendel, eds., *Stationen. Dem Philosophen und Physiker Moritz Schlick zum 125. Geburtstag* (Wien-New York: Springer Verlag, 2009), 75.

---

[56] AEH to EBH (Paris, September 30, 1935).

[57] Grete Haas to EBH (Vienna, December 12, 1935).

iii. See Anton Pelinka, *The Dollfuß/Schuschnigg Era in Austria: A Reassessment* (New York: Routledge, 2017) and Emmerich Tálos, Wolfgang Neugebauer (eds), *Austrofaschismus —Politik —Ökonomie—Kultur, 1933–1938* (Wien: Lit Verlag, 2014).
iv. Delikatessenladen, Wiener Adressbuch von 1934, accessed January 31, 2021, https://www.digital.wienbibliothek.at/wbrobv/periodical/pageview/212236.
v. J. Blackmore, R. Itagaki, S. Tanaka, eds., *Ernst Mach's Vienna 1895—1930*, 278.
vi. Juliane Mikoletzky, ed., *Die Technik und Musen—Kunst und Kultur im Umfeld der Techischen Universität Wien* (Wein: Böhlau Verlag, 2016), 114–115.
vii. Karl Sigmund, *Sie nannten sich Der Wiener Kreis—Exaktes Denken am Rande des Untergangs* (Wiesbaden: Springer Spektrum, 2015), 243.
viii. Michael Wiescher, *Radioaktivität – Betrachtungen eines Naturphänomens* (Darmstadt: WBG, 2021).
ix. Suzanne Wolff, "Meine sogenannte Emigration," in *Erinnerungen—Lebensgeschichten von Opfern des Nationalsozialismus*, ed. Renate S. Meissner, Nationalfonds der Republik Österreich für Opfer des Nationalsozialismus (Wien 2013), 156–157.

Part IV
University of Notre Dame Emigration and Living in America 1936–1942

Chapter 19
Bowdoin College

Arthur Haas finally saw himself successful in the fall of 1934, when he had nearly reached the goal of his desires. Although not quite a highly paid professorial position at one of the well-known elite universities of the United States, it was a first step. Haas had secured a one-year position at a small but well-known elite college in Brunswick, Maine.

From its inception, Bowdoin was known for educating the political and military elite of the United States (Fig. 19.1). Therefore, Bowdoin required students to have a certificate of "good moral character," a basic classical education in Latin and ancient Greek, geography, algebra and the study of the important works of Cicero, Xenophon, Virgil and Homer. Bowdoin has also played a role in the history of American literature; Harriet Beecher Stowe began writing her influential anti-slavery novel *Uncle Tom's Cabin* in Brunswick while her husband, himself a Bowdoin graduate, taught at the college. Bowdoin was also the training ground for two of the most famous American writers of the nineteenth century, Nathaniel Hawthorne and Henry Wadsworth Longfellow.[1] In the twentieth century, the school saw the need to improve in the mathematical and scientific disciplines and the appointment of Arthur Haas, with his broad classical and scientific education and knowledge, provided a unique opportunity for the college.

In the early 1930s, physics was represented by Noel Charlton Little[2] who, as chair of the department, had invited Arthur Haas to come as a visiting faculty member.

---

[1] An impression of life at a liberal arts college like Bowdoin is given in the description of fictitious Winsted College, albeit set thirty years later in the 1960s. The story was written by David Adams Cleveland, a grandson-in-law of Arthur Haas. David A. Cleveland, *Times Betrayal* (Burlington, Fomite Press, 2017).

[2] Noel Charlton Little (1895—1978) graduated from Bowdoin College in 1917; his family had been an integral part of the school for generations. He was a successful physics professor at Bowdoin and was also instrumental in its development and expansion. Little published a successful textbook on physics and, during the war, he directed radar training for US Naval Officers. In 1967 he received an honorary doctorate from Harvard University for his contributions to physics education.

M. Wiescher, *Arthur E. Haas—The Hidden Pioneer of Quantum Mechanics*, Springer Biographies, https://doi.org/10.1007/978-3-030-80606-4_19

Fig. 19.1 Bowdoin college on a contemporary postcard

Besides Little, Boyd Wheeler Bartlett[3] was an associate professor, who was primarily dedicated to teaching.[i] In addition, George Baker Welch[4] was hired in 1934 as a young assistant professor. Teaching in physics was limited to a basic lecture with special lectures on classical physics topics such as mechanics and heat, light and optics, and electricity and magnetism, all topics from the classical physics portfolio of the nineteenth century. Young Professor Welch was given the task of introducing students to modern physical concepts including gas discharges, radioactivity, X-rays, electron theory, crystal and atomic physics, etc. So, Arthur Haas had the role of introducing the theoretical foundations of the new physics to Bowdoin and complementing the primarily experimental program. It was clear that Professor Kolbe's prediction was true; Haas would have to teach at a much more basic level than he had in Vienna.

Arthur Haas' visiting professorship at Bowdoin was funded by the Tallman Foundation.[5] Tallman funds had been available for Bowdoin College since 1928 to support guest-lecturer postings of up to one year, with no restrictions on the subject matter of the lectures. Haas was elected by the Tallman Faculty Committee to teach at Bowdoin

[3] Boyd Wheeler Bartlett (1897—1965) also completed his education at Bowdoin College in 1917. After a short military career, he received his doctorate from Columbia University in 1933 and was assistant to Arnold Sommerfeld in Munich from 1934 to 1935. He was appointed to Bowdoin College in 1935.

[4] George Baker Welch (1897—) graduated from Bowdoin College in 1922 and returned as an assistant professor after completing his doctorate at Yale University.

[5] Frank Gifford Tallman (1860—1938) was a mechanical engineer and worked as an executive with several different companies before joining E.I. du Pont de Nemours & Company in 1905; he served as head of purchases and president of several subsidiary companies. After his retirement, Tallman turned his interests to philanthropy and established the Tallman Foundation at Bowdoin College.

during the academic year 1935—1936. As described, he accepted the offer immediately in the hope of using this stay as a stepping stone to a longer-term position in the United States rather than remaining in increasingly hostile Austria.

Haas arrived at Bowdoin College in time for the beginning of the fall semester of 1935. His classes were scheduled for Tuesdays, Wednesdays and Thursdays, which set certain constraints on his lecture schedule outside the college. He had planned these lectures in order to further build his reputation but the schedule forced him to give his first external lectures locally in Maine:

> I read "A Survey of Modern Physics" for three hours in front of about twenty-five students, two hours of supplements to it, and "Wave Mechanics of Atomic Nuclei" for one hour in front of members of the staff. On October 19th I give an "address" to the Physics Teachers of the State of Maine and on October 26th, I give a lecture at a meeting of the American Physical Society in Providence.[6]

Besides physics, Haas also took care of a house that he had rented cheaply through the college, to accommodate his family:

> I am overjoyed to be here and if I will have you here too, I will not miss anything for happiness. Life and work are so easy [...] The people are all indescribably nice to me, with an unparalleled cordiality. Especially the physics professor Little and his wife, with whom I lived in the beginning [...] Then there is the very nice professor for German, Kölln, who is a Hamburg native (a Nazi enemy); he is our neighbor. [...] In the afternoon I visited the apartment. [...] It is a jewel in every respect, so spacious, so practical and so beautiful. [...] Prof. Kölln has found this treasure, for which I am very grateful to him. [...] I have become a completely different person here, cheerful, always full of work, much younger and in every respect a man of order and care and full of confidence in the future.[7]

His family joined him in the last days of October. Haas had no intention of returning to Austria and immediately began an intensive search for a more permanent teaching position in the USA. To this end he once again turned to a series of invitations that he had received through the IIE with the proven help of Ms. Waite, since the country's economic situation had improved and colleges and universities again had the necessary financial resources to invite guest lecturers. The series of lectures was originally intended to be part of Haas' work at the Drexel Institute but he did not hesitate to make use of Bowdoin's most generous travel opportunities in order to meet these obligations.

Bowdoin's President Sills gave Haas several weeks' time at the very beginning of his stay, freeing him for lecture activities in January and early February 1936. By that time Haas had already received numerous invitations, especially from colleges in the western United States—Nebraska and Idaho, Texas and California, as well as Washington and Wyoming—all of which he intended to accept for February. For the nearby lectures in the East, he worked out a complicated itinerary for visiting as many institutions as possible in the hope of obtaining a permanent position.[8] Within

---

[6] AEH to EBH (Brunswick, October 13, 1935).

[7] AEH to EBH (Brunswick, October 10, 1935).

[8] IIE to AEH (New York, October 16, 1935); AEH to IIE (Brunswick, October 17, 1935); AEH to IEE (Brunswick, October 20, 1935).

a few days, he managed to put together a complete program and accept numerous invitations.

In the first week of his trip west, he planned to visit the universities of Missouri, Kansas, and Omaha, and from there turn south to give lectures at Texas State College, the University of Southern California and Stanford University. During the third week, lectures were scheduled for San Francisco, Reed College in Oregon, the University of Washington in Seattle, Whitman College of Walla-Walla in East Washington, as well as the University of Idaho and the University of Wyoming. The last week was to be spent travelling back from the Far West, but he also planned to give lectures at Iowa State College in Ames, Pennsylvania State College and Swarthmore College. It was a murderous bus and train journey in the deepest winter and the coldest month of 1936.

During the weekends of the November and December months, he planned to visit nearby institutions from which he had received invitations, such as the Drexel Institute, Yale University, John Hopkins University, University of Vermont, Wilson College and McGill University, each engagement involving many hours of rail travel.

By the end of his first October weeks in America, Haas had successfully completed his plans for the lecture trip and he traveled to New York to pick up his family, who arrived on October 29th with the R.M.S. *Berengeria.*

During their first visit to the big city, the children were especially impressed by the fact that the houses were bigger and higher than the church towers, a circumstance that they had not been familiar with in Austria. Leaving New York, the family traveled north by train via Boston to Brunswick.

> After the train went through New England, the countryside looked a little more like what I was used to seeing, i.e., trees, meadows and individual homes. We finally arrived at the little town of Brunswick, Maine, where Bowdoin College is located and it reminded me of a typical small rural town in Austria. We drove to our house on Longfellow Street [Fig. 19.2] (which Dad had rented through the College) and settled in.[ii]

The children started school immediately, at Longfellow School right next to the house, where they had extra English lessons. (Fig. 19.2)

After only a few weeks the family had adapted to life in the small American town, which became especially exciting when the first snow fell in mid-December and covered the entire countryside with a white blanket. Emma seemed happy and wrote to Arthur:

> The children are very well-behaved and study hard. Today they went skiing for the first time. It was a hustle and bustle. I always have the feeling that it is already Christmas. Outside the snow and inside the house, the wonderful peace. I recover every hour.[9]

Arthur had started his first lecture tours, traveling to Wellesley College and particularly to Wilson College where he met Dorothy Weeks whom he held in high esteem, but nothing is known about this visit.

Over Christmas 1935, the family was together in their new home but just after Christmas, Arthur left for Boston. He would then take the long train ride to St.

[9] EBH to AEH (Brunswick, December 12, 1935).

Fig. 19.2 The new home on Longfellow Street next to Longfellow elementary school in Brunswick

Louis from where he would begin his lecture tour of the western United States. But from Boston he sent the joyful message that he had been invited by Harlow Shapley (1885—1972)[iii] to be a special speaker at the 300th anniversary of Harvard University, as seen in Fig. 19.3 . Haas was to give a lecture on his ideas about the early universe, alongside Einstein, Bohr, Heisenberg, and Eddington.

But before this important celebration took place, he still had a busy time ahead of him; there were numerous visits and lectures on which Arthur Haas had again placed all of his hopes.

At the time of his departure on December 29, 1935, the family was already firmly integrated into the social circle of professors at Bowdoin College. In particular, Emma Haas had found a new friend in Professor Little's wife who made it easy for the family to feel at home. Emma received invitations from all sides and was soon considered the center of society:

> Yesterday I was photographed from every angle for the Portland newspaper, as a Brunswick-society star. Today I was supposed to go to a Republican lunch in Portland, but I cancelled. On February 7th, I have to give a lecture on Austrian education. It is an evening lecture for the Brunswick Teachers' Association.[10]

While Arthur traveled to the West, Bice managed the household in Brunswick and, in addition to the normal schooling, gave the children lessons in German and Latin since there was still no certainty of a permanent stay in the United States. Her days were full and she seems to have been a little overwhelmed, writing:

> Yesterday I always hoped for a telegram that would say: Stop lessons for children, I received permanent professorship in Los Angeles, but since nothing arrived, our hope was dashed and we continue to study diligently.[11]

This was also the exact hope that sustained Arthur Haas since being approached by the University of Southern California, who seemed to have a great deal of interest in

[10] EBH to AEH (Brunswick, January 8, 1936).

[11] EBH to AEH (Brunswick, January 15, 1936).

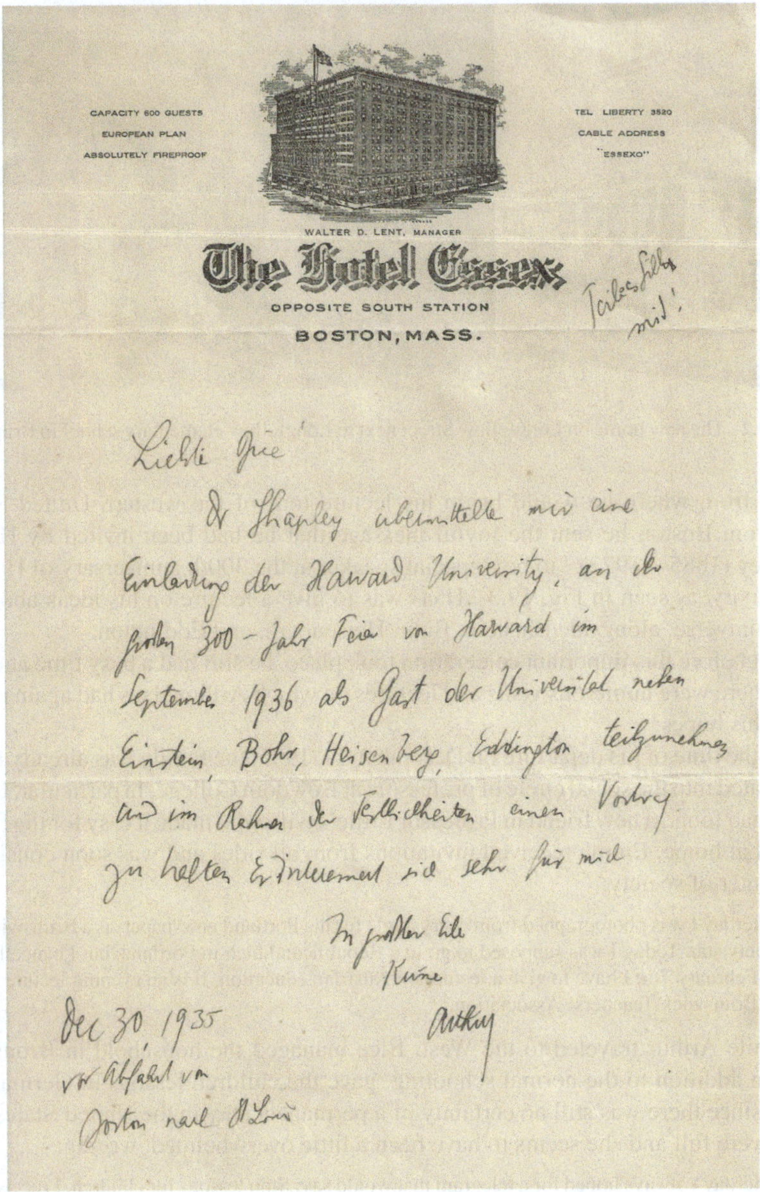
CAPACITY 600 GUESTS
EUROPEAN PLAN
ABSOLUTELY FIREPROOF

TEL. LIBERTY 3520
CABLE ADDRESS
"ESSEXO"

WALTER D. LENT, MANAGER
The Hotel Essex
OPPOSITE SOUTH STATION
BOSTON, MASS.

Teiles Silber mit!

Liebste Grie

Dr Shapley übermittelte mir eine Einladung der Harvard University, an der großen 300-Jahr Feier von Harvard im September 1936 als Gast der Universität neben Einstein, Bohr, Heisenberg, Eddington teilzunehmen und im Rahmen der Festlichkeiten einen Vortrag zu halten. Er interessiert sich sehr für mich

In großter Eile
Küsse
Arthur

Dec 30, 1935
vor Abfahrt von
Boston nach St Louis

Fig. 19.3 Letter from Arthur Haas to his wife in which he happily announces his invitation to the 300th anniversary, 1936 Centennial at Harvard University offering him a permanent professorship. USC had, therefore, made a special effort to invite him through the IIE. The visit was probably a great success and the hoped-for telegram from Los Angeles arrived in Brunswick on January 16th:

> Enormous success — long conference with vice president — University Southern California now expects a vacancy — will then offer me permanent professorship at highest-ever paid salary of six thousand — decision within two months — keep absolute silence. Here is paradise.[12]

Obviously, he had changed his views about Los Angeles and the "hell" of his last visit had now become a "paradise." Despite the heat he made numerous tours, the local professors trying very hard to make the university, city and surroundings appealing to him.

It was a great success, as Haas described in longer letters:

> Will it now become reality? It does not depend on the university, which is firmly committed, but solely on whether a local mathematics professor who wants to enter into retired life will actually make the transition in the coming weeks. But that seems almost certain.
>
> The enthusiasm for me here is increasing from hour to hour; it sounds tasteless but, the fact is, I have really taken everyone's heart by storm.[13]

The next day he was invited to Pasadena to give a lecture. The reception and thus Haas' impression were each much better than four years ago when he had first visited Caltech:

> Yesterday I was in Pasadena at the California Institute of Technology, escorted by professors from Southern California where Millikan, the Nobel Laureate, showered me with honors, which again increased my nimbus with the people at USC. At the same time, yesterday in Los Angeles, in my absence, there were consultations about my employment. The president Kleinsmidt, like the vice president Touton, is fired with enthusiasm for my permanent position and offered me an initial visiting professorship for 1936/37, regardless of my decision on the question of permanent employment. He left me as much time as I need to decide. [...] I would not have hoped that my situation would improve so rapidly, but the danger of having to return to Vienna as early as 1936 has been overcome. [...] Farewell Europe — never to be seen again except as a "tourist" perhaps. [...] Atzi's Latin no longer so important![14]

The rest of the journey was shortened because the longed-for destination seemed to be reached. Haas visited only San Francisco and Berkeley, where he also gave lectures. After that, he visited Reed College in Portland, Oregon and Spokane in Washington followed. From there he traveled back East. He wrote:

> Yesterday I spent three hours in Spokane, a city whose name I had never heard until recently but which has huge skyscrapers and a population of 150,000. Now I am on the thirty-six-hour ride from Spokane to Laramie.[15]

---

[12] AEH to EBH (Los Angeles, January 16, 1936).

[13] AEH to EBH (Los Angeles, January 15/16, 1936).

[14] AEH to EBH (Los Angeles, January 17, 1936).

[15] AEH to EBH (Railcar, January 26, 1936).

Even though the future seemed assured, other interested parties came forward who wanted to see Haas as a member of their physics departments. These were, in particular, the University of Rochester in New York that "suddenly became very rich by inheritance from Kodak Eastman", as Arthur wrote to his sister Grete on March 5th, and also a small Catholic institution, University of Notre Dame, that wanted to expand its scientific reputation and therefore its physics department in South Bend, Indiana near Chicago.

The young professor George Collins,[16] who received his doctorate from John Hopkins University, had started to build a new accelerator at Notre Dame[iv] together with young physics instructor Edward Coomes.[17] Haas, with his theoretical knowledge and practical interests, seemed to be the right man to back up this new venture with ideas and suggestions. Collins had met Haas in St. Louis when he gave his lecture there on January 9th and invited him to come directly to Notre Dame. Haas had, however, intended to make the two visits to Rochester and South Bend in succession since he did not expect much from these universities. He had his eyes entirely on Southern California, apart from opportunities at Harvard that he still imagined from time to time. What convinced him to go to California was the icy winter in Brunswick, which he did not like at all:

> The boys enjoy life here immensely. They ski almost daily. They and Bice really like the weather here. I think it is terrible. The whole city and the whole campus are one glacier field [...] what a problem it is to cover the five meters to the car and from the car, the twenty meters to the gate of the institute. There is not the slightest thing happening to clear away the glaciers. [...] In general, the passiveness of the population here is in complete contrast to everything one thinks of America. It is far more of Mohammedan fatalism.[18]

He knew that he would not have this problem in Los Angeles where he had bathed in the ocean during his last visit in January. Nevertheless, he faithfully spent the weeks after his return giving classroom and public lectures in both Brunswick

---

[16] George Collins (1906—2001) studied and graduated in physics, earning his doctorate under R.W. Wood at Johns Hopkins University before being hired as a young professor of physics at the University of Notre Dame. There he worked on nuclear excitation and nuclear decay. In 1935 he was responsible for the construction of a nuclear accelerator for the Laboratory of Nuclear Physics at Notre Dame, where he worked with Dr. Caparo. In 1935, the accelerator was the largest nuclear physics project at Notre Dame to date. The accelerator was the source for several decades of research in nuclear physics, much of which was led by Collins. In 1941, Collins joined MIT where he supported the war effort at the Radiation Laboratory by developing radar systems for military purposes. After the war he taught at the University of Rochester and was responsible for the construction and operation of the 240 meV synchrocyclotron. In 1950 he moved to Brookhaven National Laboratory and was chairman of the Cosmotron department in Brookhaven. From there he moved to Virginia Tech in 1968, where he worked for the newly founded Fermi National Laboratory and devoted himself more and more to physics education. See John R. Ficenec, "George Briggs Collins," *Physics Today* 56/1 (2003): 64.

[17] Edward Coomes (1909—1992) was born in Louisville, Kentucky and moved to South Bend to attend the University of Notre Dame in 1927. He graduated in 1938 but moved to Boston for work at MIT, developing radar for the war effort. After the war he returned to Notre Dame as a young faculty member in Engineering. Coomes retired in 1974 from Notre Dame after forty-two years of teaching and research. He died in March of 1992 in South Bend.

[18] AEH to Grete Gerstenberger (Brunswick, March 5, 1936).

and neighboring Lewiston. He also corresponded with his friend Parke Kolbe at the Drexel Institute and spent time trying to organize lecture tours for spring, as soon as the "terrible" weather improved.

Finally, Haas was able to travel to Rochester on March 24th, where he stayed for three days until Friday, March 27th; the visit was, however, a disappointment:

> I spent the whole day at the University of Rochester where I was very warmly received but there was obviously no intention of winning me over. I lived in dormitories, was the guest of colleagues for lunch and dinner, but without all the hospitality of the family.[19]

He had become used to the attentive reception of California and so, Rochester was no longer an issue. As for Notre Dame, after long negotiations Haas had arranged a visit for the same weekend so that he could get to know the campus. He took the night train and arrived on Saturday morning, March 28th. A quite different reception was reported in the same letter:

> I spent the night in the sleeping car, Rochester - South Bend. I arrived there at 8 o'clock in the morning and was taken to Notre Dame in a car by Prof. Collins (a very young professor), it took only a quarter of an hour. The golden dome of Notre Dame with the golden statue of the Mother of God shone out from afar in the most glorious sunshine. Here I live in the splendid guest rooms of the university, have a large apartment like in the Hofburg, the ceilings three times as high as normal. There is a beautiful church, the most beautiful modern one I have ever seen, as part of the university.
>
> I am extraordinarily honored here and had my first (about two hours) scientific lecture at 10 am. Huge success. Most of the professors are clergymen but they are all very comfortable gentlemen. The president is a splendid man, a clergyman himself, full of energy and youthfulness (he always swims in the cold lake).[20] [...] I wonder if they will offer me something permanent here? I like it here very much, I feel very comfortable here in the strongly religious atmosphere, I would probably have very grateful students - but Southern California is something else. If I am offered a professorship, I will ask for three weeks to think about it and I will not accept immediately. Now I should be hearing from Los Angeles soon.

Haas wavered. Notre Dame's atmosphere and warm reception obviously appealed to him; he wrote two more letters to Brunswick that were again full of praise for the university and its hospitality [Fig. 19.4]. However, after a cold winter in Maine he longed for the warm climate of Southern California. He had forgotten how much he had complained about the heat and the "hideousness" of the city just five years earlier. He expected an offer of $7,000 annual salary and in his letters, he speculated that life was cheaper in California and that even if USC would offer him only $4,500, he would still be financially better off there.

On Monday the 30th, he gave a last, big public lecture at Notre Dame, which he considered an extraordinary success and immediately afterwards he traveled back to

---

19 AEH to EBH (Notre Dame, March 28, 1936).

20 President Father John Francis O'Hara C.S.C. (1888—1960) was a member of the Congregation of the Holy Cross and served as Notre Dame's president from 1934 to 1939, during which time he expanded the Graduate School's doctoral program in a number of disciplines including philosophy, physics, mathematics and political science. In 1939 he became Military Bishop of the US Army, Bishop of Buffalo, New York and finally Archbishop of Philadelphia.

VIENNA PHYSICIST VISITS NOTRE DAME.

The University of Notre Dame Saturday morning welcomed Dr. Arthur Haas, professor of physics at the University of Vienna, who is making a three-day stay at the university to give two lectures, one yesterday and one Monday night. Shown are Rev. Christian Murphy, C. S. C., Dr. Haas and Prof. George Collins, of the physics department at Notre Dame. —Photo by Tribune Staff Photographer.

Fig. 19.4 Arthur Haas between the young physicist George Collins (right), who had constructed the Notre Dame accelerator system and who also hoped to appoint an experimental physicist to develop a research program with the new machine, and the young priest Father Murphy (left) who was interested in cosmological questions in Catholic scholasticism. They reflected the two poles of Haas' interests

Rochester by night train, where he made connections to Boston and then Brunswick so that he would be home in time for his lecture.

Arthur Haas heard from neither Notre Dame nor USC, which increasingly worried him—the semester would end in May and without a firm offer, he would have to return to Vienna and to the Academy. To be on the safe side, he wrote to Vienna and asked for an extension of his unpaid leave, until the end of 1936. Although the leave was granted, his financial situation was very tight and in the Haas' household every dollar was saved.

Haas sent further applications to the University of California Berkeley and the University of Cincinnati, as well as to two other Catholic institutions, Fordham University in New York and Villanova University in Pennsylvania, but USC and Notre Dame remained his preferred choices. Despite the agonizing wait, Haas felt safe when he learned from his friend Herzfeld at John Hopkins University that Notre Dame's president had asked for a letter of recommendation. Haas also heard from

Los Angeles that his name was included for the provisional lecture list. But then on May 13th came the letter that destroyed his hopes:

> It is with sincere regret that I must write you that, upon his return to our campus from an extended eastern trip, President KleinsSmid has reached the conclusion that it will not be possible for us to expand our budget for the next academic year to justify an invitation to you to come to work for a year at the University of Southern California. Should you go back to Vienna, it may be that we will be able at some future time to invite you to spend a year on our campus.[21]

This was a terrible setback for Haas, who immediately wrote to Professor Kolbe:

> An hour ago, I received the truly devastating news that Los Angeles will not appoint me. [...] I am trying to keep my calm but unfortunately the news of the tragedy had a terrible effect on my wife. I hope to receive good news from Notre Dame soon.[22]

Haas also contacted Robert Millikan at Caltech, who had welcomed him two months earlier, but Millikan's brief, matter of fact reply did not give any hope and pointed out that Haas' salary expectations were too high compared to the many highly qualified young faculty candidates on the job market. In the spring of 1936, it became clear that Millikan's gloomy predictions were proving to be correct. None of the hoped-for positions had been offered, the main reason being a lack of funding for new senior-faculty positions, which was explained by a decline in university revenues in the spring of 1936.[23]

Notre Dame was also hesitating despite an excellent letter of recommendation from Herzfeld. The reason was neither Haas nor his credentials but, rather, internal resistance that came from the dean, Dr. Francis Wenninger, C.S.C.[24] who felt coerced by the brash persistence of young Professor Collins. In addition, Wenninger saw little benefit in appointing a theorist to Notre Dame.

Haas had developed good internal contact with a German member of the C.S.C., Rev. Julian Sigmar[25] of Notre Dame, who gave him important feedback and advice:

---

[21] Vice President Touton to AEH (Los Angeles, May 11, 1936).

[22] AEH to Parke Kolbe (Brunswick, May 13, 1936).

[23] Private universities in the United States finance a large part of their running costs with the interest from their invested capital, which is gained from donations and other allocations (endowment); the second source of income is tuition fees. In the case of Notre Dame, the income from college football at the newly built stadium was an additional source of funding. The US capital market, after its low in 1932/33, rose until January 1935, after which there was a small slump from which the market recovered quickly. During this period, university administrations acted cautiously after the experience of the Great Depression and they tried to delay hiring Haas until the following year.

[24] Reverend Franz Wenninger (1888—1940) was Austrian by birth but came to America with his parents as a child. He attended seminary at Notre Dame and taught there as professor of biology from 1916—1926 and 1928—1940. He spent the years from 1926 to 1928 in Vienna to complete his doctorate. Notre Dame Archives, accessed February, 2021, archives.nd.edu/findaids/ead/html/WEN.htm.

[25] Reverend Julian Sigmar (von Fersen) (1886—1963) was professor of theology and philosophy at the University of Notre Dame. His civil name was Julian Schimikowski, under which he had received his education in Germany. He studied education in Berlin and psychology in Hamburg, then moving to Tübingen and Munich where he received his doctorate. He went to Notre Dame in 1932, where

> The invitation should have been discussed with the Dean himself in advance and should have been presented to the President through him. [...] F. W. believes that the "physicists" may have acted on their own. Rev. Wenninger's idea is that N.D.'s interests should be in the promotion of "Atomic Physics" with its relations to electrodynamic problems. "Is theoretical Physics of any importance to us?" he asked me. I replied to him: "Well, the experimenters get their ideas and directions, and often even the problems, from the theoretical physicist."— "Well, you may be right in that," he answered.[26]

Arthur Haas was not aware of any such internal intrigues and tensions during his visit, which he found very gratifying. However, in further letters Sigmar intensified his warnings about the Order's internal intrigues and advised Haas to turn to other Catholic schools such as Boston College and Fordham University.

> Given the academic evaluations of your colleagues, there should actually be no doubt about you deserving the position. However, this means little or nothing to Notre Dame, where science is seen as a parade for the ecclesiastical politics of the Congregation of the Holy Cross, where the influential offices are in the hands of members of the Congregation who have no idea about science and university. Do not be frightened, dear professor; it is not worth mentioning. Especially not in the case of a negative outcome. Would you feel happy to sit under cadavers, just to be able to sit?[27]

Undoubtedly not a glowing recommendation, which seems to soon turn into opposition:

> Personally, as I said before, I believe that Notre Dame, notorious for its sterile attitude in matters of science, would not satisfy you. [...] President O'Hara does not understand anything about scientific goals and methods. This is often the case in schools that are owned by a religious order. He has been made president by the superiors; he is not even designated by the Fathers of his community. (It is a perfect Hitler system here)[28] [...] Perhaps you will ignore my words and warnings, people who come from Europe often do so because they see everything with European eyes. Well, I leave it to you.[29]

Strong words from a catholic priest; Rev. Julian Sigmar was apparently not happy at Notre Dame—or vice versa—and left the school a year later in 1937.

---

he also taught sports education and physics. He seemed to hold opinions similar to Haas', as shown by a number of posthumous writings and comments in the archives of Notre Dame. He left Notre Dame in 1937 to teach at Duquesne University from 1938 to 1942, at the University of Dayton from 1942 to 1946, and at St. Louis University after 1946. In the years around 1940, he appeared for the first time under the name of a Baltic-Swedish noble family, Julian Sigmar von Fersen. At the beginning of the war, he distinguished himself through his writings of anti-British commentary for various magazines. Cf. Clement S. Mihanovich, ed., *The American Catholic Sociological Review* 7 /3 (1946): 174–199,

here, 196 and Sigmar-Fersen (Schimikowski) Julian (1886—1963), in Henryk Mross, "Słownik biograficzny kapłanów Diecezji Chełmińskiej wyświęconych w latach 1821—1920," *Pelplin* (1995): 308.

[26] Rev. Julian Sigmar to AEH (Notre Dame, April 14, 1936).

[27] Rev. Julian Sigmar to AEH (Notre Dame, April 17, 1936).

[28] Undoubtedly strong words even for 1936 when the development of the Nazi system could not yet be completely foreseen.

[29] Rev. Julian Sigmar to AEH (Notre Dame, May 21, 1936).

Despite warnings and information regarding the specific circumstances at Notre Dame, Haas sent another official application to the dean on April 16th, to follow the official route as recommended by Sigmar. He also wrote to Albert Einstein, asking him for help and a letter of recommendation:

> Now I face, in true desperation, an uncertain future with my wife and my sons of eleven and ten years. You will understand my desperation when you follow the developments in Austria. Quite apart from the fact that I have only a low-income position at the Academy of Sciences in Vienna and cannot hope to supplement 200% of this income with author's fees from Germany as I did in earlier years, I must also fear very much that I will lose my modest position in Vienna as a non-Aryan, as a result of the seizure of power by the Nazis or a Nazi-like party through general legislation. [...] My family is so happy in America [...] and now the return to barbarism with the prospect of starvation! One thinks only of Germany and forgets that the Nazi spirit was born in Austria. [...] I applied for a professorship at the University of Notre Dame [...]. I seem to have quite a chance. A short letter of recommendation from your side could trigger a favorable outcome.[30]

Einstein wrote in German to President O'Hara immediately after receiving the letter from Arthur Haas, on May 20, 1936 (Figs. 19.5 and 19.6) .

> I beg your pardon, if I may send you a few words in favor of the theoretical physicist, Prof. Dr. Arthur Haas, without having been asked. The immediate reason for this is that I have heard that there is a certain possibility that he might find employment at the University of Notre Dame.
>
> Arthur Haas is an original mind, but especially a teacher of rare talent. In today's confused state regarding the fundamentals of physics, when one is rarely able to distinguish with certainty the essential and lasting from the ephemeral, such a man is of particular importance for an academic teacher. I especially recommend his books to young people who strive to gain an overview of the essential methods and results of theoretical physics. I am convinced that Arthur Haas is able to do an excellent job as a teacher and stimulator, and that precisely this gift is rare among his usually extremely specialized peers. I would be delighted if he would be given the opportunity at your university to use his teaching talent for the benefit of the students.[31]

O'Hara responded immediately and promised that he would do his best:

> I am pleased to have your fine recommendation of Dr. Haas as a theoretical physicist and teacher. [...] the chief difficulty has been that our most urgent need is for a practical man in nuclear physics, and the eminence of Dr. Haas is rather in the field of theory.
>
> Within the next week we will probably be able to reach a decision.[32]

Einstein's letter of recommendation brought movement to the decision-making process. The internal resistance against the theorist Haas, in the interest of hiring a nuclear physicist for experiments at the new accelerator facility, was quickly and flexibly overcome. Notre Dame simply hired for both positions; however, in the interest of the university's tight financial situation, Haas had to accept a significantly reduced salary. Already on May 26th, the matter was arranged and O'Hara wrote the following to Brunswick:

---

[30] AEH to Albert Einstein (Brunswick, May 17, 1936).

[31] Albert Einstein to John Francis O'Hara C.S.C. (Princeton, May 20, 1936).

[32] John Francis O'Hara C.S.C. to Albert Einstein (Notre Dame, May 22, 1936).

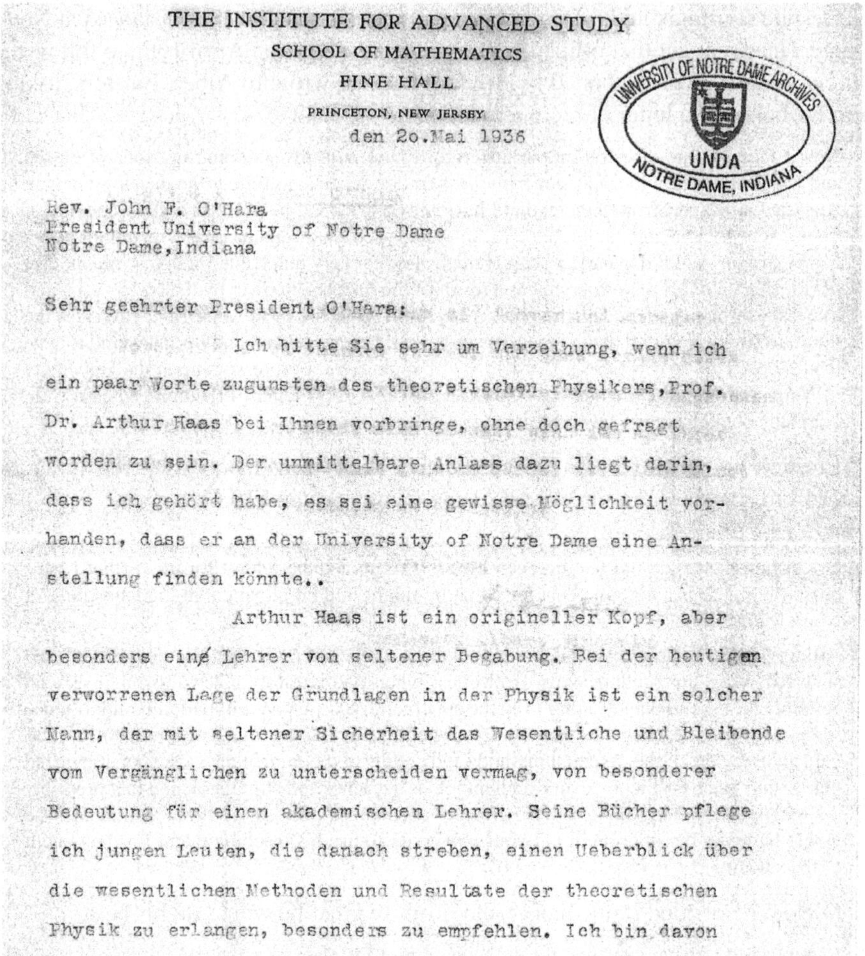

THE INSTITUTE FOR ADVANCED STUDY
SCHOOL OF MATHEMATICS
FINE HALL
PRINCETON, NEW JERSEY

UNIVERSITY OF NOTRE DAME ARCHIVES
UNDA
NOTRE DAME, INDIANA

den 2o.Mai 1936

Rev. John F. O'Hara
President University of Notre Dame
Notre Dame, Indiana

Sehr geehrter President O'Hara:

Ich bitte Sie sehr um Verzeihung, wenn ich ein paar Worte zugunsten des theoretischen Physikers, Prof. Dr. Arthur Haas bei Ihnen vorbringe, ohne doch gefragt worden zu sein. Der unmittelbare Anlass dazu liegt darin, dass ich gehört habe, es sei eine gewisse Möglichkeit vorhanden, dass er an der University of Notre Dame eine Anstellung finden könnte..

Arthur Haas ist ein origineller Kopf, aber besonders ein Lehrer von seltener Begabung. Bei der heutigen verworrenen Lage der Grundlagen in der Physik ist ein solcher Mann, der mit seltener Sicherheit das Wesentliche und Bleibende vom Vergänglichen zu unterscheiden vermag, von besonderer Bedeutung für einen akademischen Lehrer. Seine Bücher pflege ich jungen Leuten, die danach streben, einen Ueberblick über die wesentlichen Methoden und Resultate der theoretischen Physik zu erlangen, besonders zu empfehlen. Ich bin davon

Fig. 19.5 Letter from Albert Einstein to the President of the University of Notre Dame, Rev. John Francis O'Hara

My dear Dr. Haas:

I have been quite distressed at not being able to extend an invitation for you to join the faculty at Notre Dame next year. All of us who met you at Notre Dame have a warm regard for your personality, and we also hold in great esteem your contributions to the Science of Physics.

It appears now that it will be possible for us to meet our other needs and still find work that you can do. Our budget limitations are such, however, that we cannot offer you more than \$3,500 for the year. We realize that this is far from being a representative salary for you, but in view of the fact that we must engage five other professors for Mathematics and Engineering, it is the best offer that we can make at this time.

I am particularly pleased by the suggestion you make of uniting your work in Science with the work in the field of Apologetics, which we hope to inaugurate this fall. I believe that

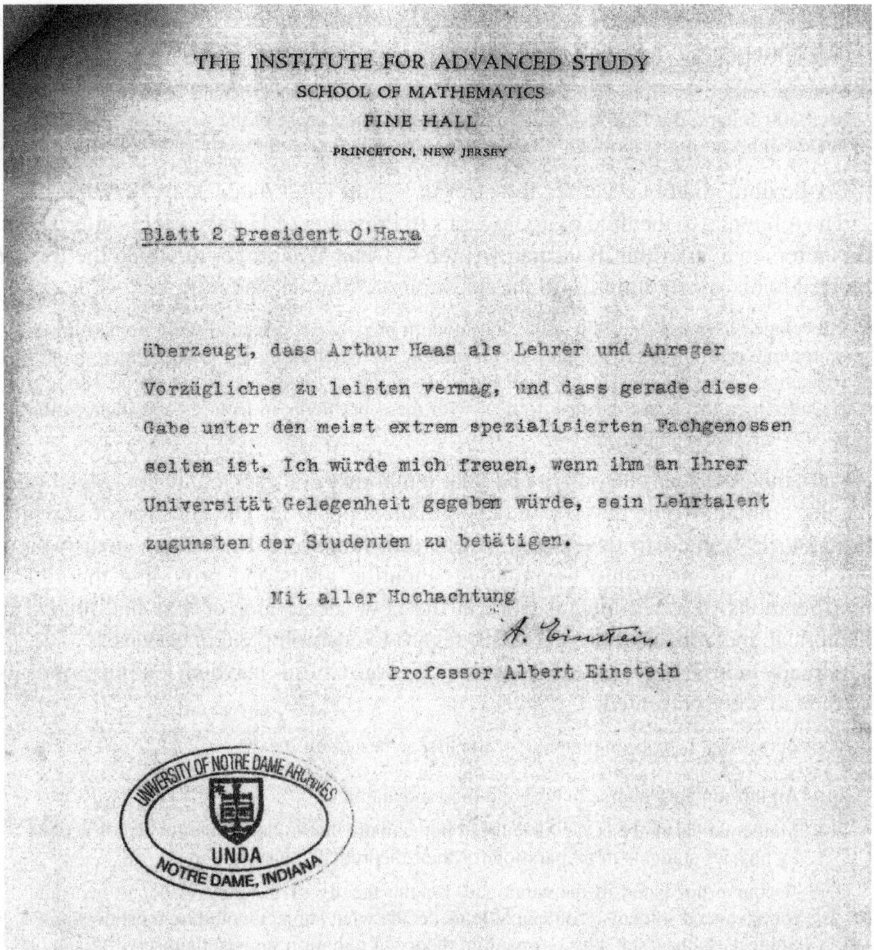

THE INSTITUTE FOR ADVANCED STUDY
SCHOOL OF MATHEMATICS
FINE HALL
PRINCETON, NEW JERSEY

Blatt 2 President O'Hara

überzeugt, dass Arthur Haas als Lehrer und Anreger Vorzügliches zu leisten vermag, und dass gerade diese Gabe unter den meist extrem spezialisierten Fachgenossen selten ist. Ich würde mich freuen, wenn ihm an Ihrer Universität Gelegenheit gegeben würde, sein Lehrtalent zugunsten der Studenten zu betätigen.

Mit aller Hochachtung

A. Einstein.

Professor Albert Einstein

Fig. 19.6 Letter from Albert Einstein to the President of the University of Notre Dame, Rev. John Francis O'Hara

> Father Murphy in particular will want to study intensively the reconciliation between the scholastic notions of Cosmology and the newer Physics theories.
>
> If you can accept a place at Notre Dame for one year on a stipend of $3,500, I will be delighted to send you a contract.[33]

This was an offer equivalent to 50% of the salary that Haas had anticipated and no doubt far less than he had hoped to receive. But he had no other choice if he did not want to return to Vienna.

[33] John Francis O'Hara C.S.C. to AEH (Notre Dame, May 26, 1936).

In a letter to his brother Otto in Vienna, Haas had described his situation as an act of desperation:

> I have no idea what Notre Dame will offer me in salary. A professor said I could easily ask for 7,500 dollars. But I am so scared to death in the truest sense of the word that everything could fail because of a too-high salary demand that I will probably only get a bad salary.[34]

So, he immediately accepted the offer in a short letter dated May 28, 1936.

The reference to the field of apologetics in President O'Hara's offer is interesting. This refers to a letter that Haas had written to Dean Wenninger in which the former referred to his special interest in the question of "physics and religion":

> I developed in some detail the view that modern physics and especially the modern ideas on the universe, on relativity and indeterminacy, are much more apt to strengthen than to weaken religious feelings. [...] Should I be so fortunate as to come to Notre Dame, I should consider it as my particular task to deal with these problems in lectures and in scientific papers.[35]

Nonetheless, Haas had only a one-year temporary contract, but he had high hopes that this contract would develop into a permanent position. The question of in which direction he wanted to develop his work did, however, still remain unanswered. Did he want to move into formulating scientific goals and providing theoretical interpretations of experimental work at the new accelerator or was his interest in theoretical, metaphysical projects with regard to cosmological questions?

Already in June he received a letter from George Collins in which lecture proposals for the fall were presented:

> The courses that have been planned for the first semester are:
>
> 1. An introductory course in heat and thermodynamics;
> 2. Mathematical analysis, a course listed in the mathematics department which is needed by physics students in preparation for their theoretical courses;
> 3. A course not listed in the catalogue, for the faculty of the physics department and for advanced students. You can help us decide what subject would be most desirable. Subjects such as statistics or quantum theory of radiation suggest themselves. Each of these would be three hours a week.[36]

Haas was delighted; he had expected general physics courses for first-year students, which were equivalent in level to the physics lessons of Viennese high schools. For the first course he chose his own book on theoretical physics as a textbook, and as a third course he suggested the Physics of Atomic Nuclei, which would again allow him to use his book on atomic physics as a textbook. With proceeds from the book sales, he hoped to improve his narrow income at Notre Dame.

He wrote to the Academy, to Secretary General Egon von Schweidler whom he esteemed, to explain the situation but also to ask for an extension of the unpaid leave from the Academy:

---

[34] AEH to Otto Haas (Brunswick, May 17, 1936).

[35] AEH to Rev. F. Wenninger (Brunswick, May 23, 1936).

[36] George Collins to AEH (Notre Dame, June 8, 1936).

Today I formally signed the contract. The University of Notre Dame has 3,500 students and was founded in 1842; it is maintained by the Catholic Congregation of the Holy Cross, but has a large number of non-Catholics both in the teaching staff and among the students, and knows no restriction whatsoever on the freedom of teaching. Some departments are very good. For example, a few months ago the head of the chemical department (Father Nieuwland) received the highest award of the American Chemical Society for his research in the field of synthetic rubber. The zoologist and the botanist are Austrians (Wenninger and Just), both students of Hofrat Molisch. In the Physics Department a large atomic disintegration facility was created a few months ago. [...] You, esteemed Mr. Secretary General, understand why I had to accept. The reasons multiply from which I must conclude that, despite all the good will of the Academy, I would no longer have a secure existence in Vienna; I and my family would always have to expect that we would be left to starve to death as a consequence of legislation similar to that of the neighboring country. If this is my fear, then - hopefully it is not justified - it is surely much better to accept the position right now than to be forced to emigrate at a later date. [...]

I ask the High Academy to forgive me my step with understanding and, if possible, to extend my leave without fees for another year, at least until the end of the calendar year 1936.[37]

Haas wrote to many friends and acquaintances to share the joy of his success in obtaining the position. He included a letter to his friend Professor Donnan who immediately responded with warmest congratulations. Donnan further announced that *Gibbs Comments*, on which they had both worked for so long within the Gibbs Committee at Yale, would finally be published as a book in September 1936. Donnan also explained that he was again active in his efforts to enable Jewish scholars to emigrate to the United Kingdom or even the United States, reporting:

I hear that Schrödinger, now in Oxford since 1933, is very likely to return to Austria, with some post in Graz or in Graz + Vienna. I am trying to do something for F. London (at present in Oxford).[38] His position is very uncertain and insecure. Born is still in Cambridge and Heitler at Bristol. I presume you know that James Franck is at John Hopkins University, and E. Teller at the George Washington University (since September 1935). Gamow is here as well.[39]

These are all names that belong to the founders of quantum physics in Germany. The "Quantum Exodus" with all of its consequences had begun.[v]

Haas had also reached his life's goal. He had left Austria, found a position in America and now felt safe from persecution. A quiet summer was ahead of him; he did not have to return to Austria. So Arthur and Emma came to a decision; she would travel back to Europe for the summer in order to liquidate the house in Vienna and arrange for all of the furniture to be shipped to South Bend by the cheapest possible means.

Arthur Haas himself remained in Maine and rented a small house for the summer, the Crystal Lake Villa in Dry Mills (Fig. 19.7); as the name promised, he spent the month directly on the shore of a small lake in the forests of Maine. The two boys

[37] AEH to Egon v. Schweidler (Brunswick, June 3, 1936).

[38] Fritz London (1900—1954) was a quantum chemist and, as a Jew, was dismissed from his position at the University of Berlin in 1933. He held temporary positions in England and emigrated to the US in 1939, where he was appointed professor at Duke University.

[39] F. E. Donnan to AEH (London, June, 19, 1936).

Fig. 19.7 Boat and bathing scenes at Crystal Lake in Dry Mills, Maine

stayed at nearby summer camp Gregory on the same lake. It was here that Haas took advantage of the opportunity and the quiet peace of the summer to write his memories—apart from his daily bathing trips and letters to Emma, he penned the notes that have been presented in the first chapter.

In his letters, filled with inner peace, he recalled the past years of their "Exodus":

> We were quite satisfied. I can say we were until the summer of 1931 when the bad years began, and they were actually shorter than the good ones. For already in 1935 in January, after the Flood, the lucky letter from Philadelphia[40] came as a dove, showing us the new way into the new world and soon after this, the redemption. [...] We are in the sign of good ascent. May it last forever now.[41]

But already terrible news from Europe was arriving. Reports of the murder of his former friend Moritz Schlick reached the Crystal Lake vacation resort:

> Today I have to add to the letter the news of a terrible tragedy that took place three weeks ago in Vienna, which I just heard about on the detour via Berkeley.
>
> Our friend Moritz Schlick has been murdered. He was the Viennese scholar that I happened to hear about on the radio, who was shot by a mad student.
>
> Helen Mackay writes us, "Wasn't it the most shocking thing you ever heard, Moritz' murder? Poor Blanche and poor children — I feel so terribly sad for them all."
>
> You'll probably hear more soon. Let me know then.[42]

Moritz Schlick was murdered in Vienna on June 22, 1936. He was killed by his former student Hans Nelböck (1903—1954) on the way to the university. Klaus Taschwer reconstructed the event in an article for the Viennese journal *Der Standard*:

> Vienna — On the morning of June 22, 1936, there is a hustle and bustle in the main building of the University of Vienna. Philosophy professor Moritz Schlick is on his way to lecture hall 41, where he plans to give his lecture. On the stairs a man he fears confronts him: Hans Nelböck had written his dissertation with him five years earlier — and had already threatened him with murder twice. "So dog, you cursed one, now you have it," Nelböck cries and fires four times at his thesis supervisor. One of the bullets from the pistol pierces the victim's heart, and Schlick dies before medical help arrives.

---

[40] The invitation from President Kolbe for a sabbatical at Drexel University.

[41] AEH to EBH (Crystal Lake, July 9, 1936).

[42] AEH to EBH (Crystal Lake, July 13, 1936).

Fig. 19.8 Reconstruction of the assassination of Moritz Schlick on June 22, 1936, on the stairs to the university

> At the scene of the crime today, a plaque is embedded in the ground to commemorate the assassination: "Moritz Schlick, protagonist of the Vienna Circle, was murdered on this spot on June 22, 1936. An intellectual climate poisoned by racism and intolerance contributed to this act." [Fig. 19.8] When the facts are examined, however, it remains controversial how influential this aspect actually was: the perpetrator had already been diagnosed with a schizoid personality disorder.[vi]

There has been much speculation about what triggered the murder. Nelböck himself put forward ideological differences as an explanation. He also brought jealousy into play as a possibility, since he and Schlick had both been interested in the same student. Nelböck himself argued that Schlick had denied him a position at the community college and that it was therefore a matter of revenge.[vii] (Fig. 19.8).

Hans Nelböck was sentenced to ten years in prison. However, after the Anschluss he submitted a plea for clemency in which he presented himself to the new regime as its champion. He argued,

> that through his deed and the resulting elimination of a Jewish teacher who disseminated doctrines that were alien to the people and harmful to them, he had rendered a service to National Socialism and had also suffered for National Socialism. Now that the ideology from which he, recognizing its correctness, committed the deed is the dominant concept of the State, he feels it would be harsh if he had to take a back seat to a great extent because of the deed born of this view.[viii]

Nelböck was released on probation and worked from then on as an employee in the state administration.

We do not have a record of Emma's thoughts and feelings, although she had been a travel companion of Moritz' and had always talked very warmly about him. For Arthur Haas, the murder was a confirmation that he had taken the right step in emigrating to escape from the poisoned atmosphere of Austria.

He was optimistic about his new future and was especially pleased about the upcoming Tercentenary Conference at Harvard University to which he had been invited by Shapley. The conference was planned as a way to bring together leading scientists from around the world, to speak in honor of Harvard and its three hundred years of existence. Arthur Haas was enormously proud that he had been invited to attend.

But he was still at the lake in Maine and in his letters to Emma, he complained that the doctors had put him on a strict diet. This was apparently with success because later, on August 3rd, he recorded that his blood pressure had dropped from 200 to 184 and his weight had dropped by six pounds. These are undoubtedly not harmless numbers and the strictness of the diet was clearly warranted.

Through his many letters, Haas was also trying to work on the dissolution of the Viennese family apartment. In his opinion, the process was a bit tedious due to the miserable postal conditions—in Maine the mail was only picked up and delivered once a day! At the very least, Haas put together long lists and plans regarding how best to arrange the packing and transport of the furniture. He conducted these negotiations primarily with and through his brother Otto in Vienna, who also acted as legal counsel in matters of financing the move.

The shipping costs were estimated at about $1,000 and Haas wanted to sell his Polish share packages with proceeds in złoty,[43] in order to finance the transport. Poland had, however, introduced a currency lock so Haas intended to circumvent this by hiring a Polish freight company, Zdenko Dworzak. He also suggested selling one of his remaining houses in Brno, in order to finance a house in South Bend. Some of the furniture was to be left behind and for these items he wanted to develop a sales strategy, which he tried to explain in his letters to Otto. All of these plans were discussed through extensive correspondence between the two brothers. For faster delivery, the new zeppelin service between the US and Germany was used, which had been made possible with the airship *Hindenburg*.[44]

But Arthur also included warnings in his letters to Otto:

> I do not believe in any immediate danger, but I do believe you would do well and wisely to get your children to speak fluent English after a year and to consider the possibility of sending Lene to a first-class American college for a year in the fall of 1937 at a cost of

[43] The Polish currency złoty had remained unstable during that period but officially, 10 złoty corresponded to approximately 1 US dollar.

[44] The Hindenburg Post, as it is referred to in the letters, transported mail by the German airship LZ 129 *Hindenburg*, beginning in March 1936. The airship guaranteed the fastest mail traffic between the American and European continents but required additional special stamps. The traffic came to a hasty end with the explosion of the *Hindenburg* in May 1937.

about $1,300 if you can afford it, if it seems necessary at that time. [...] Principiis obsta, sero medicina paratur.[45]

In looking to the future, Haas started correspondence with Professor Just,[46] an Austrian at Notre Dame, to find an apartment in South Bend. More immediately, Haas planned to spend the summer in Maine and a few days with his sons on the coast of Cape Cod. They would then travel to Boston for the Tercentenary Conference at Harvard. From there he planned to arrive in South Bend with his sons on September 10th, at the beginning of the fall semester.

Emma would arrive from Europe a few weeks later. She was to organize the final shipment of furniture from Vienna to South Bend. Another reason for Emma's extended stay in Vienna was that she wanted to finish her college diploma, which she had started with so much hope. Arthur thought that this was no longer necessary, but she insisted because she wanted to be able to continue her studies.

Emma was, however, not yet in Vienna at the time but had first traveled to London where she visited Professor Donnan. From there, she traveled via Hamburg to Berlin and Leipzig to see the publishers; as usual, they were either delayed in paying or had withheld Arthur's royalties because no export license could be obtained. Next, she wanted to go to Munich to visit her brother and also to relax at the Villa Grau on Lake Ammersee before traveling to Vienna. There she would take her exams and finally dedicate herself to packing and shipping the Viennese furniture.

These travel plans led, however, to some disagreements with Otto and his wife Greti in Vienna. The owner of the house pressed for a faster vacating of the apartment and turned to Otto as his brother's legal advisor. He and his wife complained that Emma was traveling in Germany while they had to make all of the preparations and arrangements for her. As a result, no one spared accusations when Emma finally appeared in Vienna and this led to some tension between the sisters-in-law.

In addition, communication between Emma and Arthur did not go well. Eagerly awaited letters from Europe did not make it into his forest seclusion and Arthur became increasingly restless. He was also not well; high blood pressure was bothering him, although he gradually managed to lose weight.

In the future I will only be allowed to eat fish and vegetables and fruits, at least until everything has become normal again. I must also avoid too many eggs and especially butter.

And in Munich, veal shanks and dark beer fill the tables. What remains for me of the joys of life? When I listen to American music, I get stomach cramps and when I watch American movies, I get stomach cramps. And there is the desolation of American summer resorts. Work

45 Arthur quotes the Roman poet Ovidius in his "Remedia Amoris": "principiis obsta; sero medicina paratur, (cum mala per longas convaluere moras)," which translates as, "Be prepared at the beginning, the cure comes too late (when the illness has advanced too far)." AEH to Otto Haas (Crystal Lake, July 18, 1936).

46 Theodor Just (1904—1960) had studied in Vienna and was recruited to Notre Dame by Rev. Wenninger in 1930. T. Just taught there for seventeen years before he became curator for the botanical department at the Chicago Natural History Museum in 1947. Obituary, "Theodor Karl Just, 1904—1960," *The American Midland Naturalist* 65/2 (1961): 1–15.

is the only thing that can help you overcome the wasteland. Hopefully I'll be successful at Harvard right away.[47]

The summer seemed quite long to him. Haas went to see his doctor in Brunswick and tried to calm him down, especially since Arthur had lost more weight. He explained the high blood pressure as being the stress of the current life situation and the nervousness caused by uncertainty.

While at the beginning of the retreat to Crystal Lake he had been able to concentrate on writing down his memories, Haas was now weighed down by other more practical thoughts of moving, furniture transport, money transfer, and the need for accommodation. In addition, his first worries about the longer-term future were already apparent in his letters. Would Notre Dame keep him? The position was only for a year and the next step in his homeless wanderings threatened. He realized that he was already fifty-two years old, as his health problems were making clear to him, and he hoped that he would find his inner peace at Notre Dame.

The summer came to an end at last and he spent the last week of August as planned, together with his sons on Cape Cod. As he wrote, the stay was delightful for all. Then together with the boys, he traveled with anticipation to Cambridge, MA for the Tercentenary Conference, which was held on the Harvard University campus from August 31 to September 12, 1936.[ix]

With this event Harvard commemorated the 300th anniversary of its founding. The university held elaborate celebrations to which tens of thousands of alumni, dignitaries (including the President of the United States, Franklin Delano Roosevelt), and representatives of educational institutions and scholarships from around the world were invited. This was also an important event in the world of science, as the event addressed the increasingly important role of the United States and its academic institutions.

Eminent scientists from across the globe were invited to speak but, as with any conference, some of the invited guests cancelled for personal reasons, such as Niels Bohr from Copenhagen, or were prevented by other, more political restrictions, such as Werner Heisenberg from Leipzig. Arthur Haas' main goal was to meet colleagues and friends, to refresh and improve these old acquaintances while making new connections, all for the support of his future in America. Meetings with Einstein and Eddington were at the top of his list; Einstein himself had assured Haas in his letter of May 15, 1936, that he would attend the conference.

But circumstances changed. Einstein had just bought and furnished a house at 112 Mercer Street in Princeton when his wife Elsa was diagnosed with circulation and kidney problems. The couple had spent the summer of 1936 in the Adirondacks to escape the sultry summer climate of Princeton but the symptoms had worsened there. He therefore had to cancel his participation out of concern for his wife's health. His wife later reported from those days, "… he walked around miserably and depressed."[x] Elsa Einstein died on December 20, 1936. Einstein's absence at the conference was a great disappointment to Haas, and his son reported in his memoirs that his father seemed very grumpy that day.[xi]

---

47 AEH to EBH (Crystal Lake, August 22, 1936).

Fig. 19.9 Arthur Haas in academic attire at the Harvard Tercentenary Conference. The right photo shows him posing with his two sons, Arthur and George

Despite everything, Haas enjoyed the conference. He proudly presented himself at the festive opening procession in American tradition, with gown and in full academic dignity. The occasion was commemorated with photos, as seen in Fig. 19.9 .

Attending this conference gave Haas the opportunity to become a full member of the American academic community. While he was previously known only through his lecture tours, he now planned to establish a scientific presence in the United States. His attempt to make new connections was successful, especially with the Harvard astronomer Harlow Shapley who supported Haas' election as Fellow of the American Association for the Advancement of Science (AAAS) in 1937; this is a very rare honor bestowed on few physicists. Perhaps more importantly, Haas renewed his relationship with Arthur Compton (1892—1962) from the University of Chicago, with whom Haas had already met during his visit in preparation of the Chicago World's Fair. Haas continued to cultivate a close intellectual collaboration with Compton in the years that followed, which often led him to Chicago.

At the conference itself on the Friday afternoon of September 4th, Haas gave a lecture in the area of cosmogony; his lecture had been organized jointly by the American Mathematical Society and the American Astronomical Society. He spoke immediately after Eddington as the conference program shows in Fig. 19.10 . It was a proud moment for Haas, and he felt at the height of his scientific career when he saw his name published along with his two scientific role models, Albert Einstein and Arthur S. Eddington.

Haas' lecture was a success, as he proudly wrote to Bice who had since arrived in Vienna:

> The lecture attracted the greatest attention and was a great success in every respect, as is evident from the fact that Shapley submitted it for publication by the National Academy of

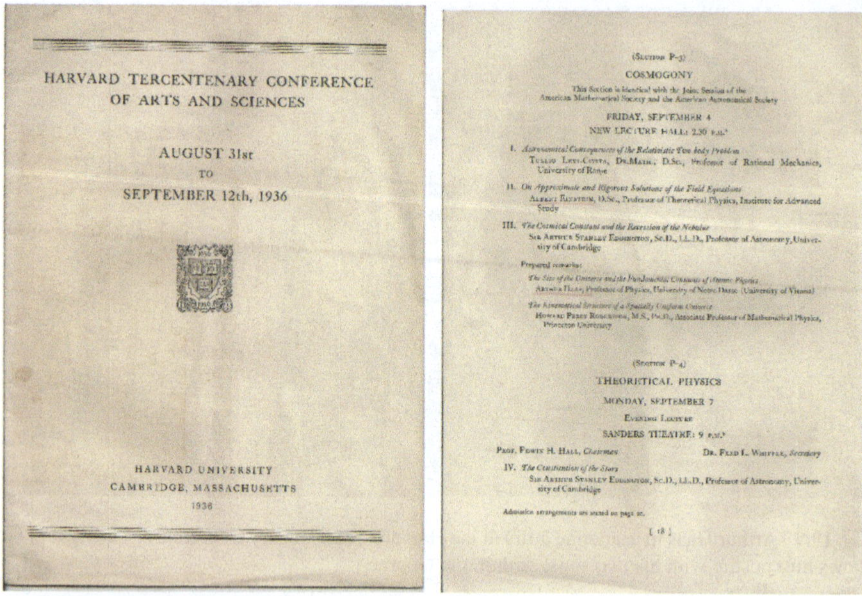

HARVARD TERCENTENARY CONFERENCE
OF ARTS AND SCIENCES

AUGUST 31st
TO
SEPTEMBER 12th, 1936

HARVARD UNIVERSITY
CAMBRIDGE, MASSACHUSETTS
1936

(SECTION P-3)

COSMOGONY

This Section is identical with the Joint Session of the
American Mathematical Society and the American Astronomical Society

FRIDAY, SEPTEMBER 4

NEW LECTURE HALL: 2.30 P.M.

I. *Astronomical Consequences of the Relativistic Two body Problem*
TULLIO LEVI-CIVITA, DR.MATH., D.SC., Professor of Rational Mechanics, University of Rome

II. *On Approximate and Rigorous Solutions of the Field Equations*
ALBERT EINSTEIN, D.SC., Professor of Theoretical Physics, Institute for Advanced Study

III. *The Cosmical Constant and the Recession of the Nebulae*
SIR ARTHUR STANLEY EDDINGTON, SC.D., LL.D., Professor of Astronomy, University of Cambridge

Prepared remarks:

The Size of the Universe and the Fundamental Constants of Atomic Physics
ARTHUR HAAS, Professor of Physics, University of Notre Dame (University of Vienna)

The Kinematical Structure of a Spatially Uniform Universe
HOWARD PERCY ROBERTSON, M.S., PH.D., Associate Professor of Mathematical Physics, Princeton University

(SECTION P-4)

THEORETICAL PHYSICS

MONDAY, SEPTEMBER 7

EVENING LECTURE

SANDERS THEATRE: 9 P.M.

PROF. EDWIN H. HALL, *Chairman* DR. FRED L. WHIPPLE, *Secretary*

IV. *The Constitution of the Stars*
SIR ARTHUR STANLEY EDDINGTON, SC.D., LL.D., Professor of Astronomy, University of Cambridge

Admission arrangements are stated on page 2c.

[18]

Fig. 19.10 Program of the Harvard Tercentenary Conference of Arts and Sciences with the announcement of lectures by Albert Einstein, Sir Arthur S. Eddington and Arthur Haas in the section on cosmogony, with Einstein's contribution cancelled at the last moment

> Sciences in Washington, the highest learned association of the United States. And as you can see from the newspaper clippings sent to you, the first newspaper in the world, the *New York Times* reported on my lecture on its first page (on September 5th).[48]
>
> My reputation is on the rise [...] In many cases, I have strengthened and deepened old relationships and never miss an opportunity to anchor my existence as much as possible.[49]

After the conference ended, Arthur Haas and his two sons traveled by train from Boston via Albany to Buffalo where they made a detour to Niagara Falls for the boys. They were impressed but not as excited as when they went shopping in Cleveland, Ohio, in a new large department store where they were confronted with escalators for the first time. The Haas men then traveled the last leg of their adventure, riding the train to Chicago (Fig. 19.11).

Notes

i. *Annual Catalogue of Bowdoin College 1934—1935* (Brunswick, 1933).
ii. George and Mary Jane Haas, "Letter to our Grandchildren," 66.
iii. Bart J. Bok, "Harlow Shapley 1885–1972 A Biographical Memoir," National Academy of Sciences, last accessed March 9, 2021, https://www.nasonline.org/publications/biographical-memoirs/memoir-pdfs/shapley-harlow.pdf.

---

[48] "UNIVERSE 'HALTED' IN PARTICLE COUNT," in special edition: *New York Times*, September 5, 1936.

[49] AEH to EBH (Cambridge, September 7, 1936).

Fig. 19.11 Trains of the New York Central Lines with which the Haas family continued on the last leg of their journey from Cleveland to Chicago

iv. Paul Chagnon, *The First Accelerator at Notre Dame*, accessed January 31, 2021, https://isnap.nd.edu/assets/186666/firstaccelerator.pdf.

v. Gordon Fraser, *The Quantum Exodus – Jewish Fugitives, the Atomic Bomb, and the Holocaust* (Oxford and New York: Oxford University Press, 2012).

vi. Klaus Taschwer, "Die Universität unter dem Kruckenkreuz," *Der Standard*, October 23, 2014.

vii. Renate Lotz-Rimbach, "Mord verjährt nicht: Psychogram eines politischen Mordes," in *Moritz Schlick – Stationen,* eds. Friedrich Stadler and Hans Jürgen Wendel (Vienna: Springer Verlag, 2009), 81–106.

viii. In Friedrich Stadler, *Studien zum Wiener Kreis. Ursprung, Entwicklung und Wirkung des Logischen Empirismus im Kontext* (Frankfurt am Main: Suhrkamp, 1997), 958.

ix. Clark A. Elliott, "The Tercentenary of Harvard University in 1936: The Scientific Dimension," *Osiris* 14 (1999): 153–175 and Wesley Fuller, "Harvard Tercentenary Conference of Arts and Sciences," *The Scientific Monthly* 43/ 5 (1936): 480–490.

x. Albrecht Fölsing, *Albert Einstein* (Frankfurt: Suhrkamp, 2020), 273.

xi. George and Mary Jane Haas, "Letter to our Grandchildren," 74.

Chapter 20
University of Notre Dame

The three Haas men, as Arthur described them in his letters, finally reached South Bend on the evening of September 11th. They stayed at The Morningside Hotel, "a good but too shabby hotel," as Arthur described it. Nor was he particularly impressed by the restaurant and food offerings of downtown; "too much pepper and paprika," he reported in his letters.

Professor Collins picked him up the next morning and took Haas to the campus where he was warmly welcomed by his new colleagues, making him feel right at home. Haas then enrolled his two sons at Madison School, which is located close to the banks of the St. Joseph River between South Bend and the independent northern suburb of Notre Dame. The school year began on September 14th, at the boy's school as well as the university.

Arthur Haas seems to have been prepared but his lecture assignments had changed somewhat:

> I have three courses of three hours each: "General Physics" for future medical students (eighty students); only for graduate students: "Mathematical Analysis" and "Thermodynamics." The first course starts on Thursday, September 17th, the other two on September 18th.[1]

In the end, he had been assigned a major lecture for undergraduate students—a not very popular duty, the training of future doctors.

The University of Notre Dame is one of the many private universities that form the backbone of the American educational system. The university was founded in November 1842 by the French religious Congregation of the Holy Cross[2] at the invitation of the Bishop of Vincennes. A whole series of Catholic universities were being founded in the United States at the time with the aim of creating higher education opportunities particularly for Catholics. This was especially important to the increasing number of Catholic immigrants from Ireland, Germany, Poland and Italy

[1] AEH to EBH (South Bend, September 13, 1936).

[2] The Congregation of the Holy Cross or Congregatio a Sancta Cruce (C.S.C.) is a Catholic congregation of missionary priests and brothers founded in 1837 by Blessed Father Basil Moreau in Le Mans, France.

M. Wiescher, *Arthur E. Haas—The Hidden Pioneer of Quantum Mechanics*, Springer Biographies, https://doi.org/10.1007/978-3-030-80606-4_20

Fig. 20.1 Left: The church and main administration building (The Golden Dome) of Notre Dame. Right: An aerial photograph of the 1934 campus picturesquely set around St. Mary's Lake (in the foreground) and St. Joseph's Lake

for whom it was just as unlikely to be accepted at the very Protestant-oriented older private universities such as Harvard, Yale and Princeton as it was for a Jew at that time.

The founding of Notre Dame coincided with the rapid rise of Irish immigrants—as a result of the great famine in Ireland in the 1840s—from whose ranks the first generation of students emerged, which explains the university's strong identification with Irish rather than French traditions. In the following decades the university grew rapidly and the few buildings of the original campus between the two lakes developed into a large campus complex.[i] Images of the lakes and administration building can be seen in Fig. 20.1.

In the 1930s, the administration developed greater ambitions than solely educating Catholic youth. The university started to evolve, developing its academic as well as its sports programs and thereby encouraging many visitors to the campus (Fig. 20.2). The university emerged from its purely educational program to become a research institution in the ranks of other private universities in the East and also in the West. Necessary funds were collected through large-scale fundraising campaigns, ensuring that growth would be secured by sufficient capital investment. In the 1920s and 1930s, much new construction took place on campus, including an impressive new Engineering Building. This development made the administration aware of the need to expand the fundamental natural sciences.[ii]

The chemistry program had already made a name for itself, particularly through Rev. Julius Nieuwland's work on the development of synthetic rubber; the patent was sold to the Dupont Company who made tremendous profits producing tires and similar products.[3] Near the end of the 1920s, physics did not yet have a comparable profile, a circumstance that was to be changed at the urging of the engineers.

---

[3] In total, Notre Dame is said to have received about two million dollars from Dupont. Cf. Christopher S. Temple, "Fostering Elite Science at an American Catholic University: The Rise of a

Fig. 20.2 The two sanctuaries of the University of Notre Dame; on the left the Football Stadium in the southeast of the campus, which has brought thousands of football fans as paying guests from all over the country to Notre Dame; on the right is the Grotto of Notre Dame, a replica of the Grotto of Lourdes. Notre Dame's grotto is located behind the church, on the hillside facing the lakes, and has attracted countless believers as visitors for silence and meditation

In the 1930s, many of the professors were still members of the Holy Cross Congregation, some with advanced degrees from prestigious American and European universities. But the number of professors who were not ordained was growing. Through an influx of European faculty members, the European conservative flair of the university developed, which particularly attracted Arthur Haas with his basic Christian and conservative mindset of the Austrian bourgeoisie.

Like other universities, Notre Dame's administration recognized opportunities for development and expansion with the exodus of European academics after the devastation of the First World War and the rise of Adolph Hitler in Germany. The decisive force in this development was Notre Dame's president, a member of the founding Congregation of Holy Cross. In the early 1920s, preparations had already begun by establishing capital investments—an endowment that was to allow for, as with large private research universities, the expansion of programs and the appointment of top scientific personnel.

President Father John O'Hara took the initiative for the second phase of the university's expansion. A graduate school was created in 1931, to allow for advanced graduate degrees, and young dynamic professors were recruited from famous schools in the East and West, and from abroad.[iii]

Prior to 1934, the Physics Department existed exclusively as a service department with professors giving only basic physics lectures to students of engineering, chemistry, and medicine. Walter Shilts taught during this period from 1924 to 1936, and he notes:

Research Culture at the University of Notre Dame, 1842–1967," (doctoral thesis, Department of History, University of Notre Dame 2020), 238–249.

> Almost all the physics students were engineers and the physics courses taught were service courses. The members of the "department" taught engineering courses, mathematics courses and astronomy courses as well. No one was interested in research. The department had grown up in this way to take care of the needs of those students who required physics as part of their undergraduate curriculum.[iv]

O'Hara, on the other hand, followed the premise that a university can only gain reputation and grow through an active research program. Although his personal interests lay in medieval philosophy and apologetics[v]—as was also expressed in the offer to Arthur Haas—he invested particularly in the development of engineering and physics. Not only were new research and doctoral programs created but also new buildings that had to be filled with new instruments.

Among the new young professors was George Briggs Collins who came to Notre Dame in 1933 from Johns Hopkins University. George Collins from the physics department, together with Joseph Caparo from Engineering and their graduate student Ed Coomes used space in the newly constructed engineering building to design and build, between 1935 and 1936, a 1 million electronvolt accelerator.[vi] Using this machine, they developed a first research program in Nuclear Physics.[vii] The accelerator itself was a copy of a machine built for the Department of Terrestrial Magnetism at the Carnegie Institute in Washington by Melvin A. Tuve.[viii]

In order to support this program and also to reduce the burden of teaching numerous courses, O'Hara appointed next to Haas another physics professor to Notre Dame, the young priest and experimenter Henry Bolger who had worked at Caltech in the Millikan group. Bolger took over as head of the department and during the next decade he developed it into an effective and widely respected research institution.

This young dynamic department, driven forward by young people, comprised the atmosphere that Arthur Haas now joined. The expectations on him were high and, with the hope of getting a permanent position, he promised to provide theoretical tools, motivation and interpretation for the young experimenters' new accelerator program. He also promised to satisfy the president's interests in apologetics through further studies in cosmology as well as to benefit from the research program of his new friend Arthur Compton in Chicago.

But first, in the fall of 1936, Haas was essentially busy settling in, accommodating the children, and making local preparations for Bice's return from Europe as well as the arrival of the furniture. Initially he took a furnished apartment, in order to get out of the uncomfortable hotel on Colfax Avenue, while he spent more time looking for an unfurnished apartment or house. At the time South Bend was a highly industrial city with a flourishing automobile industry—the Studebaker factories extended along the southern rim of the city center. In addition, there were several machine factories and other industrial enterprises that were just beginning to recover from the Great Depression and so were recruiting and hiring new workers. The housing market was tight and life was far more expensive than Haas had imagined.

> The housing problem was very difficult. Rental prices are skyrocketing. I am offered very nice furnished apartments at $100, but I can't possibly afford that. (A few months ago the price was only $50 per month!) The music professor of whom I wrote to you, who wanted

Fig. 20.3 The new Haas residence at 847 Forest Avenue: left in the winter of 1937 with George and Arthur Haas and right, the house as it looks today

to leave the apartment to me, is again vacillating about whether he should travel to Europe at all, because after the Nuremberg speeches he believes that war will break out soon.[4]

But the music professor decided to travel after all and Arthur Haas took over the house for $70 a month. The house is located at 847 Forest Avenue, in a quiet residential area near the boys' school and not too far from the university. The boys and their new home appear in Fig. 20.3. It is a typical American house on a tree-shaded street near the St. Joseph River between the university and downtown.

Taking over the house was not without problems. Several pieces of furniture that were left behind had to be moved to make room for the Viennese furniture and, in addition, it was very hot and humid during these early fall months. Haas was sweating and felt very uncomfortable but the boys didn't seem to mind:

Of course, furnishing the apartment and the many errands were a lot of effort on my part, since I have a lot of work to do at the university and the temperature is over 90 F during the day. I swim in the university lake almost every day, sometimes the boys do it too.[5]

He was settling in and the letters to Bice were becoming more positive :

You will like it very much. We have here, staying in a cottage, the amenities of a larger city — and you can get to Chicago in two hours on an electric train [Fig. 20.4]. The round-trip fare is just over $3, and a little over $2 on Sunday.[6]

---

[4] AEH to EBH (South Bend, September 16, 1936).

[5] AEH to EBH (South Bend, September 23, 1936).

[6] Ibid.

Fig. 20.4 A contemporary car of the South Shore Railway. The much-expanded South Shore still provides inexpensive transportation from South Bend to downtown Chicago

Or:

> Our lives flow quietly, peacefully, contentedly and without excitement. I'm sure I'll have a lot of success here and have the feeling of complete security as far as the future is concerned. My work takes place almost exclusively in the office. And I do not work in the morning before I go to the office or in the evening when I return for dinner. I usually leave the house at 7:30 and return home at 6 in the evening. By streetcar it takes me –thirty to forty minutes from house to office [Fig. 20.5].[7] From the house to the city center it takes about fifteen minutes by streetcar. The children have the school right next to the house.[8]

Haas also fit in well at the university, as he reports in the same letter:

> I am in the center of interest and every word I say is circulating. [...] The dialect of the less educated is, however, terrible and most of the time I don't understand a word, but then I simply pull myself out of the affair by saying, "You bet." That is always right.

Bice's arrival was approaching. She had successfully passed her final exams and Arthur was extremely proud, especially since she did very well in Physics and Chemistry although she was a little less successful in Mathematics, with a "Sufficient" (Fig. 20.6).

And already opportunities were opening up in America for academic advancement. By chance, Arthur had met the Sisters of St. Xavier College; they remembered Emma's earlier stay in the convent community and offered her a position as an instructor of German. The journey, though a bit long, was as Arthur calculated it financially feasible since the day ticket for the South Shore Trolley would cost only

[7] It would, however, be just as fast if not faster on foot, as the author knows from many personal experiences.

[8] AEH to EBH (South Bend, October 4, 1936).

Fig. 20.5 The streetcar at the Notre Dame Avenue terminus, with the buildings of the university in the background

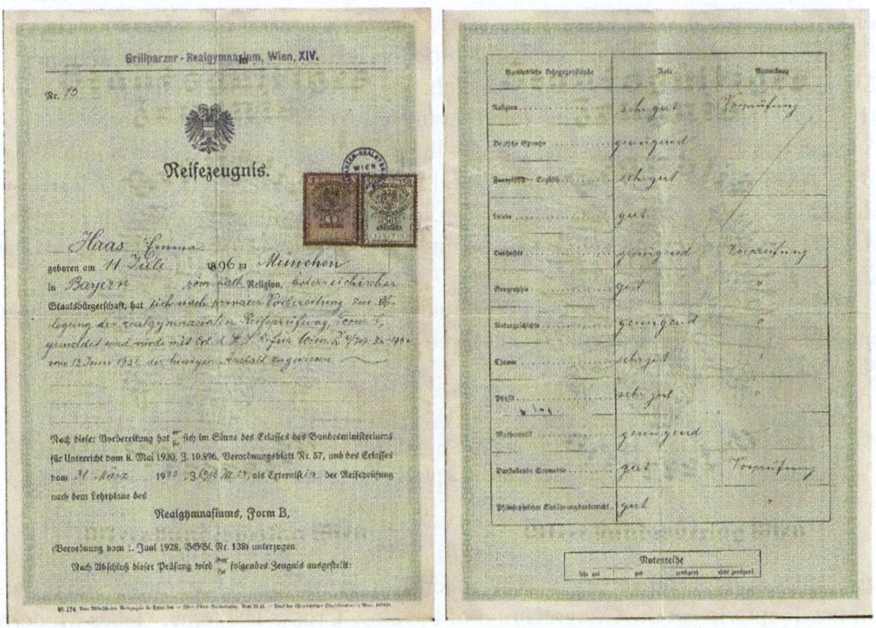

Grillparzer - Realgymnasium, Wien, XIV.

Nr. 13

Reifezeugnis.

Haas Emma

geboren am 11 Juli 1896 zu München in Bayern röm. kath. Religion, österreichischer Staatsbürgerschaft, hat sich nach privater Vorbereitung zur Ablegung der realgymnasialen Reifeprüfung, Form B, gemeldet und wurde mit Erl. d. [illegible] für Wien [illegible] vom 13 Juni 1932 der hiesigen Anstalt zugewiesen

Nach dieser Vorbereitung hat er/sie sich im Sinne des Erlasses des Bundesministeriums für Unterricht vom 8. Mai 1930, Z. 10.896, Verordnungsblatt Nr. 57, und des Erlasses vom 31. März 1933, Z. [illegible], als Externist/in der Reifeprüfung nach dem Lehrplane des

Realgymnasiums, Form B,

(Verordnung vom 1. Juni 1928, BGBl. Nr. 138) unterzogen.

Nach Abschluß dieser Prüfung wird dem/der folgendes Zeugnis ausgestellt:

Verbindliche Lehrgegenstände	Note	Anmerkung
Religion	sehr gut	Vorprüfung
Deutsche Sprache	genügend	
Französisch – Englisch	sehr gut	
Latein	gut	
Geschichte	genügend	Vorprüfung
Geographie	gut	"
Naturgeschichte	genügend	"
Chemie	sehr gut	"
Physik	sehr gut	"
Mathematik	genügend	
Darstellende Geometrie	gut	Vorprüfung
Philosophischer Einführungsunterricht	gut	"

Notenreihe

sehr gut	gut	genügend	nicht genügend

Fig. 20.6 Emma Beatrice Haas' final-exam certificate, which she earned on October 10, 1936, one month before she moved to South Bend

$3.30. There also seemed to be opportunities for Emma at the girls' college next door to Notre Dame, St. Mary's College, and Arthur was full of hope for her chances.

> The physics professor from St. Xavier visited me the other day. I heard from him that they probably want you to stay at the college two consecutive days a week (or three?), where you can stay and live. You are to teach "seniors," i.e., read and discuss German books with them. Of course, I haven't touched the question of your fee with anyone but, based on my general knowledge, I estimate that it will be between $50 and $80 per month (10 months per year), $20 of which will be travel expenses.[9]

Besides this information Haas also sent numerous letters of instruction to Vienna, mostly by zeppelin mail; he wanted to have a say in the furniture transport that was being negotiated between Emma, Otto, and various transport companies. The main issues were the selection of the most important heirlooms and whether they would fit into the small American house, and of course there were questions about shipping costs. Finally, it was done. The Viennese furniture was packed and on its way—albeit at an overvalued price, as Otto complained. It was now time for Bice to travel back to America.

While it was originally Bice's adventurous plan to travel eastward to the USA, through the Indian Ocean and then by way of Japan to cross the Pacific Ocean in order to reach the West Coast from where she would travel from San Francisco to South Bend by train, Arthur knew how to stop this plan with financial arguments. She did, however, travel elegantly enough; she booked a cabin on the *Queen Mary* (Fig. 20.7) and the ship was due to arrive in New York on November 16th.

Arthur also gave strict instructions for the journey from New York to South Bend. Bice was to take a train from Grand Central Station on the New York Central Railway, the *20th Century Limited*, depicted in Fig. 20.8 that would arrive in South Bend the next day at 5 pm. He insisted on a sleeping-car compartment for this last leg of the trip, which would cost $30.

Thanks to Arthur's planning in the choice of transportation, the journey time was minimized to ten days including the three days' journey from Vienna via Paris to Southampton.

> Here everything looks forward with great excitement to Bice's appearance. She is only supposed to diligently practice "Mid-West" English, especially the always appropriate phrase "You bet" and to speak it while sucking on a piece of candy. The English then sounds more like Midwest.[10]

Bice's speedy arrival was important to Arthur as he was originally scheduled to give a lecture to the university on November 12th, in the presence of the president and all the dignitaries of the Congregation of the Holy Cross. The topic, "Does Modern Physics Contradict Religion?" was a subject that particularly fascinated President O'Hara with respect to his interest in apologetics:

---

[9] AEH to EBH (South Bend, October 22, 1936).

[10] AEH to Liddy Haas (South Bend, October 27, 1936). Liddy Haas was an unmarried cousin, daughter of the pharmacists Arnold Haas and his wife Paula in Brno.

Fig. 20.7 The R.M.S. *Queen Mary* was one of the great luxury ocean liners of her time. She had only been put into service by the Cunard Line in May of that year and had already won the Blue Ribbon for the fastest Atlantic crossing, in August 1936. The *Queen Mary* was a ship famous for her Art Deco furnishings and her popularity with passengers

> My upcoming lecture [...] is already now, before I have even given it, the great event, and Father O'Hara, who has seen only the "abstract," has studied it eagerly and presented it to the bishop.[11]

However, the event was postponed:

> My big public lecture has been postponed from November 12th to November 19th, so that Bice can embellish the lecture with her presence.

The Haas family was warmly welcomed by the university and the town, as seen in the newspaper clippings in Fig. 20.9. And so, Arthur Haas' big, much-awaited lecture also became the welcoming reception of Emma Haas to Notre Dame Society, which would become her home and community for almost ten years. She was immediately absorbed by the professors' wives and won over for social clubs and commitments, while Haas was increasingly taken up by the priestly world of the university; this world met with the ideas and wishes that he had sought to fulfill with the Leo Society of Vienna.

Haas' lecture took place in Washington Hall, the largest lecture hall of the university, and students, professors, administrators and priests attended in large numbers. It was a success and the local press reported under the headline,

> Calls Science Atheism's Foe

[11] AEH to EBH (South Bend, October 10, 1936).

Fig. 20.8 The *20th Century Limited* was a record-setting train of the New Central Railway that needed only sixteen hours for the distance from Grand Central Station in New York to LaSalle Station in Chicago, leaving behind all competition on this route that was hotly contested by all of the passenger lines. The train was famous for its powerful locomotives, metal-clad in the Art Deco style and shown here at the entrance to the South Bend station with the Studebaker plant in the background

Vienna Family Comes To South Bend

New residents in South Bend are George Haas, left, Professor Arthur Haas and Arthur, jr., above, living at 847 Forest avenue. Professor Haas is in the physics department at the University of Notre Dame. George, 10, and Arthur, 11, are enrolled in Madison school. Mrs. Haas, now in Vienna, Austria, home of the Haas family, will join her husband and sons in American soon. (News-Times Photo.)

RETURNING TO U. S. ON QUEEN MARY

Mrs. Arthur E. Haas, wife of Professor Haas of the University of Notre Dame, is expected to arrive in South Bend about November 15, from Vienna. She will arrive in the United States on the Queen Mary, and will be at home at 847 Forest avenue.

Fig. 20.9 The Haas family in reports by the South Bend press, in the fall of 1936

> "Deeper scientific knowledge is the best defense against atheism," according to Dr. Arthur E. Haas, former Professor of Physics at the University of Vienna, Austria and now a member of the faculty of the University of Notre Dame who lectured Thursday night in Washington hall on the university campus.
>
> In speaking on "Does Modern Physics Contradict Religion?" Dr. Haas pointed out that the rapid strides of science during the last twenty-five years have dissipated the 19th century attitude of many scientists who scoffed at the idea of God, the creator.
>
> "Now most real scientists realize that there are many inexplicable phenomena — so far as science is concerned — that belong in the realm of spiritual things, and that can be attributed to God alone. Few real scientists in the world today are conceited enough to feel that they can explain the existence of the universe without admitting the existence of the deity," he said.
>
> "A new path from physics to religion seems to be open in such a way that physics of itself sees the necessary limits of its own claims. If this is the case, then from the very outset the conclusion of those writers, advocating atheistic ideas based on physics, are meaningless."[ix]

Here Haas seems to have made a decisive turn to apologetics. Although he gave another lecture in Washington Hall two weeks later on the topic "The Atom and the Universe," this lecture was also determined by philosophical and cosmological explanations. He seems to have turned his interest back to cosmology.

Yet, there was another project that kept his attention. At Bowdoin College, Haas had begun a new book project, a textbook on experimental physics that was geared more to American students and their level of knowledge but would also be useful to students in Germany and Austria, who were minoring in physics. He had learned a great deal of English during that first year in America but he began writing the book in German anyway, since at the time his future in the new country still seemed uncertain. The book covered the whole field of classical physics but also took up topics on modern atomic physics and nuclear physics. He completed the work during his first semester at Notre Dame and submitted the manuscript to Julius Springer Publishers in Vienna during January 1937 under the title *Elementary Physics*.[12] Since Haas was considered a successful author, the book was accepted and published immediately.

Besides teaching responsibilities (Fig. 20.10), it was the hope of the young experimenters around Collins and Coon that Haas would provide theoretical guidance for the experimental program of their new accelerator. In addition. they were anticipating that Haas would present and explain interpretations of the experimental results, in the context of quantum physics of the atomic nucleus. This hope was entirely justified since in his last book, on the "transformations of the chemical elements" that had just been published in Vienna in 1935, Haas had described in summary form the latest experimental developments and methods, and had also explained in detail the new possibilities that were available now that accelerators were a means of "atomic disintegration."[13] He had described the role of Russian theorist George Gamow in the development of the Cockcroft-Walton accelerator at the Cavendish Laboratory and the importance of the cyclotron, developed by Lawrence at Berkeley, for atomic

---

[12] Arthur Haas, *Elementare Physik* (Wien: Julius Springer Verlag, 1937).

[13] Arthur Haas, *Die Umwandlungen der Chemischen Elemente*, 45–66.

Fig. 20.10 The university's Science Building built in 1883, where Arthur Haas taught and today serves as the La Fortune Student Center

disintegration. So it was not unreasonable that Collins expected Haas to come up with important ideas and proposals.

The accelerator at Notre Dame (Fig. 20.11) was a machine for accelerating electrons, while the aforementioned machines used protons and deuterons; it was this niche in the competition for new scientific results that Collins hoped to exploit. The machine Collins built was based on the Van de Graaff principle[14] but was a copy of the design by Merle Tuve at Carnegie Institute, who used it to accelerate protons.[15]

Collins used the machine to generate intense, accelerated electron beams for nuclear physics experiments. Until then, electrons had hardly been used in this application. The reasons for this decision were simple: First, it was much easier to produce electrons than protons and second, no other research group had done it before.[16] In 1937, one year after Haas had come to Notre Dame, the machine was ready to perform the first experiments.

However, Haas seemed to have little interest in participating in this work or in supporting it with direct theoretical calculations. With the assumption of his position at Notre Dame and with the direct support of President O'Hara, his interest had turned away from practical laboratory and experimental questions and was again focused on fundamental questions of cosmology. Therefore, his contribution to the

[14] R.J. Van de Graaff, "A 1,500,000 V Electrostatic Generator," *Phys. Rev.* 38 (1931): 1919.

[15] M.A. Tuve, L.R. Hafstad, and O. Dahl, "High Voltage Technique for Nuclear Physics Studies," *Phys. Rev.* 48 (1935): 315–337.

[16] William K. Collins, C.S.C., "The Early History of Nuclear Physics at Notre Dame," (doctoral thesis, University of Notre Dame, 1977), 76.

Fig. 20.11 The new accelerator—at that time mostly called a generator—for Notre Dame. A copper shell mounted around a wooden frame stood on three insulating legs (plastic pipes that had just been developed for drainage systems at the time). Electrical charge was transferred to the copper sphere by a rotating insulated belt, which allowed for attaining up to one million volts. At high humidity the charge could not be held and the high voltage was discharged in a flash along the insulating legs, as shown in the right figure. Inside the copper sphere was an electrical filament in a glass vacuum vessel—similar to a light bulb—in which the electrons were released and then accelerated, by the voltage difference between the sphere and the earth, to bombard different materials. The resulting radiation, as the effect of atomic disintegration or excitation, was then measured

laboratory was essentially to bring the young Hungarian physicist Dr. Eugen Guth[x] to Notre Dame.

Guth was originally from Budapest, received his doctorate from Hans Thirring in Vienna in 1928, and was at that time also integrated in the discussion group of the Vienna Circle.[17] Haas knew Guth from these days, if not from the discussion groups of the Vienna Circle then at least because Guth had provided him with substantial help in completing the fifth edition of his textbook in theoretical physics:

[17] "The circle held weekly sessions on philosophy of science, which were very interesting, but sometimes quite confusing for me. For example, there was a discussion about a book by Herman Weyl, 'What is Matter?'. There was the expression 'there is.' I remember a lively discussion about the possible meaning of this expression. As a young student of science, I did not appreciate the importance of such semantic discussions at that time. Later I realized that the exact meaning of statements can be very important in philosophy. Nevertheless, I always remembered a saying by Goethe in *Faust*: 'Where terms are missing, a word comes at the right time'..." in, Eugen Guth, "Lectures on the History of Physics," Center for Theoretical Studies, University of Miami and Oak Ridge National Laboratory (January 1967).

> It is a pleasant task for me to thank Dr. Eugen Guth [... et al.] for the valuable help they gave me in reading the corrections for the new parts of this volume.[18]

After his doctorate, Guth was employed as an assistant in the group of Wolfgang Pauli (1900–1958) at the ETH in Zurich and then moved to work with Heisenberg in Leipzig before returning to Vienna in 1932, to complete his habilitation. On the advice of his doctoral supervisor Thirring, Guth joined the group of the newly appointed professor of chemistry Hermann Mark (1895–1992), who attempted to establish a center for polymer physics and chemistry in Vienna.

Guth began working with Mark and his group on the development of a formalism to describe rubber elasticity and viscosity of liquid solutions and suspensions. During this time in polymer chemistry, Eugen Guth again helped Arthur Haas, this time in the final drafting of his book on the transformation of chemical elements, as again noted in the preface. Guth was thus well acquainted with the questions and problems of the young field of experimental nuclear physics.

Already before and during his habilitation period, Guth had been successful in publishing several papers on the topic of nuclear physics before submitting his habilitation thesis on the viscosity of suspensions and solutions in 1936. The decision on this habilitation application was delayed because Guth was not an Austrian citizen. In the meantime, tired of waiting and concerned about the political situation in Austria, he accepted the offer from Notre Dame.

In order to push through this offer, Haas had argued with university superiors that Guth could continue the work of the late Father Julius Nieuwland (1878–1936), who had been instrumental in the development of synthetic rubber. Indeed, in the following years Guth built up a very successful research program in polymer physics at Notre Dame. In the early years, however, Guth was also directly involved in all of the projects of the experimental nuclear physics group, from the generation of Cerenkov radiation to the disintegration of atomic nuclei by electron irradiation, contributing to both theoretical prediction and analysis of the data. He had thus taken over the role originally intended for Haas within the physics group.

In addition to the second theorist Guth, the university also brought the young experimenter Bernard Waldman (1913–1986)[xi] from New York to Notre Dame. With these additional young enthusiastic researchers, Notre Dame was soon considered a center for electron and X-ray research in the United States. This dynamic group is pictured on the steps of the science building in Fig. 20.12.

With the successful implementation of the graduate school, the number of students wishing to do their Master's or PhD work with the physics department's new program grew and within a few years Notre Dame was considered a center in the new field of nuclear physics in the United States. As a result, Notre Dame was later directly involved in the Manhattan Project, which we will not discuss further in the context of our topic.[xii]

---

[18] Arthur Haas, *Einführung in die theoretische Physik mit besonderer Berücksichtigung ihrer modernen Probleme* Bd 2, (Berlin-Leipzig: De Gruter, 1930), Preface to the fifth and sixth editions of the second volume, IV.

Fig. 20.12 A photo by George Collins showing the successful young physics group at the entrance of the science building in August 1939. Arthur Haas is in the back to the right of Fr. Bolger; Bernie Waldman is in the next lower row, second from the left with Ed Coomes on the far right. Eugen Guth stands on the next lower step left from Coomes. The other people are mostly young doctoral students of the group

Between 1937 and 1941, work with the Notre Dame accelerator led to several scientific publications that were very much within the range of the possibilities and interests that Haas had described in his book. However, there is no conclusive evidence that Haas contributed directly to these experiments or the interpretation of the data. He left this to the young Eugen Guth who was thus able to make a name for himself at his new university (Fig. 20.13). Only two examples shall be mentioned here. The first work was on the disintegration of beryllium by electron bombardment,[19] to which Guth provided the theoretical predictions.[20]

[19] George B. Collins, Bernhard Waldman, Eugen Guth, "Disintegration of Beryllium by Electrons," *Phys. Rev.* 56 (1939): 876–880.

[20] Eugene Guth, "Theory of Electrodisintegration of Beryllium," *Phys. Rev.* 55 (1939): 412.

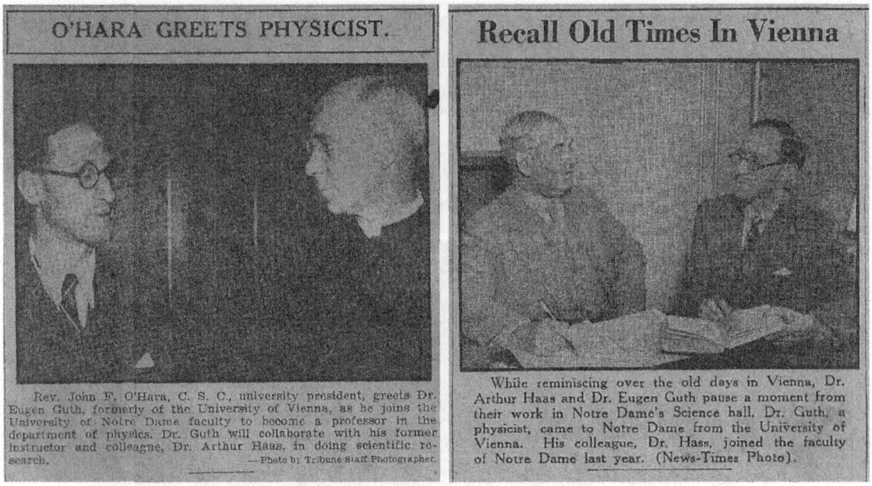

Fig. 20.13 Welcome for Eugen Guth to Notre Dame, a press event

This was a first example demonstrating that electrons could be used for nuclear transformation. It also provides an example of the close cooperation between experimenters and theorists that made nuclear physics so successful in the following decades. In addition, several papers were published demonstrating the excitation of various atomic nuclei, from indium to lead, by the energy transfer of electron radiation. Again, Guth had provided the theoretical calculations to interpret the results.[21] Another important work within the scope of Arthur Haas' interests was on pair formation,[22] where positrons were measured directly as anti-particles of electrons; this was a possibility propagated by Arthur Haas again and again as a means to produce mass from energy.

A series of experiments with the new accelerator and also with the successor machine completed in 1942[xiii] produced the basic data on which scientific knowledge is based. However, when it came to the constant and painstaking work of experimental procedure, Haas did not see himself engaged with the implementation of his ideas. Consequently, none of these works have any indication of Arthur Haas' intellectual or directly active participation. His interests had evolved in other directions, directions that, in his opinion, were more likely to address the fundamental questions of human existence.

Notes

i. Thomas J. Schlereth, *The University of Notre Dame—A Portrait of its History and Campus* (Notre Dame: University of Notre Dame Press, 1976).

ii. Thomas E. Blantz, C.S.C., *The University of Notre Dame—A History* (Notre Dame: University of Notre Dame Press, 2020).

[21] Eugene Guth, "Interaction of Electrons with Nuclei," *Phys. Rev.* 57 (1940): 349.

[22] Ibid.

iii. William K. Collins, C.S.C., "The Early History of Nuclear Physics at Notre Dame," (doctoral thesis, University of Notre Dame, 1977), 4–19, accessed on January 31, 2021, (https://isnap.nd.edu/assets/186671/earlyhistory_nd_nuclearphysics_small.pdf) and Christopher S. Temple, "Fostering Elite Science at an American Catholic University: The Rise of a Research Culture at the University of Notre Dame, 1842–1967," 217–233.

iv. *Bulletin of Information, University of Notre Dame* XXIV, no. 3 (1929), 222.

v. William K. Collins, C.S.C., "The Early History of Nuclear Physics at Notre Dame," 4–19, accessed on January 31, 2021, https://isnap.nd.edu/assets/186671/earlyhistory_nd_nuclearphysics_small.pdf.

vi. Paul Chagnon, *The First Accelerator at Notre Dame* (1985), accessed on January 31, 2021, https://isnap.nd.edu/assets/186666/firstaccelerator.pdf.

vii. Michael Wiescher, *The Nuclear Structure Laboratory at Notre Dame* (2005), accessed on January 31, 2021, https://isnap.nd.edu/assets/186667/lab_history.pdf.

viii. Louis Brown, *Centennial History of the Carnegie Institution of Washington – Volume II, The Department of Terrestrial Magnetism* (Cambridge: Cambridge University Press, 2012), 81–88.

ix. *South Bend Tribune*, November 20, 1936.

x. Daniela Angetter, Michael Martischnig, "Biografien österreichischer [Physiker]innen," Austrian State Archives Vienna (2005), 42.

xi. Cornelius P. Browne, "Bernard Waldman," *Physics Today* 40/12 (1987): 108.

xii. Philipp T. Wiescher, "Early Days of Nuclear Physics at Notre Dame and the Manhattan Project," (2004), accessed on January 31, 2021, (https://isnap.nd.edu/assets/186668/nd_manhattan_project.pdf) and Christopher S. Temple, "Fostering Elite Science at an American Catholic University," 314–328.

xiii. Michael Wiescher, *The Nuclear Structure Laboratory at Notre Dame.*

Chapter 21
Church and Cosmologies

Arthur Haas established close and personal contact with Notre Dame's president Father O'Hara through their common interest in apologetic questions around the commonality and mutual affirmation of religion and science. While O'Hara was originally and primarily a teacher of economics and economic theory, his interest in medieval scholasticism and the resulting apologetics had developed through his vocation as a priest, while Haas, coming from the Protestant approach of performing science for the betterment of the world, nurtured Catholic attitudes through his political conservatism, as attested by his involvement with the Leo Society. However, the philosophical questions he had posed at the beginning of his career in Vienna and Leipzig were also essential to his thinking and were reinforced by his focus on cosmology as well as his growing conviction regarding the parallel worlds of atom and universe, with a higher being reflected in this symmetry.

Already in his 1907 article on cosmology, Haas had argued for a finite universe using physical arguments and concluded that such a finite universe required a higher being.[i] He postulated,

> the assumption of a being whose actions are not subject to the general laws of nature, and whom, independent of nature, stands and rules over it.[1]

Haas maintained a separation between the laws of physics, which help to describe and understand the universe, and the eternal Creator who rules over it—whatever "ruling" means in this context. In the decades that followed, prior to becoming more absorbed in the thoughts and concepts of cosmology, Haas had left God alone. During these years, Haas followed the reasoning of enlightenment and rationality—best expressed in Laplace's answer to Napoleon, regarding why he had not mentioned God in his book on the cosmos: "Sire, Je n'avais pas besoin de cette hypothèse-là" (My lord, I see no need for this hypothesis). For Haas, this perspective was probably fostered by the influence of the Viennese Circle where he tried to present all of his

---

[1] Arthur Erich Haas, "Die Physik und das kosmologische Problem," *Archiv für Systematische Philosophie* 13 (1907): 511–525.

M. Wiescher, *Arthur E. Haas—The Hidden Pioneer of Quantum Mechanics*, Springer Biographies, https://doi.org/10.1007/978-3-030-80606-4_21

arguments in the positivist approach, seeking to understand and explain nature by correlating the results of observations and mathematical concepts.

Nevertheless, Haas seems to have kept the "hypothesis" in mind and this idea was strengthened when he again turned to cosmological thoughts and concepts. In addition, Haas was now influenced by the impressions raised by the uncertainty principle resulting from quantum mechanics. With this he basically recognized a reduction in the role of causality:

> The sharply pronounced causality that is revealed in macroscopic-physical events has been contrasted by the latest physics with the indeterminacy of atomic processes; it assumes a determinacy only for statistical quantities, which it assigns to the individual, elementary processes of physics. Of course, such a view also opens up new perspectives for the evaluation of physical laws as such. In the center of interest in natural philosophy, the question of whether there is any other law in nature than a purely statistical one is becoming more and more important...[2]

He took precisely this uncertainty and indeterminacy as a sign of a Creator who can never be observed or experienced exactly. In so doing, Haas took a position in contrast to Albert Einstein who with his famous and much quoted words, "God does not roll the dice,"[3] vehemently rejected statistical uncertainty and never really accepted the results and statements of quantum mechanics.

These thoughts had taken shape for Haas in Vienna under the impression of numerous personal conversations with his friend, Professor Petritsch. Petritsch was a member of the board of directors and in 1936 was even vice president of the Leo Society, which he had motivated Haas to join as well. There, between 1932 and 1934, Haas met the Spanish priest Father Antonio Romañá Pujó S.J.[4] who performed research as an astronomer at the Vienna Observatory and used the opportunity of his two-year stay in Vienna to translate Haas' book on cosmology into Spanish.

The Jesuit priest Antonio Romañá Pujó was also a member of the Leo Society in whose annual reports he published some of his thoughts. Haas made friends with the priest and the two men had many discussions on questions of physics and religion, on the role of God in the universe. Father Antonio Romañá held the view that Catholicism and the idea of a God were absolutely necessary for progress in science. After his stay in Vienna, Father Antonio Romañá returned to Barcelona and published several writings on this subject, especially after the end of the Civil War and during the early period of the Franco dictatorship. He observed:

[2] Arthur Haas, *Materiewellen und Quantenmechenik* 4th and 5th editions (1934), 276.

[3] This is the frequently quoted but incomplete formulation that Einstein had used in a letter to Max Born on 4.12.1926: "Quantum mechanics is very respectful. But an inner voice tells me that it is not yet the real thing. The theory delivers much, but it brings us hardly closer to the secret of the Old Man. In any case, I am convinced that he does not play dice.".

[4] P. Antonio Romañá Pujó (1900–1981) was an astronomer and mathematician born on March 21, 1900 in Barcelona; he died on October 13, 1981 in San Cugat de la Vallés, Barcelona. In 1917 he entered the Jesuit Order. After his licentiate and doctorate, he was for two years at the Vienna Observatory where he probably worked until 1934. Until his death, he was at the Ebro Observatory in Roquetas (Tarragona) and was its director after the Spanish Civil War, from 1939 to 1971. He was the first president of the UGGI (International Union of Geodesy and Geophysics). He published more than 150 studies, most of them on the interaction between the Earth and the Sun.

> Those who, in the name of Science, have dared to deny it, [...] have wanted to use knowledge acquired superficially and not well assimilated, to support their vain arguments. One statement is always true: "Much science leads to God; little science moves away from God."[ii]

These conversations with Antonio Romañá seem to have had a great influence on Haas' thinking; two years later in a letter to Dean Franz Wenninger of Notre Dame, Haas stated:

> I developed in some detail the view that modern physics and especially the modern ideas on the universe, on relativity and indeterminacy, are much more adapted to strengthen than to weaken religious feelings.[5]

This indeed seems to reflect the influence of Father Antonio Romañá. Haas followed this view further at Notre Dame, most likely out of conviction but probably also due to the Catholic thought that dominated university life and so appealed to him. This is reflected in his letters to his wife in which he reported on these convictions and conversations, even if only briefly:

> Father O'Hara continues to honor me [Fig. 21.1]. Yesterday, when the most prominent leader of the Catholic movement, Fr. Gillis, editor of *Catholic World* from Washington, was visiting, I was invited by President O'Hara to take together with Fr. Gillis the two seats of honor on his side.[6]

The Rev. James Gillis (1877—1957)[7] had been invited to Notre Dame from Washington for five lectures. Meetings and connections with such people were regarded as a personal honor and Haas was flattered to be included. He visibly tried to live up to his promise: to unite the thought worlds of religion and science, and to observe and record connections and parallels between the two disciplines.

Gillis was socially progressive, advocating social reform and support for President Roosevelt's New Deal in the early 1930s, but politically Gillis was conservative and rejected any cooperation with the Soviet Union. He played an important role in political developments and decisions but was less interested in the scientific issues of the time[iii]; there was probably little overlap in the interests of Haas and Gilles.

President O'Hara also introduced Haas to various other guests and visitors from abroad who were interested in questions of apologetics and had come to Notre Dame for that reason. One of these was the British writer Arnold Lunn (1888—1974) who was a convert to Catholicism (Fig. 21.2). Lunn had traveled through Spain, during the Civil War, before coming to Notre Dame as a visiting associate professor in the newly founded Department for Apologetics.[8] Like Haas, Lunn saw the Catholic Church as

[5] AEH to Dean Franz Wenninger (Brunswick, May 23, 1936).

[6] AEH to EBH (Notre Dame, October 22, 1936).

[7] James Martin Gillis was a member of the Pauline Fathers Order and a close confidant of Cardinal Spellman. He was editor of the magazine *The Catholic World* from 1922 to 1948. In the 1930s he was known in the United States through his widely broadcast radio program *The Catholic Hour*. His ideas and suggestions were incorporated into the curricula of many of the Catholic universities in America. Cf. Richard Gribble C.S.C., *Guardian of America: The Life of James Martin Gillis, C.S.P.* (Mahwah: Paulist Press, 1998).

[8] Sir Arnold Henry Moore Lunn was a famous mountaineer and skier, considered the inventor of the ski slalom. Lunn studied in Oxford at Balliol College and established himself as an agnostic writer,

Fig. 21.1 One of the many occasions Haas was featured in the South Bend press, in a joint photo with President O'Hara in 1936

the only rock in the flood of new atheist movements. In his view, this included Communism as well as National Socialism and he did not necessarily exclude other churches or liberal movements, as he argued in the following article:

> From the steadiness of the rock [Catholic Church], one sees the swirling world in a new light. One sees the stream of human thought and action, with the Communists and the Nazis swimming with strong strokes in the front. One sees the Dean of Canterbury paddling desperately to keep up with the stream, and Mr. Bernard Shaw diving like a dolphin in and out of it. One sees the Liberals trying to persuade themselves that there is no stream, and to build a permanent house upon the shifting waters. But, with the exception of those who are anchored, consciously or unconsciously, to the rock, all are heading toward a secular organization of society, the sea where all must drown.[9]

Lunn's political views at that time were similar to those of Rev. Julian Sigmar of Notre Dame, who saw a visionary in Mussolini and the devil in Hitler. Lunn's views

mostly writing about sports and society. In 1933 he converted to Catholicism. Lunn supported the Franco movement in Spain, which he traveled through before going to Notre Dame in the fall of 1936. He stayed at Notre Dame until the summer of 1937, where he gave several lectures. He then set out on a lecture tour to various American universities where he presented his views on the Spanish Civil War and on Catholic fascism as a bulwark against atheist movements, particularly communism and Nazism. After this lecture tour he stayed in Italy for a longer period of time, as he also held positive views on Mussolini. He spent World War II in England. See Elisabeth Hussey, *Biography of Sir Arnold Lunn: 1888–1974* (Independent Publishing, 2014).

[9] Arnold Lunn, "Religion," *The Catholic Mind* XXXVI, no. 863 (1938), 468.

N. D. ENGAGES FOREIGN TEACHERS

Two professors, both Europeans, who have been added to the University of Notre Dame faculty this scholastic year are shown with Rev. John F. O'Hara, C. S. C., president of the university, who is in the center. At the left is Dr. Arthur Haas, of Vienna, Austria, internationally famed physicist, and at the right is Arnold Lunn, of England, writer and lecturer, who arrived Tuesday afternoon. Mr. Lunn will teach in the new graduate apologetics department for laymen.

Fig. 21.2 Reception of two new European professors, Arnold Lunn (right) and Arthur Haas (left), by President O'Hara (center) in October 1936

also ran parallel to Father Antonio Romañá's, who saw the salvation of the Church in the Franco regime, and to those of Arthur Haas who hoped to save Austria from Nazism though the fascist Vaterländische Front.

In an article in the Notre Dame magazine *Scholastics*, Lunn formulated his thesis:

> But to return to my premise. The Catholic who knows his Europe does not make the silly mistake of equating Italian Fascism with German Nazism. The former has its roots in the old culture of Catholic Europe, which incidentally was as authoritarian as modem Italy, whereas

the latter, like Communism, is bitterly hostile not only to Catholicism but to the Catholic culture.[10]

At Notre Dame, Lunn gave six lectures, two of which attracted the attention of Haas, "The Joy of Controversy" and "Science and the Supernatural." It is not known to what extent Lunn, who was a ski enthusiast and mountaineer, and Haas, who was more inclined to the leisurely life if one disregards his swimming, understood each other or exchanged ideas beyond the meeting with O'Hara. Lunn's presence on campus did, however, reflect the mindset of the university at that time.

But apart from bringing visitors together, President O'Hara had already suggested in Haas' letter of appointment that he collaborate with Father Christian Murphy. Therefore, Haas possibly recognized this work as a professional obligation, particularly since he felt compelled to make a good impression in his first probationary year.

It was in this spirit that Haas developed his special series of lectures, which opened on November 19, 1936 with the aforementioned lecture, "Does Modern Physics contradict Religion?" In this, he developed the history of the idea of God from the perspective of developments in modern physics over the previous fifty years, explaining at the outset:

the harmony, which manifests itself in physical phenomena, must be traced to a supernatural Intelligence as founder of this harmony.[11]

He based his arguments on two points: First, the universe is not eternal and infinite, in contrast to the atheistic point of view:

A concept which is fundamental to atheistic literature is the idea of the universe as being infinite in space and time. Man's desire for the existence of the infinite is so deep that even the incredulous person who does not believe in a deity of infinite power seeks a substitute for it by endowing the universe with infinity.[12]

As his second point, Haas argued that the physical laws are not universally valid:

The laws of nature, the materialistic substitute for a deity, appear in modern physics as no longer absolutely valid; they only appear as rules based on statistical knowledge, so that deviations from the norm do not contradict the basic principles of modern physics. If, however, laws of nature are essentially statistical, then it seems meaningless to consider, from the standpoint of physical laws, unique events, such as a creation of a universe. Processes so unique that no second of its kind can be thought of, cannot be the subject of the statistical and, therefore, not of physical consideration.[13]

---

[10] Arnold Lunn, "Arnold Lunn writes a Letter—Itinerant Apologist jumps from Football to Fascism," *Notre Dame Scholastic* 73/12 (1939): 6, 23.

[11] Arthur Haas, "Modern Physics and Religion," *New Scholasticism* XII (1938): 1.

[12] Ibid., 4.

[13] Arthur Haas, "Modern Physics and the Supernatural," lecture notes, University of Notre Dame Archives, 1936—1940.

The first point he had already made in 1907. It is, indeed, the old scholastic dialogue of the Middle Ages—between the representatives of either a finite or an infinite universe—based on an interpretation of Aristotle's philosophy.[14]

In his further explanations, Haas concentrated on the limitations of the laws of physics. He discussed the development of thermodynamics and the concept of statistical systems as opposed to the concept of predictable single-particle behavior that does not lead to statistical description and the unpredictability of quantum–mechanical systems. From this he concluded that physical laws are only statistically valid and therefore not applicable to unique systems, like the universe and the deity. From this he deduced the limitations of human knowledge and the limits of human ability in the exploration of the unfathomable:

> Very remarkable is the number of important discoveries, among which one of the most significant is the discovery of boundaries to man's knowledge of the physical. The modern physicist is happy in the progress that has been made in the exploration of the explorable, but he is at the same time quite conscious of the natural limitations to his science.[15]

Thus, Haas did not attempt to extract a direct proof of God from cosmology, as Pope Pius XII[16] did a few years later on November 22, 1951.[17] Rather, Haas only attacked what he considered to be the fundamental assumption of atheists—the universal validity of physical laws in an "Eternal and Infinite Universe." In the exact derivation of how the existence of a God results from this, he remained relatively vague and indefinite. It becomes clear that he was more interested in the physical than in the philosophical-theological aspects of cosmology, as he notes:

> Physics cannot be expected to solve the riddles of philosophy nor to answer the fundamental questions of the philosophers. All that physics can do is to open new viewpoints for the consideration of philosophical problems and to establish connections between fundamental problems of philosophy.[18]

---

[14] The first group followed the teachings of the Arab philosopher Averroes and represented the doctrine of the eternal and infinite world; they were later called "atheists" by Aquinas. The second group followed the teachings of the Jewish philosopher Maimonides who held the image of a finite world created by a Creator. See Charles Singer, "Historical Relations of Science and Religion," in *Science Religion and Reality* ed. Joseph Needham (London: Sheldon Press, 1925), 116.

[15] Arthur Haas, "Modern Physics and Religion," 8.

[16] It should be noted that Cardinal Eugenio Pacelli, who later became Pope Pius XII, received his honorary doctorate from the University of Notre Dame on October 25, 1936, and on that occasion also discussed with President O'Hara the question of bringing science and religion together. "Eugenio Cardinal Pacelli is Awarded Honorary Doctorate of Letters," *Notre Dame Scholastic* 73/12 (1936): 3, 23.

[17] The Pope's conclusion regarding the Big Bang thesis of the Jesuit priest Georges Lemaître—"Hence creation took place in time, therefore there is a creator. Therefore God exists"—has been the subject of countless debates since then. See Giuseppe Tanzella-Nitti, *The Pius XII—Lemaître Affair (1951—1952) on Big Bang and Creation*, accessed on January 31, 2021, http://inters.org/pius-xii-lemaitre, and George V. Coyne, "Georges Lemaître: Science and Conviction," in *Georges Lemaître, Life, Science and Legacy*, eds. Rodney D. Holder, Simon Mitton (Heidelberg: Springer Verlag, 2012), 69–74.

[18] Arthur Haas, "The Limitations of Physical Knowledge," lecture notes, University of Notre Dame Archives, 1936–1940, 4.

It is interesting to note that Haas based his claim for a limited universe in time and space on the view that the redshift of light from distant star systems is not due to an expansion of the universe but results from a fatigue of light, which manifests itself in an increased wavelength with increasing distances.[19] In his book, he had rejected such a phenomenon.[20] However, under the influence of the information and arguments that he had heard at the Harvard Tercentenary Conference on redshift—that the fatigue of light could be caused by Compton scattering or by the influence of gravity—he seems to have found for himself a much more convincing thesis than that of the continuous expansion of the universe. Therefore, he quickly returned to the idea of a temporally, spatially limited and singular universe. It was not until two years later with the model of the "Primeval Atom" of Belgian priest Georges Lemaître[21] that Haas seems to have been convinced to accept the idea of an expanding universe, as will be discussed later.

The singularity of the universe as seen by Haas and his contemporaries is being questioned today. The Standard Model of the Universe is formulated, but there are still inconsistencies that point to physics beyond the Standard Model. These inconsistencies have triggered a search for further evidence in the microscopic world, beyond the existence and the origin of our universe. Modern physical theory, with its concept of the multiverse, has raised the possibility of alternative or parallel universes in coexistence.[iv] This concept is comparable to the discovery, made 100 years ago through observation, of galaxies outside our own galaxy. But, the multiverse is still a hypothesis that has not been confirmed by either experiments or observation. According to our physical laws, we may not be able to experience or observe alternative universes.

Arthur Haas confirmed his opinion of a finite universe in a series of four public lectures that he gave over the following months. In these he reflected on the development of modern physics during his lifetime. Starting with his fundamental question, "Is the universe infinite?" he then continued with a lecture on the "World of Atoms," a concept that he explained historically in the subsequent third lecture, "Physics and Philosophy," in which he described the development of physical concepts from a positivist point of view. His fourth lecture, "Modern Physics and the Supernatural" deals with the gap between materialism and religion that emerged in the nineteenth century. In his opinion, this gap lost its force with the development of modern physics. Haas' discussion started by repeating the representations of his first lecture and concluded with the opinion he had developed in discussion with Antonio Romañá Pujó, S.J.:

---

[19] This idea was presented in 1929 by the Swiss astronomer Fritz Zwicky at Caltech, whom Haas had probably met at the Tercentenary Conference at Harvard. See Fritz Zwicky, "On the Red Shift of Spectral Lines through Interstellar Space," *Proceedings of the National Academy* 15/10 (1929): 773–779.

[20] Arthur Haas, *Kosmologische Probleme*, 16–30.

[21] Georges Henri Joseph Édouard Lemaître, (1894—1966) was a Belgian Catholic priest, mathematician, astronomer, and professor of physics at the Catholic University of Louvain. He earned his PhD with Arthur Eddington and today is considered the father of the Big Bang theory, albeit he didn't coin the name.

> Recent findings, as to the finite size of the universe and apparent tendencies of cosmic processes, seem to strengthen rather than to weaken religious sentiments of anyone studying such problems.[22]

Arthur Haas held this view. The discussion about "Matter and Spirit in the Universe" or "God and the Multiverse" is still controversial today. Some see no necessity for God's existence in the physical observations and laws:

> Currently God is an additional hypothesis not required by the data. If he were, he would be included in the set of premises that constitute scientific theories. He is not. While we certainly do not know everything, we know of no observed fact that requires the existence of God, indeed many observed facts that are inconsistent with the God hypothesis serve to prove beyond a reasonable doubt that a God who plays an active role in the Universe and in human lives does not exist. God has left no footprints...[v]

While others in their consideration of the anthropic principle contend that by virtue of the fact that we exist, there is proof of God or the footprint of God—that physical conditions and laws have been so finely balanced, in the billions of years since the beginning of the universe with the Big Bang, that they have led to the creation of man against all probability:

> A universe with the requisites for complex life was almost infinitely improbable, given the immense range of theoretically equally plausible alternatives in which life could not arise. So why this one? Well, because we are here [...] our presence here "explains" the universe.[vi]

Although in his lectures Haas had expressed concerns about the interpretation of the spectral redshift as expansion—and although the effects of Compton scattering had always been a topic of preference in which he could easily imagine the loss of energy or the fatigue of light—he did, none the less, gradually come to recognize the advantages of an expanding universe for the Catholic cosmological world view. In particular, he appreciated the ideas of Georges Lemaître, as these were in line with his convictions of finite space and finite time.

In 1927, Lemaître had succeeded in combining the mathematical formalism of relativity and the observational results of Edwin Hubble to create a new cosmological model of an expanding universe. Then in 1931, Lemaître postulated a "Primeval Atom" as the universe's origin.[vii] He derived this idea by consistently applying the equations of Einstein's cosmological model, which fit perfectly with Hubble's observation of an expanding universe. Even Einstein was convinced to abandon his idea of a constant and unchanging universe after a personal conversation with Lemaître.

Lemaître is still considered the spiritual father of the Big Bang theory, according to which the universe expanded from a single event at a single point in time. This was called the primeval or primeval atomic model; the term *Big Bang* was only later introduced by Fred Hoyle (1915—2001) in a radio interview, in order to describe the model in comparison to the "Steady State Model"[viii] that he advocated.

Arthur Haas used his influence with President O'Hara to invite Georges Lemaître to Notre Dame for a sabbatical year (Fig. 21.3). This was probably not particularly

[22] Arthur Haas, "Modern Physics and the Supernatural.".

Takes Post

CANON LEMAITRE

Added to the faculty of Notre Dame is Canon Georges LeMaitre, priest-professor of mathematics and theoretical physics at the University of Louvain. He will assume duties when the university opens September 16 for its 97th year.

Fig. 21.3 First announcement of Georges Lemaître's arrival for the fall semester of 1937, but the visit was then delayed by several months

difficult since already in 1933, the university journal *Notre Dame Scholastic* had published a positive article about Lemaître's apologetic opinions:

> Lemaître pointed out the heart of the misunderstanding between religion and science. "It lies," he said, "in the fact that many scientists" believe the Bible pretends to teach science. When they find the Bible's scientific references wrong, they repudiate it utterly. Should a priest reject relativity because it contains no authoritative exposition of the doctrine of the Trinity?[ix]

Notre Dame made Lemaître a generous offer for the six-month stay: $3,500 salary, free lodging and a $400 travel allowance. He accepted this offer immediately and offered a course on cosmology for the spring semester of 1938. Arthur Haas was a great admirer of Lemaître and was enthusiastic about the idea of bringing him to Notre Dame as another cosmologist, if only for a limited time. Especially regarding the alternative possibility of explaining the redshift by the scattering energy loss of radiation over long distances, Haas expected an interesting debate with his new colleague.

Georges Lemaître arrived in New York aboard the S.S. *Normandy* in January 1938 and traveled by train directly to South Bend, where he was welcomed by Haas. Apparently the two cosmologists complemented each other well over the following semester. Haas used the opportunity of Lemaître's visit to organize a conference on cosmological issues toward the end of the semester, bringing together the leading experts in this field.

In addition to the possibilities regarding the redshift of light from distant galaxies as observed by Hubble—whether due to energy loss through Compton scattering or due to gravitational effects, instead of an expanding universe—Haas was particularly interested in the question of the origin of mass and how this influenced the energy balance in the universe, since the net equilibrium between gravitational energy and total energy had to be maintained. In stars, mass is transformed into energy that is emitted as light, which he had already carefully calculated in his book. But how is energy converted back into mass? For years, Haas favored the idea of generating mass through the process of pair formation by high-energy gamma radiation, the latter in turn being produced by inverse Compton scattering of photons by relativistic electrons in cosmic rays.

Lemaître, on the other hand, was at that point in his career more interested in new observational results that could further substantiate his mathematical models. These observations included not only the increasing redshift of the spectra of ever more distant galaxies—which was for him a clear sign of the expansion of the universe—but also other signals, such as the cosmic rays discovered by Viktor Hess in 1912, which Lemaître considered to be the echoes of his primeval explosion.

The analysis of this radiation had shown that these were not only high-energy photons but also high-energy, partly relativistic particles. In Lemaître opinion, these particles had been hurled from the primordial explosion with high energy and in all directions, and had been moving unhindered through the empty expanding universe in order to eventually bombard the earth as petrified sparks of the primordial fireworks.[23] As he explained:

> The evolution of the universe can be linked to a display of fireworks that has just ended—some few wisps, ashes and smoke. Standing on a well-chilled cinder, we see the fading of the suns and try to recall the vanished brilliance of the origin of the worlds.[24]

Haas, too, was very interested in this kind of radiation and had described it in his book in great detail, discussing the newly identified corpuscular nature of this radiation and its energy distribution. He did not, however, commit himself to the question of the origin of this radiation but referred to Lemaître's thesis,

> according to which the corpuscles of the rays, as the tiniest explosive particles, originate from the cosmic "explosion" that initiated the expansion of the universe.[25]

For the planned Notre Dame conference, these different but also complementary interests of the two cosmologists dictated the selection of invited speakers and the topics of their lectures. The question of the nature and origin of the high-energy cosmic rays, and their connection with the Big Bang as advocated by Lemaître, determined the topics as can be seen from the program in Fig. 21.4. The conference took place on May 2 and 3, 1938 in the newly built Cushing Hall of Engineering, shown in Fig. 21.5, which also housed the particle accelerator.

Arthur Haas invited his friend and Nobel Prize winner Arthur Compton from Chicago to be the main speaker at the conference. He was to talk about the nature of cosmic rays and his latest research results on the question of the intensity dependence between energy and latitude. In so doing, Compton was to also emphasize the importance of the inverse Compton effect; he had just carried out a large-scale

---

[23] Georges Lemaître, "Contributions to a British Association Discussion on the Evolution of the Universe," *Nature* 128 (1931): 704.

[24] Georges Lemaître as quoted in Martin Rees, *Before the Beginning* (Reading: Addison-Wesley, 1997), 47.

[25] Arthur Haas, *Kosmologische Probleme*, 51.

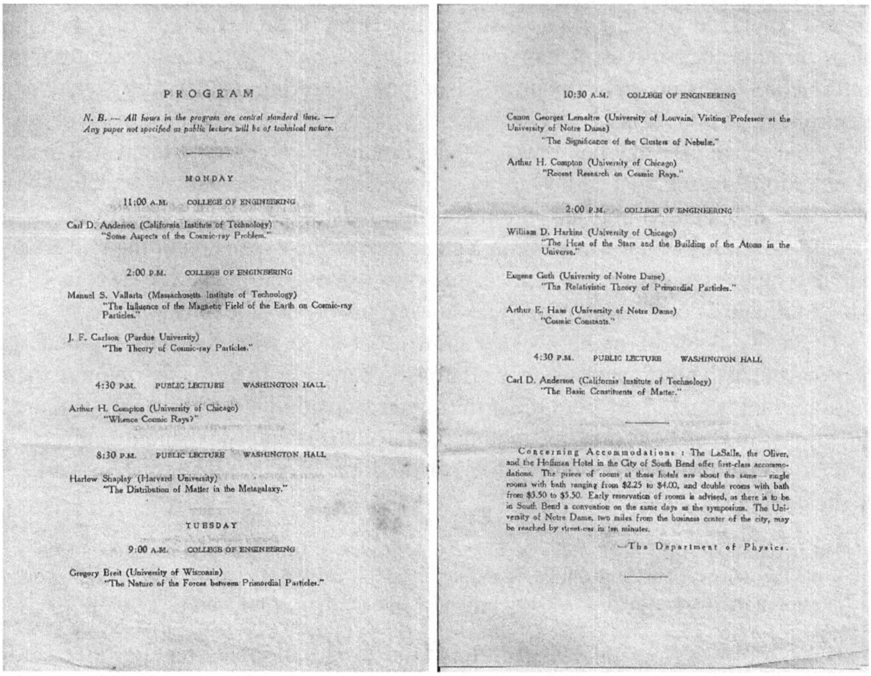

PROGRAM

N. B. — All hours in the program are central standard time. — Any paper not specified as public lecture will be of technical nature.

MONDAY

11:00 A.M. COLLEGE OF ENGINEERING

Carl D. Anderson (California Institute of Technology)
"Some Aspects of the Cosmic-ray Problem."

2:00 P.M. COLLEGE OF ENGINEERING

Manuel S. Vallarta (Massachusetts Institute of Technology)
"The Influence of the Magnetic Field of the Earth on Cosmic-ray Particles."

J. F. Carlson (Purdue University)
"The Theory of Cosmic-ray Particles."

4:30 P.M. PUBLIC LECTURE WASHINGTON HALL

Arthur H. Compton (University of Chicago)
"Whence Cosmic Rays?"

8:30 P.M. PUBLIC LECTURE WASHINGTON HALL

Harlow Shapley (Harvard University)
"The Distribution of Matter in the Metagalaxy."

TUESDAY

9:00 A.M. COLLEGE OF ENGINEERING

Gregory Breit (University of Wisconsin)
"The Nature of the Forces between Primordial Particles."

10:30 A.M. COLLEGE OF ENGINEERING

Canon Georges Lemaitre (University of Louvain, Visiting Professor at the University of Notre Dame)
"The Significance of the Clusters of Nebulæ."

Arthur H. Compton (University of Chicago)
"Recent Research on Cosmic Rays."

2:00 P.M. COLLEGE OF ENGINEERING

William D. Harkins (University of Chicago)
"The Heat of the Stars and the Building of the Atoms in the Universe."

Eugene Guth (University of Notre Dame)
"The Relativistic Theory of Primordial Particles."

Arthur E. Haas (University of Notre Dame)
"Cosmic Constants."

4:30 P.M. PUBLIC LECTURE WASHINGTON HALL

Carl D. Anderson (California Institute of Technology)
"The Basic Constituents of Matter."

Concerning Accommodations: The LaSalle, the Oliver, and the Hoffman Hotel in the City of South Bend offer first-class accommodations. The prices of rooms at these hotels are about the same—single rooms with bath ranging from $2.25 to $4.00, and double rooms with bath from $3.50 to $5.50. Early reservation of rooms is advised, as there is to be in South Bend a convention on the same days as the symposium. The University of Notre Dame, two miles from the business center of the city, may be reached by street-car in ten minutes.

—The Department of Physics.

Fig. 21.4 Program for the conference, "The Physics of the Universe and the Nature of Primordial Particles," which took place at Notre Dame on May 2 and 3, 1938

Fig. 21.5 Built in 1933, Cushing Hall for Engineering Sciences housed the physicists' newly constructed accelerator

study of cosmic rays in the previous years and found that the intensity increased considerably with increasing latitude, i.e., toward the poles.

Haas also invited Millikan from Caltech, who, however, could not come; Gregory Breit (1899–1981) from Wisconsin, who together with John Wheeler (1911–2008) had theoretically predicted the mechanism of pair formation, was also invited. This topic harmonized directly with the interests of Haas since the electron-positron pairs produced in this process had also been detected in cosmic rays by Carl Anderson (1905–1981). And so to demonstrate the experimental evidence for pair formation, Carl Anderson from Caltech was invited; Anderson had just received the Nobel Prize in Physics in 1936 for his discovery. He shared the prize with Victor Hess (1883–1964), the discoverer of cosmic rays, who was an Austrian friend and fellow student of Haas.

John Franklin Carlson (1899–1954), a former student of J. Robert Oppenheimer, had completed his doctorate in 1932 at Berkeley on the question of the origin of secondary cosmic rays. The Carlson-Oppenheimer theory of cosmic-ray showers had laid the foundation for extensive theoretical study in the field of cosmic radiation and its interaction with the atmosphere; Lark Horovitz invited Carlson to Purdue University to establish the field of nuclear physics. For the conference, Carlson gave a lecture on the theoretical interpretation of cosmic rays.

Another friend and good acquaintance of Haas', Harlow Shapley from Harvard University was also invited and he talked about the "Distribution of Matter in the Metagalaxy," discussing the claim that the stellar nebulae that are far away and newly discovered by Hubble comprise independent distant galaxies in the Universe. Shapley had originally been an opponent of this interpretation but under the pressure of observational results, he had abandoned his view of the Milky Way as the solely existing galaxy in the universe:

> The earth's galaxy is but a small sub-division of the system of stars as a galaxy. Hitherto only about 50,000 galaxies have been catalogued. But at Harvard we have been taking a census of the meta-galaxy—meaning, all the galaxies in the universe—and this census has already revealed 300,000 new ones and calculations show that before our survey is completed an additional 300,000 galaxies will be disclosed.[x]

Shapley proposed the term *meta-galaxies* for what is now called a supercluster.

The most important speaker was of course Lemaître, who spoke not so much about his original idea of the primeval atomic model itself but, rather, about the "significance of clusters of nebulae" for understanding and interpreting the expanding universe. Lemaître's friend and close collaborator Manoel Vallarta (1899–1977) from MIT spoke about "The Influence of the Earth's Magnetic Field on Cosmic Ray Particles." Vallarta and Lemaître proposed that the influence of the Earth's magnetic field deflects charged cosmic particles from their orbits and guides them along the magnetic field lines into the polar zones. This concept actually explains the dependence of cosmic-ray intensity on latitude.[26]

---

[26] Georges Lemaître, Manoel Vallarta, "Latitude Effect of Cosmic Radiation," *Phys. Rev.* 43 (1933): 87.

Other speakers included Eugen Guth on the "Relativistic Theory of Judgements," with a special focus on the possible interactions of relativistic electrons. Haas himself gave a lecture on his ideas regarding the "Cosmic Constants," which he had already represented in his book.

William Harkins from Chicago reported on "The Heat of Stars and the Structure of Atoms in the Universe." This was a new topic that showed the connection between stellar energy generation in stars by nuclear reaction and the resulting change in the elemental abundances. This last lecture led the focus from a purely cosmologically oriented conference to the remotely related question of the origin of the elements. The topic had been discussed in March of the same year at a conference organized by George Gamow at George Washington University.

The 1938 conference at Notre Dame, organized by Arthur Haas, was the first meeting of the scientific community to be devoted entirely to the question of cosmology and related phenomena. The scientific journal *Science* took note of the event and published a short summary.[xi] The campus' press also commented on the event's success:

> In addition to the scientists presenting papers, scholars from colleges and universities throughout the country attended the sessions and participated in the discussions which followed the lectures.[xii]

However, despite great interest and the participation of many leading scientists in nuclear physics and cosmology, this meeting did not have the long-term impact that Haas may have hoped for. While the Gamow-organized meeting in Washington motivated Hans Bethe to focus his interest on the synthesis of the elements and to quantitatively formulate the reaction mechanisms of the CNO cycle and the pp chains,[27] the cosmology conference at Notre Dame did not have a comparable motivating effect on the scientific community.[xiii]

This was due to a number of factors. To begin with, cosmology was still considered a somewhat esoteric field at that time; the close connection between the cosmological development of the universe and particle physics was only to become clear forty years later.[xiv] Also, the communication between the scientists involved was not mature enough to reach mutual understanding. Lemaître did not realize from Compton's presentations that the cosmic rays could not be the product of his primeval explosion, and the nuclear physicists present did not connect the observations of the expanding universe with the idea of element synthesis. This idea would be proposed years later by Gamow in his famous Alpher-Bethe-Gamow paper.[28]

---

[27] George Gamow, *The Birth and Death of the Sun – Stellar Evolution and Subatomic Energy* (New York: Viking Press, 1949), 111–116.

[28] Hermann Alpher, Hans Bethe, George Gamow, "The Origin of Chemical Elements," *Physical Review* 73 (1948): 803–804.

George Gamow[29] represented a new young generation of brilliant scientists and he recognized the need for interdisciplinary scientific communication when organizing his conference. By skillfully selecting his speakers, he was able to initiate and facilitate a conversation about the traditional dividing lines between disciplines.[xv] He succeeded not only in drawing and theoretically formulating important quantum-mechanical conclusions that were to become decisive for the field of nuclear physics and astrophysics, but he also wrote numerous popular-science books, such as the Mr. Tompkins series that brought fundamental scientific ideas of the time closer to the general public.

For Haas' part, it is unclear whether he had given up the idea of the energy loss of radiation, which he had so cherished. If so, then this could only have come about with difficulty; the idea of a possibly open dynamic universe was foreign to him and the idea of energy loss via Compton scattering was dear and familiar to him. Although he followed the ideas put forward at the Notre Dame conference, he also turned away from cosmological questions in the following two years and followed the suggestions of William Harkins from Chicago, who seemed to recognize the cosmological importance of microscopic phenomena in nuclear physics. However, personal worries and commitments did not allow Haas to complete these ideas.

Notes

i. Helge Kragh, *Matter and Spirit in the Universe*, 60.

ii. Antoni Romañá, *Los Atrayentes Problemas de la Moderna Astronomía* (Madrid: Redención, Madrid, 1940), 166–167 quoted in Matteo Realdi, "Cosmology, religion, and cultural hegemony: The scientific apostolate of Antoni Romañá in early Francoist Spain," in: *Cultural hegemony in a scientific world: Civil society and scientific culture*, eds. Badino, Massimiliano, and Pietro Omodeo (Leiden, Brill Academic Publishers, 2020).

iii. George Quitman Flynn, "Franklin D. Roosevelt and American Catholicism, 1932–1936," (doctoral thesis, Louisiana State University, 1966).

iv. Brian Greene, *The Hidden Reality: Parallel Universes and the Deep Laws of the Cosmos* (New York: Vintage Books, 2011).

[29] George Gamow (1904–1968) was born in Odessa in czarist Russia. He studied from 1923 to 1925 under Alexander Friedmann, one of the early theorists of cosmological thought at the University of Leningrad. In the following years, Gamow worked in Göttingen, Copenhagen and Leningrad on important topics in quantum and nuclear physics, developed the theory of α-decay by quantum tunneling, and formulated the first ideas on element synthesis in stars. He motivated the construction of the first accelerator at the Cavendish Laboratory and in 1932 helped to develop the first Russian cyclotron in Leningrad. In 1933 he left the Soviet Union and moved to George Washington University where he not only carried out fundamental work with Edward Teller on β decay, the so-called Gamow-Teller Decay, but also organized a successful series of conferences in the style of the Copenhagen Meetings. In the following years, he became a strong supporter of Lemaître's Big Bang theory, especially regarding the question of the element synthesis associated with it. In 1954 he moved to the University of Colorado in Boulder. In these postwar years he published numerous popular-science books before turning to questions of biology and DNA research. George Gamow, *My World Line: An Informal Autobiography* (New York: Viking Press, 1970) and Beverly Orndorff, *George Gamow—The Whimsical Mind behind the Big Bang* (Self-published, Monee, Il, 2013).

v. Victor J. Stenger, *God and the Multiverse - Humanity's expanding View of the Cosmos* (Amherst: Prometheus Books, 2014), 371.
vi. Ernan McMullin, *Evolution and Creation* (Notre Dame: Notre Dame Press, 1985), 44 and Helge Kragh, *Matter and Spirit in the Universe*, 254–255.
vii. Jeremiah P. Ostriker, Simon Mitton, *Heart of Darkness: Unraveling the Mysteries of the Invisible Universe* (Princeton: Princeton University Press, 2013), 66–78 and Helge Kragh, *Cosmology and Controversy* (Princeton: Princeton University Press, 1996), 22–79.
viii. See Jeremia P. Ostriker, Simon Mitton, *The Heart of Darkness: Unraveling the Mysteries of the Invisible Universe*, 84–88 and Helge Kragh, *Cosmology and Controversy*, 142–201.
ix. Louis Hruby, "Einstein and Priest one on Fundamentals – Science and Religion Never Antagonistic—Lemaître," *Notre Dame Scholastic* 66/18 (1933): 12.
x. Quote in George Haithcock, "Dr. H. Shapley Reveals new Facts about Galaxies in Symposium Lecture," *Notre Dame Scholastic* 71/6 (1938): 5.
xi. "Symposium at the University of Notre Dame," *Science* 87 (2257), (1938): 294.
xii. George Haithcock, "Dr. H. Shapley Reveals new Facts about Galaxies in Symposium Lecture," 21.
xiii. Helge Kragh, *Cosmology and Controversy*, 96.
xiv. Steven Weinberg, *Die ersten drei Minuten* (München: Piper Verlag, 1979).
xv. See Michael Wiescher, "The History and Impact of the CNO Cycles in Nuclear Astrophysics," *Physics in Perspective* 20/1 (2018): 124–158.

Chapter 22
German Publishers

"German publisher scum" was the term that Arthur Haas used for his publishers in Germany. He often felt he was being passed over and duped, a feeling that became increasingly strong in the 1930s when strict exchange controls by the Nazi regime made it increasingly difficult to transfer his royalties, which always came late; Haas blamed the publishers and not the new German regulations.

Haas' popular science writings had provided him with necessary extra financial income during the previous decade, without which he probably would have barely been able to survive the subsequent difficult years of emigration between 1935 and 1937. Especially his book *Physik für Jedermann*, published by Julius Springer in its series *Verständliche Wissenschaft* [Understandable Science], was an enormous success. By the summer of 1937, more than 4,200 copies had been sold and the publisher was pressing for a new edition:

> To conclude, it will be sufficient to complete the manuscript for the 2nd edition by April 1, 1938.[1]

These letters came from the Berlin office of Springer publishing house. The enclosed reviews were full of praise for Haas' clear presentation of the subject matter and only suggested an expansion in some aspects.

Arthur Haas worked on the new edition during the summer and fall of 1937, so that the revised and updated manuscript could be submitted in January 1938. This would allow enough time for the book to be completed and distributed by the publisher on schedule in April 1938. Distribution was to be handled by the Vienna Depéndénce, a division of the publishing house of Julius Springer. This division was under the direction of Otto Langer who represented German-speaking Jewish authors, in order to avoid political difficulties in Berlin. On November 25, 1937 another reminder from the publisher came, which assured the distribution of the second edition.

In February 1938, however, Arthur Haas received a brief letter (Fig. 22.1) from Springer Verlag in Berlin:

---

[1] Springer Verlag to AEH (Berlin, July 28, 1937).

M. Wiescher, *Arthur E. Haas—The Hidden Pioneer of Quantum Mechanics*, Springer Biographies, https://doi.org/10.1007/978-3-030-80606-4_22

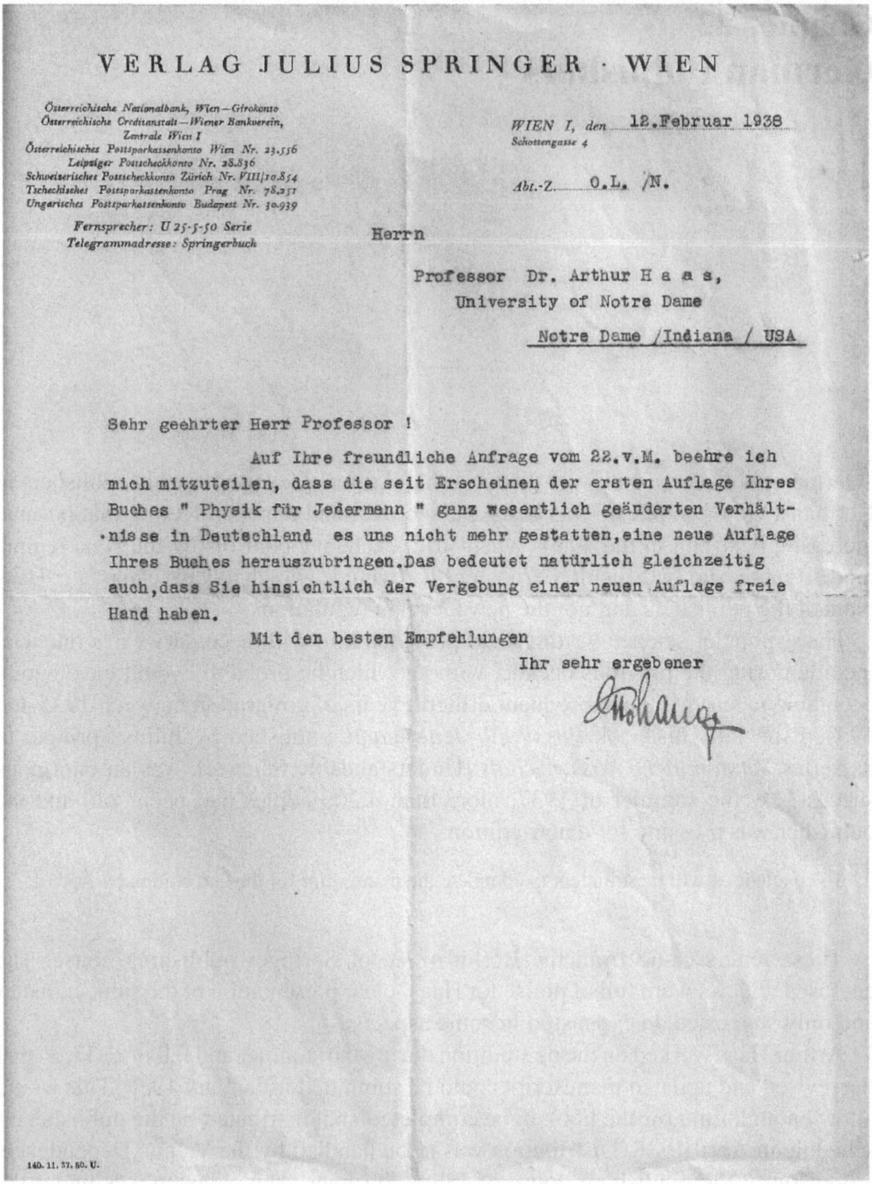

VERLAG JULIUS SPRINGER · WIEN

Österreichische Nationalbank, Wien—Girokonto
Österreichische Creditanstalt—Wiener Bankverein,
Zentrale Wien I
Österreichisches Postsparkassenkonto Wien Nr. 23.556
Leipziger Postscheckkonto Nr. 28.836
Schweizerisches Postscheckkonto Zürich Nr. VIII/10.854
Tschechisches Postsparkassenkonto Prag Nr. 78.251
Ungarisches Postsparkassenkonto Budapest Nr. 30.939

Fernsprecher: U 25-5-50 Serie
Telegrammadresse: Springerbuch

WIEN I, den 12.Februar 1938
Schottengasse 4

Abt.-Z. O.L. /N.

Herrn

Professor Dr. Arthur H a a s,
University of Notre Dame
Notre Dame /Indiana / USA

Sehr geehrter Herr Professor !

Auf Ihre freundliche Anfrage vom 22.v.M. beehre ich mich mitzuteilen, dass die seit Erscheinen der ersten Auflage Ihres Buches " Physik für Jedermann " ganz wesentlich geänderten Verhältnisse in Deutschland es uns nicht mehr gestatten,eine neue Auflage Ihres Buches herauszubringen.Das bedeutet natürlich gleichzeitig auch,dass Sie hinsichtlich der Vergebung einer neuen Auflage freie Hand haben.

Mit den besten Empfehlungen

Ihr sehr ergebener

140. 11. 37. 50. U.

Fig. 22.1 Letter of February 12, 1938 from Julius Springer Verlag in Vienna to Arthur Haas

> I regret to inform you that, in the meantime, I have come to the decision that I will not publish a new edition of your book *Physik für Jedermann* [Physics for Everyone] as Volume 20 of the *Verständliche Wissenschaft* series.[2]

When Haas asked for the reason and demanded an explanation for the sudden cancelation of the project, he only received a brief confirmation from Vienna stating that changing conditions in Germany would no longer permit a new edition. Obviously, the political situation now determined actions for the Vienna branch of Springer as well.

These were the last months of the Austrian state. The annexation of Austria by the German Empire, forced by pressure from Germany, would not take place until two months later in March 1938; existing legal provisions or regulations could therefore not have been the trigger for the decision. Question arises about the reason for this step by the publishing house, which had published with good profit, only one year earlier in Vienna, the aforementioned new textbook *Elementare Physik*. This had been a relatively simple textbook that was not directly suitable for physics students, designed to give medical and chemistry students the necessary basic knowledge of physics. To this end, however, the book also covered modern fields such as X-rays and radioactive decay.

The withdrawal had probably resulted from internal difficulties the publishing house itself was having with the National Socialist regime. Julius Springer himself had a lineage that included three Jewish grandparents and so he was considered a Jew under the racial laws. As such, he had already been forced to resign as a partner in 1935, in the interest of the continued existence of the company.[3] His successor was Tönjes Lange, brother of the Viennese publishing-house representative. In addition, the Munich Higher Regional Court (Oberlandesgericht) had issued a ruling in 1935 that allowed all publishing houses to terminate existing contracts with Jewish authors as shown in Fig. 22.2.

However, Springer Verlag—as a scientific and technical publisher—did have a large number of Jewish authors until well into the 1930s. These included Dane Niels Bohr in Copenhagen, the Germans Max Born and Richard Courant from Göttingen, as well as its most famous author Albert Einstein from Berlin, and many more. While most of these authors had emigrated to the United States after 1933, many of them continued to write their books in German and these were published by Julius Springer. A number of these scientists also continued to publish in the scientific journals of German publishers. In November 1937, Arthur Haas had also published an article on the question of absolute constants in the journal *Naturwissenschaften,* which was published by Julius Springer.[4]

---

[2] Springer Verlag to AEH (Berlin, February 12, 1938).

[3] Julius Springer was arrested in 1938 and spent some time as a prisoner in the Sachsenhausen concentration camp. See Heinz Sarkowski, *Springer-Verlag: History of a Scientific Publishing House* (Berlin Heidelberg: Springer Verlag, 1996), 325–385.

[4] Arthur Haas, "Zur Frage der physikalischen Weltkonstanten," *Naturwissenschaften* 25 (1937): 733–734.

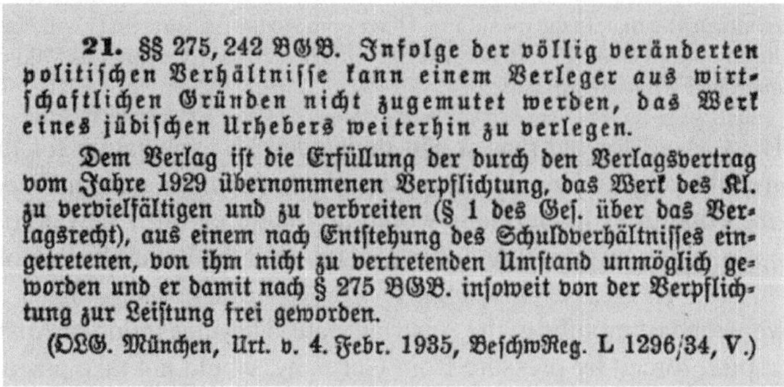

21. §§ 275, 242 BGB. Infolge der völlig veränderten politischen Verhältnisse kann einem Verleger aus wirtschaftlichen Gründen nicht zugemutet werden, das Werk eines jüdischen Urhebers weiterhin zu verlegen.

Dem Verlag ist die Erfüllung der durch den Verlagsvertrag vom Jahre 1929 übernommenen Verpflichtung, das Werk des Kl. zu vervielfältigen und zu verbreiten (§ 1 des Ges. über das Verlagsrecht), aus einem nach Entstehung des Schuldverhältnisses eingetretenen, von ihm nicht zu vertretenden Umstand unmöglich geworden und er damit nach § 275 BGB. insoweit von der Verpflichtung zur Leistung frei geworden.

(OLG. München, Urt. v. 4. Febr. 1935, BeschwReg. L 1296/34, V.)

Fig. 22.2 Decision of the Munich Higher Regional Court stating that publishers cannot be expected to publish Jewish authors for economic reasons

Between 1935 and 1939, in the years following Julius Springer's "retirement," all Jewish authors were removed from the Springer publishing program. Due to the increasing difficulties of publishing Jewish authors in Germany itself, the publishing house first transferred publication of new editions to Vienna in order to continue serving other German-speaking countries. This procedure came to an end, however, with the annexation of Austria. Nonetheless, it is not clear what caused the sudden decision to remove Haas as an author; it seems to have been part of a gradual process to reduce the high percentage of Jewish authors to whom no specific explanation was necessary when contracts were broken. New editions were canceled and the existing stock of these author's books was kept and sold for the publishing house's profit.

The signs for the end of the Austrian state were clear. First Austria's protector the Italian fascist leader Benito Mussolini signed—with Adolf Hitler in November 1937—the Anti-Komintern Treaty against the Soviet Union, followed by an economic treaty with Hitler in December. The price for this was the sovereignty of the small buffer state of Austria. In the course of these developments, Austria's chancellor Schuschnigg was forced to lift the ban on National Socialists and admit the latter as members of the Völkische Front. With this, Catholic fascism had failed in its attempt to stop National Socialism in Austria. The Anschluss was already foreseeable in January 1938 and the publishing house had taken its own steps, in recognition of this inevitability and in the interest of political and economic survival.

Arthur Haas showed little appreciation for these publishing policy considerations; he saw himself betrayed and complained to his brother about the loss of anticipated payments from Springer:

> Since I do not expect any payments from Vienna (the company—from which, as you may know, I expected a large payment for 1938—has turned out to be a horde of word-breaking pigs with whom I broke off contact), my business relations with Vienna appear to have come to an end for good.[5]

[5] AEH to Otto Haas (South Bend, April 31, 1938).

Until the summer of 1938, Arthur Haas tried to place his revised edition of *Physik für Jedermann* with various German-language publishers, in Switzerland (Orell Füssli Verlag, Birkhäuser Verlag) and the Netherlands (Martinus Nijhoff Verlag), but without success. His inquiries remained unanswered.

With the annexation of Austria in March 1938, other German publishers—including De Gruyter in Berlin and the Akademische Verlagsgesellschaft of Leo Jolowicz and his son Walter Jolowicz in Leipzig, which were also expropriated in 1938—also stopped publishing Arthur Haas' books. It was a severe financial setback, even though his annual income at Notre Dame had in the meantime increased to $4,500.[6]

With these events, Arthur Haas gave up his career as a scientific writer; only two attempts for US publication are recorded from his time in the United States. He contacted the Alfred A. Knopf publishing house in 1937, proposing an English edition for his last book, *Elementary Physics* that had been published by Julius Springer in Vienna in 1937. Knopf turned the request down because the topic did not fit into their program:

> I am sorry to inform you that we are declining it. We doubt very much that we would be able to find an adequate American market.[7]

The E. P. Dutton publishing house in New York took a different view; they accepted the manuscript in August 1937 and it was distributed in the English version under the title: *Elementary Survey of Physics; a Non-mathematical Presentation with a Special Supplement for Pre-medical Students.* The book was very positively reviewed by the influential science journal *Nature*.[8]

Yet in contrast to the reception in Germany, the English version was not a success and Arthur Haas did not write another book. He had other worries that demanded his care.

---

[6] Based on the IRS tax forms of Arthur Haas, 1938.

[7] Alfred A. Knopf to AEH (New York, June 30, 1937).

[8] "There is a tendency today to impart a general acquaintance with physical science to students outside this field; even those in the college of liberal arts are now taught the elements of physical science as part of their general culture. A serious difficulty arises from the antipathy on the part of most students, including pre-medical students, to the application of mathematical methods. Prof. Haas has presented a survey of physics without introducing mathematics except of the most elementary nature. In spite of the absence of mathematics, the field covered is comprehensive, and the explanations are so lucid that no one need be deterred by the fear of failure to understand the physical picture of Nature delineated by the author." M.D., Review: *Nature* 144 (1939): 744.

Chapter 23
Vienna Exodus

In the years following the move to South Bend, the family's restless life normalized. The boys had easily become used to the new language, the new environment, their new school and friends. Bice had quickly adapted to the social world of the professors' wives at Notre Dame, which revolved around the university itself, and she was generally popular and appreciated. At St. Mary's College, she had resumed her study of art history and architecture, together with some teaching duties that led to further connections.

Arthur himself was busy preparing his lectures and conferences as well as making the occasional trip to Chicago to confer with his physics colleagues there. He was happy at the university, whose conservative Catholic atmosphere appealed to him and whose pursuit of scientific success harmonized with his own ambitions. He had close connections with the university administration and the university president, which he knew how to use. Life became quieter and more relaxed; the new house was accompanied by a new car, which could be used for tours and excursions in the city and its surroundings.

Part of this new normality, which developed gradually in 1937, was the summer retreat. No longer traveling to the mountain lakes of Austria, the family now visited the countless Ice Age lakes of the Midwest, which stretch like a string of pearls along the hilly terminal-moraine landscape. The family spent the summer of 1937 at Christy Lake, about 70 km north in Michigan. (Fig. 23.1) They had rented a small log cabin through friends and enjoyed uninhibited bathing in the lake, as son George Haas reported in his family memoirs. (Fig. 23.2) However, he also told of the inner restlessness and anxiety that seemed to fill his parents:

> At Christy Lake we were virtually isolated from the outside except for a portable battery-operated radio (a rarity in those days), which Dad turned on about twice a day. While he spent most of the time working on the manuscripts for his new book, he would never miss the latest news broadcasts. Although he felt a strong gratification in what he was able to accomplish for his family during the past year, I think he and Mom felt that the storm clouds in Europe were becoming even more ominous than anticipated and strong concern was felt for family and friends still left in Europe.[1]

M. Wiescher, *Arthur E. Haas—The Hidden Pioneer of Quantum Mechanics*, Springer Biographies, https://doi.org/10.1007/978-3-030-80606-4_23

Fig. 23.1 The log cabins along the shores of Christy Lake where the Haas family spent their summer vacations with new friends and acquaintances from South Bend

Fig. 23.2 The two sons, Arthur Gustav and George at Christy Lake in Michigan, which for four years provided an unspoiled vacation destination for the Haas family and their swimming passion

It was around this time that Arthur Haas began talking to Viennese colleagues, including Eugen Guth and others about the potential availability of a position at Notre Dame and encouraging them to apply. Haas was not the only one who saw and feared future developments in Europe and an increasing number—especially of Jewish professors in Vienna—were looking for alternative opportunities with American universities. Many of these institutions had, however, closed their doors because they had already hired Jewish professors from Germany in the years after 1933; there were not many job opportunities left for the remaining colleagues from Austria.

Arthur Haas' successes—especially the explicit mention of Notre Dame in newspaper reports about the Harvard Tercentenary Conference—had, on the other hand, made a big impression on the university administration. In addition, President O'Hara saw a way to act on his plan to improve the status and visibility of the university by appointing European scholars to the faculty. O'Hara saw Haas as a welcome advisor in his plans and Haas took the opportunity. For the Faculty of Mathematics, he recommended Dr. Karl Menger of the University of Vienna, who was considered one of the outstanding mathematicians of the time.

Karl Menger, who initially studied physics with Hans Thirring in Vienna, had switched to mathematics. Menger received his doctorate in 1924 while working with the Viennese mathematician Hans Hahn (1879—1934) as his advisor; Hahn was one of the actual founders of the Vienna Circle. After a few years in Amsterdam, Menger returned to Vienna as a professor in 1927 and built up the mathematical institute for the university. Through his work, Menger made Vienna a center of mathematical topology, logic, and also economic mathematics—an area of interest presumably inherited from his father Carl Menger (1840—1921), the founder of the Austrian national economy. The younger Menger became a member of the aforementioned Vienna Circle and was thus well known by and friends with Moritz Schlick. In addition to his participation in the discussions and disputes of the Vienna Circle, Menger formed a Mathematical Colloquium in 1928, which brought together the most outstanding mathematicians of the university and included the young Kurt Gödel, Rudolf Carnap and Norbert Wiener.[ii]

Menger had had some experience of American life, having been a visiting lecturer at Harvard University and at the Rice Institute in 1930 and 1931, respectively. Amidst growing hostility toward Jewish scholars, as well as the fear of violence—such as the murder of Schlick, an act that was initially considered politically motivated, and the subtle gloating that occurred in reports about it in the Viennese press—Menger saw himself increasingly as an outsider in the more and more radicalized Vienna. As he would later recall:

> Then in June 1936, Schlick was shot dead in a university building by a psychopathic former student of his. Still under the impact of the tragedy, I described the situation to some friends at the International Congress of Mathematics in Oslo. A few weeks later, I received a cable offering me a professorship at the University of Notre Dame in Indiana. I found it very hard to give up the Colloquium, but in Vienna the time for meetings of the type of ours was obviously running out, while I hoped that I might have the opportunity to develop a similar group in America. And it was very sad to leave Gödel, Wald, and many others, but I trusted that before long I should see them in the free world.[iii]

The offer from Notre Dame was initiated by Arthur Haas, who was well known to Menger, and came at the right time; Karl Menger accepted immediately.

For the first year of his stay at Notre Dame, Menger was considered “on leave of absence” in Vienna. But immediately after the Anschluss he sent a telegram to the University of Vienna, in April 1938, announcing his resignation in order to avoid the certain enforced retirement that, as a Jew, was to be expected.

Menger arrived in South Bend for the spring semester of 1937, where Haas gave him a warm welcome (Fig. 23.3). While they already knew each other well in Europe, the two Austrian families formed a closer bond in the new exile community of South Bend, as George Haas reports:

> Thanksgiving that Fall [1940] was spent with Professor Menger and his family, as we had done for the previous two years. It seems that we all had a lot to be thankful for and Professor Menger especially, who was thankful to Dad for helping him bring his family to America. For this reason, we were always invited to their home for Thanksgiving.[iv]

At Notre Dame, Menger established the doctoral program in the Department of Mathematics and expanded the program by hiring more emigrant mathematicians. He also organized a mathematical publication series, the *Notre Dame Mathematical*

Fig. 23.3 Arthur Haas (left) introduces Karl Menger (center) to the Chair of the Physics and Mathematics Department, Fr. Henry Bolger, C.S.C. (right)

Lectures which quickly became known and served to establish the fame of the young institute. Menger also began a Mathematical Colloquium (modelled on the Vienna Colloquium) and published the related reports in a second series, *Mathematical Colloquium* which appeared between 1938 and 1946. World War II did, however, also affect academic life in the United States and the success of the Mathematical Colloquium was limited. Menger left Notre Dame after the end of the war to become a professor at the Illinois Institute for Technology in Chicago, where he remained until his retirement in 1971.

Menger and Haas succeeded in appointing another mathematician to Notre Dame, Emil Artin (1898—1962)[v] who had already heard lectures by Haas as a student in 1919 in Leipzig and had therefore remained closely associated with him.[vi] (Fig. 23.4) At the time, Haas had given several personally dedicated copies of his books to Artin.[1] Artin originally grew up in Austria and, after military service during the war, had spent his first semesters at the University of Vienna before he went to Leipzig. There he enrolled for several lectures with Haas. From Leipzig, Artin moved to Göttingen in 1921 to study with Hilbert and Courant. In 1925 he was offered a professorship in

[1] The dedication read: Emil Artwin, in sincere friendship, Arthur Haas.

Fig. 23.4 Kurt Gödel in 1939 at Notre Dame (Notre Dame, Archives) and a photograph of Emil Artin from 1938, taken by his wife Natascha Artin (Estate of Natascha Artin Brunswick, Inventory number P2005)

Hamburg where he remained for the next twelve years despite various offers. In 1937, he was dismissed from his professorship because of his Jewish wife and thereupon accepted the offer of a professorship at the University of Notre Dame.

Once assembled, this young group of mathematicians achieved many milestones together. One of these was a mathematical symposium, on the "Algebra of Geometry," which Menger and Artin organized for February 11 and 12, 1938. Lemaître had just arrived and he spoke on the topic "Algebraic Details of the Relativity of Protons and Electrons." The quality of their collaboration is reflected in the successful second volume of *Notre Dame Mathematical Lectures*, which documents Artin's lectures on Galois theory. Artin did, however, take an appointment with Indiana University in Bloomington only one year later. The Catholic atmosphere at Notre Dame did not appeal to him despite the good working conditions. Nevertheless, he traveled regularly, once a week from Bloomington to Notre Dame to continue to give lectures and work with Menger. After the war, in 1946, Artin transferred to Princeton University to continue his career there.

In the spring of 1939, Menger succeeded in getting his former friend and colleague Kurt Gödel to come to Notre Dame for a guest semester.[vii] Menger was dismayed when Gödel insisted on returning to Vienna even with the National Socialists' seizure of power and despite the fact that he had received a good offer from the university. Menger commented on Gödel in his memoirs:

> But I must confess that it was not easy for me to feel the same warmth for him as before.[viii]

But Gödel did not last long in National Socialist Vienna; in particular, a troop of Nazi henchmen beat him because he appeared to be a Jew. When war broke out, Gödel fled via Poland to the Soviet Union, then on to Siberia on the Trans-Siberian Railway. From there he reached Manchuria, then occupied by Japan, from where he took passage via Japan to San Francisco. He then continued by train to Princeton where Albert Einstein had procured him a position at the Institute for Advanced Studies.[ix]

Notes

i. George and Mary Jane Haas, "Letter to our Grandchildren," 75.
ii. Karl Menger, *Reminiscences of the Vienna Circle and the Mathematical Colloquium* (Dordrecht: Springer Verlag, 1994).
iii. Ibid., 215.
iv. George and Mary Jane Haas, "Letter to our Grandchildren," 81.
v. H. Zassenhaus, "Emil Artin, his life and his work," *Notre Dame Journal of Formal Logic* 5 (1964): 1–9.
vi. Karin Reich, *Emil Artin—Mathematiker von Weltruf. In Das Hauptgebäude der Universität Hamburg als Gedächtnisort—Mit sieben Porträts in der NS-Zeit vertriebener Wissenschaftlerinnen und Wissenschaftler*, ed. Rainer Nicolaysen (Hamburg: Hamburg University Press, 2011), 144, 158–159.
vii. Karl Menger, *Reminiscences of the Vienna Circle and the Mathematical Colloquium*, 200–236.
viii. Quoted in Karl Signund, John Dawson, Kurt Mühlberger, *Kurt Gödel—Das Album* (Wiesbaden: Vieweg Verlag, 2006), 170 and Karl Menger, *Reminiscences of the Vienna Circle and the Mathematical Colloquium*, 225.
ix. Matthias Baaz, Christos H. Papadimitriou, Hilary W. Putnam, Dana S. Scott, Charles L. Harper, Jr., *Kurt Gödel and the Foundations of Mathematics—Horizons of Truth* (Cambridge: Cambridge University Press, 2011), 88–91.

Chapter 24
Family Matters

Although Arthur Haas felt satisfaction in the successful recruitment of Viennese physicists and mathematicians, in the face of political developments he became increasingly concerned about his own family—his brother Otto Haas and his sister Margarete Gerstenberger, and their children.

The relationship between Arthur and his siblings appears to have been close on business terms but personally, things were at times a little tense. This can perhaps be explained by the loss of their joint parental property, which had evaporated during the years of economic collapse and the subsequent inflation throughout Austria. There were also tensions between the brothers-in-law and sisters-in-law, which were sometimes expressed in their letters. Neither Arthur nor Otto showed great appreciation for their brother-in-law Edi Gerstenberger in Sattendorf, nor did the two sisters-in-law, Emma Haas and Grete Haas, seem particularly close, which sometimes manifested itself in sensitivities and criticism. The latter became particularly clear when frictions arose between Emma and Otto over questions regarding the Vienna household dissolution.

Otto Haas, as a lawyer, represented the business and financial interests of his siblings and, until 1935, the correspondence with Arthur had mainly concentrated on budget statements and accounting issues. This is understandable since private matters could certainly be discussed at the various family meetings. Arthur's relationship with Grete in Sattendorf was perhaps closer since his family regularly traveled to Sattendorf, to spend their summer vacations, or to the Kanzel Mountains overlooking Sattendorf where they even spent their winter skiing vacations. As a result, closer friendships had developed between Arthur's sons and Grete's three daughters, Maria, Christl, and Erika.

Personal correspondence with Otto only developed after Arthur moved to Brunswick, Maine. This was, however, mostly limited to travel and success stories from Arthur's side, which were followed by complaints about the search for a permanent position, first in Southern California and then at Notre Dame.

The families supported each other, so the brothers and sisters helped with the care of the house on Baumgartenstraße when Arthur and his family were in Maine during

M. Wiescher, *Arthur E. Haas—The Hidden Pioneer of Quantum Mechanics*, Springer Biographies, https://doi.org/10.1007/978-3-030-80606-4_24

his sabbatical year of 1935, and they also had to help a great deal with the dissolution of the estate when Arthur decided to stay in America.

In particular, Otto Haas had to carry this burden. The apartment had to be liquidated, the subtenants had to be sued due to neglect and damage to the apartment and furniture, insurance policies had to be cancelled—Arthur expected reimbursement of his payments, an expectation that Otto immediately identified as absolutely illusional—as well as all of the many other organizational tasks that came with the responsibility. Otto's wife (Greti) and sister (Grete) were also involved in the work. Arthur and Bice wanted to transfer most of the furniture from Vienna to the US because they did not have a high opinion of the quality of American furniture:

> The American furniture is so miserable, not at all durable, and so tasteless that my most urgent wish would be to have, apart from my and the children's bedroom and the kitchen, only Viennese furniture.[1]

Emma spent three months in Europe but she was mostly in Germany. This caused dissatisfaction among the local sisters-in-law, who thought that this reflected a lack of interest in the moving effort. For his part, Otto was dissatisfied with Emma's quick decision on what he considered to be an overly expensive moving company.

The main problem was, however, the financing of the move. To do this, Arthur demanded that Otto sell not only the Polish mortgage bonds but also the Brno houses, which were still owned by the family as a whole. Otto had to make it clear to Arthur that, firstly, his profit expectations were too high and, secondly, the rules for currency exchange would prohibit any transfer of the profit to the United States. As an alternative, an additional mortgage on the houses was agreed upon and the sale was postponed to the following year.

Otto was also instructed to claim outstanding royalties from the German publishers, which was not an easy task either since at that time Austria was subject to special financial boycotts from National Socialist Germany. With a lot of effort, which included personal trips to Brno, Otto did manage to release funds and finance the shipment, as well as transfer the remaining amounts to Arthur in the US:

> According to a letter sent to me here Dris. Weigl [the Czech financial representative] of 7.1. d.M., the cz.-sl National Bank has already authorized the transfer from Kč 7.000.- and Weigl has already ordered it. It may still take about three months until the money arrives in Vienna. But I have no doubt that the financing of your Viennese money needs in autumn will be possible on one side or the other, in the worst case by way of an advance payment by me. [...] Actually, I can't report anything good about myself either: The excitements and over exertions of July have brought me down with my nerves to such an extent that I am far from being on my normal level.[2]

So, it was a constant writing back and forth via sea and *Hindenburg* mail. There were numerous misunderstandings and corrections, with accusations and complaints, which understandably strained the relationship between the families.

[1] AEH to EBH and Otto Haas, Leaflet No. 7 on relocation (South Bend, September 18, 1936).

[2] Otto Haas to AEH (Vienna, August 13, 1936).

Nevertheless, Arthur Haas was worried about his family in Austria and tried several times to convince Otto to at least send the children—Lene, Dorle, and Gustav—to America or at the very least teach them good English:

> I do not believe there is any immediate danger here. But I do believe you would do well and wisely to get your children to speak fluent English after one year, and to consider the possibility of sending Lene [Otto's eldest daughter] to a first-class American college for one year in the fall of 1937, at a cost of about $1,300 if you can afford it, if it seems appropriate at that time.[3]

Otto disregarded these urgings; his head was full of business worries and burdens, and he saw no direct political danger. Neither did sister Grete, who also did not react to Arthur's suggestions; she had other worries at the time, mainly with her family. Her husband Edi had—despite his Jewish wife and despite his daughter Christl who was mentally retarded and therefore considered unworthy of life by the National Socialists—actually just joined the Party. Grete applied for a divorce from Edi Gerstenberger but he would not give his consent at first, constantly building up new objections and barriers. Nevertheless, Grete did manage to get the divorce finalized in the spring of 1937.

Further, since Arthur Haas finally saw himself settled in the United States in 1937, he was keen to liquidate as much as possible of his remaining Austrian property and shares. Therefore, he urged his brother to sell all assets as quickly as possible and as lucratively as possible and to transfer the money to America. Otto, on the other hand, saw the difficulties in implementing his brother's plans at that time:

> Apart from the fact that I can only consider your explanations [...] as a product of your fantasy, which is working more lavishly than ever in the Midwest, they also seem to me to be quite pointless; if we succeed in selling the houses ourselves, we will not be able to transfer the equivalent value to you in the foreseeable future. At most, we may be able to invest it in Czech shares, but I believe that we have long since missed the most favorable moment for this as well.[4]

He also advised against the quick sale of the shares for similar reasons, as the transfer of money had become more and more difficult. However in the more private part of his letter, in response to Arthur's Christmas wishes, Otto reported on his increasing political fears and the family's worries. Otto also wrote about his sense of hopelessness regarding his situation:

> Even an outlook on the year 1937 [...] does not suggest any improvement, so that my basic mood must remain rather cloudy, especially in view of the general and the specifically Austrian political situation. What you write about the conditions there, on the other hand, may fill us with envy here but it does not permit any practical application, at least not to my person. I think you keep overlooking the fact that what was possible for you is not possible for me. Unfortunately, I am not such an impostor as to suddenly, after a twenty-three-year break, strive for a professorship in paleontology in the States, even though I might even be able to obtain one in a country of unlimited possibilities and unlimited acceptance; I have taken too much after our grandfather who, according to our father, told a contemporary

---

[3] AEH to Otto Haas (Dry Mills, July 18, 1936).

[4] Otto Haas to AEH (Vienna, January 20, 1937).

> Rothschild when he offered a professorship in internal medicine at the University of Vienna: "Baron, I know too little for that!" And I do not see any other basis for emigration, wherever. [...]
>
> As far as the situation itself is concerned, I do not really believe in a new world war; the events of the very recent past seem to me to prove that the well-founded fear of it is still too great on all sides, at least for the time being, to allow the certainly all too plentiful cannons to go off by themselves one day. As far as our country in particular is concerned, there is no doubt that it can be overtaken on any day by a "Gleichschaltung" [enforced political alignment on all levels] of one form or another (the "pénétration pacifique" [peaceful penetration] is undoubtedly making slow but noticeable progress), as soon as Italy, for whatever reason, suddenly abandons its interest in Austrian independence; as long as it does not do so, the new "Berlin-Rome Axis," so often referred to in recent times, will, according to the best informed voices in the world press, have the effect of strengthening Austrian independence rather than the opposite.

Thus, Otto recognized the political conditions and dangers but with his trust in the world press, he failed to recognize the degree of danger. In addition, he saw little hope for himself in emigration; he needed to provide not only for his wife and children but also for his sister and her children, who left Sattendorf after the divorce and had moved to Vienna:

> With Grete's divorce, we too, i.e., Greti and I [...] have now had to go through about six very exciting weeks. The degree of meanness with which our brother-in-law has proven himself capable is almost unimaginable. Until yesterday, Grete had lived with us for about 4½ weeks, the last 2½ weeks also Maria [the oldest daughter], while the other two children had to be accommodated in the children's home Trazerberg during that time.[5]

On May 15, 1937, Whit Monday, Uncle Fritz—who had been an important unifying element of the family and the last representative of the family's older generation—died at the age of eighty. Concern for Uncle Fritz had played an important role in Otto's perseverance. He now needed to take over the distribution of the old man's estate, which lead to further correspondence with Arthur. These letters revolved around financial issues and overall, Otto sometimes showed impatience with his brother's demands and inquiries:

> By the way, I consider all your moaning to be grossly exaggerated and I am convinced that if not 500, then certainly the missing 200 dollars will be easily obtained over there by summer. By the way, I also cannot understand why it is so completely impossible [...] to take an advance from the university before raising such a hue and cry and accusing God and the world.[6]

As demonstrated here, impatience with his older brother's concerns had begun to sound through. Otto's frustrations can also be understood as part of his growing concern about his family and the gradually worsening situation in Austria.

The Schuschnigg regime was constantly forced to give in to the economic pressures of National Socialist Germany; the thousand-mark ban—a law requiring that every German who traveled to Austria must pay the German Reich a fee of 1000

[5] Otto Haas to AEH (Vienna, May 4, 1937).

[6] Otto Haas to AEH (Vienna, April 20, 1937).

reichsmark when crossing the border—had led to a collapse of the tourism industry. The lifting of the ban in July 1936 was accompanied by substantial concessions to the National Socialists who had, until then, been banned from Austria. The Party was subsequently able to take on more and more of the key roles in the country, slowly undermining the National Front and the government. This development also undermined Otto's hopes, as he wrote to his brother:

> In particular, news I received from various quarters during my stay in Sattendorf had a deep impact on me, suggesting that the current [Schuschnigg] regime will not last much longer. As I was told, under the seal of the deepest secrecy, the Anschluss is expected to take place in the course of spring at the latest. I have therefore thought long and hard about finding a way out that would at least allow the children to finish their studies abroad.[7]

In the following, he offered to Arthur the entire fortune that was invested abroad. In return Otto asked that his younger children, from three to eight years, be taken in and provided for by Arthur and Emma in America. There was no reply from Arthur. The letter shows Otto's growing concern based on his observations of conditions in Germany; it was clear to him as a Jew that his children would be denied a higher school and university education in the event of Anschluss. It was also clear to him that he personally might not be allowed to continue as a lawyer. Over the previous year, his income had already decreased considerably and he was pessimistic about the future.

In February 1938 the sale of the Brno houses finally worked out. With the help of his friend in Brno, the businessman Hans Beth, Otto Haas succeeded in transferring a portion of the proceeds from the sale, moving funds from Czechoslovakia to Austrian and American accounts. He managed this just in time; in March 1938, German troops, with the help of Nazi sympathizers in the country, removed the barriers at the borders; annexation to the German Reich was completed in a coup d'état and the Austrian population cheerfully welcomed—at least in large part—the German occupiers as seen in a newspaper clipping, Fig. 24.1.

Although this development had seemed inevitable for months and Otto had spoken of it several times in his letters, it still came as a shock to him; despite warnings and admonitions from his brother, he was not prepared for anything. His eldest daughter Lene had not yet graduated from high school and her plans for a college degree evaporated; the completion of schooling for the two younger children, Dorle and Gustav, was also in question. Otto's professional license as a lawyer was not immediately revoked but it was limited, and he was ordered to vacate the premises of his law office on Stubenring within six months. It was an economic and emotional catastrophe for Otto Haas, which he could barely grasp.

Even Arthur could not quite grasp the extent of the catastrophe. While Otto begged him to look for ways to bring the entire family to the US as soon as possible and to create affidavits for them—legally binding carte-blanche letters with an obligation to provide accommodation and livelihood for the respective persons if they could not find a job—Arthur hesitated. Contrary to better knowledge, which he himself had already stated in many letters, he did not want to believe in the long-term dangers

---

[7] Otto Haas to AEH (Vienna, October 23, 1937).

Fig. 24.1 The German conqueror waves to Austrian girls on his way along the Danube to Vienna. Some raise their arms in a Hitler salute, but one turns away and hides her face in her hands

for his brother. A letter from Hans Beth—who informed Arthur about the situation and his brother's plans, and who probably also tried to calm Arthur down—was interpreted as an affirmation that the situation was not as dangerous and uncertain as Arthur himself had always believed. Beth stated:

> He [Otto] himself wants to gradually liquidate his assets in Austria, in order to get at least a few cents in return for the emigrant's mark! As a war veteran, he remains admitted to the bar, but the legal profession certainly does not give the admitted Jew enough to live on. His means are such that he cannot think of retirement, with a family to care for… but a modest income would help to slowly increase his capital income enough that he might not need to attack the capital itself.[8]

Although the letter asked for help in preparing affidavits for Otto and his family, Arthur felt reassured. The letter showed that Otto was doing sufficiently well, that he could continue to work and that he probably had sufficient assets abroad—so the situation was only half as bad as Arthur had thought and he felt free to think about alternatives that seemed better to him, in view of the situation. Stay calm and wait, this was the advice that Arthur had to give:

> As for O., there are two options. I do not believe that the current regime, which determines O.'s fate, will last very long. Therefore, with several years of perseverance, he could enter a period where he could maintain himself with his old profession—he will of course be even older then and the question will be whether, after the great cleaning of O's hometown, it will still be in as good a condition as it was in 1919.[9]

---

[8] Hans Beth to AEH (Amsterdam, April 21, 1938).

[9] AEH to Hans Beth (South Bend, April 30, 1938).

This view, that the Nazi ghost would pass quickly, was held by many of Vienna's Jews but it was based more on wishful thinking than on a realistic analysis of the situation. Arthur, too, was addicted to this wishful thinking; in his economically limited situation, he saw himself unable to take in and provide for his brother Otto and his family and possibly also for Grete and her daughters. He blocked the idea with the excuse that Otto, as a war veteran and lawyer with financial resources, could not be so badly situated. Arthur refused to believe his brother and cited as an excuse that only one affidavit was allocated to him. He explained all of this in the same letter to Hans Beth.

> I have already sent an affidavit to Otto. Whether it will be of any use I do not know. The fact that I myself have three dependents (people I have to provide for, namely my wife and two children) and the number of dependents has to be stated, is a serious obstacle, as well as my low income and the fact that I am not yet an American citizen, and I am apparently stateless.

Arthur didn't recognize the fear and panic that Otto was going through and did not want to see the danger. Arthur was looking at the situation from his position and his experience, remembering the difficulties that he had had with his job search in the USA. The United States had since moved into another economic recession and European lawyers such as Otto, without knowledge of the American legal code, were not even making it onto the list of people who could apply for a visa. One remaining hope was Otto's second field of study, paleontology, which he had also completed with a doctorate before needing to take over his father's law practice. But who wanted to hire a fifty-one-year-old paleontologist without any professional experience? These were Arthur's concerns and worries.

They were understandable concerns. In addition, many of the crimes perpetrated by the National Socialists, while they are known today, had not yet entered the public consciousness. Hope was closing the eyes of many citizens on both sides. For his part, Otto did not want to immediately leave the country. Rather, he hoped to close his business and financial affairs properly, in order to be able to take as much money and property as possible with him from Austria. He considered this an additional defense, reassuring himself that there was no need for haste.

So too, Arthur wanted to wait and see how things would develop. Efforts toward an immediate immigration of the entire family to the US did not seem necessary to him.[10] Moreover, his conference on cosmological issues was rapidly approaching; he had worked on this for a long time and it required all of his attention. Waiting seemed to be the most sensible way to proceed.

But in the meantime, he remained under the pressure of these family concerns. Arthur Haas first focused his efforts on transferring Otto's eldest daughter to the America. Secondly, Arthur worked on finding a permanent position for his brother, so that he could emigrate with financial security.

---

[10] That Arthur was not entirely wrong in his assessment of the situation is proven by an analysis of the emigration of older Jewish emigrants from Austria and their difficulties in adapting to their new American living conditions. See Melissa Jane Taylor, "Family Matters: The Emigration of Elderly Jews from Vienna to the United States, 1938—1941," *Journal of Social History* 45/1 (2011): 238–260.

Arthur wrote countless letters to people he knew at universities and museums, exploring every possibility for his brother; the answers were negative and, as expected by Arthur, gave little hope:

> Dear Dr. Haas:
> I am sorry to have to write you that we have no vacancy in our department of paleontology at the present time. It is also unfortunate that should there be one we would have to give a competitive examination under the U.S. Civil Service, which is open only to Americans.[11]

No position, no examination, no US citizenship—that was the typical answer Arthur Haas received from the various museums and institutions he approached about a position for his brother. Arthur tried to reassure his brother and make him aware of the difficult situation in the USA. However, he still assumed that the situation was less bad than described and that financial means were available as support.

> One problem that is always being discussed vividly in the USA is the desperate problem of job creation for domestic people who have already passed the age of forty. Almost all of them are doomed to unemployment once they have lost their jobs and do not have special skills. [...] If the unbelievable miracle should happen that you could even get a job in the USA, it would be so badly paid that you could hardly live on it alone, let alone with a family. Just keep in mind that the salaries of the most famous scholars are far below the average income of a man with a college education, and again you can only count on a small fraction of the income of such a scholar.[12]

No doubt, this was not a particularly encouraging letter that could give hope to the desperate Otto. It probably was not meant to be encouraging; from his own experience, Arthur considered any chance of Otto getting an academic position in paleontology to be absolutely hopeless. Arthur saw the family's hope in Lene, the eldest daughter, who would have the best chance of moving to the US as an Au-Pair or college student. She could then use the college years to gain a foothold, marry if possible and then later move the rest of the family. He recommended concentrating exclusively on this possibility because he saw it as the best chance for success.

Regarding Otto, Arthur recognized no danger. As a decorated front-line fighter in the world war, he considered Otto safe and recommended, in the same letter, that he wait in Austria with the rest of the family:

> You should do it with your wife and not-yet-grown-up children, as long as you can work in Vienna or live from your investment capital, at least until Lene can arrange for you to emigrate. But if you have the possibility to work for any European company (Czech, Hungarian, Polish) in Vienna, do it with all your means. [...] Above all, you should keep a calm head and not rush into anything rash. Work on Lene's flight with all your energy and sacrifices, which might prove to be the best investment. After all, only 24,000 are expected to emigrate annually and you are in a privileged position several times over, mainly because of your front-line earnings.

This was now the plan on which Arthur placed all of his efforts. He used all of his connections, including connections from his IIE days, to find a place for Lene

[11] Smithsonian Institute to AEH (Washington, DC, May 31, 1938).
[12] AEH to Otto Haas (South Bend, May 8, 1938).

at an American college, if possible free of charge. For this reason, he had issued an affidavit for Lene, to give her a better chance.

Otto also talked to all kinds of acquaintances in Vienna who, like Arthur ten years before, told him about the golden opportunities in the USA. They gave him countless positive examples from his circle of acquaintances—how one could easily immigrate to the US and just as easily get a permanent position.

This led to misunderstandings and misinterpretations between the brothers. Arthur pressed for documents, certificates and forms—for both Lene and Otto. For his part, Otto worried about the financial aspects and hoped for immediate relocation with quick employment. Also, as a father he had little sympathy for Arthur's frequent reference to a possible quick marriage for Lene, who was described by all sides as beautiful. This drove Arthur to apply renewed pressure:

> With respect to Lene, I would like to advise you once again to do everything, and if you have to sacrifice almost everything to give her a year of study at an American university, you should do so. College. If she applies for a job here personally—from a distance it's not possible at all—with a European high school diploma, she can become a parlor maid, maybe a nanny, waitress and the like [...] If she applies for a job after she has been in a college for one year, she has incomparably better chances not to mention that she creates a wealth of relationships in college; with brilliant academic success comes the possibility of a free continuation of her studies and moreover, she acquires excellent chances of getting married. You unfortunately never had the ability to make generous decisions, have always thought too much about details, about hours instead of years—don't miss this opportunity that never returns.[13]

Otto wrote to various American aid organizations, including the Institute for International Education, and sent application papers, while Arthur continued to write directly to museums and other institutions to ask for at least a temporary position for his brother as well as, if possible, a free place to study for his niece. Both options bordered on impossibility, as he knew. He wrote to his brother more often, always reviewing the politely formulated refusals.

Finally, Wellesley College responded with an offer; the good connections that Arthur had fostered there paid off. The college administration agreed to consider an application from Lene and, after Arthur's repeated reminders, the application was submitted on June 17, 1938. A month later the college's approval arrived, guaranteeing her a scholarship for one year. On July 26th, she was sent a visa so that she could arrive at Wellesley in time for the beginning of the semester on September 12th. The plan was for Lene to travel to Holland and then continue to the USA.

Now it was time to concentrate on Otto who had also written to various institutions but, due to his lack of academic references, was only met with rejection and incomprehension regarding his request to be assigned to a museum.

Betty Drury, the secretary of the Emergency Committee in Aid of Displaced German Scholars wrote to him:

> The difficulty, however, as far as our own committee is concerned, is this: with the best will in the world, we cannot attempt to "place" you in any sort of position, whether academic or

---

13 AEH to Otto Haas (South Bend, May 19, 1938).

> curatorial. We are rigidly restricted by our rules and by those policies which it has deemed necessary, in view of conditions in this country, to adopt, from trying to find positions for any German scholars whatsoever.[14]

Rev. Joseph Ostermann of the Committee for Catholic Refugees from Germany also wrote:

> We regret that you are of the opinion that a position has to be waiting for you upon your arrival. Please be informed that it is very difficult to secure a position for anybody that is still residing abroad.[15]

It was therefore a matter of restoring Otto Haas' scientific reputation in the field of paleontology. So, Arthur asked Otto to write at least one or two papers, demonstrating that he was still scientifically active:

> Publish a paleont. paper as soon as possible, however short it may be, and have Prof. Julius von P.[16] immediately send it to the Academy of Sciences, of which he is a corr. member, for the *Anzeiger* [...] since a work sent in by a corr. member is automatically considered accepted, [...] if you can do this with two papers, even better. [...] as soon as one or two new papers of yours are accepted by certain journals, thus proving that you are really not rusty, I will ask in Pittsburg and Chicago if they would like to offer you a job.[17]

The growing stress of the situation resulted in continued, mutual accusations and yet, everyone worked and wrote to the point of exhaustion. As a Jew in Vienna, Otto Haas was under constant threat both officially and privately. He was also under increasing official pressure to clear out his office. In addition, as part of the harassment inflicted by the authorities, Otto had to appear in person for each administrative step and request. This involved going to several offices each time and thus exposing himself to humiliating conditions and treatment. All of this while writing several letters every day, often asking his sister for advice because of his imperfect English. At the same time, Arthur was still organizing his cosmology conference while keeping up his scientific work and also writing daily letters to help his brother.

Communication between the brothers became more irritable and impatient and the misunderstandings and misinterpretations became more frequent. The quarrels went so far that Otto's wife Greti asked Arthur for his help in a personal letter. Asking Arthur for moderation, she wrote:

> You have no idea how much I worry about your brother. Please write everything unpleasant, hopeless, etc. that you have to tell him either to me and I will tell him in a way that does not hurt him so much, or write it to him, if it is possible for you, as one treats a sick person who is totally down with his nerves. I must speak very clearly, otherwise you may not understand me either; on many occasions your B. asks me to give my consent to end his life! When a

[14] Emergency Committee in Aid of Displaced German Scholars to Otto Haas (New York, NY, July 12, 1938).

[15] Committee for Catholic Refugees from Germany to Otto Haas (New York, NY, July 28, 1938).

[16] Julius von Pia (1887—1943) had been since 1927, Associate Professor and since 1937, Professor for Paleontology. He served from 1913 to 1943 in the Geological-Paleontological Department of the Natural History Museum.

[17] AEH to Otto Haas (South Bend, July 6, 1938).

> letter comes from you and he has read it, this request becomes more and more serious and urgent.[18]

Greatest misunderstanding arose from the fact that Arthur had lost all hope in Otto; as expressed in various letters, Arthur saw no hope of his brother making a living in the US. This may explain his insistence that Otto look for a job that Arthur himself considered impossible to find. He expressed his pessimistic views quite clearly in a letter to Otto's school friend Hans Beth in Brno, who had found refuge in Holland and was looking for emigration and investment opportunities in Australia, France, or even the United States. Although Lene was to stay with Beth before she left for the US, Arthur wrote regarding Otto:

> I no longer have the slightest hope that O. [Otto] will be able to build a new future in the USA, especially since his flexibility, agility and adaptability, which had never been great anyway, has diminished immensely under the impact of recent events. [...] Since I do not believe (which would be possible) that he has a chance in another country, and since I do not believe that L.'s [Lene's] mother could become fit for work, I unfortunately believe in no other possibility of salvation than for L.'s parents to spend the rest of their lives as modest pensioners [...] A life annuity for L's parents, with the greatest modesty and living in a very cheap country (Dalmatia?), would require a capital of about $30,000, corresponding to a monthly annuity of $100, which I believe is enough in Dalmatia, for example.[19]

Arthur Haas tried to build up the capital for such an account, in order to first finance Lene's studies and to then support his brother in distant Europe for the longer term. Hans Beth had agreed to pre-finance Lena's trip and also offered to help transfer funds, that Otto had invested, to this account in the US. Arthur wrote,

> As I already wrote to you, I will [...] have the account opened in the name of L. (other possibilities would be L. and her younger sister together or L. and O. or O. alone) and there will be me in the first place and in the second place my colleague Dr. Karl M. (who will also be the cash auditor) as trustees. As I already wrote to them, I will invest the money, depending upon the earnings, on account or in savings bank deposits (1½% per annually) or in the form of $1,000 divisible U.S. government securities (interest approx. 2¼%).[20]

Arthur had, however, underestimated his brother. As he wrote to Arthur in September, Otto Haas succeeded in sending his two younger children to England,

> where, thanks to the helpfulness of local friends, they move into schools, Dorle to the Girl's High School of the Convent of Our Lady of Dolours, Cullompton in Devon and Gustav to the Norwood School, Pennsylvania, Exeter. [...] Just at the moment I am writing this, Lene is expected to arrive in Rotterdam to embark on the *Statendam* from there. So we are scattered all over the world, "gone with the wind"![21]

---

[18] Greti Haas to AEH (Vienna, August 26, 1938).

[19] AEH to Hans Beth (South Bend, August 11, 1938).

[20] AEH to Hans Beth (South Bend, August 26, 1938). It is interesting to note the reference to Karl Menger as a participant in the investment account, pointing to the friendly bond between Haas and Menger.

[21] Otto Haas to AEH (Vienna, September 9, 1938).

In addition to Otto's two children, Grete's eldest daughter Erika was also successfully sent to England.[22] It seems that Otto could then think about not spending his days as a pensioner in Dalmatia, since he resumed writing letters on his own behalf. In addition to his brother, he wrote to various American aid organizations looking for a way to make his move possible. The last letter from Vienna is dated October 29, 1938.

During the night between November 9th and November 10th, the pogroms against the Jewish population in the German Reich began with the so-called Reichskristallnacht.[23] Officially, the pogroms were declared in response to the assassination of the German Legation Secretary in Paris, Ernst von Rath, by the Polish Jew Herschel Grynszpan. The Reichskristallnacht manifested in brutal riots against the Jewish population. Jewish stores and synagogues throughout the Reich were shattered but the destruction was carried out with particular brutality in Vienna. More than 4,000 stores were destroyed. Thousands of apartments were looted, Jewish residents beaten, expelled, and arbitrarily arrested, while synagogues and prayer houses were robbed and set on fire. Almost 4,000 of the Jews arrested in Vienna were deported to the newly built concentration camp in Dachau. The perpetrators were HJ members, SA and SS units, as well as citizens of the city who joined the lust for robbery and violence.[i]

The events in the city were recounted by Carl Zuckmayer (1896–1977) in his autobiography, which was quoted by André Heller (b. 1947) for the 80th anniversary of this event on March 12, 2018; the commemorative event was organized by the Austrian federal government in Vienna:

> That evening all hell broke loose. The underworld had opened its gates and unleashed its lowest, most hideous, most unclean spirits. The city was transformed into a nightmare painting by Hieronymus Bosch: lemurs and half-demons seemed to have crawled out of dirty eggs and risen from boggy holes in the ground. The air was filled with an incessant shrieking, wild, hysterical shrieking, from the throats of men and women, which continued to shrill for days and nights. And all the people lost their faces, resembled distorted grimaces; some in fear, others in lies, others in wild, hateful triumph. [...] What was unleashed here had nothing to do with the 'seizure of power' in Germany, which outwardly appeared to be taking place legally and was received by part of the population with alienation, with skepticism, or with a clueless national idealism. What was unleashed here was the uprising of envy, resentment, bitterness, blind malicious vindictiveness - and all other voices were condemned to silence. [...] Here nothing was let loose but the dull mass, the blind destructiveness, and its hatred was directed against everything ennobled by nature or spirit. It was a coven of the mob and a burial of all human dignity.[ii]

[22] Such transports allowed and organized the departure of over 10,000 children, who were considered "Jewish" in the sense of the Nuremberg Laws, from Germany to Great Britain. In this way, primarily children from Germany, Austria, Poland, the Free City of Danzig, and Czechoslovakia could escape into exile. While the children were able to leave the country, most of them never saw their parents again. In the case of the family described here, their lives were happier because the parents were able to follow.

[23] *Kristallnacht* or the "Night of Broken Glass," named for the many windows that were shattered all over the German Nazi Empire.

How Otto Haas and his wife survived the night and the following weeks is not known. They were Catholic and therefore probably did not correspond to the expected Jewish stereotype, but this was only short-term protection. It very quickly became clear that a privileged position, as assumed by Arthur Haas, would not guarantee long-term protection for Jewish veterans.

There are no letters between the two brothers for the months of November and December 1938. During these months thousands of Jewish families tried to leave Vienna. Many received help from the Jewish Community Vienna (IKG), which tried to coordinate the emigration storm and provide assistance.[iii] However, Otto Haas and his wife Greti were Catholic and not members of the community. They did not consider themselves Jewish, or they at least saw themselves as culturally far superior to the Galician Eastern Jews, and it must have been an additionally humiliating shock to suddenly be reclassified as Jews.

Many Viennese Jews joining together through family connections and social networks did, however, successfully organize the exodus of several thousand Jews from Austria. Twenty-five thousand Jews applied to emigrate in the first days after the Reichskristallnacht. The shock had its effect and they were now ready to go anywhere, pay any price and leave everything behind in order to ensure their safety. Within a few months, a fully functioning system of mass emigration was in place, requiring the support of foreign aid committees and the full cooperation of Viennese Jews regardless of their religious affinity.[iv]

In Otto's case, it was probably through his association with Dr. Fritz Demuth (1876–1965)—who was the head of the Notgemeinschaft Deutscher Wissenschaftler (Emergency Association of German Scientists) in London and a close acquaintance of Margarethe Haas'—that the needed help was provided in obtaining the necessary papers. This connection was not so much a result of his earlier scientific career but because of his connections to the Catholic Church.

Otto Haas did not make any mention of this and he sketched these months only briefly and in a business-like tone:

> As a result of the upheaval of 1938, my wife and I sent our eldest daughter to the U.S.A. and our two younger children to England, where we succeeded them on April 2, 1939.[24]

Nonetheless, these must have been desperate months of waiting in National Socialist Vienna; privileged Jews who could afford to and who had the necessary connections abroad were able—if only with difficulty and through these connections—to obtain the necessary papers and affidavits to escape from Vienna to England. England was the preferred, easier destination since she maintained a liberal immigration policy; the United States, especially the State Department, made it increasingly difficult for Jewish emigrants and refugees to enter. In Austria this

[24] Otto Haas, "Kurze Lebensgeschichte," (unpublised Typoskript, Reno, 1969).

exodus met with few barriers from German authorities until 1941, as long as property and assets remained in Germany. Otto and Greti Haas only succeeded in leaving Austria by leaving behind the majority of their possessions.[25]

In England, Otto Haas initially had only very limited financial means, although he had invested some money in Swiss accounts. He was, however, lucky or he had good connections because already in May 1939, he found a temporary position as a paleontologist at the British Museum in London. Although a short-term project, he helped to catalog a collection of Eocene nautilids from British Somaliland.

The immediate danger was over but war seemed imminent, and the Haas family was particularly concerned about an invasion of England by German troops. In fact, the Gestapo (the secret police in Germany) had compiled long lists for the arrest of all former German citizens after the successful occupation of the British Isles. Therefore, the important thing was to flee further to America which was believed to be safe.

Even though Otto and Greti Haas did not travel to England until April 1939, Lene had written to her uncle Arthur as early as December 1938, saying that she had successfully obtained an affidavit for her mother. Arthur then applied for an additional affidavit for his brother's entry into the USA:

> I am very happy that you were able to procure an affidavit and that you were able to get your mother a position for after her arrival. Hopefully everything will work out fine. [...] Also, you should have told me exactly whether the affidavit is for your father or your mother, since I now want to do something myself.[26]

But things were delayed again because, in the beginning, there were no employment opportunities in the United States that offered the family economic security. Finally, Otto and his family were granted a visa and in December 1939, several months after the outbreak of war, they managed to reach the US. As Otto recalled,

> In December 1939, we moved to the United States, where we landed on December 30th. From January to August 1940, I worked at the Museum of Comparative Zoology of Harvard in Cambridge, Mass., then in September of the same year, at the Field Museum of Natural History, Chicago. On October 1, 1940, I began my service at the American Museum of Natural History, New York, initially as a Research Associate.[27]

Otto wrote to Arthur several times in April 1940, mainly to tell him about the gradual adjustment of the family and also to inform him of a possible offer of a permanent position at the American Museum of Natural History in New York, since the position at Harvard was also limited. The New York position was delayed and Arthur managed to secure employment for Otto through his colleague Theodor Just, working for a month in Chicago at the famous Field Museum.

This provided the only opportunity for the brothers to again meet. Otto spent a weekend with Arthur and his family at the Indiana Dunes, an extensive dune area

---

[25] The mostly personal material that was successfully shipped only reached Rotterdam where it remained due to the outbreak of war and was then destroyed in the German bombing of the city and port in May 1940.

[26] AEH to Lene Haas (South Bend, January 2, 1939).

[27] Otto Haas, "Kurze Lebensgeschichte."

Fig. 24.2 Arthur Haas and his brother Otto with the young Arthur on the left side and the tall George Haas on the right side, in September 1940 on the steps of the house at 934 Riverside Drive

located between Chicago and South Bend along the southeast coast of Lake Michigan. Otto also traveled to South Bend to visit Arthur's place of work. On this occasion, the last picture of the two brothers, together with Arthur Haas' two sons, was taken on the steps of his house, Fig. 24.2.

Contrary to his brother's original expectations in 1938 and despite the dramatic changes in his life, Otto was to have an extremely successful career as a paleontologist during the next two decades.

Unlike Otto, Arthur was never to see his sister Margarethe Gerstenberger again. She spent the turbulent days in the autumn of 1938, after her eldest daughter Erika had been sent to England, with her two younger daughters, Christl and Maria. They stayed in Sattendorf under the protection of her former husband but then managed to leave Austria with her brother and his wife in early April 1939. Once in England, Margarethe had to make a living on her own. Arthur initially tried to support her by means of bank transfers (Fig. 24.3), yet it was probably a traumatic new beginning for the doctor of history. She had to earn her living in the first few years by doing housework and cleaning, without her own means of support from other assets.

Notes

i. Christa Zöchling, *Novemberpogrom in Wien. Keiner der Brandstifter und Mörder wurde nach 1945 vor Gericht gestellt*, accessed on January 31, 2021, https://www.profil.at/oesterreich/novemberpogrom-in-wien-keiner-der-taeter-wurde-vor-gericht-gestellt/400889639.

ii. Carl Zuckmayer, *Als wär's ein Stück von mir – Horen der Freundschaft* (Frankfurt: Fischer Verlag, 1969), 71.

iii. Melissa Jane Taylor, "Family Matters: The Emigration of Elderly Jews from Vienna," 242.

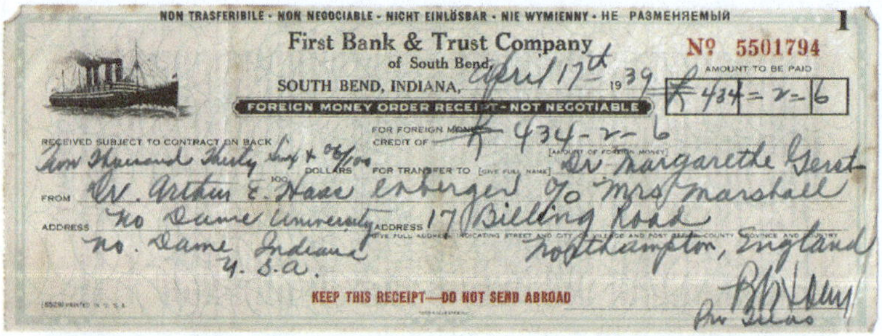
NON TRASFERIBILE · NON NEGOCIABLE · NICHT EINLÖSBAR · NIE WYMIENNY · НЕ РАЗМЕНЯЕМЫЙ

First Bank & Trust Company of South Bend, SOUTH BEND, INDIANA, April 17th 1939

Nº 5501794

AMOUNT TO BE PAID £ 434 = 2 = 6

FOREIGN MONEY ORDER RECEIPT - NOT NEGOTIABLE

RECEIVED SUBJECT TO CONTRACT ON BACK Two Thousand Thirty Six + 06/100 DOLLARS

FOR FOREIGN MONEY CREDIT OF £ 434 - 2 - 6

FOR TRANSFER TO Dr. Margarethe Gerstenberger c/o Mrs Marshall

FROM Dr. Arthur E. Haas

ADDRESS No Dame University No. Dame Indiana U.S.A.

ADDRESS 17 Billing Road Northampton, England

KEEP THIS RECEIPT—DO NOT SEND ABROAD

Fig. 24.3 Transfer of £ 434 from Dr. Arthur Haas in South Bend to Northampton in England where Dr. Margarethe Gerstenberger and her three daughters had found accommodation

iv. Ilana Fritz Offenberger, *The Jews of Nazi Vienna, 1938—1945*, *Palgrave Studies in the History of Genocide* (Cham: Palgrave Macmillan, 2017), 31–209.

Chapter 25
Last Plans

In the summer months of 1938—after the successful conclusion of his cosmology conference in May and the departure of Lemaître back to Leuven, and while under constant strain, uncertainty and worry about the fate of the family in Vienna, with the emotional pressure of bringing his brother and family to the US—Arthur Haas buried himself further in his work.

He had found a new goal worthy of his attention. His doubts about the validity of the expansion of the universe, which he had developed and advocated since the Tercentenary Conference in 1936, had been dispelled by Lemaître's visit and reinforced by the statements and results of the conference at Notre Dame. The explanation of redshift as a light that is weakened by Compton scattering had been an interesting idea, which he had particularly appreciated in order to maintain the concept of a closed static universe. But now this idea had to give way to expansion. In addition, the lectures and discussions at the conference had led Haas back to the microscopic world that he had left behind with his focus on cosmology. He was to return to the field of nuclear physics, motivated by the lecture of the nuclear chemist Harkins during the conference.

William Harkins[1] was regarded as the leading nuclear physicist or nuclear chemist during the first decades of the twentieth century, at a time when most American physicists still had no understanding of or appreciation of the quantum-physical phenomena of atomic nuclei. As early as 1917, Harkins argued for the synthesis of elements in stars based on his analysis of meteoric materials. At that time, he had

[1] William Harkins (1873—1951) enrolled in 1896 to study chemistry at Stanford University, graduating in 1900. From 1898 to 1900 he was an assistant, then lecturer of chemistry while at Stanford. For the next twelve years, Harkins taught as professor of chemistry at the University of Montana in Missoula but he spent considerable time on research at the University of Chicago and Stanford University. For this work, he was awarded his doctorate from Stanford in 1908. He conducted research in Germany in 1909 and worked as a research assistant at the Massachusetts Institute of Technology from 1909 to 1910. In 1912, at the age of thirty-nine, Harkins left Missoula and went to the University of Chicago where he spent the remaining thirty-nine years of his life doing research. Robert S. Mulliken, "William Draper Harkins – 1873—1951," (National Academy of Sciences, Washington, DC, 1975).

M. Wiescher, *Arthur E. Haas—The Hidden Pioneer of Quantum Mechanics*, Springer Biographies, https://doi.org/10.1007/978-3-030-80606-4_25

also discovered the odd-numbered effect in the elemental abundance distribution. He first tried to build up the more complex heavier elements from basic components like hydrogen and helium, to establish the systematics for a new periodic table.[2] In modern nuclear physics, this would be described as a mixture of the single-particle and alpha-cluster models of atomic nuclei. He based his arguments on results concerning the mass ratios as well as on the decay chains of the heaviest elements, most of which decayed by the emission of particles.[3] In later work, after the discovery of neutrons, he searched for the correlation between nuclear stability and the number of protons and neutrons in light and massive nuclei. He discovered a correlation to the β decay phenomenon since after the α-decay in the decay series, subsequent β-decays convert excess neutrons in the atomic nucleus into protons.[4]

Haas was fascinated by the regularities in the structure of nuclei, which Harkins had pointed out in his lecture at Notre Dame. Harkins ordered nuclei by isotopes, characterized by the difference between the number of protons ***P*** and neutrons ***N***, the so-called isotopic number I = N-P or I = A-2Z, where ***A*** is the mass number and ***Z*** the charge number of an atomic nucleus. This observation reflects the quantum number *isospin* predicted in 1932.[5] Haas built on this and developed his own sorting scheme for nucleons in nuclei, with even mass number in the form of two nucleonic clusters, $\alpha = {}^4\text{He}$ and $\sigma = {}^6\text{He}$; he then considered and sorted all even isotopes according to a simple scaling relationship

$$A = r \cdot \alpha + s \cdot \sigma$$

where ***A*** is the mass number and ***r*** and ***s*** are integers, where ***r*** is between 1 and 22 and ***s*** is between 0 and (***r***-1). This means that he was able to construct each of the then known atomic nuclei[6]—by a configuration of ***r*** ^{4}He nuclei or particles, and ***s*** ^{6}He nuclei—as particles coupled with two neutrons; in today's language, α particles with a neutron halo.

This idea led to the cluster model of atomic nuclei. For isotopes with even protons and odd neutron numbers or vice versa, the so-called even–odd nuclei, Haas added only a single nucleon to the series of clusters—which today can be described in the context of a single-particle model. With this principle, all 300 isotopes[l] observed at that time could be sorted into a simple mathematical series and many more atomic nuclei not observed at the time could in principle be predicted empirically. The latter was particularly important because the quantum physics of nuclei, with the

[2] William D. Harkins and R. E. Hall, "The Periodic System and the Properties of the Elements," *American Chemical Journal* XXXVIII/2 (1916): 169–221 and William D. Harkins, "The Nuclei of Atoms and the New Periodic System," *Physical Review* XV (2), (1920): 73–94.

[3] William D. Harkins, "The Periodic System of Atomic Nuclei and the Principle of Regulatory and Continuity of Series," *Physical Review* 38 (1931): 1270–1288.

[4] William D. Harkins, "The Nuclear Exclusion Principle and the Neutron-Proton Pattern," *Physical Review* 52 (1937): 472.

[5] Werner *Heisenberg, "Über den Bau der Atomkerne," Zeitschrift für Physik* 77 (1932): 1–11.

[6] For the discovery of the different nuclear isotopes see Michael Thoennessen, *The Discovery of Isotopes* (Cham: Springer International, 2016).

complex interactions between the individual nuclear components, made it almost impossible to directly predict the different nuclear structures and also the nuclear reaction parameters; still today, one has to rely largely on empirical observations and the correlations derived from them.

With respect to this scheme, Haas observed periodicities in the properties of nuclei ranging from light to heavy mass, which are correlated with mass and stability as well as the frequency of stable nuclei and the half-life of unstable nuclei. He interpreted these correlations as a reflection of some quantum relationships that are now known as closed shell configurations, or others that lead to the formation of α clusters or pair effects that ensure greater stability of particle pairs rather than of single nucleons in nuclei.[7]

In a second part, Arthur Haas concentrated especially on this systematic with respect to the half-lives of the β^- decay of neutron-rich isotopes and was the first to extend this argument to the positron or β^+ decay of neutron-deficient isotopes. He demonstrated the effect of pairing between two nucleons, which leads to a higher stability of even-nuclei compared to nuclei with odd nucleon numbers. He also concluded that the pairing interaction between two neutrons is stronger than that between two protons.

Harkins was the reviewer of this work and he sent, directly to Haas, his comments on some of the isotopes Haas was interested in, isotopes whose identification or stability were under discussion during these years, such as ^{5}He:

> I doubt that helium-5 has been proved to be stable, although some theorists have claimed it is.[8] Also a radioactive form He-5 does not seem to have been produced as yet. I wonder therefore in table I, where you indicate the addition of a neutron to He-4, if you might not want to mark this doubtful.[9]

This undoubtedly was a necessary correction because we know today that, for nuclear structure reasons, there are no stable nuclei with the mass number 5 or 8. They are too weakly bound and decay almost immediately.

Harkins also questioned the premature identification of the element Z = 43, which was labeled masurium and is now called technetium:

> Many scientists doubt that Masurium was actually discovered, and this applies also to Illinium. Therefore, you might wish, if you include it to indicate under $_{42}$Mo^{98} that p is doubtful.[10]

---

[7] Arthur Haas, "On Some Periodic Properties of the System of Isotopes," *Proceedings of the National Academy of Sciences* 26 (4), (1940): 305–312.

[8] Harkins possibly refers to Carl Friedrich von Weizsäcker who, based on his empirical mass formula, considered the possibility of a long-lived ^{5}He and, on this assumption, postulated a first explanation for the production of the heavy elements by simply adding protons or neutrons to ^{5}He, proposing a nucleosynthesis path up to the heavy masses. See Michael Wiescher, "The History and Impact of the CNO Cycles in Nuclear Astrophysics.".

[9] William Harkins to AEH (Chicago, February 18, 1940).

[10] The original discovery of the element 43 was based on the work of two German chemists, Ida Tacke and Walter Noddack who named the element masurium after Noddack's birthplace in the Masurian Lake district of East Prussia. However, this discovery could not be confirmed at first and

Despite these and other critical and not always justified remarks, like the latter on the discovery of masurium, Harkins strongly recommended the work for publication:

> Almost every paper published on this subject during the last fifteen years has presented nothing original, but has only changed the variables of relations known. Professor Haas, however, has presented here some entirely new relations, which are helpful in connection with the problem of how nuclei are built up ... It is an excellent paper.[11]

In fact, this paper led Harkins himself to his predictions of the closed shell phenomenon, where a decade later he provided further observational evidence for magic numbers in nuclei.[12]

Haas' study—on the coupling of pairs of single nucleons to clusters—represented a first step toward the Alpha Cluster Models of nuclei. This work eventually led toward the Interacting Boson Model (IBM) , which in its earliest version was based on valence proton and neutron pairs coupled at spin 0 and 2, and which describes the nuclear structure in terms of algebraic symmetries rather than helium configurations. This opened a new path which uses a similar empirical system for the study of nuclear properties and nuclear structure. The IBM approach of benchmark symmetries and the evolution of the nuclear structure between these symmetries uses a quantification for the onset of collectivity by the N_pN_n scheme that considers the total number of valence p-n interactions for a given nucleus or the P-factor,[ii] which represents the average number of such interactions. A similar approach was found in the advanced IBM-2 model by using the concept of the F-spin quantum number, defined by the difference of neutron and proton bosons, to characterize the nuclear structure.

In the midst of these scientific successes and the esteem with which Haas was held by his colleagues in Chicago, including Arthur Compton and William Harkins, the two years from 1939 to 1940 were characterized by the private difficulties just described under the shadow of the outbreak of war in Europe. On the other side, economic conditions for the family in America improved considerably as a result of the gradual economic upswing brought about by economic development as, with the developments in Europe, the country gradually shifted to war production.

With the improved standard of living, an automobile was purchased, a 1938 Plymouth that allowed for car tours and even vacation trips. Emma was probably

---

element 43 was not artificially produced until 1937, by neutron activation of molybdenum at the Berkeley cyclotron. Because it seemingly could only be produced by technical means, it was called technetium (Tc). In fact, there is no stable isotope of technetium; it occurs in minute amounts in molybdenum-containing rocks produced by natural neutron capture. These spurious amounts might have been detected by Tacke and Noddack. Twenty years later, technetium was also discovered in the spectra of red giant stars (Paul Merrill, "Spectroscopic Observations of Stars of Class S," *Astrophys J.* 116 (1952): 21–26), a first direct evidence for the production of radioactive elements in stars. Today ^{99}Tc is produced in large quantities for medical applications. See Roberto Zingales, "From Masurium to Trinacrium: The Troubled Story of Element 43," *Journal of Chem. Education* 82/2 (2005): 221–227.

[11] William Harkins to National Academy of Sciences (Boston, February 18, 1940).

[12] William D. Harkins, "The Effect of Nuclear Shells upon the Pattern of the Atomic Species," *Physical Review* 76 (1949): 1538–1540.

Fig. 25.1 The house at 934 Riverside Drive, with a view of the St. Joseph River from the backyard

the only one who used the car regularly, as Arthur continued to do his work-related travel by tram and train.

In the summer of 1939 the music professor, whose house Arthur Haas had rented, returned to South Bend. The family did not have to move far, only a few blocks away to a much larger house at 934 Riverside Drive where the garden overlooked the St. Joseph River, as seen in photos (Fig. 25.1). The younger son George reports about these years in his memoirs:

> It was a beautiful home and with Dad's permanent professorship, life couldn't have been better. I know Mom and Dad were very grateful for everything that America had to offer. Dad was really looking forward to receiving his American citizenship papers and I recall one time when I woke up shortly after midnight and heard the late news on the radio downstairs in Dad's library (where he was usually working on some of his papers). I sneaked to the top of the stairs just as the station was getting ready to go off the air and they were playing the National Anthem. I looked down and there was Dad standing at attention, every bit as proper as if he were an honor guard at West Point.[iii]

War had long been expected by Haas, but no direct comment is recorded about the occupation of Czechoslovakia and his old hometown of Brno by the Germans. Nor is there word about his thoughts when the war actually started in August 1939. There are only a few letters from this period and they mainly deal with private and professional developments and not with world events. However, the newspaper articles and caricatures that were cut out and preserved (Fig. 25.2) indicate that these world-political events were read and discussed daily. The son's memories reflect some of the events of those years and help in understanding later developments.

> It was hard to interpret any personal news coming out of Europe at that time. We received word that my Uncle George [Fig. 25.3] was dead—much to the shock and anguish of Mother. The word was that he died of a heart attack—but he was only forty-five and seemed in good health. My mother always maintained that he met his demise at the hand of the Nazis (probably for not cooperating with them in the road-construction business that he had).[iv]

The reasons for George Huber's early death in 1940 are unknown; the official death notification declared that it was a work-related accident, when he fell off of

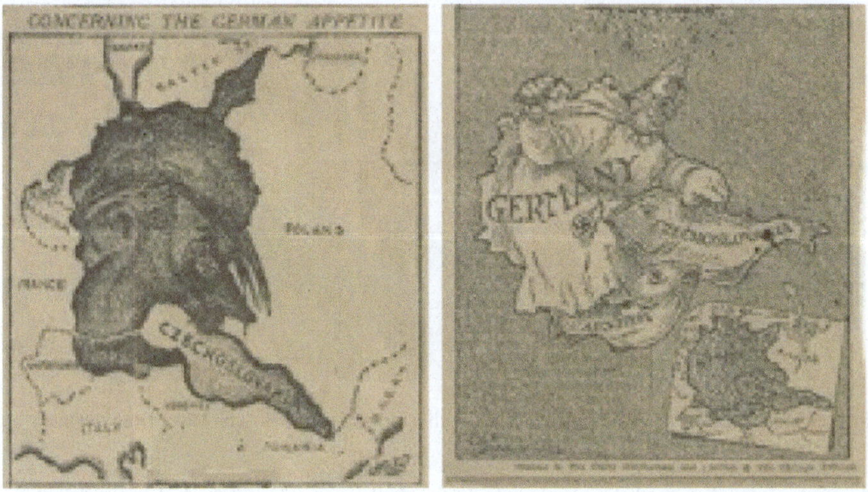

Fig. 25.2 Newspaper articles and caricatures cut out by Haas, which reported on the encirclement and possible occupation of the Sudetenland and on the dissolution of Czechoslovakia

Fig. 25.3 George Huber in his younger years. He had taken over his father's company and despite numerous political difficulties, George always tried to maintain his sister's share and ensure that she received her portion

a roof. The mere fact that he was not a roofer but in the business of constructing and paving streets convinced Emma Haas that he was killed as a result of arrest and torture by the Gestapo.

> The general news from Europe in the summer of 1940 became ever more and more grim. The British troops were forced to retreat out of Europe via Dunkirk, and soon France surrendered. Now all of Europe was controlled by the Axis Powers (i.e., the Rome-Berlin axis, or the combined war effort of Hitler and Mussolini). —However, there were still a few light moments for us at Christy Lake. It was to be the fourth (and last) summer that we were to spend there as a family.[v]

Arthur Haas probably followed the news on the radio, while Emma was desperate over the loss of her brother. So, Arthur hid in work and traveled from conference to conference. Emma found her own ways, building close relationships with the Catholic community, the priests and women of Notre Dame. The small Austrian community at Notre Dame supported each other—the families of Haas, Menger and Guth, with Gödel having already left for Vienna. Without doubt, they discussed the turbulent developments in Central Europe—first the rapid occupation and annexation of the homeland, and then the successful assault on the Sudetenland and Czechoslovakia, which was sacrificed in the hope of peace.

Austrian refugees passed through South Bend and George Haas reports of visits to his family's home by the Trapp family and even by the imperial heir to the throne, Otto von Habsburg:

> I recall two families in particular. One of these was Archduke Otto von Habsburg, "Pretender to the Throne of Austria" - I don't know what he was "pretending," but the Habsburg Monarchy was gone long before even Hitler came around. - The other family was not nearly as "pretentious" and to me, seemed much more friendly - they were the Trapp Family Singers - including the Baron - who were on a singing engagement with some college campuses. They were at St. Mary's and Mom was their official guide.[vi]

The Trapp family may have impressed George more in retrospect since a few years later, in 1965, their story and escape were portrayed in the film *The Sound of Music*. Arthur was probably more impressed by the imperial visit of the heir to the throne, Otto von Habsburg (1912—2011) —the eldest son of the last emperor Karl I of Austria who had died in 1922, three years after his abdication—and the Empress Zita. Otto von Habsburg had never given up his claim to the imperial chair and this is reflected in the letter of thanks to Professor Haas that was sent by the imperial adjutant Franz, Count Trautmannsdorf:

> His Majesty the Emperor has instructed me to thank you very much for your kind words and to tell you in his own name as well as in the name of his brother Ehzg. Felix how pleased His Majesty was to meet you and the other Austrian gentlemen who represent the Austrian intellectual culture at the University of Notre Dame. - His Majesty would also like to thank you and your wife for the kind invitation to your house.[13]

One has to consider what such a letter and such a visit must have meant to Arthur Haas, to him who, in his youth, had been invited to the receptions of Emperor Franz Joseph, Otto's grandfather. For as much as Haas valued with gratitude his new home, he nevertheless remained conservative and loyal to the Emperor in his heart for all of his life. Haas regarded the end of the Habsburg monarchy and the collapse of its empire as a catastrophe for the bourgeoisie—the empire represented his origins and his homeland, whose destruction had allowed the rise of the "mob," the Socialists and National Socialists.

Further visits and opportunities were also documented. Haas received a letter from Arthur Compton in June 1940, which offered a new opportunity to strengthen the scientific profile of Notre Dame as well as the small Austrian scientific community:

---

[13] Franz, Count Trautmannsdorf to AEH (Ritz Carlton, New York, April 14, 1940).

> I have just received a wire from Professor Tolman of Pasadena, stating:
>
> "Schroedinger would like immediate summer appointment in United States. If interested cable Schroedinger. 45 Victoria Road, Clontarf, Dublin, Ireland."
>
> It has occurred to me that Notre Dame might be interested in making such an appointment. Our summer program in Chicago is already complete.[14]

It was typical for Schrödinger to make such a request at such short notice. Schrödinger, with his wife Anny and his lover Hilde March, had only been half a year in Dublin at Trinity College but the financial situation was tight.[vii] He had lost his salary and pensions from Vienna and Graz, and therefore had only very little financial support. He would not have been paid over the summer without giving lectures and a well-paid visit to America was an attractive idea. This idea may also have been strengthened by the lightning-fast victories of the German armies against Belgium and the Netherlands in the spring of 1940.

It is not clear whether Schrödinger would have chosen Notre Dame over Caltech in Pasadena but his somewhat unconventional personal circumstances with two equally treated women, while ignored and tolerated by the Catholics in Ireland,[15] would likely have led to complications in the more puritanical Midwest. But regardless, despite a brief hope on the part of Arthur Haas that he would be able to welcome his old friend to Notre Dame, the question was resolved with the founding of the Dublin Institute for Advanced Studies where Schrödinger found accommodation and a decent salary.[viii]

So, Schrödinger stayed in Europe, in neutral Ireland and Arthur Haas was never to see him again.

During these two years from 1939 to 1940, the Haas sons Arthur Gustav and George grew up; they were successful in school and transferred to Central Senior High School in South Bend, which was also easily accessible by bicycle from their new home. Arthur and George became American teenagers and enjoyed the last years of their family's time together. However, 1939 saw the last year of their joined summer vacations; the traditional family vacations, in the small cabin with friends at Christy Lake, were replaced by separate vacations and trips.

By early June 1940, Bice was already traveling alone; she did not seem well, possibly due to the news of her brother's death. She spent some time at the Mayo Clinic but nothing was found there. Arthur was looking for further answers and a resolution:

> We should be glad that you are not missing anything organic, thank God. Of course, we have to work together so that the nervous causes of your weakness are quickly eliminated.[16]

Arthur expressed his hope to see Emma again soon after this trip but he left when she returned, to spend a few quiet days at Canyon Inn in Spencer, Indiana. Perhaps

---

[14] Arthur Compton to AEH (Chicago, June 3, 1940).

[15] Erwin Schrödinger himself described this situation from experience in his previous positions: "In Germany, if a thing was not allowed, it was forbidden. In England, when a thing was not forbidden, it was allowed. In Austria and Ireland, whether allowed or forbidden, they did if they wanted to." Walter Moore, *Schrödinger – Life and Thought.*, 357.

[16] AEH to EBH (Spencer, June 21, 1940).

spontaneous or maybe motivated by an argument, it was a trip in the tradition of short swimming vacations at Lake Oissach or in the Salzkammergut, which had thus continued in the US. He had already been to the Canyon Inn for a week the previous year and he wrote:

> It is very lonely here and I am all alone with myself. The nature is beautiful and I am also in the water a lot, have worked quite a lot so far. [...] I'm recovering quite well here. I also wrote a manuscript and sent it off, but the European events break my heart and won't let me sleep.[17]

Events weighed on him; he used the time to look back and to long for a "better" past, which now seemed so much better to him despite all the hardships and complaints:

> I have thought about everything, especially on long lonely walks, and I can say with the purest heart and conscience that I am convinced that my feelings are still the same as they were in Portorose, and this quite honest assertion is in no contradiction to statements that may have sometimes been made in uncontrolled rage.
>
> You too should realize this and consider if I haven't done more in our sixteen years than just drop toothpaste in the wash bowl or leave a lamp burning.
>
> I have firmly resolved not to take part anymore when you turn mosquitoes into elephants, then there should be no more friction.
>
> And also think about whether a return to the feelings of 1924 is so impossible on your part, if for no other reason than your health. It would be so easy to fill family life with a spirit of better family love if you would only finally free yourself from your obsessions that make you see everything as wrong. And then true peace would almost immediately put an end to your nervousness and its consequences.[18]

Whether his plea was successful or not, the time of the summer retreat came and was spent separately. Arthur spent it with his sons one last time in the silence of Christy Lake and Bice traveled with the Guth family to New Mexico for health reasons. They spent a few weeks there, first two weeks at Laguna Vista Lodge in Eagle Nest near Taos followed by another two weeks at the nearby Chase Ranch in Cimarron, NM. Bice was hoping to recover from her migraine pain and nervousness in the clear cool mountain air at the almost 3,000-m-high locations. She stayed in the high-altitude area during the hot August days and into September, returning to the sultry Midwest and her family near the beginning of the fall semester.

In the fall of 1940, the lectures at Notre Dame began as usual in the last days of August. The United States had previously remained neutral in the European war, watching with concern the German successes on the war front as one country after another was overrun. Europe was transformed into a German fortress and prison.

As described above, Otto Haas and his family had barely managed to get out and Otto visited his brother and sons in South Bend in the September days, as described. Emma was still in New Mexico at the time and returned from Santa Fe only a few days after Otto had moved to Harvard.

The autumn days went quietly, except for the disturbing news from Europe. Arthur Haas gave his lectures and worked on the completion of his systematics on the

---

17 AEH to EBH (Spencer, June 12 and 19, 1940).

18 AEH to EBH (Spencer, June 22, 1940).

Fig. 25.4 Arthur Haas at his desk in the fall of 1940—the last photograph of him

structure and properties of radioactive isotopes. (Fig. 25.4) He was to speak about this work in November, in Chicago at the autumn meeting of the American Physical Society, at the invitation of William Harkins.

These were exciting days in Notre Dame's physics department. The quiet department that Arthur Haas had joined just four years ago had become a dynamic research facility where students could be trained and awarded doctorates. The new accelerator was producing results that were published in rapid succession. Plans had begun for the construction of a new machine based on the Van-de-Graff principle, whereby the entire facility was to be installed within a large pressure vessel in order to avoid the electrical discharges that made the operation of the first machine so difficult. Construction of the facility began in 1940 and an annex was added to the Science Hall to provide the necessary space for the new machine.

Arthur Haas stood somewhat outside of these experimental events and Eugen Guth had also turned his attention to polymer physics in those days, although he still published with the young experimenters. A new generation had taken over the department, in particular the young dynamic Bernie Waldman (1913—1986) who was to take over the leadership of the experimental group from Collins in 1941 and who pushed the construction of the new accelerator forward:

> In line with recent installations of atom-smashing machines elsewhere, the University has expanded its own facilities for research. During the past summer, a basement addition was built on the Science Hall. Here a steel tank weighing twenty-one tons rests on wheels. To the uninitiated it suggests a gift from Nazi skies. [...] This generator is designed to give seven to eight million volts.[19]

This machine was designed to produce maximum voltage of eight million Volts, a record at the time. It was completed in 1942 but with the engagement of the United States in the war after Pearl Harbor, the new facility was immediately used in research for the US Manhattan Project and toward the building of the first nuclear weapons; the accelerator played an important role in determining the radiation levels of critical materials, which was important for the development of the plutonium bomb.[ix]

[19] Ibid.

Fig. 25.5 The young physicists' group at Notre Dame. Left is a meeting in the control room: In the center is the chair Fr. Bolger; standing are Bernie Waldman and Charlie Mullin; in the foreground is the engineer for the development work, and sitting on the right is Eugen Guth. The rest of the group were students. Arthur Haas was not present on such occasions. On the right, the interior of the new accelerator as it had been developed by the young team

Thus, unrest and construction noise had filled the building since the summer of 1939 and so Haas often worked at home in the fall of 1940. It also seems that, in contrast to Guth, he completely detached himself from experimental activities and had little interest in the experimental successes and developments of the young physicists' community assembled in photo Fig. 25.5.

This was the autumn semester with its official end at Thanksgiving, considered one of the most important holidays in America. Thanksgiving fell on November 21st that year and as previously described, a tradition had already emerged among the Austrian emigrants at Notre Dame. As George remembered of this particular year:

> Thanksgiving that fall was spent with Professor Menger and his family, […] I know we always ate much more than we should and then lingered (barely being able to move) from couch to easy chair and so on. Dad, as I recall, had to leave early since he had to get ready for a scientific meeting in Chicago.[x]

The next morning, November 22, 1940, Haas took the train to Chicago in order to attend the fall meeting of the American Physical Society at the Windermere Hotel. He was scheduled to deliver a lecture on "Some Rules Concerning Radioactive Isobaric Nuclei,"[20] the topic that had occupied him for the past two years. On leaving the hotel after his lecture, he suffered a stroke and collapsed on the steps; he was taken to Cook County Hospital.

Emma was alerted in South Bend and hurried to Chicago where she waited at his hospital bed. Arthur woke only occasionally; his speech center was damaged and he could only remember the classical languages that he had learned in his youth, whether he perceived his surroundings remains uncertain. Emma informed Otto who arrived in Chicago by plane, hoping to exchange a few last words of reconciliation

---

[20] Arthur Haas, "Some Rules Concerning Radioactive Isobaric Nuclei," *Physical Review* 59 (1941): 107.

with his brother; his trip was in vain. Otto wrote a somewhat bitter letter after his return:

> The urgent request I received at midnight on November 27th - five days after his first stroke - to fly to Chicago immediately, I could only understand as news that Arthur had an urgent - though understandable - wish to speak with me again. In fact, he was never able to do so during my nearly forty-eight-hour stay in the hospital, and perhaps never has been able to do so since.[21]

Arthur was still alive but the hope that he would recover was diminishing. His son George remembered the last few weeks:

> Dad would occasionally come out of the coma and then was talking to Mom, reciting both Latin and Greek from his early classical studies - which sounded fantastic, but Mom admitted that this was only sporadic and that he had no comprehension of present or recent past. After a few weeks it seemed from tests that the brain damage was so extensive that he would never be able to resume his teaching profession. ... He died on Feb. 20, 1941 (actually of pneumonia).[xi]

Arthur Haas was buried on February 22, 1941 in the Cedar Grove Cemetery of the University of Notre Dame. The pallbearers were his friends and colleagues, Karl Menger and Eugen Guth from Vienna, the family friend in South Bend, Joseph Caparo, Professor of Engineering, and the biologist Theodor Just, also from Vienna, as well as the physicist Robert Anthony who worked with Guth on polymer physics. Haas' long-time friend and colleague Eugen Guth wrote the obituary for the physics community.[22] Dr. Arthur E. Haas is commemorated in the university archive with the photo in Fig. 25.6.

Notes

i. John J. Livingood, Glenn T. Seaborg, "A Table of Induced Radioactivities," *Review of Modern Physics* 12 (1940): 30–46.
ii. Richard F. Casten, *Nuclear Structure from a Simple Perspective* (Oxford: Oxford University Press, 2001).
iii. George and Mary Jane Haas, "Letter to our Grandchildren," 78.
iv. Ibid., 80.
v. Ibid.
vi. Ibid., 78.
vii. Walter Moore, *Schrödinger – Life and Thought* (Cambridge: University of Cambridge, 1993), 355–363.
viii. Ibid., 365–366.
ix. See Philipp T. Wiescher, "Early Days of Nuclear Physics at Notre Dame and the Manhattan Project," and Christopher S. Temple, "Fostering Elite Science at an American Catholic University," 269–288.
x. George and Mary Jane Haas, "Letter to our Grandchildren," 81.
xi. Ibid.

---

[21] Otto Haas to EBH (New York, January 20, 1941).

[22] Eugene Guth, "Arthur Erich Haas, 1884–1941," *American Journal of Physics* 9 (1941): 198.

Fig. 25.6 Arthur E. Haas' archive photo for the University of Notre Dame

Chapter 26
De Mortuis Nil Nisi Bene

Arthur Erich Haas died unexpectedly. While the topic of his last lecture corresponded to the scientific interests of his last two years, this topic also fit into his broader interests and his search for the correlations and systematics of physical systems; he observed phenomena and tried to construct correlations between them to subsequently put his observations into mathematical form. This approach has always been of great importance in physics, a kind of scouting work for exploring new directions and possibilities, an approach that provides the great mathematical theorists the motivation and information for opening new directions and fields in physics. Haas did his work by comparing and correlating apparently or ostensibly incompatible formal approaches. In these correlations he sought deeper meaning or the shaping hand of his creator.

Through upbringing and propensity, Haas was a highly educated man. He not only underwent the rigorous training of the classical-humanistic gymnasium, which measured its ideological and moral values against those of antiquity, but he also received a fundamental and rigorous education in all other humanistic-philosophical and also scientific-mathematical disciplines. He shone in his striving for knowledge and cognition and, from his youth on, he received the intellectual tools necessary to recognize and utilize hidden relations.

Growing up in the wealthy bourgeoisie society of the Habsburg Empire, Arthur Haas as well as his brother Otto followed the classical German educational canon that formed their understanding of the world. This canon considered Goethe as the representative and guide in identifying these hidden connections to a deeper understanding of the world in all its complexities. This way of thinking was fundamentally ingrained in the educated middle class and characterized especially classical physicists, as illustrated in the following quotation:

> At the Gymnasium Goethe had been his favorite author, and since then he had read Goethe's works beginning to end several times. To this day, he left physics books behind when being on vacation, but he always took *Faust* or Goethe's writings on Italy and classical architecture …[i]

M. Wiescher, *Arthur E. Haas—The Hidden Pioneer of Quantum Mechanics*, Springer Biographies, https://doi.org/10.1007/978-3-030-80606-4_26

In the case of Haas, this is perhaps most evident in his lifelong considerations about cosmology. In his *Physik für Jedermann*, he described cosmology as one of the most interesting contemporary questions in physics and noted:

> But probably for no branch of physics the words of Goethe apply more than for cosmology, that it is the highest happiness of thinking man to have explored the explorable and to quietly worship the inscrutable.[1]

In the close parallelism of the worlds, forces and forms, large and small—Haas thought that he could recognize the hand of a creator, the inscrutable. He also referred to this basic idea when he concluded his book on cosmology, discussing his vision of the close relationship between, on one hand, the microscopic cosmos of the atomic world and, on the other hand, the macroscopic cosmos of the universe:

> Among the unsolved questions of physics, one of the most interesting, but also one of the most daring, is that of the possible relationship between the cosmological and atomic physical constants. In Goethe's *Proverbs and Rhymes* we find the beautiful words,
>
> "If in the *infinite* you want to stride,
>
> Just *walk* in the *finite* to every side."
>
> How strange would it be if it might really be possible to recognize the size of the universe by stepping mentally through the unimaginably tiny atom.[2]

Haas quoted only a fraction of this long poem, *Gott, Gemüt und Welt* by Johann Wolfgang von Goethe (1749—1832), in which the harmony of the world and its inner understanding result from trust in God. In his comparison, Haas could also have continued by quoting the following two lines of the poem as his guideline:

> If the whole is ever to gladden thee,
>
> That whole in the smallest thing thou must see.[ii]

Haas' comparison of microcosm and macrocosm, and the correlation he sought to find between the ruling constants, seems to reflect the thoughts Goethe formulated in his "plant morphology." In this Goethe recognizes a fundamental similarity of natural phenomena, the "essential form with which nature always seems to play, creating by play the various forms of life." Heisenberg observed that Goethe pushed the thought further and noted, "Adopting this model ... one can invent plants to infinity, which, even if they do not exist, they could exist because of their inner truth and necessity."[iii]

Arthur's brother Otto also recognized in Goethe's plant morphology a guideline, however, not toward an understanding of the universe but rather toward an understanding of the evolution in this very universe. He saw this evolution in the numerous artifacts in his paleontological collections and, like his brother, could not help but justify his scientific discoveries with the thoughts and conclusions of Goethe, which he also cited as evidence. From the *Metamorphosis of the Plant,* Otto notes:

> Here doth Nature close the ring of her forces eternal;
>
> Yet doth a new one, at once, cling to the one gone before,

---

1 Arthur Haas, *Physik für Jedermann*, 266.

2 Arthur Haas, *Kosmologische Probleme der Physik*, 109.

So that the chain will be prolonged forever through all generations.[3]

Goethe was a driving force in the development of Haas' scientific thought but he was not alone in this. Goethe, or rather Goethe's *Faust*, was the guide and driving force for German science and scientists throughout the entire nineteenth century and well into the early twentieth century—from Alexander von Humboldt (1769—1859) to whom Goethe still conveyed his ideas personally, through Hermann von Helmholtz, the initial doubter and critic, to Werner Heisenberg who repeatedly attempted to unite Goethe's theory of colors with the modern concept of "dividers and separators" in the modern sciences, rejected by Goethe.[iv] They all followed Faust's drive for knowledge and understanding as expressed in the first scene, in Faust's first monologue:

That I may know what the world contains
In its innermost heart and finer veins.

In addition to the Greek classics, Goethe's *Faust* was an inspiration for the young Haas in his quest to understand the world. The German and equally the Austrian system of classical education emphasized Goethe and elevated *Faust* to a canonical text, a text known to every schoolboy, a text that was to accompany the German troops to the front in their rucksacks, a text to lead the physicists in their striving for knowledge and truth. There is no doubt that Goethe was omnipresent.

Haas wrote his first essay on quotations from Goethe at the Gymnasium in Brünn. His academic role model and teacher, Boltzmann, quoted precisely these lines of Faust's monologue without any reference, in his lectures on Maxwell's theory of electricity and light published in 1891 and 1893, knowing full well that any student would recognize them. This was what Goethe and his Faust had instilled in the German scientists of the nineteenth century: the search for a single, coherent picture of the world that encompasses all phenomena, constants, laws and concepts.[v] This also inspired Haas in his passionate lifelong search for the connections between constants and concepts.

But Goethe's influence reached still further because Goethe also represented the spirit that defied the path of modern science, a science that sought to understand nature by breaking it down into its smallest components. This is a path that, in Goethe's view, would lead to missing the fundamental phenomenon of the whole. He warned against this path:

Even if a fundamental phenomenon is found, there still remains the danger that one does not want to recognize it as such, that we seek behind it and above it for something more, instead of admitting our limitations of vision. The naturalist should let the primary phenomena exist in their eternal rest and glory.

Humboldt accepted this idea and saw nature and its phenomena in their beauty and wholeness. As reflected in his *Cosmos,* he was the first natural scientist to describe nature as a connection between all being, as a network of ecological interactions that must be seen as a whole and not categorized and catalogued into independent units. Helmholtz, initially a critic of Goethe's view of nature, expressed an ever-deeper

[3] Quoted in Otto Haas, "Goethe and Evolution," *Osiris* 10 (1952): 35–42.

appreciation of Goethe's view in his own work on physiology and perception, later giving public lectures on Goethe's scientific ideas. This was also taken up by Haas; in particular, he addressed this in one of his first works on the historical analysis of the energy principle, advocating the idea and development of the "physical unity idea" in "the new Energetic."

During his years in Göttingen, the young Haas had numerous conversations with his friend and early mentor Jellinek, who held the opinion that modern thinking and spiritual life were in crisis:

> Our intellectual life has fallen into anarchy by the fact that the individual branches of science have become too specialized, and no attempt is made to balance them; further by the fact that no intellectual bond is woven between science, art and religion, but rather that there are gaping rifts between these disciplines.[vi]

Haas saw this as a conflict of modern life—the danger that modern science could lose itself in an abstract formalism and lose sight of the overall picture and the deeper meaning. Was it this, perhaps, that motivated him to focus initially on the history of science and not on its future—in order to protect the fundamental secrets of nature? In his first essay on cosmology, "Is the World Infinite in Space and Time?" Haas asks to what extent we can understand and interpret the world: Is it accessible? Is it reality? Are there generally valid physical laws or should we listen to Goethe's warnings and accept Faust's insight,

> Mysterious in open day,
> Nature retains her veil, despite our clamors:
> That which she doth not willingly display
> Cannot be wrenched from her with levers, screws, and hammers.

Haas agreed with this view but followed only partially the Protestantism of Jellinek, who later joined the anthroposophy of Rudolf Steiner. Haas instead took, probably after his return to Austria and under the influence of political developments, a conservative turn to the Catholic Church in whose thinking and dogma he found his home.

This was the intellectual basis that motivated his later striving and with which Haas fueled his research work and visions, yet he seems to have had little success in implementing this work and in achieving broader recognition. He was one of the pioneers of modern physics, not as successful as some of his contemporaries but through his work and his books he became a household name in the scientific community. He had been trained as a classical physicist with a love for ancient Greek thought and culture, and he certainly shared the view that classical physics was a result of the European spirit which was based on classical Greek heritage.

> When he thought of classical physics, he did not see it so much as a defining way of understanding nature, but rather as a value system, a certain way of thinking and personality. The classical physicist knew that for him there was a close connection between physics and classical education, although it might sound strange. Classical philologists tried to grasp the spirit of ancient thought and life and the physicist the inner essence of nature. Classical education was based on the kinship of ancient and modern spirit - like physics on the similarity of mental (model ideas) and natural processes.[vii]

Arthur Erich Haas had lived both: He had captured the spirit of ancient thought with his work on the understanding of nature in antiquity and he tried to understand the inner essence of nature with his work on modern physics. He recognized the inner connections between the microcosm and the cosmos, that both worlds were deeply connected and conditioned each other. In this sense Haas contributed thoughts and ideas to the model of the atom as well as to the model of the universe, and this work may have contributed to our present knowledge of nature as known through the great communication field of science. Today these early ideas may sound naive but they were thoughts and ideas that were generally represented in the world of the new science; it has only been through further work that they have reached the complexity of today's model of the world. As demonstrated by his many personal letters and writings, Haas tried to remain at the forefront of science until his early death, contributing to fundamental understanding and progress.

But he did more than just erect buildings of thought in the ivory tower of the theoretical physicist. Through his books, Arthur Haas motivated young scientists and inspired many of his readers to the ideas of the new science. He recognized that science cannot be done in an ivory tower alone, that the results and findings must also be made accessible to the general public. In this he followed the thinking of his great mentor Ludwig Boltzmann whom Haas, like his friend Erwin Schrödinger, tried to emulate in public relations. Haas also went beyond this by trying to reach everyone, not only to demonstrate the effects of physics on his personal life but also to oppose alternative models that were not based on scientific thinking, models based on political ideology and myth. This work seemed successful but was ultimately swept away with the upsurge of National Socialist madness.

However, the great scientific breakthrough was denied him. Personal weaknesses as well as political conditions of the early twentieth century may have prevented this achievement for Haas. Presumably an interplay of multiple factors was responsible for the failure. The adverse circumstances of the time, the outbreak of war that prevented him from making progress and banished him to the horse stables of the home front; the growing anti-Semitism in Vienna that denied him the Austrian faculty position; the inflation that robbed him of his fortune; the economic crises that crashed his many hopes of the longed-for professorship in America. All of these factors stood in his way and could be considered sufficient reason for the lack of success in the development of his early ideas. There are, however, also other factors of a more personal nature. Besides the adverse circumstances of the time, one can also recognize in his continual hesitation and indecisiveness a certain personal weakness that kept him from bringing "the great idea" to fruition and achieving "the great goal." This indecisiveness was especially noticeable when Haas felt doubt about himself and his abilities.

These tendencies became evident on many occasions—such as in his decision to side-step difficulties with his projected doctoral thesis on thermodynamics when Boltzmann unexpectedly died. Instead of turning to unloved professors such as Franz-Serafin Exner, whom Haas felt were inferior and from whom he sensed criticism, he simply chose a philosophical-historical topic for his work.

This approach also came to light in his second, physical habilitation thesis. When Haas applied the electromagnetic laws to the interpretation of quantum systems, he had developed as a young scientist a remarkable approach. Yet when this fundamental first step toward unifying classical and modern physics seemed unacceptable to his examiner Hasenöhrl, Haas shrank away from the necessary next step. He did not follow the topic any further and it was Niels Bohr who provided the decisive conclusion on the explanation and quantification of atomic systems, while Haas moved on to Leipzig and the history of physics.

In his cosmology, Haas saw the possibility of a correlation between the laws of the macrocosm and the microcosm. This was a comparison that may seem naive to us today but in which Haas was not alone. It was the search for the great unifying solution linking the mystically mysterious numerical values of the fundamental constants that define the forces of nature with their corresponding range in our universe. Haas was following Eddington in his search for a description of the world through the phenomenology of numbers, while Einstein and Heisenberg hoped to find the solutions in their respective world formulas of the all-encompassing unifying theory that encompasses everything and describes everything.

Haas recognized the importance of redshift, acknowledging the possibility of an expanding universe from an origin, but he hesitated to build on this because he fundamentally doubted the feasibility of a dynamic expanding universe, which contradicted his classical Greek education and his traditional Catholic understanding of a closed world. He considered a universe that is finite in space and time to be consistent with his Catholic view of the world and further, he saw the basic thesis of atheism in the observation of a universe that is unlimited in space and time.[viii] Therefore he avoided the idea of an expanding universe. Although this concept could have included the idea of divine origin, Haas still maintained the physical explanation of redshift, which maintained not the expansion of a dynamic universe but rather, alternative, traditional physical models such as the energy loss of photons due to Compton scattering and a resulting redshift.

In an unpublished version of a biography of Haas, Armin Hermann recognized this weakness in Haas' resolve, sighting an earlier episode associated with Felix Klein's seminar in Göttingen. As described in the first chapter of this book, Haas shrank from the honor of being selected as the main speaker on a scientific topic and left this task to his fellow student Max Born. For his presentation, Born received the 1905 Göttingen Prize for the best work on this topic and continued to develop it into his later dissertation. Haas, despite his theoretical talent, withdrew into the supposedly safe world of experiment as Hermann notes:

> The seminar lecture on "elastic lines" was the first opportunity for Haas to measure his abilities against those of Born, to whom he was so similar in age, origin and education. The younger one, Haas had been given the opportunity of probation before Born, yet he could not realize it.[ix]

Hermann identifies this as a symptomatic problem for Haas that, despite all his abilities and talents, prevented him from making the big, decisive breakthrough. Haas was aware of his self-doubt but blamed it on his origins, his sheltered upbringing and the lack of challenge associated with it. In his life review, written at Chrystal Lake in Maine, Haas regretted his father's original financial independence, which had been immensely damaging for self-confidence: "I lacked the strong drive to work that would otherwise be the necessity of obtaining paid employment."

Whatever the reason may have been, inner indecisiveness or lack of self-motivation, he certainly could not be blamed for the latter in his later years. Arthur Haas was a scientist with vision but he also had shortcomings. The perpetual theme reflected in his letters is his thriftiness. He was a penny-pincher who counted every cent, shilling, or dollar individually, a trait that met with criticism and incomprehension from the financially much more generous Bice and undoubtedly led to many a marital row. This behavior is explained by the sudden loss, through war bonds and inflation, of the inherited wealth previously considered inexhaustible.

This financial catastrophe affected not only Haas but the entire bourgeois class and it sent him into shock. Interestingly, it has been noted that Erwin Schrödinger had similar financial behavior; he too grew up in a wealthy home and was similarly plunged into financial ruin by inflation. He traded and bargained with the institutions that supported him for every mark and every pound, and with the villagers for every penny. As said of Schrödinger:

> He never achieved financial security until it was too late to enjoy it, but his insecurity was always more psychological than economic.[x]

One can see the similarities in financial behavior and economic fears.

Economic fear and the aforementioned indecisiveness are probably also the keys to understanding Haas' hesitation to help his brother leave Austria after the annexation by Germany. Issuing an affidavit for his brother and his family would have placed on Haas the financial responsibility of providing for several family members. In his mind, he simply could not bear this burden. In his letters to friends and also to Otto, he embellished Otto's situation in Vienna and denied the constantly growing threat. The unquestionable legal and administrative difficulties, in descriptions of Arthur's efforts to find a job for his brother, grew in equal measure. Without a position through which Otto could support himself and his family independently, Arthur saw no possibility for Otto in America and in some letters, he did not even consider this possibility. He could not bring himself to make a decision and so he avoided the decision. He never atoned for his alienation from his brother, a development that both were aware of and never discussed, as Otto recalled:

> This [...] was only due to the tension that unfortunately existed between us in the last few years.[4]

To this extent, Arthur's refusal to provide direct help cannot be understood and excused by financial worries alone and only by psychological fears in an extreme time. Arthur Haas died too early to be able to reconcile with his brother.

Arthur Erich Haas left behind a rich portfolio of published work; his textbook on theoretical physics has been reprinted several times in ever-improving form, as have

[4] Otto Haas to EBH (New York, January 20, 1941).

his books on atomic physics, which have been translated into many languages. In addition, there are his booklet on cosmology and many other smaller contributions to the consideration of general and modern physics. He wrote clearly and understandably and his books were considered a great success. Ban, emigration, war, and his ultimate, untimely death prevented the printing of any new editions after the war. With this, Arthur Haas and his contributions to physics have been largely forgotten.

Notes

i. Russell McCormmach, *Night Thoughts of a Classical Physicist*, 35.
ii. Johann Wolfgang Goethe, "Gedichte, Ausgabe letzter Hand. 1827," from: *Gott, Gemüt und Welt,* trans. Stefania Degaetano-Ortlieb.
iii. Werner Heisenberg, "Das Naturbild Goethes und die technisch-naturwissenschaftliche Welt," Vortrag auf der Hauptversammlung der Goethe-Gesellschaft zu Weimar (1967).
iv. Richard Friedenthal, *Goethe, Sein Leben und seine Zeit* (München: Pieper Verlag, 1982), 452–468.
v. Helmut Carl, *Geheimnisvoll am lichten Tag. Physik und Physiker im 20. Jahrhundert* (Frankfurt am Main: Scheffler Verlag, 1963) and Gino Segrè, *Faust in Copenhagen – A Struggle for the Soul of Physics* (New York: Viking, 2007).
vi. Karl Jellinek, *Das Weltengeheimnis* (Stuttgart: Verlag Ferdinand Enke, 1922), preface.
vii. Russell McCormmach, *Night Thoughts of a Classical Physicist*, 147.
viii. Helge Kragh, "Eight Physicists and Astronomers: Biographical Portraits" in *Biographical Encyclopedia of Astronomers*, ed. Thoma Hockey, (New York: Springer Verlag, 2007, 2014), accessed on January 31, 2021, https://arxiv.org/ftp/arxiv/papers/1208/1208.3114.pdf.
ix. Armin Hermann, "Arthur Erich Haas – Eine Biographie," (unpublished typoscript, 1965), 3. Published as Armin Hermann, "Arthur Erich Haas–eine Biographie," in *Arthur Erich Haas, Der erste Quantenansatz für das Atom*, ed. A. Hermann, (Stuttgart: Ernst Battenberg Verlag, 1965), 7–25.
x. Walter Moore, *Schrödinger – Life and Thought* (New York: Cambridge University Press, 1993), 365.

Part V
Post Mortem 1943–2017

Chapter 27
Margarethe Gerstenberger

Not much can be reported here about the young and talented Margarethe Gerstenberger née Haas. (Fig. 27.1) She was successfully awarded a doctorate in history at the University of Vienna, presumably to take up the teaching profession. However, after the war she married the young Eduard Gerstenberger and moved to Sattendorf in Carinthia, where Edi initially worked as a minor civil servant before turning to business. She never followed her academic ambitions; she had three daughters and lived peacefully in Sattendorf where she earned an additional income by renting rooms to tourists. She initially refused to follow Arthur's admonitions to emigrate to America:

> After all, not all of Central Europe can emigrate, everything from the mediocre downwards just sits here and awaits the things that are to come, as well as the millions of fellow sufferers. That the situation of the mixed blood is particularly unfavorable and the embarrassment is also undeniable. I personally can't do anything else but say "carpe diem" - every day that I still walk around freely and unhindered - and let my children learn English, which they are doing intensively this summer.[1]

From this it is clear that she had at least listened to Arthur's warnings and had taught the children English. But life was soon to change when the ban on the Nazi Party was lifted in 1937. Edi, who had not been very successful so far, joined the Austrian Nazi Party because like so many others he saw hope for change and a new, albeit radical beginning. He did not see that this radical beginning was at the expense of his wife and children but Grete recognized this and, despite difficulties and protests, filed for divorce. This was finally accomplished in 1937 but her life was not easy financially, as she wrote to Arthur:

> Nevertheless, of course, I have not yet succeeded in achieving a balance between income and expenses, because I have too many extraordinary expenses, mainly due to the constant harassments by my divorced husband. [...] What my husband performs is nerve-racking, the worst backstairs novel. He tries with all means, partly dangerous threats (which earned him

[1] Margarethe Gerstenberger to AEH (Vienna, August 13, 1936).

M. Wiescher, *Arthur E. Haas—The Hidden Pioneer of Quantum Mechanics*, Springer Biographies, https://doi.org/10.1007/978-3-030-80606-4_27

Fig. 27.1 Otto Haas and his sister Margarethe during the last years of the First World War

> three weeks in the regional jail) partly through the intervention of all kinds of third parties, to get me to take him back.[2]

Through English acquaintances she managed to have the eldest daughter Maria sent to England and then fled from Sattendorf to Vienna to seek protection and legal counsel from her brother. Edi begged her to return to him with the two younger daughters Erika and Christl. Grete not only didn't return but wanted more, working to also at least send Erika to England or the USA. It was even more difficult with Christl who was mentally retarded and needed constant care and attention; she could not be passed on to English foster parents. Grete herself was also thinking of emigrating. She considered Czechoslovakia, back to Brno, or even to South Tyrol in Italy. This had less to do with the growing threat of Nazism and was more because of her husband's unending threats. But Edi Gerstenberger, who was a ranking Party member, finally agreed in 1938 that it would be better to send the children to England.

Arthur paid for Erika's accommodation at the Catholic Convent of Cullompton in Devon, England, and so in August the second daughter managed to travel to England. In the files of the *Israelitischen Kultusgemeinde*, who largely organized the transport of children from Austria to Great Britain, there are no records of Erika; her emigration may have been possible through private connections, such as those with Grete Haas' former educator Miss Norah Barnett or through connections to the Catholic Church. Grete hoped to eventually get a job as a German teacher in England but also considered the possibility of leaving Europe altogether for New Zealand:

[2] Margarethe Gerstenberger to AEH (Vienna, November 4, 1937).

> Of course, I am very skeptical and do not give in to any hopes, but of course, I am also making constant efforts. Because to rely ad infinitum on Eduard's alimony and to lay my hands in my lap for the time being would probably be irresponsible.[3]

In addition to teaching German, she also had plans to set up a handicraft business with her daughter Maria and to manufacture jewelry. Her last letter to Arthur is dated December 1938, but it took another four months before she could leave Austria with her brother Otto and sister-in-law Greti to settle in England.

These must have been difficult years since Margarethe Gerstenberger had to earn her living by cleaning and doing other household chores. After a few months serving as maid in London, she moved and lived as a subtenant of a Mrs. Marshall in Northampton at 17 Billing Road, a small Victorian villa. Northampton was a small town in the Midlands of England, located between Manchester and London. It was probably the Catholic diocese in the town that attracted her. Whether she ever managed to find a job as a teacher is unknown. Her later fate is also forgotten; there is only a small note about her death in 1971 in England.[i]

Margarethe's former husband Edi Gerstenberger remarried and lived as a businessman in Vienna. The Party had presumably not turned out to be the bearer of betterment that he had hoped for in his life. He apparently began to have doubts, especially as final victory became increasingly unlikely. On December 22, 1944, he was arrested by the Gestapo for favoring the enemy.[ii] According to the indictment, he had contacts with resistance fighters in Carinthia and had helped a group of several parachute agents. Three young Austrians—Ernst Unger, Fritz Gertner and Johann Pascher—under the command of an American officer Jack Taylor parachuted into the area of Lake Neusiedl[iii] with Gerstenberger's assistance. He was lucky and was released on February 14, 1945. Unger, Gertner, and Pascher were sentenced to death by a court-martial for war treason, espionage, and desertion but they were able to flee from their planned execution by firing squad in Döllersheim. Captain Taylor was transferred to the Mauthausen concentration camp and he, too, was lucky enough to survive.[iv] Whether the charges against Gerstenberger were not confirmed or whether his connections to the Party helped him is not known. Only a few were able to escape unscathed from such an accusation in these last days of the war.

The Gerstenberger's second daughter Erika found shelter in the convent and stayed, joining the order. As a Sister of the order, she traveled a lot and occasionally visited the United States where she met with the families of her two uncles. She died in the years after 2001 and the time of their last meeting, but the exact date of her death is unknown. Her sister Maria married Edward Timothy Udall and lived with him in Bury St. Edmunds in Suffolk; she was considered a strong open-hearted person in the village. The couple had no children but they cared for sister Christl. Christl Gerstenberger died around 2010, her sister Maria died a year later in 2011 as the result of a traffic accident. Tim Udall then lived alone but already in 2016, the Catholic parish newspaper of Bury St. Edmonds requested prayers for him as he

[3] Margarethe Gerstenberger to AEH (Vienna, October 15, 1938).

suffered from an incurable disease. He died in February 2019 in Bury St. Edmonds. A requiem was given in his honor in the local Catholic Basilica.[v]

Notes

i. Werner Röder, Herbert A. Strauss, eds., *Biographisches Handbuch der deutschsprachigen Emigration nach 1933–1945* (München: Saur Verlag, 1999), 5, 443.
ii. Dokumentationsarchiv des österreichischen Widerstandes, accessed on January 31, 2021, https://www.doew.at.
iii. Wolfgang Neugebauer, "Mit der Waffe in der Hand … Partisaninnen in Österreich 1938—1945," in *Bananen, Cola, Zeitgeschichte: Oliver Rathkolb und das lange 20. Jahrhundert* (Wien: Böhlau Verlag, 2015), 383–396.
iv. Wolfgang Neugebauer and Peter Schwarz, *Der Wille zum Aufrechten Gang* (Wein: Czernin Verlag, 2005), 128.
v. St. Edmund's Catholic Parish, accessed on January 31, 2021, http://www.stedmundkm.org.uk.

Chapter 28
Otto Haas

Contrary to the expectations of his brother Arthur, Otto Haas quickly and with great success established himself in his new American home. He seemed to miss neither his old profession as a successful Viennese lawyer nor his Viennese home where he had lived and worked for forty years. On the contrary, after decades of legal and economic activity, he had finally reached his secret goal in life: to be able to work scientifically as a paleontologist. He was considered a specialist for ammonites, shellfish which for 340 million years from the Devonian to the end of the Cretaceous period formed an important species with an enormous diversity in the biological development of our planet. This diversity is reflected in the rich shape and form of fossil ammonite shells. Ammonites were the subject of Otto Haas' doctoral thesis on "The Fauna of Middle Liassic of Ballino in South Tyrol" under his advisor Carl Diener (1862—1928) in 1914. Otto was acknowledged as a well-known expert in this field throughout the remainder of his life.

Otto was more practically oriented than his brother. Although less concerned with fundamental questions about the origins of the universe, Otto did share a concern with his brother regarding questions about the evolution of life, which he, like Arthur, associated with the thinking of Goethe.[1] Otto was interested in the systematics of species order, which he pursued in detail during the following decades.

He knew his preferences and wished nothing more than to be able to work in a museum where he could organize and catalog the species of the different classes of ammonites. This was his wish and hope after his doctorate and military service during the First World War. Yet this hope was not fulfilled since he had to take over the law practice of his late father. The wish was, however, eventually fulfilled after he fled Austria twenty years later. Arthur had seen inflexibility in Otto's way of life, considering him therefore as unfit for the dynamic American ways. But it was duty not inflexibility that had determined Otto's choices. He felt it his duty to fulfill the promise he had made to his father. Otto also had an obligation to his young family after the war—he had married Margarethe Hirsch in 1919—to provide a sufficient standard of living. For these reasons, Otto Haas had put his dream of paleontology behind him and devoted himself to law. The escape from Vienna and

M. Wiescher, *Arthur E. Haas—The Hidden Pioneer of Quantum Mechanics*, Springer Biographies, https://doi.org/10.1007/978-3-030-80606-4_28

its obligations gave him a new opportunity and new hope. Perhaps Arthur had also motivated and ultimately aided him with admonitions to give up jurisprudence and turn to the science of paleontology.

Shortly after his successful escape to England in March 1939, Otto was offered a temporary position at the British Museum. There he worked on Eocene Nautiloids[1] from British Somaliland while arrangements were being made to join their eldest daughter Lene, as she was now known in the United States. Relocation to Boston was accomplished in December 1939 with the help of affidavits issued by Helene and Arthur Haas, so Otto's family was finally reunited with Helene near Wellesley College.

Through his connections, Otto found a temporary position at the Harvard Museum of Comparative Zoology. Afterward, he spent the summer of 1940 at the Field Museum in Chicago and was able to visit his brother in South Bend. In October of that year, Otto found a permanent position at the American Museum for Natural History in New York where he remained for the next fifteen years, first as a research associate and later as curator of the fossil collection of invertebrates. With this, he was then able to devote himself full-time to the study of Mesozoic ammonites or cephalopods as reflected in his nearly seventy scientific articles from the time. Otto's scientific knowledge and importance were immediately recognized and already in 1941, he was awarded the A. Cressy-Morrison-Prize in New York for his extensive work, "The Vernay Collection of Cretaceous (Albian) Ammonites from Angola."[ii]

He and his family adapted well in the United States and he resumed his quiet way of life, which had already characterized him in Vienna and had caused his brother to doubt him. He worked systematically in the archives and collections of the museum and he traveled regularly with his family during the summer; it was no longer the lakes of the Salzkammergut in Austria that they visited but, rather, the coast of Maine and the mountains of New Hampshire.

The connection to his brother's family was formal after his visit to Arthur's sickbed but individual members of the family did meet on rare occasions, such as in 1953 during Otto's summer stay in Maine where the newly married George Haas saw his Uncle Otto again after almost twelve years. A photograph from this time is shown in Fig. 28.1.

In the ten years of his tenure as curator at the American Museum for Natural History, Otto Haas worked steadily and published a total of about seventy articles in the museum's *Museum Novitates* or in other scientific journals such as the *Bulletin of the Geological Society of America*, the *Journal of Paleontology*, and even the leading more general science journals such as *Science* and *Nature*. He systematized his results which were mainly based on investigations of ammonite finds from Africa, North America and the Middle East. Throughout this classification work, Otto never lost sight of how his results and interpretations could fit into the overall picture of the evolution of cephalopods.

[1] Nautiloids are a large and diverse group of marine cephalopods (mollusca) that emerged in the Late Cambrian Age.

Fig. 28.1 Otto Haas in the summer of 1953 in Maine

Although Otto Haas reached retirement age in 1955, he stayed at the museum until spring 1956 to complete some ongoing catalogue work. Then his final retirement from the museum was followed by another surprising change in his career.

Instead of spending his retirement years peacefully in New York, Otto was drawn to new ventures, demonstrating a flexibility in his life circumstances that brother Arthur would never have believed possible. Otto moved to South America to work as a geological consultant for the International Petroleum Company of Colombia, for which he was also appointed to the faculty of the National Mining School in Medellin. He stayed in Colombia for one year but continued teaching after his return to the United States. He spent one semester as a visiting professor at Vassar College and then three years at Hofstra University on Long Island from where he moved to Reno, Nevada in 1960. In Reno, Otto became Distinguished Visiting Professor at the Mackay School of Mines where he taught for the next six years; he was then appointed Curator of Paleontology and dedicated his energy to the University's fossil collections. He also did not shy away from technical innovations and performed a very important service when he committed himself to changing the fossil catalog to a punch-card system that he had also designed; this was a first step in the computerization of data catalogues.

Fig. 28.2 The family of Otto and Margarethe Haas in 1968 in Reno, Nevada. The children on the left are Lene and Gustav and the younger sister Dorle is on the right in her nun's habit. In the foreground is Caron, Lene's daughter, born in 1956

While Otto Haas and his wife Margarete enjoyed Reno and the surrounding landscape of the American Southwest, Otto became increasingly frustrated with the lack of research equipment at the college and especially with the limitations of his library in terms of paleontological holdings. He requested loans from his old place of work, the American Museum in New York, and established a very broad international correspondence through which he built up an enormous database, enabling Otto to register and catalogue the latest results in his field of research. He published seventeen volumes in the years after 1957, the last only a few months before his death in 1976.

Figure 28.2 shows the last family photo. Only one year later, in 1969, Margarete and Otto were hit by a heavy blow of fate. Tragedy struck when their younger daughter Dorothea or Dorle was killed in her sleep. As a member of the Little Sisters of Jesus Order, she worked and lived in the Chicago ghettos with members of her order as part of her missionary work. On her last evening in Chicago before her transfer to the administration in Washington, D.C., a stray bullet from a gang war entered the nun's quarters. Sister Dorothea was killed by this cross shot, as reported in the clipping in Fig. 28.3.

Otto survived his daughter by seven years; he died in Reno in 1976, outliving his brother by thirty-five years. Despite the increasing weakness of his eyesight, Otto worked at the university until the end, well regarded and considered with respect by his colleagues, an immaculate figure from the past:

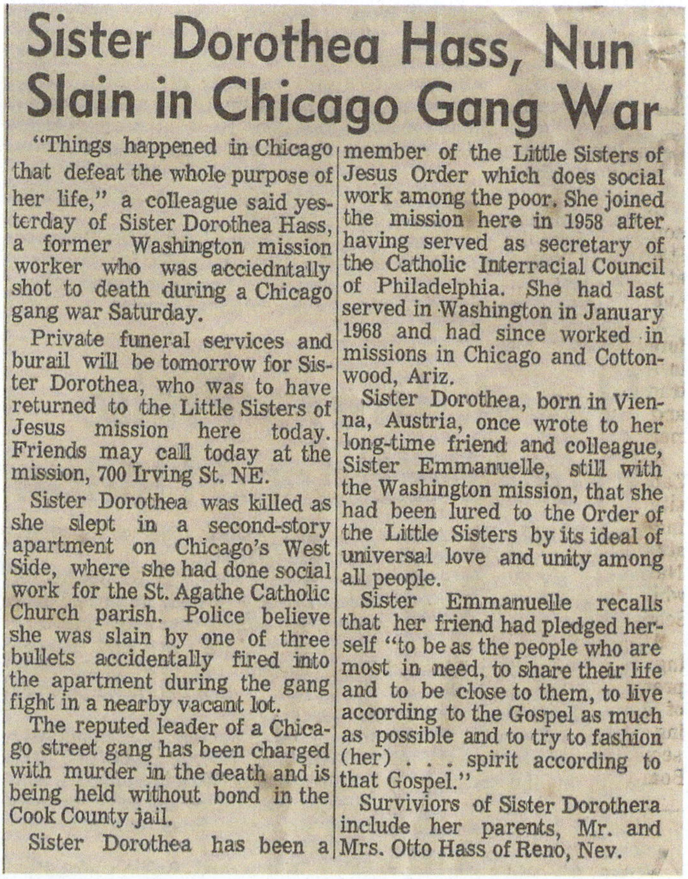

Sister Dorothea Hass, Nun Slain in Chicago Gang War

"Things happened in Chicago that defeat the whole purpose of her life," a colleague said yesterday of Sister Dorothea Hass, a former Washington mission worker who was acciedntally shot to death during a Chicago gang war Saturday.

Private funeral services and burail will be tomorrow for Sister Dorothea, who was to have returned to the Little Sisters of Jesus mission here today. Friends may call today at the mission, 700 Irving St. NE.

Sister Dorothea was killed as she slept in a second-story apartment on Chicago's West Side, where she had done social work for the St. Agathe Catholic Church parish. Police believe she was slain by one of three bullets accidentally fired into the apartment during the gang fight in a nearby vacant lot.

The reputed leader of a Chicago street gang has been charged with murder in the death and is being held without bond in the Cook County jail.

Sister Dorothea has been a member of the Little Sisters of Jesus Order which does social work among the poor. She joined the mission here in 1958 after having served as secretary of the Catholic Interracial Council of Philadelphia. She had last served in Washington in January 1968 and had since worked in missions in Chicago and Cottonwood, Ariz.

Sister Dorothea, born in Vienna, Austria, once wrote to her long-time friend and colleague, Sister Emmanuelle, still with the Washington mission, that she had been lured to the Order of the Little Sisters by its ideal of universal love and unity among all people.

Sister Emmanuelle recalls that her friend had pledged herself "to be as the people who are most in need, to share their life and to be close to them, to live according to the Gospel as much as possible and to try to fashion (her) . . . spirit according to that Gospel."

Surviviors of Sister Dorothera include her parents, Mr. and Mrs. Otto Hass of Reno, Nev.

Fig. 28.3 Report of the *Evening Star*, Washington DC, September 2, 1969

He was always immaculately turned out, and in later years when he and I were associated, even in his field clothes he possessed an air of distinction. His love of good food and fine wine was well known, yet he maintained a trim, athletic figure through constant exercise all his life. In manner he was courteous and somewhat formal, but never sufficiently so to scare away a young and timid student.[iii]

Notes

i. Otto Haas, "Goethe and Evolution," *Osiris* 10 (1952): 35–42.

ii. Otto Haas, "The Vernay Collection of Cretaceous (Albian) Ammonites from Angola," *American Museum of Natural History Bulletin* 81 (1942): 1–224.

iii. Joseph Lintz, Jr., "Memorial to Otto Heinrich Haas (1887—1976)," *Geological Society of Amerika*, 1976.

Chapter 29
Emma Beatrice Haas

Emma Beatrice Haas, Arthur's wife may have had the most interesting destiny of the family. Nonetheless, there is little documentation and the little that there is remains mostly under lock and key and therefore not accessible despite several requests. Emma Beatrice had already lived a relatively independent life in the community of professors and priests at the University of Notre Dame and neighboring St. Mary's College in the two years before her husband's death. The sons were increasingly independent and the couple had grown somewhat apart; Arthur Haas was absorbed in his work and Emma Beatrice took care of the social and artistic life of the Notre Dame community, as seen in Fig. 29.1.

The common vacations had come to an end and the mood was irritable on both sides, as can be seen from the letters quoted above. Arthur went to his meetings or to lonely forest lakes in Indiana State Parks where he swam, hiked and thought about his plans and books. Emma traveled with friends to more exotic places, from Texas to New Mexico, where she also hiked to free herself from her chronic nervousness and irritability. This unbearable kind of living together could have continued for years, because divorce was unimaginable on both sides, but death came suddenly.

With the sudden death of Arthur Haas, a financial catastrophe befell the family. Arthur had not been employed at the university long enough to accumulate sufficient capital, the Austrian assets had disappeared and the money invested by Arthur in Switzerland had been largely used up by years of moving around. The education of the two sons required additional funds, especially since the change from the local high school to an expensive college was imminent.

Thanks to her good connections with Notre Dame and her excellent language skills, a solution was found by securing employment in the university archives where she was responsible for translating French, Latin and German texts into English. Since the university was originally a French foundation and had brought countless members of the order from German-speaking countries to South Bend, a considerable history had accumulated. In addition, the archives of Notre Dame were also considered the official repository for all correspondence between Catholic missionaries in the US

M. Wiescher, *Arthur E. Haas—The Hidden Pioneer of Quantum Mechanics*, Springer Biographies, https://doi.org/10.1007/978-3-030-80606-4_29

Fig. 29.1 Emma Beatrice Haas in a long evening gown

and their motherhouses in Europe in the eighteenth and nineteenth centuries. There was a rich field of information for forty-five-year-old Emma to work through.

The university archive was located in the old library (now Bond Hall, Fig. 29.2). To accommodate the archivists, a miniature elevator brought them to the premises located in the center of the large classical building. It was a relatively large room that mainly housed the card files and several tables. The offices of the archive staff were hidden behind filing cabinets and large metal cupboards. Emma Beatrice was based there with Fr. Thomas McAvoy C.S.C. (1903—1969), Professor and Chairman of the Department of History and Head of the University Archives.[i]

Through her work in the archive (Fig. 29.3), she made many new friends and soon became indispensable as many of the letters left behind by staff and archive

Fig. 29.2 Bond Hall, the former library building of the University of Notre Dame

Fig. 29.3 Emma Beatrice Haas at her workplace in the archives of the University of Notre Dame located in the old library building of the university

users demonstrate. In particular, she became close friends with McAvoy who greatly appreciated her collaboration.

Emma Beatrice helped McAvoy transform the University of Notre Dame archives into a useful research tool by organizing and cataloguing over 500,000 documents. As

Elected.

DR. YVES SIMON.

Dr. Yves Simon, professor of philosophy at the University of Notre Dame, is among three new members elected Monday by the pontifical Roman Academy of St. Thomas Aquinas, the Associated Press reported in a dispatch from Vatican City.

The academy is a society composed of the outstanding Catholic philosophers of the world. Dr. Simon has been a member of the Notre Dame faculty since 1938. He came from the Catholic university of Lille, France, where he taught philosophy. He is a native of Cherbourg, France.

Fig. 29.4 Left, the priest and historian Thomas McAvoy; right, the philosopher Yves Simon, both of whom taught at the University of Notre Dame in the 1940s and developed a close friendship with Emma Beatrice Haas

a priest and historian, McAvoy was particularly concerned about the so-called American Catholic dilemma of living in an often-hostile puritanical Protestant world.[1] In this sense he got along very well with Emma Beatrice and the two developed a close relationship. (Fig. 29.4).

[1] As an historian, McAvoy broke new ground with his interpretative approach that focused on accurate documentation, the rejection of generalizations, and an impartial judgment on the history of the Catholic Church. He wrote in many historical fields but made his most significant contribution to American church history and educational history by creating a composite sketch of American Catholicism in history, especially through his unique perspective of looking at all events through the lens of Catholicism's minority status. Richard Gribble, "Thomas T. McAvoy, CSC: Historian, Archivist and Educator," *American Catholic Studies* 115, 1 (2004): 25–43.

Emma Beatrice also got along well with the French philosopher Yves Simon (1903—1961), who had come to Notre Dame in 1938 as a visiting professor from Lille. Due to the outbreak of war, he could not return to France and remained as a professor at Notre Dame until 1948. Afterward he went to the University of Chicago to head the Committee on Social Thought.[2] The acquaintance was possibly made by Karl Menger as early as 1938, since Simon attended Kurt Gödel's lectures on mathematical logic with him. Later, Simon also joined with Father McAvoy in planning the journal *The Review of Politics*. The Catholic academic and priestly world at Notre Dame represented a closely knit community. From 1942 to 1948, Yves Simon and Emma Beatrice had an intensive correspondence in French but only his letters have survived. The relationship seems to have been particularly close from his side for he wrote extremely frequently, especially in the years from 1945 to 1948, whereas replies were rather sparse. This was a regrettable habit that her late husband Arthur E. Haas had already lamented throughout his life.

In addition to these contacts on campus, Emma was also very active in South Bend at the community level, as described in Fig. 29.5. She had graduated from St. Mary's College and was now lecturing in local schools on art and education. She also took the opportunity of her new job and experience to give public lectures on the importance of archives, under the title "Adventures in Archives." She must have been an entertaining speaker since her presentations were highly lauded in the local press.

Besides speaking on these academic topics, she also felt a strong responsibility with respect to her civic duties. Shortly after Arthur's death, their joint application for citizenship was granted and Emma took very seriously the task of informing the local community—in clubs, schools, and church centers—on the dangers of Nazism and the duty to defend democracy. Much of this was driven by the loss of her brother who she maintained was murdered by the Gestapo. In her personal conversations Emma asserted that she saw the defeat of Nazism as her mission in life.

Life also changed for her two sons in 1941, in the months following the death of their father. After the summer vacation, which they spent again with friends at Christy Lake, Arthur Gustav began his last year at Central High School in South Bend, the plan being that he would go to college at Notre Dame. George still had time to explore; his interests were mainly focused on electronics and radio tinkering, but he also played a lot of sports and was an especially passionate swimmer in family tradition. Arthur Gustav was a bit more reserved but George had many friends in the community:

> One weekend in the winter of '41, we were visiting with a group of kids in the afternoon. When we were ready to leave, an older brother of one of them offered to take us home in his car since the weather was a bit nasty. As we were dropping off one of the kids, the regular radio broadcast was interrupted and we heard the news of Pearl Harbor. We were stunned

[2] Yves Simon retired in 1958 due to illness and in his last letters to Emma Beatrice Haas in 1949, he already complained of frequent indisposition. Even during his time in Chicago, he lived in South Bend where he died of cancer in 1961. Walter Nicgorski, "Yves R. Simon: A Philosopher's Quest for Science and Prudence," *The Review of Politics* 71 (2009): 68–84.

Widow of Notre Dame Scientist Haas Has Vital Job in University Archives

BY JAMES V. CUNNINGHAM.

A smiling lady with a soft, European accent, whose greatest pride is her new American citizenship, holds one of the most highly specialized positions in the archives of the University of Notre Dame.

Mrs. Emma B. Haas is a learned woman, whose proficient knowledge extends to six languages. She is the widow of the late scientist, Dr. Arthur Haas, who was a professor of physics at Notre Dame and an internationally known experimenter into the nature of atoms.

Translating French, German, Italian and Latin, historical letters and documents, most of them over 100 years old, this delicately featured woman works at a large desk in a many windowed office on the top floor of the Notre Dame library.

Lauds American Progress.

The sun throws off a dull glow from her graying blond hair, and when she speaks it is in a low, cultured voice. She talks of the United States: "I think it is the most wonderful country in the world. In art, in music, in architecture, in all fields of science America has come as far in 100 years as Europe has in the last 500."

Mrs. Haas came to the United States with her scientist husband and two young sons in 1935, after many years of learning and happy living in colorful and peaceful pre-Hitler Vienna, where there were dozens of plays, operas and concerts playing to full houses every night. The trip was made with some misgivings but along with her husband she was soon captivated by the American way.

A wide background of education has prepared Mrs. Haas for her job in the archives, a position for which only one person in many thousands can qualify. University authorities consider her an invaluable asset, for in addition to linguistic ability she has studied thoroughly the history and literature of the 19th century with which her work deals.

As a little girl in Munich, she learned French and English along

Continued on Page Eight, Column Four.

Mrs. Emma Haas pauses at her desk in the Notre Dame library.
—Photo by Tribune Staff Photographer.

Fig. 29.5 Newspaper report in the *South Bend Tribune* describing Emma Beatrice Haas and her employment in the archives of the University of Notre Dame

> and we also knew that now we were definitely going to be part of the global conflict. - And so we were, as President Roosevelt would announce to Congress the following day.[ii]

At first not much changed but with entry of the US into the Second World War, the campus became increasingly empty as many of the students and younger professors were drafted into the military. This caused great concern for the university administration since without students and the associated tuition fees, the university could not continue to operate. But the US Navy approached the administration suggesting to its board of directors that the university be used as a training facility for young cadets. This so-called Reserve Officers' Training Corps (ROTC) began as a small test program but was to expand during the war.[3] The program also seems to have

[3] On the implementation of the ROTC Program of the Navy at the University of Notre Dame see Thomas E. Blantz, *The University of Notre Dame – A History* (Notre Dame: University of Notre Dame Press, 2020), 321–347.

attracted the two Haas sons, especially since the Navy would bear the costs for their academic training.[4]

> After school let out in June 1942, Arthur started at Notre Dame and I spent part of the summer working at Camp Eberhart on Magician Lake in Michigan. (I worked in the supply store for which I got my room and board. Among other things, we sold stationery supplies and made sure the kids would write home regularly.) I was looking forward, in my senior year, to returning to Central since I had a good chance to become the Captain of the swimming team - but this was not in the cards.
> Mom was going to be sent to St. Edward's College in San Antonio, Texas for some courses in Library Science to help her in her advancement at the Notre Dame Library, and since Arthur was a resident R.O.T.C. student at Notre Dame, I would be the only one home. Therefore it was arranged that I should go to a boarding school just outside Chicago called Marmion Military Academy.[iii]

Before her departure in the winter of 1942, Emma Beatrice had a breakdown and was hospitalized for some time. Was it her health or was it the sudden separation from her sons who were increasingly being drawn into the war? The training at St. Edward's was postponed until January and she left after Christmas. While she was away, Thomas McAvoy faithfully reported to Emma about how things were developing on campus, especially as more and more students were being drafted.

> Until Sunday, no one seemed to know anything new. But Saturday there were several announcements which are not yet too clear. One is that members of the army air corps reserve have to go into service. This means about 150 boys. Then there was the announcement that Navy would use Notre Dame as one of its schools. ...[5]

In addition, many of the older professors, especially in physics and engineering, were recalled to special military duty. They reported to MIT for work on the development of radar—such as George Collins and Edward Coomes—or they simply disappeared into secret research facilities in Tennessee or New Mexico—where men including Bernie Waldman were enlisted to places that existed only as vague rumors due top secrecy. The accelerator laboratory was also used for classified scientific work. Each day, groups of scientists from Chicago came with the South Shore railway to use the machine for secret experiments that we now know involved studying the behavior of different materials under strong radiation. All of this together became known as the Manhattan Project.

These were major changes to the university campus. The quiet, comfortable seclusion of the university with its priestly world was no more. With the introduction of the Navy's ROTC program, life and education on campus became militarized. New courses, such as Naval Gunnery, Seamanship, Naval History, and Navigation were offered. For a while, the university staff debated the possible consequences. Thomas McAvoy in particular feared for the Catholic character of the university, as he considered the church to be a persecuted institution in the United States. This is evident in the letters he wrote to Emma Beatrice in the winter months of 1943:

---

[4] For sons of university employees, university fees were traditionally waived but the incidental costs for room and board, study materials, and related costs were still required.

[5] McAvoy to EBH (Notre Dame, IN, February 15, 1943).

> Boys are leaving the school in droves. This last call to the air service has caught many of them and those left behind are very restless. Added to this is the "scuttlebutt" that the Navy is to take over Alumni and Dillon [university buildings] and maybe the whole place. One rumor is to the effect that the Navy wants the whole place but will keep the lay faculty. They do not want any of the priests to teach. …[6]

One of the drafted students, Robert Schorsch, had helped in the archive and he wrote letters to Emma Beatrice about his training as a pilot and paratrooper in Texas, about his new friendships in the military, and about his training in skydiving. He fell, only twenty-four years old on September 24, 1944 as First Lt. Robert S. Schorsch in the battle of Arnhem.

As Emma Beatrice recovered on the quiet campus of St. Edward's, McAvoy continued to write and to remind her to think about her health, encouraging her to stay in the warmer climate:

> Remember that you and I have one thing in common. Neither of us were supposed to be alive at this day and date and no matter how poorly we do our work, we are just that much ahead of the game. Unless you take time now to get really well, and incidentally leave your worries to the Lord, you will not be able to do anything well later.[7]

She liked it in Texas; she was learning to ride and recovering from the dark November and December months in Indiana.

These are efforts that she continued after returning to Notre Dame in the fall of 1943; she worked in the archives, gave German lessons for the few remaining students and continued to be invited as a lecturer, to talk about her work but also about her experiences in Austria and its educational system before annexation by Nazi Germany.

In 1944 a new turning point came. Emma Beatrice had ambitions; she wanted to do more than archive work in the library of Notre Dame. Her sons were out of the house and she felt the emptiness around her. Thomas McAvoy had observed this tendency: "You are always trying to do another job and merely spoiling the one in front of you."[8] Nevertheless, with his support, she applied for a scholarship to work at the National Archives in Washington where she wanted to expand her training as an archivist; whether she had more far-reaching career plans in mind remains unclear.

She received a six-month scholarship; Yves Simon drove her to the train station in Niles, Michigan from where she departed for Washington, DC. He expressed his worries about being lonely without her, but she assured him that she would be back before his loneliness would be noticeable. He reminded her about this promise in his many letters asking her to come back. While working in DC, she lived in an apartment in nearby Alexandria, Virginia. She was once again very successful and thanks to her language skills, her position turned into permanent employment. From the purely archival work, she moved to a new department where she was entrusted with the translation of current German documents.

---

[6] McAvoy to EBH (Notre Dame, IN, February 17, 1943).

[7] McAvoy to EBH (Notre Dame, IN, January 8, 1943).

[8] Ibid.

Fig. 29.6 Miss Josephine Cobb, Mrs. Beatrice Haas, and Mr. James Roberts listening to Memovox recordings of World War II broadcasts from enemy stations. This is a duplicate of 64-NA-1-78 (Records of the National Archives and Records Administration, 1789—ca. 2007)

Other departments within the government also became aware of her and, like many of the former German immigrants, she was recruited by the Office of Strategic Services (OSS), the US espionage department founded in 1942. (Fig. 29.6) She was to provide information about Germany and the conditions there, as her son recalled:

In 1944, she was sent to the National Archives in Washington D.C. for some additional archival training. The people at the National Archives liked her so well they offered her a permanent position in Washington, which she eventually accepted. One of her assignments was translating some of Hitler's papers, for which she also spent some time going over O.S.S. documents (now called CIA).[iv]

This information is part of the family's record, however neither the name Haas nor her maiden name Huber are mentioned in the available documents on members of the OSS. This does not necessarily mean that Emma's work with the OSS and later the CIA as described here does not correspond to the facts. There may have been special circumstances that do not allow access to further information. Yet even this is speculation, although family information is clear:

As before, the CIA liked her work so well, she was asked to become a permanent member of that organization. She was sent to Europe in 1950 and worked undercover in Frankfurt. As I understand it, one of her duties was to alert the Army to former Nazis, looking for positions at the Army installation in Frankfurt. As an undercover agent she played the part of a German widow (under an assumed name) and thus could find out about the background of many of these people.[v]

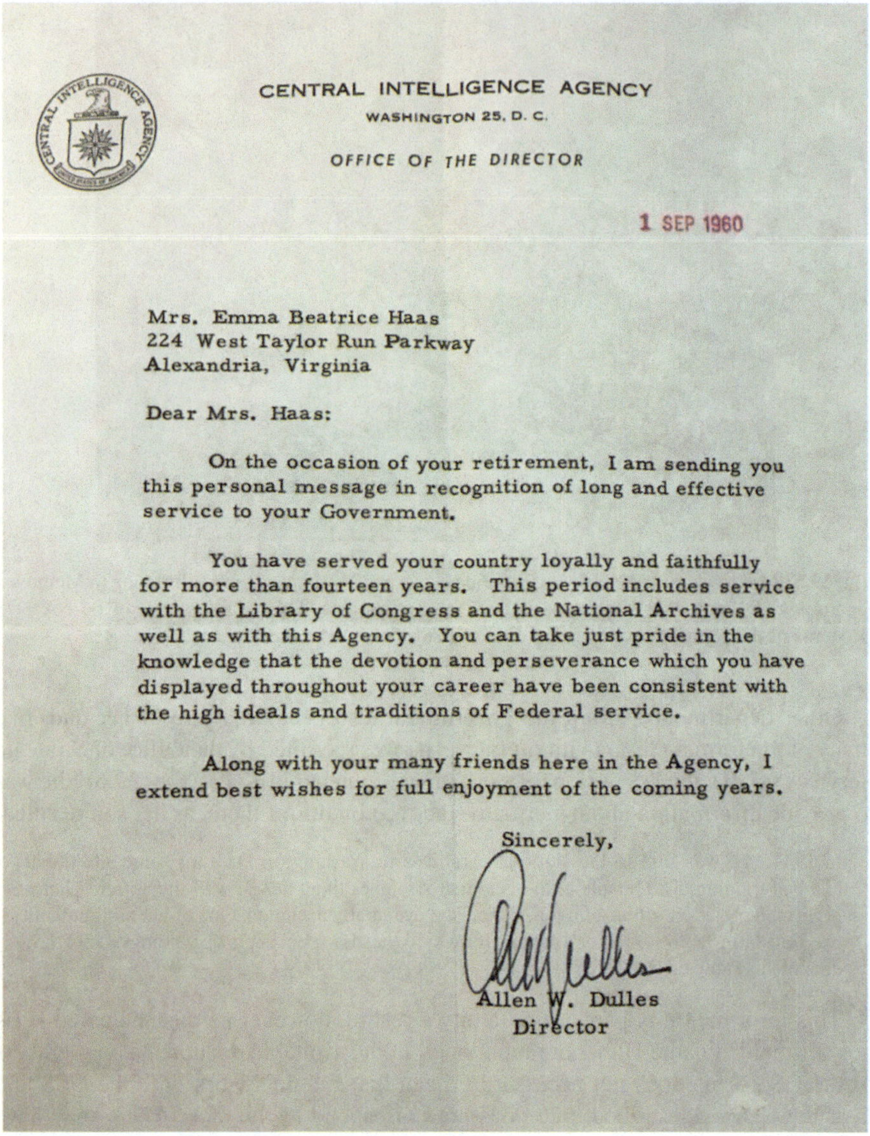

CENTRAL INTELLIGENCE AGENCY
WASHINGTON 25, D. C.

OFFICE OF THE DIRECTOR

1 SEP 1960

Mrs. Emma Beatrice Haas
224 West Taylor Run Parkway
Alexandria, Virginia

Dear Mrs. Haas:

On the occasion of your retirement, I am sending you this personal message in recognition of long and effective service to your Government.

You have served your country loyally and faithfully for more than fourteen years. This period includes service with the Library of Congress and the National Archives as well as with this Agency. You can take just pride in the knowledge that the devotion and perseverance which you have displayed throughout your career have been consistent with the high ideals and traditions of Federal service.

Along with your many friends here in the Agency, I extend best wishes for full enjoyment of the coming years.

Sincerely,

Allen W. Dulles
Director

Fig. 29.7 Allen W. Dulles, CIA Director on the retirement of Emma Beatrice Haas in September 1960

As her release papers document (Fig. 29.7), Emma Beatrice worked as an agent with the CIA for the following years. According to family tales, her position was first possible thanks to her knowledge of German. Since she spoke with an unadulterated Bavarian dialect, she was infiltrated into the remaining right-wing extremist splinter groups, most of which were successful in smuggling high-ranking Nazi party members and SS officers out of Germany and into South America through connections with the Catholic Church and the Vatican. Emma Haas' excellent connections to Catholic officials as well as her good knowledge of the Italian language would have undoubtedly helped her in this mission.

Both assets were important since especially the Catholic Church was anxious to lead former National Socialists back to church and faith as fighters against communism. It has been documented that many former Nazi members took this step:

> National Socialism had turned out to be a false doctrine, and the church celebrated every converted soul as moral victory. It came to a kind of moral crusade for souls. In the planned re-Christianization of Europe, the Protestant and Catholic churches were in competition.[vi]

However, this was not only an action on behalf of the souls of the imprisoned and condemned. National Socialist party members as well as SS representatives were in many cases actively seeking a means of escape and sought assistance through church channels:

> In the first post-war months, the escape of SS members in Italy was still very much improvised. In addition to SS comrades, Catholic clergymen in particular helped the fugitives. Church circles saw SS men primarily as anti-Communists, and "godless Communism" always seemed to have been a much greater enemy than National Socialism. As upright anti-Communists, SS members could count on support. Organized structures of flight took shape in the summer of 1946.[vii]

According to family documents Beatrice Haas, as a new member of the OSS and the subsequent CIA organization founded in 1947, was part of this hunt for former SS officers. However, this mission soon acquired a new undertone. With rapidly growing anti-Communism in the United States, and the resulting difficulties with the Soviet Union as a former ally, the CIA recognized the benefits of finding former SS officers and so developed a new recruitment plan. This was a complete change in American procedure; the systematic search for Nazi criminals and their escape organizations that had been conducted until then was quietly discontinued:

> The National Socialists' escape channels were not just uncovered by the US Americans, but were soon utilized for their own purpose. Soon after the end of the war, the Americans had a great interest in SS secret service agents with anti-Communist know-how. These secret service experts were now to be "recycled" and used against the "common enemy." For this reason, the Americans supported the escape of SS members from Europe starting in 1947. Those who had learned their trade did not become unemployed. In 1947, the US Counter Intelligence Corps (CIC) took over the church's escape route for National Socialists via Italy to overseas for its own agents. The Americans called this escape route the "rat line."[viii]

Whether Emma Beatrice Haas ever took part in such actions is unknown.[9] It is doubtful, since she remained convinced that the Gestapo had killed her brother George. It is known that she was in a position that today is called an analyst, for

---

[9] Inquiries by the author and various family members to the CIA - based on the Freedom of Information Act (FOI) - about the person and area of responsibility of Emma Beatrice Haas during her work there between 1945 and 1960 were turned down or ignored.

which she was to sift through incoming data from the various crisis areas, translate it if necessary and analyze it for possible consequences.

> CIA was a small organization in 1950 and lacked a robust collection and analytical capability. Formed by the National Security Act of 1947, it had only 5,000 employees worldwide in late January 1950, with just 1,000 of them employed as analysts and only three operations officers in Korea before the June 1950 invasion.[ix]

Perhaps she had duties beyond an analyst, given the small number of CIA employees in those early years. Emma's personal involvement as the alleged wife of an SS officer in hiding in right-wing extremist circles was not without danger, which undoubtedly speaks for more than a desk job. The rich collection of postcards picturing cultural and historical monuments from the various crisis zones of the world, which she carefully preserved,[10] also speaks for this. As one can see, it was precisely in these early 1950s that she undertook a long journey to Japan, where she spent most of her time in Nagano and Kobe. However, this probably had less to do with the sudden outbreak of the Korean War than with the reappraisal of Japanese German documents that demonstrated the cooperation between Japanese industry and other interest groups and the Nazi regime.[x]

Emma also traveled extensively in Central Europe, from Norway to Italy as her correspondence with Yves Simon documents. Furthermore, she stayed in French speaking Algeria for a long time and lived in the Hotel Transatlantique in the oasis Bou Saâda between the Atlas Mountains and the Sahara during the time of the Algerian war. Rumor claims that Bou Saâda was a secret CIA station in that country which was fighting for its independence. In addition to Algeria, the Middle East was also the destination of Emma's further travels; she was in Egypt which, in the fifty years under socialist-minded Gamal Abdel Nasser, overthrew the monarchy and carried out far-reaching land reforms. These included the nationalization of the Suez Canal, which led to the Suez Crisis. Residences for Emma are also documented in Jordan, Lebanon and Syria as well as in the newly founded state of Israel. Although all of these trips were of an official nature, the postcard collection reflects her great interest in the history of Christianity and the Catholic Church.

From these trips she wrote to Yves Simon several times to inquire about the fate of her sons. Yves Simon still lived in South Bend where George studied physics at Notre Dame. But Simon worked and taught in Chicago where Arthur Gustav studied history as a doctoral student. Simon wrote long reassuring replies, recounting her sons' academic crises and successes:

> I saw Arthur yesterday during the lunch hour. I found him much more mature and manly with plenty of sense. Plainly, he was greatly moved by your difficulties and his concern about you had pulled him out of the childishness that you and I have noticed. Such reaction to your troubles is a very fine evidence of good heart and good judgement. ... Believe me, I was very favorably impressed.[11]

---

[10] According to the family, these trips were business trips although there is no direct written evidence for this. It is, however, unlikely that such extensive travel could have been financed from the family's private coffers; the age of cheap, global mass tourism was still forty years away.

[11] Yves Simon wrote nothing about Emma Haas' special difficulties, which were probably a matter of professional decisions that she had to make. Yves Simon to EBH (South Bend, May 19, 1950).

Fig. 29.8 Emma Beatrice Haas in 1958 after her return from Europe, pictured here with grandchild Susan and daughter-in-law Mary Jane, who was called Nana, the wife of George Haas

George seems to have caused fewer worries, as Simon writes:

> George himself seems to be in top shape: Tall, big, smiling, very friendly. He seems to enjoy his job [at the physics department]. I met Fr. Bolger and another physicist who both told me that he is doing his work very well.[12]

Thus Yves Simon watched over the welfare of Emma's sons until 1953, the year Arthur Gustav Haas received his doctorate in Chicago. In his letters, Yves Simon repeatedly asked Beatrice to return to South Bend to revive their friendship:

> Dear friend, will you not soon come home? I search my memory and use associations to give a date to your last visit here: It was in spring 1951, almost two years ago. Many miss you over here.[13]

After Arthur Gustav's final exams in the summer of 1953, there is only one more letter; Yves Simon thanked Beatrice for her postcards from Spain and Vienna and wrote that Arthur had applied for a scholarship in Germany to write his doctoral thesis on Metternich's politics.[14] In this last letter, Simon considered:

> with Arthur in Germany, I perhaps should wish that you should stay over there for another year. Yet I would be happy if the rumor of your coming back this fall was confirmed.[15]

Yves Simon had often complained about an illness and it was diagnosed as cancer. Five years later in 1958, he retired from the University of Chicago and he died in South Bend in May of 1961.

Emma Beatrice Haas did not return to Washington from Europe until the late 1950s. She remained in the Alexandria area for some time (Fig. 29.8) before retiring

[12] Yves Simon to EBH (South Bend, November 13, 1950).

[13] Yves Simon to EBH (South Bend, March 16, 1953).

[14] Arthur G. Haas, *Metternich, Reorganization and Nationality 1813—1818* (Wiesbaden: Steiner Verlag, 1963).

[15] Yves Simon to EBH (South Bend, October 3, 1953).

in 1960 at the age of sixty-four. Her merits were undoubtedly recognized as the personal letter of the then CIA Director Allen W. Dulles underlines. Unfortunately, details were not disclosed.

Emma Beatrice Haas returned to Notre Dame and helped in the university archives for several years until the sudden death of her old friend Thomas T. McAvoy, Director of the Archives, in July 1969.

> The end came suddenly in his office in the University Archives, where, as was his custom, he was working outside of office hours. He had just completed correcting the page proofs of his general history of American Catholicism.[xi]

With this, Emma retired for good; she was seventy-three years old.

As she grew older, she eventually needed more and more assistance. Her son George, who was living in Alexandria at the time, convinced her in 1973 to give up her independent but increasingly stressful life in South Bend and return to Virginia. She moved to a Catholic nursing home run by nuns, feeling as she did in her early days as a novice in the convent of St. Xavier. She lived there under good care for another twelve years but the memories of her eventful life disappeared. She died in 1985 in her sleep at the age of eighty-nine. She was buried at Notre Dame's Cedar Grove cemetery at her husband's side.

Notes

i. Philip Gleason, "Thomas T. McAvoy, C.S.C.: Some Informal Recollections," *American Catholic Studies* 115.4 (2004): 59–68.
ii. George and Mary Jane Haas, "Letter to our Grandchildren," 84.
iii. Ibid.
iv. Ibid., 46.
v. Ibid., 47.
vi. Gerald Steinacher, *Nazis auf der Flucht – Wie Kriegsverbrecher über Italien nach Übersee entkamen* (Frankfurt am Main: Fischer Verlag, 2010), 175.
vii. Ibid., 179.
viii. Ibid. and https://www.cia.gov/library/center-for-the-study-of-intelligence/csi-publications/csi-studies/studies/97unclass/naziwar.html (September 6, 2020).
ix. "The Korean War Controversy: An Intelligence Success or Failure?" accessed March 9, 2021, https://www.legistorm.com/stormfeed/view_rss/663785/organization/95196.
x. Freedom of Information Act Electronic Reading Room, accessed on January 31, 2021, https://www.cia.gov/readingroom/document/519cd81a993294098d515f12.
xi. M. A. Fitzsimons, *Obituary Thomas T. McAvoy*, accessed on January 31, 2021, https://www.cambridge.org/core/services/aop-cambridge-core/content/view/628ECA91F89E32BC67746F27035A8CA3/S0034670500011761a.pdf/thomas-t-mcavoy-c-s-c-1903-1969.pdf.

Chapter 30
The Next Generation

The sons of Arthur and Emma Beatrice Haas both entered successful careers. Like so many of his fellow students, Arthur Gustav joined the US Navy ROTC at the University of Notre Dame in 1943. The Navy recognized his gift for languages and assigned him to the Oriental Language School program to study Japanese at the University of Colorado and the University of Oklahoma. In 1945, he was named a Lieutenant Junior Grade. After the end of the Second World War, Arthur was given the choice to either continue studying Japanese or to go and stay in occupied Japan. He decided that it would be more exciting to go to Japan and in fact he could tell many interesting stories about his six months of service there. At the end of his service, he returned to Notre Dame to complete his official college degree. Arthur received a Bachelor of Naval Science in 1945 for his ROTC training and a Master of Arts in 1949 for his study of history at Notre Dame.

After completing his degrees, he had probably had enough of Notre Dame but Arthur did not go far. He applied to the University of Chicago and was accepted into the graduate school's doctoral history program. He chose as his doctoral advisor the German-Jewish historian Hans Rothfels[1] who was an outsider in the Department of History in Chicago at the time. However in 1951, Rothfels accepted an offer to return to Germany and teach at the University of Tübingen. Arthur Gustav followed him to Germany after finishing his graduate classes in 1954. He had received a research

[1] An interesting choice because Hans Rothfels (1891—1976) was a nationalist German-Jewish historian, initially a well-known expert in nineteenth century German history who, despite his request to be classified as an Aryan on an honorary basis, had to leave Germany in 1939. Rothfels went via Oxford to Chicago where he worked as a professor between 1941 and 1951. He wrote important works on Clausewitz and Bismarck and after the war authored influential works on the history and interpretation of the conservative resistance of July 20th against Hitler. In 1951 he returned to Germany and taught at the University of Tübingen where he headed the Institute for Contemporary History. Since Arthur Gustav expressly refers to him as his doctoral supervisor in the foreword to his work, it can be assumed that the two historians also maintained contact through the later 1950s. Cf. Jan Eckel, "Hans Rothfels: An Intellectual Biography in the Age of Extremes," *Journal of Contemporary History* 42/3 (2007): 421–446.

M. Wiescher, *Arthur E. Haas—The Hidden Pioneer of Quantum Mechanics*, Springer Biographies, https://doi.org/10.1007/978-3-030-80606-4_30

fellowship for several years in Mainz at the Institute for European History.[2] The Institute had been founded in 1950 on the initiative of the French military government with the aim of conducting research. The Institute's charter defines the primary goals of the IEG as follows:

> Research on the religious and intellectual traditions of Europe, their development and crises, and particularly on religious differences, their effects and the possibilities of overcoming these differences.

In the years that followed, the institution promoted cooperation between scholars from many countries on this topic.

The city of Mainz had been largely destroyed in numerous air raids by the British and Americans and lay in ruins. The city is considered one of the oldest inhabited places in the Rhine Valley and had been an important trading center even before the arrival of the Romans. The experience Arthur had in Mainz with its countless ruins of Roman and medieval buildings was particularly fascinating for him as he had always had a keen interest in the history of ancient Rome. He searched and found many artifacts from Roman times in the shattered city. While visiting Roman ruins in Germany, Arthur Gustav met his future wife Roswitha Huchting from Bremen, who had studied chemistry in the war-torn city of Münster from 1951 to 1962 and now worked as a scientific assistant in Mainz.[3]

Despite his interest, Arthur Gustav did not, however, work on a topic of Roman history. Under the guidance of the institute's director Martin Göhring,[4] Arthur turned his attention to the diplomatic history of Habsburg Austria. He focused on the period from 1813 to 1818 when Austria was represented by Prince Metternich at the famous Congress of Vienna, where the destiny and future of post-Napoleonic Europe was decided.

---

[2] Today the Leibniz-Institute for European History. At its inception, the French and American military governments collaborated with the young state of Rhineland-Palatinate. The French promoted facilities to aid in the "re-education" of Germans toward democracy as well as rapprochement with France. The US also endorsed institutions that should strengthen the young democracy in Western Germany. It is very likely that the CIA was also involved, funding many European cultural and scientific centers in order to stabilize the continent against the perceived threat of Soviet Communism.

[3] Roswitha Haas worked as a chemist at the Oak Ridge National Laboratory in the US and gave chemistry lectures at the University of Tennessee in Knoxville.

[4] Martin Göhring (1903—1968) is another interesting choice. He was considered an expert of the French Ancient Regime and of France under Napoleon. He had joined the National Socialist Party in 1934. In 1939, he was a lecturer for middle and modern history in Halle, but from 1940 onward he was called up in occupied France to work on files of the French Foreign Ministry. From 1942 to 1944 he was associate professor of political foreign studies and Western European history at the National Socialist elite university Reichsuniversität Strasbourg. After the war, Göhring first became a lecturer in modern French history at the University of Tübingen and a visiting professor at the Technische Hochschule Stuttgart. In 1951, he became director of the Department of Universal History at the Institute for European History in Mainz. Cf. Heinz Duchhardt, ed., "Martin Göhring (1903—1968). Stationen eines Historikerlebens," *Veröffentlichungen des Instituts für Europäische Geschichte Mainz* 64 (Mainz 2005).

Arthur Gustav stayed in Mainz for several years but often visited Vienna, where he used the inexhaustible resources of the Vienna House, Court, and State Archives. In 1961, Arthur received a doctorate from the University of Chicago for his work. His official supervisor in Chicago was Louis R. Gottschalk.[5] His selection of professors is significant, especially the initial choice of Hans Rothfels as doctoral supervisor; in his writings, Rothfels played down the persecution of the Jews and portrayed National Socialism as an aberration through mass hysteria comparable to Communism. Rothfels argued that the National Socialists, like the Communists, should be considered "an extreme consequence of the secularization movement of the nineteenth century."[i] He further maintained that the resulting values of modern mass civilization caused most people to forget the basic values of the West, which in Rothfels' opinion were based on respect for God and the individual. These are undoubtedly views that Arthur Erich Haas also held in his opposition to the National Socialists and in his hope for the clerical fascist movement of the Vaterländische Front in Austria. One wonders if the academic outsider Rothfels was chosen as a doctoral supervisor because his views were similar to those of Arthur Gustav's father.

After several years in Germany, the successful completion of his doctorate, and the publication of his book on Metternich, Arthur Gustav Haas was offered a position in the History Department of the University of Tennessee at Knoxville in 1963. There Arthur Haas taught a series of courses over the next thirty-five years, ranging from the annual compulsory course on the history of Western civilization to special topics on recent Austrian and German history, most of which he represented in doctoral seminars. In addition, he provided topics for doctoral students that were reminiscent of his own doctoral supervisor Rothfels. These included a topic on the development of Adolf Hitler as a politician as well as on the question of political resistance in the so-called Kreisau Circle of conservative officers against Hitler; the latter was a topic to which Haas' student Thomas Childers was particularly devoted.[ii]

Many students have remembered Arthur Gustav Haas. Even forty years later, one student recently wrote:

> My love of history was born in Dr. Haas' lectures on Western civilization. Dr. Haas had a great gift as a storyteller, making information fascinating and engaging.

Another student noted:

> Dr. Haas had such a positive influence on my intellectual processes and critical thinking.[iii]

Arthur Gustav Haas was living in Knoxville when he died at the age of ninety from Parkinson's disease in 2016.

Arthur Gustav's younger brother George Arthur Haas embarked on a completely different career, following his father's tradition and choosing to study physics. As a teenager he had been very interested in technical instruments and developments. After graduating from the Marmion Military Academy, he moved to Notre Dame

[5] Louis Gottschalk (1899—1975)was an American Jewish historian who taught at the University of Chicago from 1927 to 1959. His area of expertise was the history of the French Revolution and the subsequent Napoleonic era as well as the impact of General Lafayette on America.

as a freshman in 1943. There he enrolled in a program to study metallurgy and engineering. He also hoped to be accepted into the ROTC program in order to enjoy the preferential treatment that came with it but admission had become more difficult and, unlike his brother, the ROTC program remained closed to him:

> Because Notre Dame was on a three-semester-per-year schedule during the war years, I was ready to start my second semester, sophomore year while I was still seventeen. Since I didn't want to be drafted into the Infantry when I reached eighteen (I had enough Infantry life at Marmion), I thought it might be good to try to become an Officer in the Air Force.
>
> I took the tests for the Aviation Cadet Training Program and passed both the mental and physical exams but as I was waiting to be called, I was told that I would have to get a special waiver approved. It seemed that for security reasons, there was a new rule which prohibited former immigrants from Axis controlled countries (even though it was "annexed" by Hitler, Austria was considered as such) to enter officer training programs unless they had been American Citizens for at least ten years. Since we had been Citizens for only three years, this clearly left me out. I also was told it might jeopardize Arthur's position in the ROTC to appeal, since no one seemed to know how retroactive this rule might be.[iv]

In order not to be drafted into the army, George applied to the Navy. After passing his exams, he was immediately hired as an electronics technician because of his knowledge and expertise. This involved a whole series of new courses as well as training in Michigan City, Chicago, and New Orleans. George was a fully trained technician in electronics and radar when the war ended in 1945, but his interest in submarines had been awakened and he volunteered for another two years. Electronics technicians were in high demand and after another short training period on various submarines, he sailed the Seven Seas. He described these travels and adventures in detail in his memoirs.

After completing his service in 1947, George Haas returned to Notre Dame to finish his studies. First, he worked as an electrical engineer installing the first radio and television antennas in South Bend, to bridge the time until the beginning of the semester. He was also questioning whether he should return to metallurgy:

> I met my old roommate Jim Sackinger again and he convinced me that in view of all my experience in the Navy Electronics Program, I should join him in the Physics Department at N.D. rather than go back into Metallurgical Engineering - and he said he'd introduce me to the head of the Physical Electronics Lab. Well, the head was none other than Professor E. A. Coomes. Ed Coomes was at the famous "Radiation Laboratory" at M.I.T. during the war and when this was disbanded, he took several people to N.D. with him to start the Physical Electronics Laboratory there. The "Radiation Lab" was the place where innovations and new improvements in radar were developed during the war, and what "Los Alamos" was to the Atomic Bomb, the "Radiation Lab" was to Electronic Warfare.[v]

It was a good decision for George. At Notre Dame in the fall of 1950, he met a young student from neighboring St. Mary's College, Mary Jane Farrell (Nana) from Alexandria, VA, the town where his mother lived at the time. The relationship developed quickly and well; soon Nana was introduced to Emma Beatrice in Alexandria on the occasion of a home visit. During the Christmas break that year, a marriage proposal was accepted and followed by a wedding scheduled for the following year.

George completed his studies within only two years, in June 1951. He then applied for the doctoral programs at Cornell, the University of Illinois, and Notre Dame. He was accepted by all of them but decided to remain at Notre Dame to earn his doctorate.

For the summer of 1951, George Haas held a position at the US Naval Research Laboratory (NRL) in Washington, DC, where he worked on electron emissivity for radio tubes. These were the main electrical control units in electronic systems from radio to radar before the development of the transistor. At the NRL he learned about materials research to improve technical developments. He was very excited to work in such a large laboratory with state-of-the-art equipment, but he returned to Notre Dame for the fall semester while Nana stayed in Alexandria to prepare the wedding which took place in Alexandria on December 27, 1951. The bride and groom were joined by the bride's family, friends and even the Guth family from South Bend traveled to Virginia. Emma Beatrice could not attend as she was on one of her assignments in faraway lands.

George worked hard and by 1952 he had passed the Candidacy Exam, the official admission exam for the doctoral program; only a year later he submitted his thesis, completed under the supervision of Edward A. Coomes, on the topic: "Experimental Study of the Schottky Effect in Thermionic Emission from Strontium Oxide on Polycrystalline Molybdenum."[6] This was a work that explored the emissivity of electrons from materials under different arrangements of external electrical fields. This not only led to the improvement of the electron sources at the Notre Dame accelerator but also had numerous other applications in electronics and material sciences. It was a technical development thesis, a direction of work that would determine his later life. What drove him on was the offer of a permanent position at the NRL. There he worked successfully for the next thirty years as a physicist and researcher. It was a career that satisfied him professionally and personally and also gave him the time to travel a lot—he and his wife traveled the world, also in retirement; he probably inherited his love of travel from his mother—and to provide for his growing family with three daughters Theresa (*1952), Patricia (*1953), Susan (*1957) and a son Christopher Arthur (*1960). (Fig. 30.1).

The physics that George Haas practiced was fundamentally different from the physics that his father had found so compelling. It was a new era and although only twenty years had passed, the war had fundamentally changed the world and the goals of science. The Manhattan Project had demonstrated the enormous power of science and scientists to create something new and unprecedented. After the war and during the early Cold War period, the continued quest for weapons technology freed up an enormous amount of research funds. The United States attracted researchers from all countries, regardless of their past, to improve and strengthen US science; the aim was to face the danger, or presumed danger, of the Communist world movement that manifested itself in the Soviet Union.

[6] The Schottky effect or field enhanced thermionic emission is a phenomenon named after Walter H. Schottky and is used in electron emission devices, especially in electron sources. This additional electric field facilitates the emission from solid state materials. Haas had the task of determining the effect in oxides and crystals.

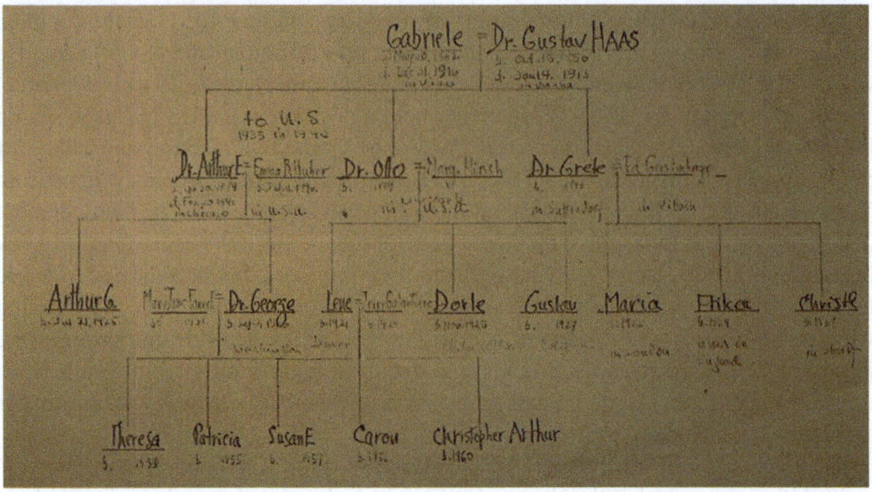

Fig. 30.1 The children, grandchildren and great grandchildren of Gustav and Yella Haas. From the marriage of George and Mary Jane Haas three daughters, Theresa (*1953), Patricia (*1955), Susan (*1957) and a son Christopher Arthur (*1960). Otto's daughter Lene and her husband Jerry Galantiere had one daughter Caron (*1956)

It was no longer the emigrants of the 1930s who had come to America begging for and seeking help; there was active recruitment of a new wave of young innovative scientists that continued into the 1990s. The search was no longer about the fundamental questions regarding the structure of the microcosm in the atomic and subatomic realms nor was it about the questions of the origin and decay of our universe. The questions were about the application of just-learned basic principles of the atom and more so, the success of the Soviet atomic and later hydrogen bomb as well as the launching of the Sputnik satellite. This drove development and funding of the US science and atomic energy program and also the surge of space and rocket science to reach and conquer the Moon and the planets.[vi]

The applied physics that George Haas practiced had the goal of improving the necessary instruments and their functioning. He worked in the field of materials research, especially in the field of so-called surface physics as well as gas plasma physics. With so-called atomic spectroscopic methods, the electronic properties of the thinnest materials were tested for their ability to emit electrons, feeding ever smaller electrical circuits essential for the development and effective operation of ever smaller electrical control components for industry and defense.

In his specific work George investigated the properties of a wide variety of materials, of which new ones were constantly being created to take advantage of electronic properties. Without these instruments there would be no space travel and no nanotechnologies, there would be no computers and no intelligent weapons technology, there would be no consumer electronics, there would be none of the multitude of electronic gadgets that govern our lives today and without these, mankind apparently feels unable to exist. With his work at the NRL, George contributed to these developments that have changed our lives over the last decades. As an applied physicist George Haas published mainly in the scientific journals *Applied Surface Science*, *Journal of Applied Physics* and in *IEEE Transactions*; he published more than 100 scientific papers and also successfully filed numerous patents.

George Haas remained at NRL for thirty years. He did not contribute to the big picture in the understanding of the universe, a desire that drove his father, but he moved science forward through his systematic testing and analysis of electronic materials and components for science applications. This he performed with a degree of patience that his father could not muster in the early days of his studies under Professor Hermann Theodor Simon, trying to understand the problem of singing arc lamps in electrical engineering. In 2010, George Arthur Haas and his wife Mary Jane moved to a retirement home for Navy personnel, Arleigh Burke Pavilion in McLean, Virginia; there she passed away in 2013 and he died peacefully in his sleep on February 10, 2018. He left behind a large and successful family.

Notes

i. Georg Iggers, *The German Conception of History* (Middletown: Wesleyan University Press, 1968), 257.
ii. Thomas Childers, "Hortensienstrasse 50. The Kreisau Circle and the 20th of July" (master's thesis, University of Tennessee, 1971), accessed January 31, 2021, https://trace.tennessee.edu/utk_gradthes/3110.
iii. *Arthur Gustav Haas*; Obituary, accessed on January 31, 2021, https://www.legacy.com/obituaries/knoxnews/obituary.aspx?n=arthur-haas&pid=177874960&fhid=5920.
iv. George and Mary Jane Haas, "Letter to our Grandchildren," 86.
v. Ibid., 99.
vi. Audra J. Wolfe, *Competing with the Soviets – Science, Technology and the State in Cold War America* (Baltimore: John Hopkins University Press, 2013).

Index of Persons

M. Wiescher, *Arthur E. Haas—The Hidden Pioneer of Quantum Mechanics*, Springer Biographies, https://doi.org/10.1007/978-3-030-80606-4

Subject Index

M. Wiescher, *Arthur E. Haas—The Hidden Pioneer of Quantum Mechanics*, Springer Biographies, https://doi.org/10.1007/978-3-030-80606-4

The manufacturer's authorised representative in the EU is Springer Nature Customer Service Centre GmbH, Europaplatz 3, 69115 Heidelberg, Germany. If you have any concerns regarding our products, please contact ProductSafety@springernature.com

Printed and bound by CPI Group (UK) Ltd, Croydon, CR0 4YY
28/09/2026
02224065-0015